AMERICAN CIVIL ENGINEERING HISTORY

The Pioneering Years

PROCEEDINGS OF THE FOURTH NATIONAL CONGRESS ON
CIVIL ENGINEERING HISTORY AND HERITAGE

November 2–6, 2002
Washington, D.C.

SPONSORED BY
American Society of Civil Engineers Committee on History and Heritage
ASCE National Capital Section

EDITED BY
Bernard G. Dennis, Jr.
Robert J. Kapsch
Robert J. LoConte
Bruce W. Mattheiss
Steven M. Pennington

 American Society
of Civil Engineers

1801 ALEXANDER BELL DRIVE
RESTON, VIRGINIA 20191–4400

Abstract: In this 150th year of the American Society of Civil Engineers (ASCE), it is fitting and appropriate to recognize the engineers and their accomplishments that have made civil engineering and quality of life what it is today. From John Smeaton, who first sought to raise the stature of the engineer to a professional level, to Benjamin Wright, the engineer and lawyer who is the father of American Civil Engineering, to Neal FitzSimons, who was instrumental is establishing the National Committee on Civil Engineering History and Heritage within ASCE and the National Historic Civil Engineering Landmarks program. These proceedings of the Fourth National Congress on Civil Engineering History and Heritage memorialize the tangible and intangible accomplishments of these men and those who they have influenced. The proceedings are a compilation of papers prepared for ASCE and presented at its annual convention from November 2 to 6, 2002 in Washington, D.C.

ISBN 0-7844-0654-5

Contents

The History of Building Materials and Methods: Cement and Steel

Historic Water Supply Systems

Preservation Case Studies

Perspectives on American Civil Engineering—A View from Afar

Introduction

The focus of the Fourth National Congress on Civil Engineering History and Heritage is the development of American Civil Engineering. This is especially appropriate as we celebrate the 150th anniversary of the American Society of Civil Engineers (ASCE). Our year-long celebration began with the October 2001 convention in Houston, Texas, and has brought us to our nation's capital, Washington, D.C., in November 2002.

To mark ASCE's anniversary, the History & Heritage Committee extended invitations to prominent scholars, not necessarily only engineers, to provide a broad view of the development of civil engineering in America. The result is an outstanding collection of original manuscripts documenting the emergence and growth of our profession.

The beginning of the profession of Civil Engineering is recounted in the paper on the first civil engineer, John Smeaton. A unique summary is presented of the engineering career of Benjamin Wright, the Father of American Civil Engineering.

The struggle for the formation and early years of ASCE is presented. Insight into the start of Britain's Institution of Civil Engineers, and the Canadian Society for Civil Engineering is also provided.

Excerpts from the diary of a young civil engineer visiting the United States in 1837 provides a glimpse of the new frontier where America's rapid growth and expanding boundaries defined a unique engineering approach not found in England or continental Europe. The career path of an engineer in the 1840s is outlined in a paper that addresses the transfer of people and ideas across our border with Canada. Early engineering achievements in Alaska are covered.

The outstanding efforts of three prominent engineers—Latrobe, Jervis, and Meigs in the design and development of water distribution systems in Philadelphia, New York, and Washington, D.C., respectively—are documented in three papers. Transportation history is covered with histories of the Potomac Canal, the C&O Canal, the National Road, and the B&O Railroad. Material science is covered with papers on hydraulic cement, reinforced concrete, and early steel rolling mills.

New engineering discoveries of portions of the lost Inca Trail at Machu Picchu are revealed in a paper by Kenneth Wright, winner of the 2000 ASCE Civil Engineering History and Heritage Award. Dating from 1500, this is truly early American civil engineering.

Preservation case studies include special insights into structural renovations to the U.S. Capitol and the geology of the monumental core in Washington, D.C.

These proceedings support the continuing efforts of the ASCE Engineering Education and History and Heritage committees to develop course material for teaching civil engineering history in universities. An expanded module of Civil Engineering Education History is offered, along with the history of analytical modeling, and the development of testing and monitoring tools.

The reader will be amazed at the way these varied papers weave a fabric of common thread in the evolution of knowledge and skills in the field of civil engineering. This volume deserves to become a key reference for educators and future generations of engineers.

The early history of how we arrived at this point is recounted in these papers, and new chapters in our history are being written even now. As we prepared for our grand celebration of ASCE's 150th anniversary, our focus was directed to events of the present—a new chapter in history, and in Civil Engineering. The terrorist attacks on the World Trade Center in New York City and the Pentagon in Washington, D.C. on September 11, 2001 will be an important marker in historic engineering records. We will long remember the heroic efforts not only of firefighters and police, but also engineers who worked tirelessly to restore order, regain control of the disaster sites, and led our country, and the world, from shock and despair to hope and recovery. In support of these heroes, the American spirit soared as we overcame the disbelief and began to rebuild. In the aftermath, civil engineers were there, are there, and will be there—the "quiet profession."

Many thanks to the authors who gave freely of their time and talents to paint a magnificent mural of our predecessors' lives, challenges, and remarkable achievements. To Roland Paxton and Alistair MacKenzie who joined in our anniversary party by providing views of early American Civil Engineering from Britain and Canada, respectively; a sincere thank you. Special thanks go to the editorial committee: Steve Pennington, Robert Kapsch, Bruce Mattheiss, and Robert LoConte, for the many hours devoted to planning, topic selection, speaker recommendations, coordination, and paper review. Each of us knew, admired, and learned from Neal FitzSimons, who inspired us to pursue our engineering heritage. This effort is devoted to his memory. Special recognition and deep gratitude goes to my wife Susan, for the many meals at committee meetings, editorial assistance, computer graphics expertise, moral support, and being there - always.

I thoroughly enjoyed reading and learning from each of these papers. I am confident you will, too. Sit back and enjoy the history of the first 150 years of American Civil Engineering.

Bernard G. Dennis, Jr.
Editor

NEAL FITZSIMONS
(1928 – 2000)

A MEMORIAL

In early March 2000 the American Society of Civil Engineers lost one of its more prominent members, Neal FitzSimons. Neal was extremely active in Society affairs at the local, national and international levels. His great legacy was the establishment of the ASCE Committee on History and Heritage of American Civil Engineering, which he helped organize and chaired for many years.

Neal became active in ASCE while a student at Cornell where he would graduate in the Class of 1950. He would continue to be active in the Society when, by the time of his death, he was both a Fellow and Life Member. In addition to ASCE Neal was also a member of the American Concrete Institute, the National Society of Professional Engineers, the American Society for Testing and Materials, the Canadian Society of Civil Engineers and the Construction Specifications Institute.

Throughout his career, Neal was filled with a passion for the heritage of civil engineering. He worked tirelessly and, to those that know him, he seemed like a "man possessed." Neal was unique in that he graduated from Cornell with both a

---

Photograph above title: Neal at Benjamin Wright's birthplace during ASCE plaque dedication ceremony, 1970.

degree in civil engineering and a degree in history. His love of history was furthered through his association with James Kip Finch, Dean of the School of Engineering at Columbia. Neal felt, as others do as well, that one of the primary goals of the Society is to instill in all its members a strong fundamental understanding of the heritage of the profession. To this end he worked both through his personal interactions with other civil engineers and through his writing.

In April 1965 Neal introduced the column: "Who Am I?" in *Civil Engineering* magazine. Each month he would present material about a prominent civil engineer. The story was spoken in first person in such a way as to entice the reader to guess the identity of the author. In the following month's installment the identity would be revealed. The following extract is a paraphrase of the article from the October 1966 issue of *Civil Engineering*.

> I was born during the month of Captain Prescott's trial for murder. This was one of the events which soon after led to a bloody eight-year war. . . Although formal schooling was not to be my lot, a kind uncle patiently taught me the rudiments of surveying. . . Under the ubiquitous English Civil Engineer, William Weston, I began my early career as a canal builder. . . The great project of the day was about to begin and I was selected as a principal engineer. . . The magnificent canal was to become the "First American School of Civil Engineering.". . . Of all the early American civil engineers, I perhaps had the longest and most diversified career. So if some believe that I deserve the appellation, "Father of American Civil Engineering," only modesty would cause me to raise a mild protest.

> Who am I?

The answer appeared in the November 1966 issue of *Civil Engineering* and revealed the author's identity as Benjamin Wright (1770-1842). The great canal project was the Erie Canal in New York.

In 1976 Neal received ASCE's Civil Engineering History and Heritage Award. Quoting from his citation:

> *(Recognition for his). . . contributions to enable civil engineers, as well as others, to have a better understanding of and appreciation for the history and heritage of civil engineering. His knowledge of the subject is boundless – his enthusiasm contagious. The profession of civil engineering has benefited significantly from his leadership in the field of Civil Engineering History.*

For the many who knew Neal over the years, his passing was a great and tragic personal loss. For both the Society and the profession as a whole, Neal has left a void that will be hard to fill.

Photograph of Neal on Bollman's Truss Bridge (Maryland) at its dedication as a National Historic Civil Engineering Landmark in 1966.

Birth of American Civil Engineering

A MAN FOR ALL REASONS
John Smeaton, FRS, The First Civil Engineer

By M. D. Morris, P.E., FASCE[1]

PREMISE

John Smeaton was not a product of his time. He qualifies as a true, quiet giant in history, who was able to alter the times in which he lived, thus altering all the times that followed. This genuine, proverbial, historic hero is underrated, unheralded, and unsung. Adam Smith published his *Wealth of Nations*[2] also in 1776, but all that socioeconomic advocacy would have been at least another century in materializing were it not for John Smeaton's efforts that same year at delineating its enabling profession, Civil Engineering.

BROAD BACKGROUND

In 1776, perhaps the most significant historic event in the life of England, the United States, and the rest of civilization, was not the American Revolution! I believe it was the hanging of a shingle outside a London office. It proclaimed, "JOHN SMEATON, CIVIL ENGINEER. "

Seventeen seventy-six may not have been a good year for red wines, but it was replete with heavy historical happening: Matthew Boulton and James Watt created power from an efficient steam engine; Josiah Wedgewood produced quality china crockery for the populace; and, there was Adam Smith's *Wealth of Nations*. All that, in addition to the Jefferson-Franklin magnum opus on "truths ... self evident." Why then should the public proclamation of a professional presence by a person of particular perspicacity take precedence over all other contemporary events?

Because in the world purview, John Smeaton established and clearly defined the hitherto unrealized and unnamed profession of Civil Engineering. Those practitioners in the public good, then became the recognized people who eventually were able to provide England (and the rest of the world) with a fully designed and constructed physical plant, power base, and means of conveyance. That is what enabled the Industrial Revolution to achieve its inevitable goal, to change the world from agricultural to industrial.

Without Smeaton and his followers, that maturation might have taken another century, or even longer. Is it not true that from the Roman era until then, very little had changed? In some areas things mechanical even had regressed. During Smeaton's 1755 trip to the Low Countries he observed a windmill at Deftshaven.

---

[1] Technical communications consultant to government and industry, worldwide, Ithaca, New York.
[2] Adam Smith, *An Inquiry into the Nature and Causes of the Wealth of Nations*, Scotland, 1776.

John Smeaton, F.R.S.
(1724-1792)

Portrait oil on canvas by George Romney, owned by the Smeatonian Society, hangs in the Institution of Civil Engineers, London. Reproduced with permission.

"Near this place I Saw a mill which turns about at the top; this Mill appears to have been built 100 Years, but is entirely the same as those of more modern date."[3]

What marks man's ascending civilization is:
1) To adjust his behavior to accommodate natural forces, (he comes in out of the rain) intellect;
2) To devise an adjustment to negate the effect of natural forces, (he makes an umbrella to keep himself dry while he's out in the rain) craftsmanship;
3) To develop an adjustment to negate a local influence of natural forces, (he builds a building complex to keep a community dry,) engineering; and hopefully,
4) To adjust the natural forces to accommodate his desires, (stop, or start the rain) future engineering.

As a humanist engineer functioning in the interface between science and society, Smeaton's efforts were the result of pondering problems, then attempting solutions based on both science/math, and in applying knowledge of properties of materials. This perspicacity is what enabled him to go, professionally, where no man had gone before.

The recording of history can only be done by contemporaries. But they should not interpret history, that is for the heirs and assigns of those events. It is equally unsound for later generations to attempt to revise or redirect history toward another meaning, unless some valid and compelling evidence has been unearthed. But logical interpretation is one of the delights of studying history. Another is, by knowing the real beginnings of a field you can acquire a true grasp of its value. Aristotle wrote, "He who considers things in their first growth and origin ... will obtain the clearest view of them."[4]

I spent fifty years of my lifetime as a professional civil engineer. The contributing force that enabled my non-mechanical mind to succeed in an area where it was technically unsuited was a compensating emotional suitability. That stemmed from a clear understanding of what I was doing and why I was doing it. Much of that came, at my father's urging, from reading history, and interpreting it. As such I believe today John Smeaton is perhaps the western world's least known public benefactor.

HISTORICAL BASIS

Many historians record events in terms of military, political, or socioeconomic events, strategies, and victories. Picón Salás[5] contends (and I tend to agree) that history should be measured in terms of cultural changes. A people's culture reflects its history. This is what I perceive is the real value of Smeaton and his effort. Details

---

[3] (N) P-28
[4] *The Politics*, Bk. 1, Ch. 2, Trns Jowitt, Benjamin, London, 1921.
[5] Picón Salás, Mariano, *A Cultural History of Spanish America From Conquest to Independence*, University of California Press, Berkeley, 1963.

being what they may (and the 1783 political separation of England from U.S. included) our way of life today is a direct outgrowth of the English Industrial Revolution. Socioeconomic, yes; political, yes; but mainly cultural.

The general historic tendency is to combine the three revolutions of the 18th century: the Industrial Revolution in England, the American Revolution from England, and the French Revolution. I feel that in the retrospective view, the French Revolution did not have quite the impact on 18th Century (or later) civilization that the other two did. I believe Smeaton's declaration hastened the evolution of the Industrial Revolution and thus enabled both England and America to survive separately.

"Revolutions are not made by fate but by men. Sometimes they are solitary men of genius. But the great revolutions in the 18th Century were made by many lesser men banded together. What drove them was the conviction that every man is master of his own salvation," wrote Jacob Bronowski in his poetic prose.[6] Yet I am astounded that that particular man, who could distill such sensitive wisdom from the historical record would, by error of omission (or design) completely ignore Smeaton. That accomplished name appears nowhere in Bronowski's otherwise brilliant and flawless *Assent of Man*,[7] while the text cites lesser men all around Smeaton: Brindley, Watt, Wedgewood, Wilkinson, etc. Why?

Though Bronowski doesn't actually call those men "cultural," the consequences of the Industrial Revolution were. Its roots were in the countryside. They grew from the people who demanded more for their efforts than a meager subsistence with Dark Ages amenities. New industry provided standard Wedgewood crockery for every table at a price a worker could afford. (Or Jasper, and bone china for a little more.)

From his Yorkshire Iron & Steel works John Wilkinson not only made swords and blades for the King's army, but he was able also to turn out flatware for the tables, and tools for farm, home, and shop. His consultant (and brother-in-law) Joseph Priestly succinctly summarized the Industrial Revolution as "the happiness of the greatest number."[8] Priestley also had much to do with oxygen, and later taught chemistry at the University of Pennsylvania in Philadelphia.

Joseph Priestley was also scientific adviser to Wedgewood who, along with Dr. Erasmus Darwin, all belonged to a scientific and industrial society. They met monthly on the full moon at various places, and called themselves the Lunar Society. Members included Boulton and Wilkinson, among others. (Wedgewood and Darwin eventually became both grandfathers of Charles Darwin.[9]) The Lunar Society is

---

[6] (C) P-259.

[7] This blunder in historical recording is matched by (Cornell and *The New York Times'*) Allison Danzig's, *The History of American Football*, (Prentice Hall, N.J., 1956). "The most detailed and accurate account of our most colorful sport," as it describes itself, completely omits mention of Rutger's two-time (first) All American, Paul Robeson.

[8] (C) P-272.

[9] Charles Darwin, *The Origin of the Species*.

noteworthy here because it extrapolates and verifies the separation of the scientific/industrial group from the civil engineers who, along with Smeaton, formed their own group, described later. Prior to this point (as the top personages had done in the Royal Society) all men of a scientific or technological bent were "of imagination all compact."[10]

Quid pro quo is generally the way with a revolution. People get some things in exchange for other things they must give up. The Industrial Revolution brought people unlimited power in the real sense but it made the workers slaves to the place and pace of the machines. It created physical pollution of the air and the streams; and social pollution in the burgeoning squalid urban centers. All cultural changes.

Before, in the relative (though mostly impoverished) tranquility of rural "cottage industry," people did piece work at their own pace in their homes. Then they took their products to market centers. The Industrial Revolution ended that by bringing people from the countryside to the work centers, and to their inevitable ghettos. They lived in cold crowded rabbit warrens with the most meager of conveniences or social graces. But the general public gained things they never had before, at mass produced prices within their reach: dishes, flatware, stoves, tools, shoes, clothing, and more machines to make more things.

To accommodate that turning over of the socioeconomic body, [11] demanded immediately, and in full scale, things from technology that until then existed only in dreamers' minds. Factory and mill buildings had to house the great new engines. Canals had to be enlarged and extended. Then roads, unchanged since the Roman *iter* had to be made all-weather and reach all places. When those two modes of transport still failed to meet the new demands, the railroads evolved. Ports, harbors, and navigation aids also had to be concocted then improved.

To keep pace with all that advancement and the rapid population shift from the soil to the cities, new water supplies, eventually waste disposal systems had to be devised and installed. Larger multiple dwellings evolved. Dams, dykes, tract drainage, bridges, and buildings came next. All that cultural change became the bailiwick of the newly emerged profession of Civil Engineering, as did the industrial plants.

James Brindley (1716-1772) built canals, improved water wheels, and consulted on such production commencing less than a decade before Smeaton. He, like Smeaton, began, as a self-taught mechanician, who went on from there. But unlike Smeaton he had not the imagination, the resourcefulness, nor the mental capacity for the top. His efforts were well built on dead reckoning and continuation of "standard practice," plus a liberal share of common sense and good fortune. Brindley's time was the transition from the artisanship before him to the engineering that followed. Also he worked well with Smeaton when they actually did, on several occasions.

---

[10] William Shakespeare, *A Midsummer's Night's Dream*, v, i, 7.
[11] "Fanshen" of the Chinese People's Revolution of 1949.

ENGINEERING, CIVIL

Humanist and Civil Engineer, Solomon Cady Hollister, Dean Emeritus and father of today's College of Engineering at Cornell University lined us in, direct to what we need to know about the "Ingenious Contriver of the Instruments of Civilization."[12]

"Use of the word *engineer* has been traced back at least to the twelfth century. In Roman days the classical Latin term for military constructor was *rchitectus militarus*; but in 1196 we find the Latin term *encignesius* used in Lombardy. In 1238 the spelling is modified to *inzegnerium*. In France the designation *maistre engingnierre* was used in 1248, modified in 1276 to the spelling *engegynnyre*. In Germany *ingenieu* appeared before the middle of the fifteenth century; this word is still in use, as it is in France. The origin of our word *engineer* is connected to the German word, and is associated with the word *ingenuity*."[13]

From that etymology it appears then that the noun *engines* evolved from the devices of the "ingenious contrivers " rather than *engineers* becoming the name for those people who operated those contrivances.

The Engineers' Council for Professional Development (ECPD) in 1962 determined:

"Engineering is the profession in which a knowledge of the mathematical and natural sciences gained by study, experience, and practice is applied with judgement to develop ways to utilize, economically, the materials and forces of nature for the benefit of mankind." [14]

Early in my college days I learned more simply, "an engineer was someone who could do for one dollar, what anyone else could do for two." It baffles me how this comparative definition has held up over the years in the light of the cost of today's constructed projects.

A succinctly more direct version of the ECPD definition came from England about thirty years after Smeaton. In 1828 Thomas Treadgold said of engineering (specifically Civil), "... the art of directing the great sources of power in nature for the use and convenience of man." [15]

Inherent in all those definitions is the notion that an engineer is not just a builder, and is vastly different from an architect. He learns scientific principles (mathematics and physics) then applies them (along with an acquired knowledge of the makeup and

---

[12] (G) Subtitle.
[13] (G) P-30.
[14] (F) P-2.
[15] (F) P-1.

behavior of materials) to design and construct devices to alter the effects and forces of nature. Smeaton's efforts harnessed wind and water to work for people.

"In the old days all engineering was military. Ever since Apius Claudius built the military roads that led from Rome, engineers have constructed impregnable fortifications—then contrived their ultimate destruction." "Since Smeaton, all the public works plus traffic patterns, tunnels, and management have become honored civil preoccupations,"[16] while the military engineer continues to practice his own skills along with allied professions (chemical, aeronautic, electronic, etc.). The 'Civil's' satisfaction comes from tangible results and their usefulness."[17]

The ancient engineers of Sumer, Egypt, and Greece brought wealth and greatness to their lands through construction of fortifications, harbors, buildings, and war machines. The Romans added roads, bridges, and aqueducts. But, as with the major works of Leonardo daVinci, it was all for conquest or defense. Just as all the machines (catapults, etc.) were devised for warfare, little if anything was done in the public interest. Hollister allows that prior to about 1600 all the materials for construction were stone, wood, and brick. And all the labor was by hand, aided only by lever, pulley, wedge, and wheel.

The old military engineers both built the fortifications, then manned them as defense directors being majors domo, so to speak, over the activities as they developed. In the observations of his visit to the Lunette, Zealand, "works that were attaqued by the french," Smeaton wrote, "Under the ravelins in many directions are chambers for the Mines, and places for the Engineers to set in, and give directions during a siege."[18]

Today's surviving structures show us that the only remnants of great efforts of the Renaissance are churches and castles— refinements of the temples and forts of the ancients. The number of homes, reservoirs, roads, or any public works is minuscule, if at all extant in isolated places.

Public works in the Low Countries consisted mainly in dykes, canals, and sluices. But again they were done principally to aid defensive military operations. Commerce was of raw materials for, and in, the products of cottage industry. The time had come to stop the driving effort for the military and concentrate on the needs and wants of the general public.

"The business of building is not new." As National Chairman I wrote in the introduction to the booklet commemorating the 1975 Golden Jubilee of the Construction Division, American Society of Civil Engineers: "Man has been erecting structures to overcome natural obstacles for over 10,000 years. What distinguishes our current efforts from those of the ancients is efficiency. The classic roof support columns of the Egyptian and Greek temples occupied more than 66% of the floor

---

[16] (K) P-15.
[17] (G) P-30, 31.
[18] (N) P-25.

space. Now we dome-cover an 87,500-seat football stadium entirely without inside support. (NFL Superdome, New Orleans, Louisiana.) Naturally the elements of time, labor, and materials have evolved considerably but the real difference today is the engineering that goes into both the structural design and the construction method."

Smeaton started us thinking of the entire system in his thoroughly complete execution of the Eddystone Light. He examined the site, analyzed the problem, and then designed the structure from his knowledge of materials. By making a composite structure applying knowledge of the properties of the materials as criteria, he replaced uneconomical building ways that relied on the skill of the craftsmen and the mass of the structure to make it stand.

JOHN SMEATON

That John Smeaton was the maker and first standard bearer of the profession that today provides us with our tangible world, is best documented by Skempton:[19]

> "The term 'civil engineer' appears for the first time in history on the first page[20] of the Minutes of our Society; the decision by Smeaton and the other founder-members to adopt this title marks the recognition of a new profession in Britain, as distinct from the much older calling of the military engineer. Obviously long before this date there had been men carrying out non-military engineering works (both civil and mechanical in the modern sense) and some of them are quite well known, at least to historians. But it was only after 1760 that the numbers of such men in England began to increase to the point that they could think of themselves as belonging to a profession."

The actual statement in the Minutes book was: "Ano 1771 Society of Civil Engineers."

Smeaton's primacy in the profession is acknowledged by Hollister,[21] Kirby, et. al.,[22] Titley,[23] Straub (indirectly),[24] Hartman,[25] Encyclopedia Britannica,[26] and of course Holmes, and Smiles.[27] Though the active practice began about 1760 and the Society was formalized in 1771, it was not until July 1776 that his own sign, "JOHN SMEATON, CIVIL ENGINEER" hung outside "Gray's Inn, his London Office."[28] He headquartered all summers at his family home at Austhorpe in Leeds and also

---

[19] (M) P-5.
[20] Ibid.
[21] (G) P-33.
[22] (I) P-167.
[23] (N) P-v.
[24] (Q) P-134 + 170.
[25] (F) P-26.
[26] (B) Ed 1959 "Civil Engineer."
[27] (P) P-93.
[28] (E) P-xiii.

hared an office with his friend and colleague John Holmes in the joint proprietorship of the water works for Deptford and Greenwich.[29]

Far more could have been known of this rough-hewn gentleman of knowledge, culture, pragmatism, and human empathy. Unfortunately, his closest kin (except for his daughters Mary and Ann, who helped with his writings and drawings) knew so little of his accomplishments, "... that so much 'rubbish' as it was termed, was found in that square tower at his death, that a fire was kindled in the yard, and a vast quantity of papers, letters, books, plans, tools, and scraps of al kinds, were remorselessly burnt."[30] Happily, all was not lost, for enough still remained to show us a clear picture of that remarkable man.

That John Smeaton had both the character and the personality to be a public personage becomes self evident on examining aspects from random samplings of his life. Smeaton tempered his forthright honest opinions[31] with an innate wry sense of humor:

> On Amsterdam's clock: "... the contrivance however of the pendulum part, with respect to measuring time is *tres miserbale*. The Bells are very good but I think inferior to those of Bruges...."[32]

> On Dutch machines, "this was the only attempt towards a dry dock I see in the low countrys and indeed from the awkwardness with which they had putt it into execution, one would not be surprised that they had grown weary of it.[33]

> On Dutch women, "...but to give the dutch their due, in this branch of Statuary they have shown great tast and Skill, which one would more wonder at, as the original living models in this country are not of the most delicate sort, especially those of the female Kind."[34]

> On Belgian machines and people, "Upon the whole this Machine is one of the worst contrived I ever say, and as badly executed: I should have thought it too despicable to have taken the least notice of it, had I not heared of it from several persons, as being
> cryed up for a most curious piece of art, which numbers had been to see."[35]

> On Dutch Protestants, "The Sunday afternoon in Holland is rather looked upon as a Holyday for making merry than a sacred abstince from Labour."[36]

---

[29] (P) P-169.
[30] (P) P-167.
[31] All entries are quoted exactly as they appeared in the original work, (sic), complete with errors.
[32] (N) P-37.
[33] (N) P-21.
[34] (N) P-30.
[35] (N) P-10.
[36] (N) P-44.

On Dutch Jews, "This morning I was at the worship at the Jews Synagogue, in which there was much Shew and Ceremony. I also heard their singing, which is not performed but once in 3 weeks or 9 month, at which time it is an extraordinary favour for Strangers to be present; the performance was by 3 persons, viz. a Bass, Tenor, and trible and the musick was of much the same kind as that used in our Cathedrals and, considering it was performed by 3 persons and the trible a grown man, the Effect was Extraodrinary."[37]

On French Security, "... and finding how jealous the french are of the English, especially at this time which is looked upon as being the Eave of a War; and fearing least my curiosity at Dunkerk might have been noticed; concerning which they are very watchful, as they are constructing works there, in Breach of the treaty of Utrecht, I was resolved to depart the next morning for Ypres and get out of the french territory as fast as I could.[38]

On Dutch Security, "This Town is in the dutch territory which being a Garrison Town I here underwent the ceremony of being Examined by the military Gentry, who handed me from one to another, was made to walk an hour in the rain and at last suffered to go to my lodging without being asked one material question."[39]

On gambling when about to lose a sizeable sum in a card game with the Duke and Duchess of Queensberry, he beseeched her, "Your Grace will recollect that the field in which my house at Austhorpe stands may be about five acres, three roods, and seven purches, which at thirty years' purchase, will be just my stake; and if your Grace will make a Duke of me, I presume the winner will not dislike my mortgage."[40]

John Smeaton's character and disposition evinced themselves in his reaction to life's most difficult impositions, frustration, and adversity. At the end of his trip to the Low Countries 9 July Smeaton tried to sail for home. He found the ship he expected had sailed without him so he used the day for more observations. The same thing befell him on 10 July and 11 July. But on the 13th he made it aboard a vessel. Then poor winds and low tides detained it until 5 p.m. of the 15th. At the English coast, again bad winds and tides kept them out of their destined port for two more days. Finally, ten days later and a considerable land distance from his destination at Leeds, he summed that great frustration with, "I gladly took this opertunity of packing up my Bag and Baggage, and was to my great satisfaction put safely on Shoar by the pilot Boat."[41]

---

[37] (N) P-40.
[38] (N) P-6.
[39] (N) P-17.
[40] (P) P-175.
[41] (N) P-60.

In adversity John Smeaton was equally as solid. He befriended a young man gone astray. He taught him to be a clerk and got him a job of trust and responsibility, for which Smeaton personally posted a bond. Ultimately the chap went awry again, committing a forgery to meet his debts. He went to jail and Smeaton had to pay the claim for the forfeit bond. Despite this emotional and financial beating, because his wife was ill at the time, John showed no outward sign of his distress until she was well.[42]

To any builder the abysmal worst is the failure of a project after completion. His only English bridge (the rest were all in Scotland), a handsome nine arch masonry structure across the Tyne at Hexham in 1777, failed in a 1782 storm. But it was not his faulty design—the foundations had scoured out under the piers. No one considered the subsoil in those days (nor even as recently as the 1930's), thus "the best laid schemes,"[43] etc.

Smeaton wrote to his resident engineer, a Mr. Pickernell, "All our honours are now in the dust! It cannot now be said that in the course of thirty years' practice, and engaged in some of the most difficult enterprises, not one of Smeaton's works had failed! Hexham Bridge is a melancholy instance to the contrary."[44] In true professional fashion he recovered from that loss, and went on.

Smeaton was perhaps spared from a second failure at Wisbeade. He designed a sluice gate for the River Nene that was shelved for lack of funding. He based his designs on his observations of a similar situation at Ostend, but again he neglected to consider the difference in soil conditions. "... subsequent experience has shown that, if executed, they would most probably have proved failures."[45]

Those two isolated instances, one probable, and one single reality, indicate the human fallibility of John Smeaton. Still, a remarkable record for a pioneer in those times, when stacked against his enormous list of achievements. His tremendous triumph, the Eddystone Light, stood for a century and a quarter only to be dismantled because the unrelenting sea had undermined and eroded the natural rock below his foundations.

John Smeaton's personal foundations were of equal integrity. The Princess Dashkoff was commissioned to make him an unconscionably lucrative offer to go to Russia in the service of the Great Empress Catherine. He politely refused because no money could induce him to leave his family, home, friends, and profession in England. The Princess told him, "Sir, I honour you! You may have your equal in abilities perhaps; but in character you stand alone. The English minister, Sir Robert Walpole was

---

[42] (P) P-176.
[43] Robert Burns, "To A Mouse on Turning up Her Nest with the Plough, November 1795."
[44] (P) P-151.
[45] (P) P-138.

mistaken, and my Sovereign has the misfortune to find one Man who has not his price."[46]

Later, "One of his largest engines went Russia, in 1775, to pump out Catherine II's naval dry docks at Kronstadt. It replaced two enormous windmills, 100 feet high, that Dutch engineers had installed in 1719. It was said that those mills had required a year for the task; the Newcomen engine built by Smeaton did the job in about two weeks."[47]   Other engineers eventually went to Russia at various occasions, John Tarey never completed the Vol. VI of Smeaton's drawings archive because of a protracted professional trip to Russia in 1819.[48]

Smeaton was a cultured man of means who traveled in the best circles, yet mingled among the construction workers with equal aplomb. He "spoke in the dialect of his native county, and was not ashamed to admit it."[49]   No connoisseur, he nonetheless did appreciate music, art, and literature, in a very human, sensitive way.   While watching cattle react to a penny whistle in Holland he noted, "I could not help thinking but the Tabor & pipe must have been the musick wherewith Orpheus charmed the Brutes and made them Dance."[50]

About a church at Bruges he wrote, "When at a small distance, I never heared anything of this kind so sweet; the great Bells were like the strong and steady tone of an Organ & the small ones like the strings of a Lute."[51]   All, comments of an intelligent traveler alive to, and aware of, all he saw. He also loved his wife Ann and his two daughters, Mary and Ann, who returned his affection with love, then cared for him in his ultimate, aged and infirm condition.

John Smeaton was, actually, "to the manner born,"[52] in his family manor at Austhorpe near Leeds, on 8 June 1724.[53]   The manse was built by his grandfather, John Smeaton, for whom he was named.   Father Smeaton was a prominent, moderately prosperous, Leeds and London barrister who really wanted young John to be so too, but John marched to a different drummer.

Because his family had some modest means, Smeaton had an esoteric upbringing. Except for grammar, he did receive a fine elementary education both at home and at the Leeds Free Grammar School.[54]   He showed an early and unusual interest in

---

[46] (P) P-171.
[47] (I) P-137.
[48] (E) P-xiii.
[49] (P) P-174.
[50] (N) P-45.
[51] (N) P-15.
[52] Shakespeare, *Hamlet*, Act 1, Scene 3.
[53] (P) P-180. The birth date, 8 June 1724, was read by Smiles from Smeaton's burial monument at Whitkirk Church, Leeds. I saw that myself, thus Titley is in error when he cites, "Smeaton was born on the 28th of May, 1724." (N) P-i.
[54] (P) P-90.

things mechanical and constructive. While his contemporaries gamboled, playing the usual children's games, pranks, and pastimes, Smeaton spent his time tinkering with tools, instruments, and devices to the extent that his schoolmates called him "Fooley Smeaton."[55] Observing workmen digging a pit or erecting a mill, he was bright and asked incisive questions. Adolescent John at one point built a pump using the design of the sump for Gasfoht Coal mines. It worked so well he pumped out the fishpond at his father's house which, of course, killed all the fish. Fortunately, his father saw the humor of the situation and scaled down the punishment.[56]

By 1742, dutifully trying to follow in his father's footsteps, he went to London to visit the Courts in Westminster Hall, and apprentice at law. It was an honest, vain attempt. Again showing warm intelligent judgement, the elder Smeaton released young John on his own recognizance. He immediately apprenticed to a philosophical instrument maker, lived a conservative, frugal existence, and became successful at what he was doing.

Not merely content with the trade of making instruments as they had always been crafted, Smeaton sought to improve them. He delved into their applications to try to employ them better, then devise new ones. That was the nature of his genius, and that was the way he had to learn. Smeaton was well schooled in mathematics and geometry at Leeds, thus had a leg-up on his fellow workers. But on the whole he was self-taught since there was no education available to him, "for there was no such thing then as the profession of a Civil Engineer."[57]

Smeaton's success came largely from three elements within him: The "bent of his genius;" "his innate love of construction;"[58] and his adherence to a maxim, "The abilities of the individual are a debt due to the common stock of public happiness."[59] He taught himself French, to read the available papers of that era and area that contained much about bridges, canals, and windmills. To know them at first hand, he made an extensive trek through the lowlands inspecting the water and land works in an area where nature had done so little. Thus a man with his meager technology then had to do so much for himself.

Smeaton is sometimes mentioned as a protege of Brindley's. This is not so because there were only eight years difference in their ages. They ran parallel courses with their professional lives until maturity, when they frequently consulted with each other on projects. There was no rivalry since Brindley worked on his canals and windmills in the west of England, while Smeaton worked in the north and in Scotland. When Brindley died in 1772, much of his unfinished work fell to the overloaded Smeaton, by then the top Civil Engineer in England.[60]

---

[55] (P) P-89.
[56] (P) P-90.
[57] (P) P-93.
[58] *Ibid.*
[59] (P) P-95.
[60] (P) P-85.

Brindley was, at age 17, a millwright at Marclesfield. In his snide, self taught way he rose above both his inherited poverty and his traditionally stifled environment to become, by androgogy and attrition, an early engineer in public works. Though he built canals of national note, he could not spell the word "navigator," thus (even today) workmen who dig trenches or canals in western England are sill called "Navvies."[61]

As Smeaton's *Reports*[62] and *Diary of His Journey to the Low Countries*[63] verify, his spelling left much to be desired. Often, he'd spell one word three different ways on the same page. Yet in truth no one ever experienced difficulty deciphering Smeaton's documentation. His expression was lively, clear, and holds up over time. He showed consistency with neither capitalization nor punctuation, a failing of the era and not of the man. When I read his original journals, written with quill and ink by his own hand, his meanings came through in vivid word pictures. The problem of perusing whole pages without the relief of paragraphs fades in the fascination of seeing his first-hand message.

Smeaton made most of his journey on foot or in the small inland canal boats of that place and time. His experiences were carefully noted in a separate *Diary of His Journey to the Low Countries*. Though Smeaton himself kept this day-book and used it extensively through the rest of his professional lifetime (especially on his drainage projects and in his great effort with Dover Harbour)[64] this historically useful tome remained hidden under dust at the Library of Trinity House, London.[65] Happily it was unearthed and transcribed in its entirety. Unchanged by these wise men (H.W. Dickinson and Arthur Titley, the editor), it retains the charm and genuineness of all Smeaton's solecisms, and the precision of his observations. Unfortunately his *Reports*[66] did not enjoy the same intelligent sensitive treatment.

"His *Reports* written in direct and homely language, are outstanding examples of lucidity that might well be followed today.
"The introductions to the *Reports* says, 'as to Mr. Smeaton's style and language, he had a particular, and in some degree a provincial way, of expressing himself, and conveying his ideas, both in speaking and writing; a way which was very correct and impressive, though his diction was far from what may be called classical or elegant.' He himself in his great folio on the Eddystone, says, that the writing of that book cost him more labour than did the building of the Lighthouse itself, and acknowledged the assistance of friends in correcting his manuscript.

---

[61] (C) P-262.
[62] (O)
[63] (N)
[64] (P) P-96.
[65] (N) P-i.
[66] (O)

"Believing that the language of a man of original genius conveys the best impression of his mentality and character, it is the writer's regret that his friends did not leave it alone."[67]

Though I quote excerpts from Smeaton's low country diary elsewhere in this work, the origin, intent, and purpose of that venture, its document, and the man, did get around to the proper places. Very soon thereafter it enabled Smeaton to rise to a challenge; exercise his genius and discipline; and, produce the great triumph of his professional life. That changed his fortunes, and our history.

Five months after Smeaton's return from his low country journey, Rudyerd's Lighthouse on the Eddystone burned to the rock. The immediate replacement of a beacon was imperative, but who could do it? One of the lessees of the Rock, under Trinity House, a Mr. Weston, wrote to the Earl of Marclesfield in 1756 requesting a recommendation. The reply as quoted by Smeaton in his *Narrative of the Building of the Eddystone Lighthouse*, says in part:

"There was one of their Body (the Royal Society) whom he could venture to recommend to the business; yet that the most material part of what he knew of him was, his having within the compass of the last seven years, recommended himself to the Society by the communication of several mechanical inventions and improvements; and though he had at first made it his business to execute things in the instrument way (without ever having been bred to the trade) yet on account of the merit of his performances, he had been chosen a Member of the Society. And that for about three years past, having found the business of a Philosophical Instrument maker not likely to offer an adequate recompense; he had wholly applied himself to such branches of mechanics as he (Mr. Weston) appeared to want; that he was then somewhere in Scotland, or in the North of England, doing business in that line; That what he had to say further of him was, his never having known him to undertake anything, but what he completed to the satisfaction of those who employed him; and that Mr. Weston might rely upon it, when the business was stated to him, he would not undertake it, unless he clearly saw himself capable of performing."[68]

And clearly see it, Smeaton did! The importance of the light on the Eddystone was immense. It was fabled in story and song. An old sea-chantey goes, "My father was the keeper of the Eddystone Light/ He slept with a mermaid one fine night/ And from that union there came three/A porpoise, and platty, and the other was--me... "
Sir Walter Scott penned:

"For in the bosom of the deep,
O'er those wild shelves my watch I keep;
A ruddy gem of changeful light,
Bound in the dusky brow of night;

---

[67] (N) P-v.
[68] (N) P-iv.

The seaman bids by lustre hail,
And scans to strike his timourous sail." [69]

The challenge was to create and situate a structure that would withstand the rigors of time and the inexorable forces of the sea, yet be buildable with the materials and methods available at that time. The comparative history and development of all four lighthouses at Eddystone Rock were best summarized in the *Encyclopedia Britannica.*[70]

Even flamboyant King Louis XIV of France realized the value of the Eddystone Light. During his war with England, a French privateer captured the crew from the Lighthouse and took them all prisoners to France. Upon hearing of it, the Sun King immediately ordered their release and for them to be returned to their station. Though he was at war with England Louis was not at war with nature nor mankind. Moreover, the Eddystone Light was so situated in the Channel to be of service to all nations, even France.[71]

The features that made Smeaton's Lighthouse superior, durable, and a true product of civil engineered construction, were that it was designed and erected by employing knowledge of composition and behavior of materials coupled with applications of mathematical principles. Smeaton personally made seven exploratory trips to the site and despite real danger walked, examined, and noted the features of the Rock. He also devised and designed ways of dovetailing the massive stone pieces of the structure so they would fit precisely, and be locked together both horizontally and vertically.

If there were a flaw in his professional character it was his grudging willingness to delegate responsibility. Smeaton made drawings, some stone cuttings, and inspections himself. Often during construction he would, even in rough seas, go out to the works to handle a difficult or exacting detail. He operated from a work-yard, and got the job done with the aplomb of a circus lion tamer. This was no mean feat in those days since each stone piece cut to his three dimensional jigsaw puzzle averaged about two tons.

Often I think of Smeaton's life as a great subject for a Hollywood movie, but its major triumph, the most significant super event comes far too early in the story.

The only Eddystone Light that withstood the onslaught of tide and time for 119 years could have gone on as a stable structure for additional centuries. The power that undermined it in the truest sense, was the eternally victorious sea. The natural rock below the foundations became so eroded, the stability of the tower then became questionable.

---

[69]   Sir Walter Scott, Pharos'Loguitur, first verse.
[70]   (B) Vol. 14, 1959, P-85 & 86.
[71]   Smeaton's *Narrative*, P-28.

Once established by the Eddystone project, fame and fortune did not rain upon John Smeaton. In modestly comfortable circumstances he could have retired to Austhorpe as a country gentleman. But his useful career had only begun and the Industrial Revolution had just become enabled.

England was poor. The people spent less and complained more because of the lack of domestic progress and the perpetual state of foreign wars. The roads were as the Romans left them, except ravaged by time. Harbors were the way nature effected, then affected them. There were a few canals. Goods still went from here to there on pack horse and continued thus to 1794.[72] The people wanted more than just a bare subsistence, to enable the King support wars in other lands.

The English Crown fought the French: Through established Sir Robert Clive in India continuously from 1748 to 1763; Through young George Washington in North America's French and Indian War from 1754 to 1763; and futilely in Europe's Seven Years War from 1756 to 1763. Those so depleted English energy and morale that when Smeaton proposed a canal, or drainage net, or a bridge, or a new harbor, the project though authorized would be shelved for lack of funds. Actually today's conditions show little change of circumstances through 200 years. The completion of the Eddystone Light in the same year as the end of the three wars with France was a happy coincidence. For Smeaton it was propinquity; for "peaceful arts," a boon.[73]

He became busy designing and building his specialties. The Newcomen Society for the Study of the History of Engineering and Technology published "A Catalogue of the Civil and Mechanical Designs (1741-1792) of John Smeaton, F.R.S. "[74] only because all the works themselves were far too voluminous to publish. The cataloguers Dickinson and Gommc have become to Smeaton's engineering what Ludvig von Koechel was to Mozart's music. Their catalogue lists design drawings in six categories:

I.   Windmills and Watermills for Grinding Corn, 137 items plus
II. Mills for various purposes and Machines for Raising Water, 171 items plus
III. Fire Steam Engines for Raising Water, 190 items plus
IV. Bridges and Buildings, 188 items plus
V.   Canal Works, Sluices and Harbours, 195 items plus
VI. Canals and River Navigations, 114 items plus

In all, some 995 plates. Many of them have items on the backs that bear the same number with a "V" and others in several leaves that have additional sketches. They are unnumbered but bear the drawing number with an "s" notation. A safe estimate is that the catalog shows about 1500 preserved design-drawings of John Smeaton's in the six classes. Smeaton also presented papers on Astronomy and Mathematics to the

---

[72] (P) P-87.
[73] (M) P-14.
[74] (E) P-xi.

Royal Society; but, choosing to keep pure science as a relief hobby,[75] he channeled his efforts into civil works.

The Romans had a trick with concrete that enabled many of their construction efforts to remain standing through Smeaton's time, on into today. Again, though Roman aqueducts, bridges, and roads were used by the public, their principal reason for existence was the operation and maintenance of military pursuit. After the Romans, through the Dark Ages nothing happened with concrete, it was forgotten. But then the need for concrete as a base, binder, foundation material for construction had resurfaced.

For the Eddystone Light and other harbor and jetty projects, Smeaton not only needed a concrete, but one that would form and set-up to hardness in sea water. Realizing the relationship of cement to concrete is like flour to bread, Smeaton reasoned concretely that the better the flour, the better the bread, ergo the cement must be improved. Thus Smeaton became the first "modern" to grapple with the question of varying hydraulic properties of different lime cements.[76]

The Romans used pozzulan, a form of natural limestone from Pozzouli, near Naples.[77] Smeaton sought in England to find a similar material. In experimenting for an underwater hardening cement for Eddystone he found the best hydraulic limes came from limestone containing a good quantity of clay.[78] On determining what he needed, he left the experimentation for others, to build on his research. It was Portland Cement, as it is generically known in today's construction world. It came from Portland, England where the natural limestone has all those required properties.

Cementing human relations was also a Smeatonian talent. Besides his being loved and respected by his family and the community at Leeds (I found he is still mentioned by St. Mary's Whitkirk clergy), Smeaton was revered by his two apprentices. One, John Holmes, wrote the first book about him.[79] Smeaton was esteemed and admired by his workers since the days of building the Eddystone Light. When there was a difficult or hazardous chore, Smeaton himself would go do it. He also moved with equanimity among the Fellows of the Royal Society, and the peerage, having been chosen as a personal friend by, among others, the Duke and Duchess of Queensbury.[80]

But his major effort, his intangible magnum opus that today has grown to outlive the Eddystone, is his founding of the Civil Engineering profession.

---

[75] (H) & (B) & (P).
[76] (B) 1959 "Civil Engineer."
[77] (B) 1959 "Civil Engineer."
[78] (L) & (I) P-196 & 197.
[79] (H)
[80] (P) P-174.

Smiles[81] writes that during the time Smeaton spent in London, "... he was accustomed to meet once a week, on Friday evenings, in a sort of club, a few friends of the same calling, canal makers, bridge builders, and others of the class then beginning to be known by the generic term of Engineers. The place of meeting was the Queen's Head Tavern in Holborn; and after they had come together a few times, the members declared themselves a Society, and kept a register of Membership..."[82]

Skempton, is in semi-agreement about the situation, but differs in the detail.[83] He writes they met at the King's Head Tavern, Holburn from 1771-1792. Quoting from the original minutes books he states they:

"Agreed that the Civil Engineers of the Kingdom do form themselves into a Society consisting of a President, a Vice President, Treasurer and Secretary and other Members who shall meet once a fortnight ... at seven o'clock from Christmas ... to the end of the sitting of the Parliament."[84]

The name "Smeatonian" Society of Civil Engineers came into being in the late 1820's and appeared first in the 1830 member roster.[85] The Society was reorganized " 'in a better and more respectable form' in 1793 by Robert Mylne."[86] John Rennie, F.R.S. the next truly prominent Civil Engineer after Smeaton, became treasurer and ran the Society without a president. "During his period of office the membership decreased slightly, partly due, I think, to the extraordinary standards which Rennie set for himself and others. At any rate, the Society tended to become rather too exclusive; and this was at a time when the numbers of able engineers in the country were increasing. Consequently the need for a more representative and more professional body was felt, and the Institution of Civil Engineers came into being in 1818. At the outset this was even smaller than the Society had been in 1771, but after Telford became President in 1820, the Institution grew rapidly and soon established for itself the position of premier engineering Institution in the world."[87]

There was neither conflict nor jealousy between the two organizations, most leading members belonged to both, and the Society gave its collection of books, drawings, and reports to the Institution in 1845.

But, as Victor Hugo wrote, this was "an idea whose time had come."[88] By 1852, the American Society of Civil Engineers had been founded[89] and the profession long in

---

[81] Samuel Smiles, a Scottish physician (1812-1904) was biographer of Smeaton, Rennie, Edwards, and other engineers of the 18th and 19th centuries. (P).
[82] (P) P-170.
[83] (M) P-5 & 9.
[84] (M) P-8.
[85] *Ibid.*
[86] (M) P-11.
[87] (M) P-15.
[88] Victor Hugo, *Histoire d'un Crime*, Paris, 1852.
[89] (A)

shadow being, was firmly established in the English speaking union, followed rapidly by the rest of the civilized world.

John Smeaton himself continued to be the patriarch and elder statesman of the Society, being personally responsible for the admittance to membership of James Watt, 29 March 1789.[90] Though Watt was a mechanical engineer, his efforts were in public works.

Smeaton's average consulting fee was "two guineas[91] for a full day's work."[92] but he limited his practice to devote time to self improvement and scientific investigations. His paper on wind and water as economical sources of power was a landmark of basics we still use. He developed a pyrometer[93] to measure the increase in length of a metal rod under increase of temperature. Still in use, today's version of this device is called the "extensometer."

John Smeaton was the first of a long line of civil engineers to appear in a court-of-law to act as what today is called an "expert witness, "a "Forensic Engineer." Two excerpt pages of 1782 court testimony[94] appear as Appendix 3. That proceeding established the idea of offering learned opinion based on existing fact, one of our current subcultures.

On that basis, "He was a frequent witness before Committees of both Houses of Parliament in support of Bills for authorizing civil engineering works."[95] And, here again is strong historical disagreement:

> "It is stated in a recent work, edited by the learned recorder of Birmingham, M. D. Hill, Esq., entitled 'Our Exemplars', that 'Smeaton was for several years an active member of Parliament, and many useful bills are the result of his exertions ... His speeches were always heard with attention, and carried conviction to the minds of his auditors.' This must, however, be a mistake, as Smeaton was never in Parliament, except for the purpose of giving engineering evidence before committees, and instead of being eloquent, Mr. Playfair says he was very embarrassed even in ordinary conversation."[96]

As John Smeaton grew older, he became somewhat hobbled by a variety of symptoms that now might be diagnosed as cancer. He also suffered a stroke that paralyzed him but left him with his intellect intact. Despite the tender, loving care from both his

---

[90] (M) P-13.
[91] About $12 in those days, around $7 today!
[92] (P) P-171.
[93] (N) P-42 & 64.
[94] Provided by Professor John W. Briscoe, Department of Civil Engineering, University of Illinois, Urbana.
[95] (P) P-169.
[96] (P) P-163.

daughters, he faded away 28 October 1792, and was interred inside St. Mary's Whitkirk in Leeds.

That was just eight years short of the 19th Century, an epoch in which his work enabled progress to burgeon beyond the imagination of even his own surviving contemporaries. A year before he died, John Smeaton officially retired from the profession he founded. With this circular he announced:

> "Mr. Smeaton begs leave to inform his friends and the public in general, that having applied himself for a great number of years to the business of a Civil Engineer, his wishes are now to dedicate the chief part of his remaining time to the Description of the several Works performed under his Direction. And he hoped that by not accepting further Undertakings he shall not incur the disapprobation of his Friends.
>
> Gray's Inn, 6th October, 1791"[97]

AFTER AFFECTS

The Newcomen "fire engine" worked so inefficiently for so long it might even have been discarded were it not for Smeaton's improvements. Those put it into a position from which James Watt evolved his efficient steam engine. That then changed the power picture forever. Watt himself acknowledged it by calling him "Father Smeaton."[98]

That engine was the development that made the Industrial Revolution possible. Plants were no longer limited to running-stream-sites. Smeaton's designs for buildings, based on engineering knowledge and practice (later carried forward by Rennie) enabled the buildings housing the steam driven plants. His work on canals and bridges made it easy for his heirs and assigns to create better transport facilities. The combining, by followers of John Smeaton's work on roads, bridges, and the steam engine provided the groundwork for the advent of the railroads.

Smeaton was the first to develop and use the diving bell, or caisson, for underwater construction.[99] Without that innovation,, the Roeblings, (father, son, and daughter-in-law) could never have begun their landmark Brooklyn Bridge, completed by 1883.[100] Nor would any of these U.S. Civil Engineering milestones have been possible, like the Philadelphia water system, begun 1799; Ascutney Mill Dam, 1834; Eads Bridge over the Mississippi (St. Louis 1870); Chesborought's Chicago Water Tower, (survivor of the fire of 1871); Wheeling, West Virginia suspension bridge 1854; New York's Croton Dam and Aqueduct System in 1842; and, the incredible 1,766 miles of U.S. transcontinental railroad accomplished within a half century of Peter Cooper's development of the steam locomotive, among many others.

---

[97] (P) P-172.
[98] (P) P-163.
[99] (Q) P-195.
[100] (A)

All those living monuments gave the 19th Century its prominent place in world history as the first century of genuine technological progress. But principally because John Smeaton created the climate for it late in the 18th.

SUMMARY

Historically, at odd times great events somehow tend to coincide. For instance, in 1187 simultaneously the world experienced: the unification of Okinawa; the frustration of the Crusaders at the gates of Jerusalem; the Russians under Prince Igor staving off the Mongols; and England muddling through without Richard. It seems little ever really changes in history. In 1492, I believe surpassing in importance the voyages of Columbus, was the unification of Spain under Ferdinand and Isabella to bring about the defeat of Boabdil III at Granada, thus expelling the direct Moorish influence in Iberia, Europe, and the world.

Accordingly, I believe John Smeaton's establishment of Civil Engineering as a bona fide unique profession, was the prime historical happening of AD 1776. His public works efforts hastened the Industrial Revolution because he amalgamated practice, experience, and theory, to displace haphazard traditional artisanship. Smeaton was an ethical human being dedicated to the improvement of the quality of life above monetary, political, or personal gain. With the Eddystone Light, not only did he erect and kindle a beacon in the open sea to offer succor to those lost in the storm, his proudly proclaimed professional practice lighted the way to public works, to the man-made physical world in which we now live. We all owe a great debt to John Smeaton, Civil Engineer.

q. e. d.

Ithaca, NY, July, 2002

BIBLIOGRAPHY AND REFERENCES

(A) American Society of Civil Engineers, ANNUAL REPORT, New York, 1975.

(B) BRITANNICA, ENCYCLOPEDIA, Vol. 20, London, 1959.

(C) Bronowsld, Jacob, THE ASCENT OF MAN, Little Brown & Co., Boston, 1973.

(D) Dept of the Army, Corps of Engineers, Civil Works Program, CIVIL WORKS IN REVIEW, (EP-1110-2-2), Washington, 1975.

(E) Dickinson, H.W.and Gomme, A.A., Eds. (for Smeaton, John), A CATALOGUE OF THE CIVIL AND MECHANICAL ENGINEERING DESIGNS, 1741-1792, Newcomen Society for the Study of the History of Engineering and Technology, Extr. Pub. No. 5, Courier Press, London, 1950.

(F) Hartman, John Paul, ENGINEERING HISTORY COMPENDIUM, J. P. Hartman, (Fla. Tech. Univ.), Orlando, 1969.

(G) Hollister, Solomon Cady, ENGINEER, Macmillan, New York, 1966.

(H) Holmes, J., A SHORT NARRATIVE OF THE GENIUS, LIFE, AND WORKS OF THE LATE MR. JOHN SMEATON, Smeatonian Society, London, 1793.

(I) Kirby, R.S., Withington, S., Darling, A.B., Kilgour, F.G., ENGINEERING IN HISTORY, McGraw-Hill, New York, 1956.

(J) Merdinger, Charles J., CIVIL ENGINEERING THROUGH THE AGES, Society of American Military Engineers, Washington, 1952.

(K) Morris, M.D., THE CIVIL ENGINEER REACHES OUTER SPACE, (Engineers' News Supplement) N.Y. Herald Tribune, NY, 26 April 1959.

(L) Pasley, Charles W., OBSERVATIONS OF LIMES,....CONCRETE, AND ON PUZZOLANAS, NATURAL AND ARTIFICIAL, J. Weale. London, 1838.

(M) Skempton, A.W., THE SMEATONIANS (Duo-Centenary Notes on the Society of Civil Engineers, 1771-1971), The Society, London, 1971.
Skempton, A.W., Ed., JOHN SMEATON, FRS, Thomas Telford Ltd., London, 1981.

(N) Smeaton, John, (Titley, John, Ed.) DIARY OF HIS JOURNEY TO THE LOW COUNTRIES, 1775, for the Newcomen Society (c.f.(e)), Extr. Pub. No. 4, Courier Press, London, 1938.

(0) Smeaton, John, REPORTS, Society of Civil Engineers, London, 1812.

(P) Smiles, Samule, LIVES OF THE ENGINEERS, (Smeaton and Rennie), John Murray, London, 1874.

(Q) Straub, Hans, (Trns. Rockwell, E.), A HISTORY OF CIVIL ENGINEERING, MIT Press Cambridge, 1964.

BENJAMIN WRIGHT
(1770 - 1842)

THE FATHER OF AMERICAN CIVIL ENGINEERING

Steven M. Pennington, P.E., MASCE[1]

Author Dedication

During the professional career of Neal FitzSimons one of his many goals was to publish three particular works. The first was an effort to further the work of James K. Finch.

Finch was a member of the Class of 1906 at Columbia University and would go on to become Renwick Professor and later the Dean of the School of Engineering at Columbia. Finch had written "The Story of Engineering" published in 1960. Upon his death, Neal was bestowed the collection of Finch history papers. Neal's hope was to someday create an expanded international history of engineering, which he jokingly referred to as his "magnum opus."

The second work was of a more personal nature. Neal's older brother, Tom, had served in World War II as a reconnaissance pilot. During the war Tom

[1] Senior Staff Engineer, Facility Engineering Associates, Fairfax, VA

made so many flights out of England over the theater of operations they were too numerous to number. Unfortunately, on April 20, 1945, Tom was killed when his plane was shot down over the Netherlands. It had always been Neal's strong belief from research he had conducted that Tom had been the last Allied pilot killed in the war. In the years prior to the war Neal had developed a special relationship with his brother and it was his desire to put that story to paper.

The third was to write a definitive biography of Benjamin Wright. Over the years Neal accumulated a large quantity of Wright material and had even established communication with Wright descendents. He would spend hours transcribing many of Wright's letters to create a computer database of resource material to form the groundwork of the work to be published. It is indeed unfortunate that his passing in 2000 came too soon for him to see any of the three become a reality.

The author feels fortunate to have had both as friend and mentor, Neal FitzSimons. It is without question an honor to prepare this paper and to further dedicate it to his memory.

Introduction

This paper will not examine, in any depth, the personal and family background of Benjamin Wright, for either his childhood or his adult life. These periods are viewed briefly to provide context in which to frame his overall character. The reader is referred to the specific works of Charles Stuart and Neal FitzSimons, listed following the text of this paper, for further detail as to both Wright's genealogy and his immediate family. It is the intent of this paper to examine the professional career of Benjamin Wright. In so doing, the significant projects from the complete body of his work will be discussed and the impact he had upon American Civil Engineering.

Family History and Early Life

America in 1770 was a country torn with the threat of revolution. In April of 1775 Paul Revere would make his famous ride and a year later the Continental Congress would sign the Declaration of Independence. Thirteen colonies would become thirteen United States. This was the time in which Benjamin Wright entered the world. Ben was born on October 10, 1770 in Wethersfield, Connecticut. Wethersfield would later become the site for the famous meeting between Washington and Rochembaeu in 1781.

Ebenezer Wright (1742-1808), Ben's father, was born in Stamford, Connecticut. His mother was the former Grace Butler. She and Ebenezer were married in 1768. While Ben was still a child Ebenezer served as an officer in a militia unit seeing action at the fateful Battle of Long Island in August of 1776.

For a period of time between 1784 and 1788 Ben lived with his Uncle, Joseph Allyn Wright in Plymouth, Connecticut. Joseph had been a major in the war. Connecticut was the scene for one of the most gruesome events of the Revolution. Details of Benedict Arnold's 1782 raid into Connecticut with the burning of New Haven and the bloody siege of Fort Griswold were still fresh in people's minds. This was the stage upon which Ben, while with his uncle, found himself learning the fundamentals of surveying. When Ebenezer moved his family from Wethersfield to Fort Stanwix, New York in 1789, Ben at age 19 left his uncle's home and rejoined his family.

Surveying and His First Work as an Engineer

With war's end, George Washington believed the best way to keep the people united was linking the various regions via routes of commerce, i.e. roads, canals and improved navigable waters. To this end he personally involved himself in projects both on the James River in Virginia and the Potomac River between Virginia and Maryland. In addition to the country's defense, improvements to the flow of commerce became a national priority. Unfortunately, the Federal government found itself without the means both politically and financially to make any of it happen. Thus, it was left to the various States to fund and build such works.

Even though Washington was pushing for expansion westward via the James and Potomac River corridors he also took the time to tour the Mohawk River valley during a trip to New York in 1783. During a tour in the company of Governor George Clinton and Alexander Hamilton, Washington encouraged the regional effort. As early as 1784 the New York legislature was discussing improvements in navigation on the Mohawk.

Territory of the
Western Inland Lock & Navigation Company
(Fort Stanwix became Rome)

The Mohawk River flows west to east emptying into the Hudson River north of Albany. At the western end of the Mohawk drainage sits the town of Rome, formerly Fort Stanwix. In 1791 General Philip Schuyler and Elkanah Watson provided the leadership in forming the Western Inland Lock Navigation Company (WILNC). The goal of the company was to improve navigation of the Mohawk and create a canal over the high ground at Rome to connect with Wood Creek to the west. In addition, the navigation of Wood Creek to its mouth at Oneida Lake was to be improved. The traveler would continue through the lake to the Oswego River and into Lake Ontario. In 1791 the New York legislature authorized the first surveys for improvements between the Mohawk River and Wood Creek.

Benjamin Wright is an excellent example of someone being in the right place at the right time. Ben re-joined his family in 1789 and while pursuing his surveying career, he assisted Major Abraham Hardenburgh in the WILNC surveys conducted during September of 1791. The following spring of 1792 the WILNC was formally incorporated with Philip Schuyler as its first President.

Work began in 1793 with a canal to bypass Little Falls. Schulyer assumed duties for the design and construction of the Little Falls canal with boats first passing through in November of 1795. In that same year William Weston was retained to fulfill duties as engineer for the company. Weston, an Englishman, was born in Oxford about 1752. He arrived in the United States about 1793 where he was hired by Loammi Baldwin to plan the Middlesex Canal in Massachusetts. He also worked on the Schuylkill and Susquehanna Navigation Company's Union Canal in Pennsylvania. In 1795 Weston joined the WILNC to continue the work for improving navigation.

During Weston's tenure with the company, the locks were rebuilt at Little Falls, channel improvements were made to the Mohawk River and canals were constructed at German Flats (1798) and at Rome (1797). After this period Weston resumed work with other projects in the U.S. including some water supply studies for the city of New York. He returned to England about 1800. About 1813 He was offered the position as Chief Engineer for the Erie Canal but declined. He died in London in 1833.

With the absence of Weston, the WILNC was in need of an engineer for the remaining work to improve the Wood Creek navigation. The traditional story has Philip Schulyer in need of an engineer to conduct the studies for Wood Creek. Weston had moved on to other projects (supposedly there were issues of money for his salary). George Huntington, a political figure in the Rome area of New York, suggested to Schulyer that a local surveyor, Benjamin Wright could handle the necessary leveling for the study. Schulyer agreed and gave Huntington the company's leveling instrument to in turn give to Wright. Wright took the level and over the next several days disassembled and re-assembled it, adjusted it and conducted level circuits over several miles and found it to be a fine instrument and agreed to pursue the work.

Wright, a resident of Rome and having assisted Hardenbrough with the initial survey, was a logical choice for the task. Wright conducted the leveling through the length of Wood Creek, about the year 1800, and prepared his report for navigation improvements. This would be Benjamin Wright's first work as an engineer. He was about thirty years of age.

The Wood Creek improvements included four sets of wooden locks constructed in 1802 and 1803. It is unclear as to whether these were of Wright's design but they were unique in that they were covered much like the covered bridges of the same era. Wright would go on to perform three additional studies and surveys along the Mohawk River. These were in 1803, 1811 and 1812 at various locations on both sides of the river. Unfortunately the WILNC did not fulfill its goal of uninterrupted navigation between the Hudson and the Great Lakes. The company would later be bought by the state of New York in 1820.

The Erie Canal

Adam Smith in his "Wealth of Nations" penned a phrase; "navigable canals are among the greatest of all improvements." There was no greater advocate than George Washington. Although interested from a commercial point of view, Washington also wanted the varied regions unified. It is an interesting commentary to note the ordeal of travel in those days prior to the railroad. In New York alone it took from two to five days to reach Albany from New York City, by river. On the Mohawk River a sailboat could make 20 miles a day with a fair wind. Travel was exceedingly slow. Things changed, however, when Robert Fulton's steamboat became a reality in 1807. Albany could be reach from New York City in less than 30 hours. The need for a canal through the interior became an even louder cry.

In 1808 at the request of Joshua Forman of Onondaga County and Thomas R. Gold of Oneida County, the New York legislature established a committee to study the feasibility of a canal linking the Hudson with Lake Erie. A survey was conducted by the New York Surveyor General at a cost of $ 600 resulting in a favorable report as to the feasibility of such an improvement. Eventually in 1810 a Canal Commission was formerly organized but it would not be until 1816 that any activity began. To this point political wrangling over financing consumed much of the time.

James Geddes (2)

In 1816 a new canal board was organized with DeWitt Clinton as chairman. Engineers were sought with an emphasis on Europe as the source. Joseph Ellicott of upstate New York (brother of Andrew

Ellicott, surveyor responsible for the boundary of the District of Columbia), along with others in upstate, urged the board to use local talent. The board of commissioners retained three men: James Geddes, Benjamin Wright and Charles C. Broadhead.

During late 1816 and early 1817 surveys were conducted, plans developed and estimates formulated. The following was the initial estimate dated March 1818 for the construction of the Erie Canal.

Western Section	James Geddes	$ 1,801,862
Middle Section	Benjamin Wright	$ 853,186
Eastern Section	Charles C. Broadhead	$ 2,271,690
Total Estimate		$ 4,926,738 (Ref. 16)

Wright's section, the middle section, was ninety-four miles in length and included that portion of the canal going "over the top" between Atlantic drainage and Great Lakes drainage, in the area between Rome and the Seneca River. The famous "Long Level", between Utica and Syracuse was within Wright's section. The "Long Level" was the longest stretch of uninterrupted canal, 69.5 miles in length. There was much debate at the time as to whether a leveling circuit could be run and even if a canal would function properly over such a great distance as "level". Wright would go on to prove differently to the skeptics.

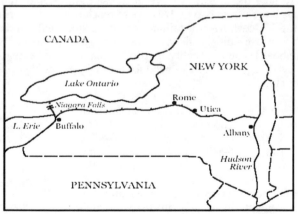

Route of the Erie Canal

The commissioners in 1817 had established dimensions for the canal prism as forty feet of width at the water line, twenty-eight feet at the bottom, and four feet of depth. Locks were to be ninety feet long and twelve feet wide. The first contract for construction was let on June 27, 1817 with ground formally broken on July 4, 1817.

Construction of the canal was done by manual labor. Pick, shovel and wheelbarrow were the implements commonly used. Once trees and undergrowth were cleared, a horse drawn winch was used to pull the stumps. The chief engineer for each section would have two or three assistant engineers under his direction, each in turn in charge of a survey party. In addition to route layout the engineers were constantly setting grade for excavation. In addition to the main line canal, feeder channels were constructed, locks, waste weirs, culverts and aqueducts carrying the canal over natural obstacles were also designed and constructed. Except for the timber lock gates, much of the structural material was stone.

Wright was fortunate to have had an excellent corps of assistant engineers. The Erie Canal is considered the cradle of American Civil Engineering for this very reason. From this enormous public works project came such men as John Jervis, Nathan Roberts, David S. Bates and Canvass White.

John Jervis (1795-1885) was born in Rome, New York. During Erie Canal construction Benjamin Wright sought out the younger Jervis to serve as an axeman in one of the survey parties. Through his own self-will he educated himself in the techniques of surveying and in time became proficient with the instruments. He had been placed under Nathan Roberts and later under David S. Bates. According to Jervis, Bates was more versed in horizontal surveying (traverse) and not as familiar with the vertical (leveling), of which Jervis was well accomplished. In a short period of time Jervis was performing engineering work.

John Jervis (3)

David Bates (1777-1839) had worked with Wright on various land survey projects prior to the Erie Canal. In 1817 Bates approached Wright for a position with the canal construction. Bates was placed as an assistant engineer in the middle section with chief responsibilities for the construction of the locks at Lockport and the aqueduct over the Genesee River at Rochester. The aqueduct was over eight hundred feet long and consisted of nine stone arches.

Nathan Roberts (1776-1852) was most noted for the design of the locks at Lockport. It was his design for the double series of five locks. Each lock had a lift of twelve feet to ascend the rock and bring the canal up from Lake Erie.

Canvass White (1790-1834) joined Wright in 1816 as an assistant engineer. His work would continue until 1824 on the Erie as well as other canals in New York. For a brief period in 1817 he traveled to Europe to study canals and

Nathan Roberts (4)   their construction. He is most noted for his discovery of hydraulic cement, which he patented in 1820. His career beyond the Erie was principally with the design and construction of canals in the United States. In a letter to W.W. Woolsey and dated June 1820, Wright talks about White's discovery.

> The specimen of Argillo-ferruginous limestone, herewith presented, is found in great abundance in the counties of Madison, Onondaga and Cayuga, in the state of New York. . . it is found to be a superior water cement, and is used very successfully in the stone work of the Erie canal, and believed to be equal to any of the kind found in any other country. . . Mr. Canvass White, a friend of mine, has obtained a patent for it when used for hydraulic purposes, and it is believed it will answer an excellent purpose for rough casting, etc. . . (Ref. 26)

Canvass White (5)

The ninety-four mile middle section of the Erie was completed on October 22, 1819 at a cost of $ 1,125,983. It is to Wright's credit that actual costs came in about 10% over the budgeted amount. The entire canal was completed in 1825 at 363 miles in length and contained 83 locks giving a total height differential between Lake Erie and the Hudson of 555 feet.

In addition to locks, culverts, aqueducts and waste weirs there were other interesting canal features. All construction was done by manual labor. Simple hoists were used for placing stone but digging was done by hand. When trees and undergrowth were cleared from the route the stumps remained for removal prior to digging. A "stump puller" was devised utilizing teams of horses or oxen to pull a large wheel hoist.

Model of stump puller (6)

When the canal opened, travel on the canal was not free. Charges were made based on tonnage for any particular boat. To that end there had to be some means in which to weigh the boats. At various locations, weighlocks were constructed for that purpose. The boats would enter the lock and the gates closed. The displacement would change the height of water in the lock. This height was correlated thus giving the weight of cargo on board.

The Erie Canal is the cradle of American Civil Engineering. It was America's first large public works project, opened the west and provided incentives for other states to do likewise. It is without doubt a significant reason why the City of New York became the important world city that it is. In 1967 the American Society of Civil Engineers declared the Erie Canal a National Historic Civil Engineering Landmark.

Weighlock (7)

The Delaware and Hudson Canal

William and Maurice Wurtz, brothers of Swiss extraction, were born in New Jersey in the period between 1783 and 1792. They were later partners in a dry goods business in Philadelphia. With hostilities during the War of 1812 all shipments of coal and fuel from England were terminated. Even with the conflict over, fuel was still in such short supply that as late as

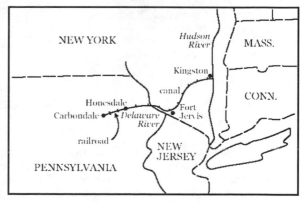

Territory of the
Delaware and Hudson Canal

1814 fuel for manufacturing was almost non-existent. Directly impacting their business the Wurtz brothers sought out other means to replace British coal.

It was common knowledge since before the Revolution that coal was present in northeastern Pennsylvania. The exact date is unclear but the Wurtz brothers came into possession of land in the Lackawanna valley for the purpose of mining coal, specifically anthracite coal. They established the village of Carbondale for their headquarters and homes for the miners. By 1822 they had managed to produce over a thousand tons of anthracite and, with a great deal of hardship, managed to get it to Philadelphia. Their real market, and the one with the greatest demand, was New York City. To reach New York required taking coal from Philadelphia by barge down the Delaware River and around New Jersey. With the success in New York State with the Erie Canal, the Wurtz brothers looked to a canal.

Benjamin Wright was retained in 1823 to conduct a study between Carbondale and the Hudson River for the purpose of a canal. Surveys were conducted and an initial report issued in 1824 giving a first glimpse as to what would be encountered. In 1825 the first board was elected for the Delaware and Hudson (D&H) Canal Company. John Jervis had joined Wright and immediately began route surveys. On May 21, 1825 Wright issued his report for the selected route of the canal. Wright was shortly appointed Chief Engineer in June 1825.

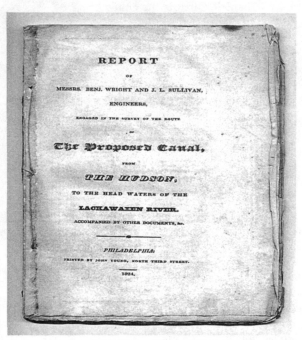

Cover of D&H Canal Report, 1824 (8)

The canal would not be an easy task. The proposed route would begin in Kingston, New York on the Hudson. The canal would travel up Rondout Creek to the divide and then over into the waters of the Neversink River. Thence down along the Neversink to its confluence with the Delaware River and up to what would later be called Honesdale (named for the first president of D&H, Philip Hone). From Honesdale a railway would reach the mines at Carbondale.

Wright would resign his position with the D&H in March of 1827, an issue to be discussed later in this paper. The first successful run of canal boats from Honesdale to the Hudson was completed in December 1828. The railroad as first proposed was to be a gravity operation. John Jervis, who succeeded Wright, proposed the use of the newly devised locomotives from England. In the years that followed the railroad portion of the operation proved troublesome. However, coal made its way to New York and the company operations were a success. It is interesting to note that Wright's original route was for the most part the route eventually utilized.

D&H Canal, Lock #51, near the Delaware River (9)

John Jervis would later resign his position with the D&H in 1830. He would go on to have a successful career with other canal activity and in the emerging railroad industry. His best-known accomplishment however, is the design and construction of the water supply system for the City of New York

Further Canal Engineering

With his successful activity at the Erie, Wright would go on to further employment with other canal companies. In 1821 he was retained to conduct route location studies for the Farmington Canal in Connecticut. The proposed route would run from New Haven on the coast up the valley of the Farmington River to a location on the Massachusetts border. Wright conducted his survey in 1822.

In 1822 Wright would become involved with the Blackstone Canal in Rhode Island and later that year he was retained by the Chesapeake and Ohio (C&O) Canal in Maryland. The C&O Canal was a continuation of the same strategy along the Potomac River valley initiated by George Washington after the revolution. The Patowmack Company had created a series of river improvements and bypass canals on the Virginia side of the Potomac. In a short period of time these elements proved insufficient and were abandoned.

Territory of the
Chesapeake and Ohio Canal

      The C&O Canal was routed along the Maryland shore of the Potomac River and sought to run from Georgetown in the District of Columbia westward and eventually cross over the mountains to reach the Ohio River. Begun in 1828 the canal encountered obstacles not the least of which was the B&O Railroad. Running from Baltimore westward the rail route hit the Potomac Valley at Point of Rocks Maryland. At first the railroad wanted joint right of way use but the C&O, through its original charter under The Patowmack Company kept the railroad out. However, the Maryland General Assembly countered and forced the canal to share right of way with the railroad between Point of Rocks and Harpers Ferry (now West Virginia) where the railroad entered Virginia.

C&O Canal at Harpers Ferry (10)

      Even with the overshadowing by the railroad the C&O Canal is a marvelous engineering feat. Wright would serve the canal between 1828 and 1831 with completion to Cumberland, Maryland in 1850 after his death. Along its route—a distance of 184.5 miles—are eleven stone aqueducts of significant size.

At the time the Monocacy Aqueduct with its seven stone arches was one of the largest masonry structures in the world. Another unique feature, and probably its most interesting, was the Paw Paw Tunnel. Built through solid rock, the tunnel was over three thousand feet long, saved five miles of canal and took twelve years to complete. The tunnel is lined with layers of brick, just under six million in all.

C&O Canal – Monocacy Aqueduct (11)   C&O Canal – Paw Paw Tunnel (12)

Wright would go on to serve as consulting engineer on various canal projects, including the Saint Lawrence Ship Canal, The Welland Canal and projects in Chicago. From his humble beginnings as a country surveyor, Wright was America's authority on canals, both their design and construction.

The Erie Railroad

In short order the railroad became the mode of transportation for the future. The Baltimore and Ohio (B&O) Railroad, America's first rail line, would bring affluence to the City of Baltimore. It didn't take long for the City of New York to acquire a sense the urgency to establish a link westward to the Great Lakes. In November of 1831 the New York legislature

Territory of the Erie Railroad

granted incorporation to a company for the construction of a rail line between the City of New York and Lake Erie. The company was The Erie Railroad.

After a couple of years obtaining financing and conducting feasibility studies, the railroad formally established itself with a board of directors and a president. The first president was Eleazar Lord. Interestingly enough, Lord received his education as a minister but never preached a sermon. He established himself in business, especially in insurance, as president of the Manhattan Fire Insurance Company. Sitting as one of the board for the railroad was Benjamin Wright. This was July 1833.

Wright no doubt held position due to his notoriety with the Erie Canal and his work on the Delaware and Hudson Canal. The political wrangling was enormous. The success of the venture relied on cooperation in all regions of the state. The New York legislature intervened and took on the responsibility for route location and authorized the sum of $ 15,000 (a small sum for the task even in those days) for the necessary surveys. Wright was retained by the state (not the railroad) to conduct the survey.

Wright's task was not an easy one. His directions from the legislature were to keep the route outside the counties with Erie Canal right of way to the north and outside the state of Pennsylvania to the south. For the study, Wright divided the 500 plus mile corridor giving the half from the Hudson to Binghampton to James Seymour and the other half from Binghampton to Lake Erie to Charles Ellet. These assistant engineers each had 2 or three survey crews conducting fieldwork for route location. Wright's report was dated January 1835

The route selected by Wright was one allowing the full use of locomotive power to pull both passenger cars and those with cargo. He estimated the steepest grade along the route to be 2% and even suggested the possibility of a tunnel under the Shawangunk Mountains. However, one element that troubled Wright was the necessity of running the railroad through the Delaware Valley above Port Jervis. This was the alignment of the Delaware and Hudson Canal and Wright expressed concern for the diversion of freight to the new railroad. Knowing that in good engineering terms the best route was through the Delaware valley, he found himself shouldering the burden of his decision.

The final route location was a decision made by the board of directors. The eastern terminus was at Piermont just north of New York City. The route did require a short portion in Pennsylvania with the western terminus at Dunkirk on Lake Erie. Wright's tenure as civil engineer concluded with his report. The road was completed to Dunkirk in 1851 at a cost of over $ 43,000 per mile. It is interesting to note that this actual cost is six times the estimate Wright had made in his report given sixteen years earlier, but that is something that could be expected considering the newness of the technology. Within a brief period of time a spur line would be constructed to Jersey City, New Jersey, giving direct access to New York City.

Wright would go on in his career with further railroad work. Along with work on the Harlem Railroad, he was involved with the Tioga and Chemung Railroad, both in New York. In 1835 he assisted his son, Benjamin Hall Wright with route selection for railroads in Cuba. It is unclear as to whether the senior Wright ever went to Cuba, but his son spent several years there.

Street Commissioner of New York

In 1832 Wright was appointed Street Commissioner for the City of New York. Wright was 62 years of age and the years of travel kept him separated from his family for months at a time. In an effort to be nearer his family he took the position in New York City. He was only there for a short period but the time was not without important events. During his tenure he oversaw studies and surveys for the water supply of the city. His report submitted in 1834 reinforced earlier reports of Canvass White and, even earlier, those of William Weston that the water source should come from the Bronx River watershed. Politics would enter and, even with Wright's reputation, pressure was strong for the source to come from the watershed of the Croton River. Eventually the Croton would become the water supply source with construction of a dam, aqueduct, and the famous Harlem River Bridge. Wright's assistant engineer from Erie Canal days, John Jervis, would bring his career to a high point with his work on this project. Jervis became Chief of the Croton Aqueduct project in 1836.

The James River and Kanawha Canal

It had long been a dream of George Washington for an all water route linking the eastern territory with land to the west. Even in those early years he was aware of the possibilities. Simply link the James River in Virginia to waters in the western part of the state that empty into the Ohio. A traveler could proceed down the Ohio to the Mississippi and from there possibly reach the foothills of the Rockies. Unification was uppermost in Washington's mind. To this end he applied himself to making it a reality. One of those projects was the James River Company.

The James River Company was chartered in 1785 for the purpose of navigation improvements to the James River. The James flows from Tidewater, through Richmond westward to its headwaters near what is now Covington, Virginia. With frequent flooding channel improvements were unsuccessful and the state purchased the charter in 1820. In March 1832 the James River and Kanawha Company was formerly organized for the purpose of creating a route linking the James River to the Kanawha River, creating a link between Richmond and Point Pleasant (in what is now West Virginia) on the Ohio.

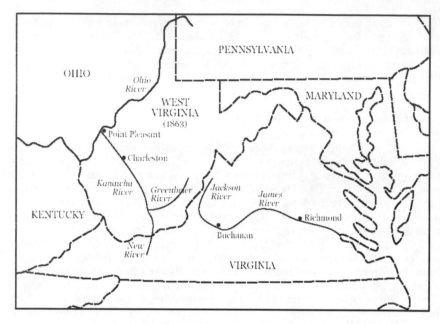

Territory of the James River and Kanawha Canal Company

As early as the spring of 1824, Claudius Crozet, the Principal Engineer for the State of Virginia, was conducting studies for a canal along the James River through the gorge where it passes through the Blue Ridge. At the request of Governor Floyd, Wright was brought in to consult with Crozet in the fall of that year. Wright reviewed surveys and the two men consulted with one another and separate opinions issued. Crozet took issue with Wright on a number of issues. The spacing of feeder canals supplying water to the improvements as well as Wright's suggested designs for masonry locks were of concern to Crozet. On the subject of a canal itself the two differed as to the size of the canal prism. Crozet wanted a canal forty feet wide and three and one half feet deep. Wright suggested a canal fifty feet wide and five feet deep.

Claudius Crozet (13)

At that point in time there were three alternatives under consideration. The first was a lock and dam configuration on the James, the second was a parallel canal and the third was an all rail route. Wright favored the canal, Crozet the lock and dam configuration. Crossing the mountains to reach the Kanawha was still an unsettled issue. Wright was not in favor of the all rail route. It is interesting to note Wright's views on these issues.

The railroad was still a technology virtually unknown to many. Wright felt that using locomotive power would require a greater mechanical skill than many Americans possessed. He also felt the railroad would be more expensive than the canal and strangely enough he felt that from a "cargo security" issue the canal boat was a safer place for cargo than a railroad car. The overwhelming consequence of the differing opinions of the two engineers was that Virginia did nothing.

The political wrangling went on for years with both sides unable to come to an agreement. The question of canal vs. lock and dam went undecided. Meanwhile Crozet would conduct further studies into the mountain crossing. It was his strong feeling there was not enough channel or water to create an economical crossing and thus the railroad was the most practical route from Covington westward. Crozet would resign his position in 1830, due to a great extent to the prioritizing of his work by the various state delegates.

One factor in all the studies, which provided a degree of uncertainty, was hydraulic cement. The cement would be necessary for lock construction and would be expensive if the material were shipped from outside the state. Eventually in 1827, John Hartwell Cocke, a prominent leader with land holdings along the James discovered a limestone source for hydraulic cement. Cocke and Wright were frequent correspondents. As can well be imagined, Cocke had much to benefit from the improvements to the James. The products of his plantation would reach their market sooner and cheaper. However, the uncertainty of the supply created uncertainty in the cost estimates formulated by each engineer. These uncertainties entered into the disagreements between the two, thus, ultimately factoring into the delay in getting anything started, an issue to be discussed later in this paper.

Eventually, Virginia made up its mind although reluctantly for some, for an all canal route to the base of the mountains. Wright took the position as chief engineer in 1835 at the age of 65. He immediately divided the task of route location and design to three assistant engineers. One of his assistants was his son Simon W. Wright, another was David Livermore and the third

James River & Kanawha Canal – Boat "Marshall" (14)

was Charles Ellet, Jr. The first portion of canal construction was that between Richmond and Lynchburg, 146.5 miles, and completed in 1840, shortly before Wright's death. A second from Lynchburg to Buchanan, 50 miles, was completed in 1851. The third from Buchanan to Covington was never built. Wright was associated with the canal operation until the time of his death in 1842.

Charles Ellet would go on to have an illustrious career. He is credited with the design and construction of the first suspension bridge across the Ohio at Wheeling. He is most noted for his elaborate studies of the Mississippi basin. His paper regarding flood control was ahead of its time and was a precursor to what would later become The Tennessee Valley Authority (TVA). During the Civil War, on the side of the Union, he put forth the idea of taking river steamboats and converting them to battering rams for naval combat on the river. He was killed in battle while commanding one such vessel.

Charles Ellet, Jr. (15)

The canal was never completed to Covington. The completed portion consisted of 197 miles with 90 locks. The railroad and possible connection with the Kanawha was not built until after the Civil War and as a different enterprise. However, as late as 1850 studies and plans had been developed for the complete length of the project including an all water link across the mountains via a series of level canals through tunnels. The state never appropriated the money and the canal was never extended beyond Buchanan.

QUEEN OF THE JAMES RIVER AND KANAWHA FLEET

This boat carried the body of General Stonewall Jackson to Lexington, Va. for burial in 1863. It was the last of the canal boats to enter Lynchburg and was beached on the south bank of James River one mile above the city and remained there until carried away by the flood of 1913. The old hull, constructed of galvanized iron is 90x14 ft. and is on display at Riverside Park. The last owner and captain was James A. Wilkinson and the last mate was Captain Wilkinson's son, James P. Wilkinson.

Canal Boat "Marshall" – Carrying the body of Stonewall Jackson, 1863 (16)

The James River and Kanawha Canal brought together two engineers from different walks of life. Crozet was a French military engineering with formal training. Wright was a country surveyor with training through experience. It is easy to see the clash of wills.

Establish a Professional Society

Civil engineering in 1838 was slowly emerging as a profession in America. It was in that year a meeting was held in Augusta, Georgia for the purpose of organizing the civil engineering profession. The details of the meeting and especially a list of those who attended are unknown. However, it is known that a direct result of that meeting was the invitation to all engineers for an organizational meeting the following year.

On Monday, February 11, 1839 at Barnum's Hotel in Baltimore, Maryland the first organizational meeting of civil engineers was held. There were forty engineers present from eleven states. Benjamin Latrobe of Pennsylvania was elected president and, from the forty engineers, two committees were appointed. The first, a committee of seventeen, was commissioned to draft a constitution, and the second, a committee of five, was to draft a proclamation to be sent to all engineers inviting them to join. On the list of forty engineers selected to draft the constitution were such names as; John Jervis, Benjamin Latrobe, Claudius Crozet and Benjamin Wright of New York.

In April 1839 the committee of seventeen convened at the Franklin Institute in Philadelphia for the purpose of drafting a constitution for the new organization of engineers. Of the seventeen, only four attended. They were Benjamin Wright, William S. Campbell, C.B. Fisk and Edward Miller. Miller was appointed secretary of the committee. With only the four, a constitution was developed. It is curious to examine this document. One stipulation of membership was a requirement that all members must publish or present a paper each year or be subjected to a fine of $ 10.00. At the conclusion of work the constitution was formerly presented to the balance of the committee for consensus.

In his letter of July 15, 1839 to the President of the society, Latrobe, Miller presents the results of the committee action. The following is an extract from his letter:

> I have the honour to inform you that the form of Constitution proposed for the Society of Civil Engineers by that portion of the Committee of Seventeen which met in Philadelphia on the 10th of April, agreeable to their appointment, is rejected. (Ref. 25)

It is interesting to note one of Miller's comments farther into the letter. He offered an idea that the extent and ordeal of travel for meetings were possible factors in the negative vote for the constitution.

Voting for the constitution were Wright, Campbell, Fisk, Robinson and Gay. Other than three abstentions, the balance was dissenting votes. Without a working structure the idea for a society of civil engineers was shelved. It would not be until after Wright's death that in 1852 a successful attempt would be made to establish the American Society of Civil Engineers. However, it is quite easy to understand how he felt about the idea. He was one of only four who made the effort to travel and take time for the organization and he voted in favor of the constitution. This shows a great deal of care for his chosen profession.

Extract of letter from Miller to Latrobe (17)

A Career Retrospective

In the examination of any prominent person's career it becomes easy to merely retell key events and important projects, to present what would amount to a historic resume of their accomplishments. Like most men, Wright was not without his faults. Any discussion of Benjamin Wright should not overlook those issues, which shed light on his true character, whether favorable or not.

The Erie Canal is without doubt America's first great public works project. It is rightfully considered the cradle of American civil engineering. Wright was fortunate to be in the right place at the right time. The canal was completed, became successful as a business venture and brought Wright a great deal of fame.

Throughout his career, Wright found himself coming back to the Erie, not in the physical sense but in his interactions with people wanting to take credit where none was due. One such involvement was with Jesse Hawley. The question was; "who was responsible for the origins of the Erie Canal?"

The traditional story is as follows. A resolution was introduced before the New York Legislature by Joshua Forman and Thomas Gold in 1808. The resolution called for the study of a canal linking the Hudson with the Great Lakes. At the time Wright was also an assemblyman. The story goes that Wright having received a copy of "Ree's Cyclopedia," which included a section on canals, spent the better part of an evening with Forman discussing the merits of just such a venture. Shortly thereafter Forman introduced his resolution. The assembly adopted the resolution and with the political push of DeWitt Clinton the canal became a reality.

Jesse Hawley, a writer from New York, put forth the notion that he in fact should take some, if not all, the credit for the canal idea. For a couple of reasons he felt he should be the one responsible. The first was a series of newspaper essays he wrote in the *"Genesee Messenger"* putting forth the notion of a cross-state canal. The articles were in fact written prior to date of the act of legislative resolution. The second was evidence that DeWitt Clinton was knowledgeable of these articles, and their message, prior to any act of the assembly. In a letter to Wright dated July 17, 1835, Hawley asks Wright for his recollections on the matter.

Wright in his response letter dated July 22, 1835 tells Hawley that he was not aware of the articles mentioned and placed Clinton in the highest of regard. He goes on to recount the events leading up to the canal including the Surveyor General's work prior to 1810. Wright's character reveals itself when he neither takes direct responsibility himself nor does he shun Hawley's attempt to share in the notoriety. Wright, in a sense, applauds Hawley's foresight and tells him that he should take pride in being among the leaders in the canal movement.

In an article written in 1866 and published in 1870, Benjamin Hall Wright, talks about the work of his father on the canal. He mentions specifically the efforts of Hawley to take credit and as would be expected gives much credit to his father. Like his father, putting self-interest second, the younger Wright says flatly that the origins of the Erie Canal belong to no one. Benjamin Wright was in a position to have captured much of the fame and glory but chose not to.

During Wright's tenure as Chief Engineer with the Delaware and Hudson Canal he was fortunate to have had John Jervis working under him as an assistant. Wright had been appointed chief engineer in 1825 and one of his first tasks was to study the route for the canal. The design was straightforward with locks and canal. However, between Honesdale and Carbondale his initial report lacked the detail as to the means of connection with the mines.

In September of 1825 George Stephenson placed into operation the first steam railway locomotive on the Stockton and Darlington Railroad in England. It was an event that would change the history of transportation. At the time John Jervis joined Wright in 1825 the connection between the canal and the mines was not yet finalized. One of the key responsibilities of Jervis was route selection. Jervis filed reports with Wright for due consideration in which he strongly advised Wright to use a railroad to link the headwater of the canal with the mines. In September 1826, in his report to the Board of Direction, Wright provides the necessary detail regarding the final link and give details for the planned railroad. What Wright failed to do in this report is give fair credit to Jervis.

The next several months proved troublesome for the two engineers. Jervis felt stepped on by his superior in both the report as well as with duties on the project itself. Wright was absent much of the time and placed a good deal of responsibility on Jervis. Jervis would repeatedly ask for Wright to come to the site and deal with issues but to no avail. Feeling frustrated Jervis challenged his superior. Wright responded to his Assistant on December 15, 1826. His letter, in its entirety is as follows:

Dear Sir;

Last week, Thursday night, I arrived here (New York) in feeble health and I continue so, attended with fever at night and a very bad cough.

On my arrival I found your favor of the 23rd. I have read it over and over again and tried to examine my own conduct for the past six month to see if I deserved the inferences which I draw from the contents of your letter. I have never, as I think, attempted to trim responsibility and particularly that worst of most unpleasant of duties, the adjustment of accounts.

My long protracted ill health and for the loss of my son (which is a severer affliction of itself) added to my professional difficulties and troubles and ought to be some excuse for my not meeting you long before this time as I wished and intended.

As I always mean to be frank with those I consider my friends I shall now open my mind freely and fully.

I have long wished and endeavored on my part to have you my friend, as I have been and still am yours. But I have perceived for many months past that a coldness, a distant non communication and reserved conduct on your part toward me gave me strong evidence that I was not numbered in your breast among your friends, I regretted it because I wished to be your friend and to show it in that frank and unreserved manner which used formerly to constitute our intercourse with each other. I tried to make it so on my part but when I could not effect it, I consoled myself with the reflection that no man can control the feeling and friendship of others and therefore although I may feel grieved and regret it I must submit.

In reflecting upon the contents of some parts of your letter where you appear to feel that the Delaware and Hudson Canal will be an injury rather than a benefit I am totally at a stand how or whence such a feeling can arise, from the manner it is introduced it would perhaps be fair to draw the inference that I am the cause of these unpleasant reflections in your breast.

In reviewing my own conduct I do not feel that I am deserving of being the cause of these feelings on your part toward me.

If you look back and see the course they have taken as to the Delaware and Hudson Company and the necessity of taking measures and plans to raise the value of the stock and thereby obtain additional subscriptions, you will there find the true cause of my remaining so long attached to the company. I know very well that you can perform all the duties of Chief Engineer for the execution of the work on the canal and its location and my good judgment as to all this and where I say that but for the causes I have hinted above nothing on my part would have prevented you such as far as my wishes are concerned, and I may add that such a course will be taken with the consent of others to bring about such an event. If it fails (of which there is I hope, no probability) you will only have to charge yourself with the blame.

I shall meet you as soon as I can and adjust the amounts of the lock contractors. But my Dear Sir let me say that if you shall continue to possess the same feeling—the cold, distant and reserved uncommunicative manner which exhibited toward me for the last five or six months—our meeting will be very illy calculated to adjust accounts as associates as it will want that frank and free expression of opinion so necessary to compare minds and come to correct results.

As there is five percent reserved until a trial of the locks I do not see any great evil in that can. . . (Ref. 29)

The letter is unfinished and unsigned but clearly in Wright's language and handwriting. Interestingly enough, it is among Jervis' collection of letters. In short, Jervis received it.

The tone of the letter is direct and tries in a diplomatic manner to put Jervis both at ease and in his place. The letter also encourages Jervis to step forward and take the responsibilities, especially if he is to further his career. It may very well be that Wright, overextended with other work and dealing with family problems (Jervis was unmarried at this point) was truly unable to attend to matters as frequently as he should. The omission of Jervis' name in the report may very well have been an oversight but it remains a question without an answer. However, in the best interest of both the friendship and the company, Wright resigned three months later.

In 1824 Wright arrived in Richmond, Virginia with the purpose of consulting with Claudius Crozet regarding the James River and Kanawha Canal. Crozet had already performed surveys and conducted studies. Three possible alternatives were present. The result of their meeting was the inability to agree on anything. They differed on many issues including the design of locks, the size of the canal prism, cost estimates, the use of railroads and the very method of the route; canal vs. lock and dam. Various political figures were behind each of the various methods and also behind each of the engineers. As a result, the lack of consensus between the engineers led state as a whole to do nothing.

In a letter to Governor Floyd dated August 1831, Wright says:

> . . . The public mind is now so unsettled in their opinions, on the comparative advantages and disadvantages between railroads and canals. And considering that it will take some little time to have the good people of Virginia satisfied, I have had doubts in my mind, whether it would be useful for me to return here again. . . (Ref. 12)

Virginia did not know what it wanted. Nor was it getting any sense of direction from the two engineers. When Virginia finally decided to take some sort of action it would be in 1850, long after Wright's death, with the canal only reaching Buchanan, short of its goal.

This period in Virginia history may very well have been a turning point for the entire nation. Wright and Crozet were two different types of engineers. Crozet was younger, the logical thinker and frustrated with politicians. Wright was older, more set in his ways and better able to deal with politicians since he had been one. Regardless of the hydraulic cement issue, had they been able to reach some sort of an agreement the state may very well have taken some sort of action much earlier. The use of hydraulic cement factored into both schemes for construction, and limestone supplies were discovered by Cocke in short order. So, if Virginia had been able to breach the mountains and reach the Ohio, could history have been different? With a route, whether rail/canal or all water, from the Ohio through the mountains to the coastal port at Hampton Rhodes, Virginia would have become a key player in American commerce. Rather than ship his crop around Florida, a planter in Alabama could ship his cotton up the Tennessee River to the Ohio and onward to the Atlantic for shipment to England at reduced costs and in much less time. This route would have connected the country and also avoid the split of the state in 1863. North and South would become more interdependent on each other. State's Rights and slavery were issues in need of a resolution but, with the increased interdependency, was war necessary?

Neither man can be said to have been either right or wrong. Their arguments were valid. They just simply could not agree. America will never know what could have been.

His Family and Personal Life

Ebenezer and Grace, Benjamin's parents were devoutly religious people. Ebenezer's father was a minister in the Congregationalist Church in Wethersfield, Connecticut. It is quite possible their strong religious beliefs were what held them together as a family during those early times as they made the transition between Connecticut and New York. Ebenezer was without work and their leaving Connecticut was to find a better life. When they arrived in Fort Stanwix (later to be named Rome) he took up farming and apparently was able to provide for his family over the years that followed.

Ebenezer's brother, Thomas married Martha Butler, Grace's sister. During the war, he and Thomas had both served together and spent some time at Fort Stanwix. Thus when the decisions were made to seek a better life Ebenezer and Thomas had some idea of where they wanted to take their families. The two Wright brothers settled in what is now Rome, New York about 1789.

Ebenezer and Grace, from the time they arrived in Rome, opened their home for religious service on Sundays. With no church building, several families would come and attend services each Sunday. In 1793, five families signed a covenant to organize what would later become the First Religious Society of Rome. Two of the families were those of Ebenezer and Thomas.

On September 27, 1798 Benjamin married Philomela Waterman of Plymouth, Connecticut, a girl he had known since his childhood. It is more than likely that Benjamin met her while he was staying with his uncle, Joseph Allyn Wright in Plymouth. Between the years 1799 and 1820 Benjamin and Philomela had nine children; Henry, Benjamin Hall, James, Simon, Albert, Joshua, George, Mary and Frances. Simon and Benjamin Hall would both choose their father's career in civil engineering. Simon assisted his father in Virginia with the canal and Benjamin Hall would pursue a career with railroad work, most notably work with the first railways in Cuba.

Philomela Waterman's father, Simon, was a minister of the Congregational Church in Plymouth. During a journey in 1800 to visit his daughter and son-in-law in Rome, Simon was asked to dedicate the newly organized church, The First Religious Society of Rome. On September 28, 1800 in the home of Ebenezer and

Grace Wright the church was formally dedicated by act of holy communion officiated by Simon Waterman.

The families would continue to meet in the Wright home. Among those families was that of Benjamin and Philomela. Eventually, with time, the church found its way to construct a church building and in 1807 the first church building in Rome was erected and simply called, "The Meeting House." According to church tradition, Benjamin Wright was one of three on the building committee and he is credited with the design.

Meeting House of 1807

"The Meeting House" (18)

Benjamin Wright died on August 24, 1842. He is buried, along with several other members of his family, in vault number 83 of New York City's Marble Cemetery. The cemetery is located between East 2nd and 3rd Streets and Second Avenue and the Bowery. In 1997 a civil engineering firm, specializing in the preservation of historic structures, conducted a condition survey of the cemetery and prepared estimates for rehabilitation. In the report under the section entitled "History and Significant Design Aspects" mention is made of prominent personages buried within. Uriah and Charles Scribner, Aaron Clark a former city mayor and James Talmadge, first president of New York University are among those buried in the cemetery. No mention is made of Benjamin Wright. The oversight was not intentional. They just simply did not know, but the irony is quite evident. As civil engineers we go to great lengths to preserve and restore those great works from our past but at the same time lack a fundamental knowledge of the history of the profession. The heritage of civil engineering is not just the great works but also the people. The two cannot be separated.

During the bicentennial of Benjamin's birth, ASCE formerly declared him as "The Father of American Civil Engineering". To acknowledge this fact a landmark plaque was placed at his birthplace in Wethersfield during ceremonies on October 17, 1970. Benjamin Wright is considered America's first civil engineer.

David Hosack, a doctor and writer of the day, was working on a biography of DeWitt Clinton. In a letter to Hosack dated December 31, 1828, Wright discusses the chronology of events leading up to the Erie Canal. In a portion of his letter he himself says it best;

> . . . In 1802 The Western Inland Lock Navigation Company determined upon improving the navigation of Wood Creek from near Fort Stanwix to a small tributary stream six miles westerly, called Little Canada Creek. In this distance there was a descent of nearly twenty-four feet and the navigation very indifferent and troublesome. The plan decided upon was by means of dams and locks, of which they constructed four in the distance above mentioned. George Huntington, Esq. of Rome, was their agent, and I was their engineer. (Ref. 28)

Acknowledgements

The author wishes to acknowledge the following;

The Principals and management of Facility Engineering Associates in allowing the time for preparation of this paper.

Dan Chung, Staff Engineer with Facility Engineering Associates for his help in the assembly and formatting of this paper.

Joan Pennington for her kind help in preparation of the map illustrations.

Photograph and Illustration Credits

1. Benjamin Wright, the Wright Collection from the L. Neal FitzSimons Library.
2. James Geddes, The New York State Library.
3. John Jervis, the Jervis Collection from the James K. Finch Library.
4. Nathan Roberts, photograph in the collection of the L. Neal FitzSimons Library.
5. Canvass White, The Smithsonian Institution, Washington, D.C.
6. Model of stump puller, photograph of model in the collection of the L. Neal FitzSimons Library, model built by Chester Williams, displayed in the Fort Stanwix Museum, Rome, New York.
7. Weighlock, copy of photograph in the collection of the L. Neal FitzSimons Library, original from the George Eastman House, Rochester, New York.
8. Cover of Engineer's Report, the Wright Collection from the L. Neal FitzSimons Library.
9. Canal Lock, Lock #51, the Jervis Collection from the James K. Finch Library.
10. The C&O Canal at Harpers Ferry, the Library of Congress, Photographs Division, Washington, D.C.
11. The C&O Canal, Monocacy Aqueduct, the Library of Congress, Photographs Division, Washington, D.C.
12. The C&O Canal, Paw Paw Tunnel, postcard in the author's collection.
13. Claudius Crozet, The Smithsonian Institution, Washington, D.C., the original in the collection of The Virginia Military Institute, Lexington, Virginia.
14. Canal Boat "Marshall," postcard in the author's collection.
15. Charles Ellet, Jr., the Wright Collection, from the L. Neal FitzSimons Library, in the possession of the author.
16. The James River and Kanawha Company, postcard in the author's collection.
17. Letter from Edward Miller to Benjamin Latrobe, copy in the Wright Collection, from the L. Neal FitzSimons Library.
18. The Meeting House, from photograph of drawing by Noel Reagan in the collection of the L. Neal FitzSimons Library, original in the Collection of the First Presbyterian Church, Rome, New York.
19. Benjamin Wright, photo of painting in the Jervis Library, New York.

Reference Material

Resource Material Used In The Preparation Of This Paper

Books

1. Andrist, Ralph K., THE ERIE CANAL, American Heritage Publishing Company, New York, 1964.

2. Burton, Anthony, THE CANAL BUILDERS, Eyre Methuen, Ltd, London, 1972.

3. Cable, Bertha Kropp, A CLOUD OF WITNESSES, A HISTORY OF THE FIRST PRESBYTERIAN CHURCH ALSO KNOWN AS THE FIRST RELIGIOUS SOCIETY OF ROME, NEW YORK, 1800-1971, Canterbury Press, Rome, New York, 1993.

4. The Committee on History and Heritage of American Civil Engineering, A BIOGRAPHICAL DICTIONARY OF AMERICAN CIVIL ENGINEERS, The American Society of Civil Engineers, New York, 1972.

5. The Delaware and Hudson Company, A CENTURY OF PROGRESS, A HISTORY OF THE DELAWARE AND HUDSON COMPANY, 1823-1923, J.B. Lyon Company, Albany, New York, 1925.

6. Couper, William, CLAUDIUS CROZET, SOLDIER-SCHOLAR-EDUCATOR-ENGINEER, The Historical Publishing Company, Charlottesville, Virginia, 1936.

7. Dunaway, Wayland Fuller, HISTORY OF THE JAMES RIVER AND KANAWHA COMPANY, Studies in History, Economics and Public Law, Columbia University, Volume CIV, Number 2, New York, 1922.

8. Finch, James Kip, THE STORY OF ENGINEERING, Doubleday and Company, Garden City, New York, 1960.

9. FitzSimons, Neal, THE REMINISCENCES OF JOHN B. JERVIS, ENGINEER OF THE OLD CROTON, Syracuse University Press, Syracuse, New York, 1971.

10. Hepburn, A. Barton, ARTIFICIAL WATERWAYS OF THE WORLD, The MacMillan Company, New York, 1914.

11. Hungerford, Edward, MEN OF ERIE, A STORY OF HUMAN EFFORT, Random House, New York, 1946.

12. Hunter, Robert F. and Dooley, Edwin L., CLAUDIUS CROZET, FRENCH ENGINEER IN AMERICA, 1790-1864, University Press of Virginia, Charlottesville, Virginia, 1989.

13. Kirkwood, James J., WATERWAY TO THE WEST, Eastern National Park and Monument Association, National Park Service Interpretive Series No. 1, 1963.

14. Larkin, F. Daniel, JOHN B. JERVIS, AN AMERICAN ENGINEERING PIONEER, Iowa State University Press, Ames, Iowa, 1990.

15. Lewis, Gene D., CHARLES ELLET, JR., THE ENGINEER AS INDIVIDUALIST, 1810-1862, University of Illinois Press, Urbana, Illinois, 1968.

16. Stuart, Charles B., LIVES AND WORKS OF CIVIL AND MILITARY ENGINEERS IN AMERICA, D.Van Nostrand Publisher, New York, 1871.

17. Wisely, William H., THE AMERICAN CIVIL ENGINEER, 1852-1974, The American Society of Civil Engineers, New York, 1974.

Manuscripts, Periodicals and Other Documents

18. Burial records, New York Marble Cemetery, Evelyn Luquer, Trustee, 1969.

19. Dedication Program, plaque dedication ceremony designating Benjamin Wright the Father of American Civil Engineering, Wethersfield, Connecticut, 1970.

20. THE ERIE CANAL AND ITS EARLY ENGINEERS, an exhibit catalog, The Rome Historical Society, Rome, New York, 1967.

21. FitzSimons, Neal, BENJAMIN WRIGHT, THE FATHER OF AMERICAN CIVIL ENGINEERING, paper presented before the society, The American Society of Civil Engineers, 1996.

22. FitzSimons, Neal, BENJAMIN WRIGHT, THE FATHER OF AMERICAN CIVIL ENGINEERING, *Civil Engineering Magazine*, The American Society of Civil Engineers, September 1970.

23. FitzSimons, Neal, "Engineer as Historian," Engineering Counsel, Kensington, Maryland, 1984.

24. Harte, Charles R., CONNECTICUT'S CANALS, paper presented before the Connecticut Society of Civil Engineers, Hartford, Connecticut, 1938.

25. Letter from Miller to Latrobe, copy in the Wright Collection, the L. Neal FitzSimons Library, in the possession of the author.

26. Letter from Wright to Woolsey, "The American Journal of Science and Arts," Benjamin Silliman, Editor, Vol. III, 1821.

27. Lord, Philip, THE COVERED LOCKS OF WOOD CREEK, *The Journal of the Society for Industrial Archeology*, Vol. 27, No. 1, 2001.

28. Misc. letters of Benjamin Wright, copies in the Wright Collection, the L. Neal FitzSimons Library, in the possession of the author.

29. Misc. letters of John Jervis, copies in the Jervis Collection, the James K. Finch Library, in the possession of the author.

30. NEW YORK MARBLE CEMETERY, SURVEY OF CONDITIONS REPORT AND REHABILITATION COST ESTIMATES, Robert Silman Associates, P.C., New York City, 1997.

31. SKETCH HISTORY OF THE WESTERN INLAND LOCK NAVIGATION COMPANY, paper presented at the Canal Museum, Syracuse, New York, 1966.

32. Thoma, Edward C., AMERICAN SOCIETY OF CIVIL ENGINEERS, ITS RISE AND GROWTH, a circular, Indiana Section, The American Society of Civil Engineers, 1946.

33. Wright, Benjamin Hall, ORIGIN OF THE ERIE CANAL, SERVICES OF BENJAMIN WRIGHT, *The New York Observer*, 1866.

The Origins, Founding, and Early Years
of the American Society of Civil Engineers:
A Case Study in Successful Failure Analysis

Henry Petroski[1]

Abstract

The story of the founding of the American Society of Civil Engineers (ASCE) is widely known. Still it bears retelling on the occasion of the sesquicentennial of the Society, for it is a story of emulation, of slow and difficult beginnings, of halting early progress, of overcoming adversity, of learning from failures, and, ultimately, of monumental achievement. In short, the story of the founding of the ASCE mirrors the story of a great engineering project.

The British Model

The origins of the ASCE, like those of America itself, are rooted in the soil of Britain. In the latter part of the eighteenth century, it was the example of pioneering British engineers like James Brindley and John Smeaton that defined the identity of the civil engineer as a member of a developing profession distinct from that of the military engineer. Establishing themselves, in modern terminology, as consulting engineers, the likes of Brindley and Smeaton asserted their intellectual and professional independence from those who hired them. These civil engineers engaged in assessment and design projects but did not necessarily serve as constructors. By 1770 there were of the order of a dozen civil engineers who were well-known among each other and throughout Britain. (Watson, 1988)

As part of their practice, these engineers found themselves often testifying before Parliament on matters relating to civil works projects. This naturally necessitated that the engineers be in London while Parliament was in session, and, at a 1771 meeting at the King's Head Tavern in Holborn, a number of them, including Smeaton, "agreed that the Civil Engineers of this Kingdom do form themselves into a Society." Fortnightly meetings ensued, at which engineers on opposite sides of issues and bills before Parliament dined together and engaged in "conversation, argument and a social communication of ideas and knowledge." Such intercourse was both "the amusement and the business of the meetings." This Society of Civil Engineers was thus a dining club rather than a learned society. (Watson, 1988)

---

[1] A.S. Vesic Professor of Civil Engineering and Professor of History, Duke University, Box 90287, Durham, NC 27708-0287.

Founder member Smeaton was one of the regulars at the Society dinners. Following his death in 1792, at age 68, the Society became inactive, but it was revived the following year and came to be known as the Smeatonian Society of Civil Engineers. It remained a dining club, however, and took on the distinct character of an old-boys' club. The original membership of the Society was aged between 37 and 63, with the average member being in his late 40s. The nature of the Society was such that its membership was limited and that it comprised the most established and famous engineers in Britain. Young engineers had little opportunity to join or benefit from activities of the Smeatonians, as the Society came to be known.

Since, save for apprenticeship, there was also no institutionalized formal engineering education in Britain through the first decades of the nineteenth century, younger engineers saw the need to establish a society whose emphasis would be not dining and conversation but information and instruction. Thus it was that 21-year-old Henry Robinson Palmer, then still an apprentice, suggested the idea of an Institution of Civil Engineers. Soon, Palmer and a group of young engineers aged 19 to 32 began to meet informally to flesh out the idea.

The first formal meeting of the Institution of Civil Engineers (ICE) took place in 1818, and at it Palmer laid out the proposed rules of the new society. These began with (Watson, 1988):

> *First*, that a society be formed, consisting of persons studying the profession of civil engineer.
> *Second*, that as much as possible to prevent reserve in the junior members, the ages of admission shall not be less than twenty years nor greater than thirty-five.
> *Third*, that the Society shall meet once every week for the purpose of mutual instruction in that knowledge requisite for the profession.

The rules were soon adopted and augmented, but within a year or so the ICE experienced a drop in interest. Dues were decreased and the age requirements for membership relaxed. A new class of membership, Corresponding Member, was instituted to attract members too far from London to attend the regular meetings. Further classes of membership, Honorary Member and Associate, were also instituted, as was the office of President. The eminent civil engineer Thomas Telford, then 62 years old, was invited to become the first president of the ICE, and thereby bolster its reputation. Telford not only accepted, he embraced the idea of the Institution and gave a collection of books to the fledgling organization. (In many regards, the ASCE would have strikingly parallel beginnings.)

In time, the ICE membership began to grow, reaching 87 in 1824, and attendance at the weekly meetings increased. Continuing expenditures did not allow the Institution's finances to grow accordingly, however, and in 1827 it was decided to establish an endowment fund, through which the future of the Institution would be guaranteed. In exploring legal options for establishing a trust to secure the Institution's assets, the option of seeking a royal charter was proposed, and it was this course that was followed. It was in the application for such a charter that Thomas Tredgold drafted his famous definition of the profession of a civil engineer as "being the art of directing the great sources of power in nature for the use and convenience of man." The charter was granted in 1828. (Watson, 1988)

Early American Efforts

The eventual success of the British Institution of Civil Engineers was not lost on American engineers. By the mid-1830s, the ICE was recognized as a leading learned society, and membership in it was a mark of distinction. In addition to its educational benefits, the entire profession of civil engineering in Britain was epitomized by the ICE, and its status elevated by the very existence of the Institution. American engineers sought a similar situation in their own country.

During the 1830s, there were several abortive attempts to begin an American counterpart to the ICE. Railroad construction at that time was a principal occupation of many civil engineers, taking them into rugged territory where they were not only cut off from cities and libraries but also presented with many a difficult engineering problem, for which the experience of others was welcome. Thus it is not surprising that engineers associated with railroads were among the early proponents of an American engineering society. In 1836, engineers associated with the Charleston and Cincinnati Railroad made efforts to establish a National Society of Civil Engineers. Two of the engineers were William Gibbs McNeill and George Washington Whistler, both American-born and West Point-educated, who like many of their fellow military academy graduates, had become prominent railroad engineers. Indeed, perhaps it was their extensive responsibilities on the Charleston and Cincinnati and other railroads that kept the society from taking hold. Another short-lived effort was organized by some engineers associated with the Baltimore and Ohio Railroad. (Calhoun, 1960; Hunt, 1897; Wisely, 1974)

The editors of engineering journals were also among proponents of a professional organization, which they hoped might in turn attract readers to "a suitable periodical to publish their proceedings," as the editor of the *American Railroad Journal* put it. Late in 1838, a "highly respectable meeting of the profession" called for Augusta, Georgia led to the announcement of a national convention to be held the following year in Baltimore. The 40 engineers who attended that meeting are believed to have represented in excess of 10 percent of all civil engineers then in the United States. The assembly adopted a resolution calling for the formation of a geographically diverse committee of seventeen charged with preparing and adopting a constitution and forming a Society of Civil Engineers of the United States. The Franklin Institute of Philadelphia, which had been founded in 1824 and was a repository for engineering literature, offered its encouragement and help. The Franklin Institute had for years been attracting civil engineers to its ranks, and it made sense for any new engineering association to affiliate with the Institute, thus taking advantage of its headquarters, library, and journal. Unfortunately, many of the committee of seventeen, elected in their absence to carry out the details of organization, were apathetic, and the effort ultimately failed.

Edward Miller, Secretary of the Committee of Seventeen, wrote in the *American Railroad Journal* of the experience. According to Hunt, Miller believed that "the causes of the failure were the appointment of a large committee, which, though democratic, was not likely to produce harmony in council." (Hunt, 1897) This condition was aggravated by the fact that most of those appointed to the committee were, in Miller's own words, "ignorant of their appointment" and some were "absolutely indifferent or hostile to the formation of any institution." Furthermore, the members were scattered geographically, thus making meeting difficult and allowing "local views, partialities and jealousies" to have undue influence on the outcome. Not surprisingly, since human beings were involved, the situation was not much different from what can be experienced today under similar circumstances.

Nevertheless, the aims of the abortive society were noble: "the collection and diffusion of professional knowledge, the advancement of mechanical philosophy, and the elevation of the character and standing of the Civil Engineers in the United States." However, the noble goals were offset by the practical obstacles. Among these were the difficulty that busy engineers scattered across the country would have in getting to meetings and the suspicion that an affiliation with a preexisting organization like the Franklin Institute would bring with it prior political associations that were unwanted by those outside the preexisting structure in the Philadelphia area. Undercurrents of competition between state-supported canals and privately financed railroads, and the engineers associated with each, further complicated the matter. The *American Railroad Journal* continued to offer itself as a national periodical for the profession, and it advocated professional society rather than government control of the training and selection of engineers for important projects, thus exercising some quality control on the profession and excluding the persuasive but incompetent. (Hunt, 1897)

An attempt to establish an American Institute of Engineers took place in Albany, New York, in 1841, but it too was unsuccessful. (Anonymous, 1890) The establishment of a local organization was naturally an easier task. The Boston Society of Civil Engineers was begun in 1848 and became the first permanent engineering organization in America. Its unique success has been attributed to a concentration of engineers in the nearby area; its rigorous membership requirements that excluded the less mature, less prestigious, and more mobile engineers; the immediate establishment of a headquarters where meetings could be held and a library maintained; and a balance between social and professional activities. Another city where there was a similar concentration of engineers and comparable circumstances was New York, and so it should be no be surprise that an ultimately successful effort at organizing civil engineers took root there. (Wisely, 1974)

Beginnings in New York

Among the founders of the Boston Society of Civil Engineers was James Laurie, who had emigrated to America from his native Scotland in the early 1830s. In America, Laurie served as associate engineer under another Scottish-born engineer, James Pugh Kirkwood, who was chief engineer of the Norwich and Worcester Railroad, and worked his way up to chief engineer and superintendent of construction. Soon Laurie was doing consulting work for railroad, canal, dam, bridge, and wharf companies. He moved his consulting office to New York in 1852, but he evidently missed the professional society he had left behind in Boston, because he soon became a driving force behind starting a civil engineering society in New York. After the idea was discussed among some local engineers, a notice was circulated to practitioners in the area (Hunt, 1897):

New York, October 23d, 1852.

Dear Sir:

A meeting will be held at the office of the Croton Aqueduct Department, Rotunda Park, on Friday, November 5th, at 7 o'clock P.M., for the purpose of making arrangements for the organization, in the city of New York, of a Society of Civil Engineers and Architects.

Should the object of the meeting obtain your approval, you are respectfully invited to attend.

Wm. H. Morell	Wm. H. Sidell
J. W. Adams	A. W. Craven
James Laurie	James P. Kirkwood

and others.

The location of the organizing meeting was in the offices of Alfred Wingate Craven, chief engineer of the Croton Aqueduct Department, which was responsible for bringing the city's water supply down the 41 miles from the dam on the Croton River. As indicated in the notice, the department's offices were located in Rotunda Park (also known simply as The Park and now called City Hall Park), so named because it was the site of the domed Romanesque building known as the Rotunda, or the Rotunda in the Park. The building, which dated from about 1818, was designed by the American neoclassicist artist and architect John Vanderlyn and contained panoramas painted by Vanderlyn himself. (His works also hung in the U.S. Capitol Rotunda.) However, the Rotunda did not remain dedicated to art. An 1846 map of the area shows the building being used as a Post Office. When a cholera epidemic hit New York in 1849, the Rotunda was used for a time as a hospital, after which the structure housed municipal offices, including those of the Croton Aqueduct Department. The New York Rotunda was located in the northeast corner of the park, near the intersection of Chambers and Centre Street, and a plaque in the park (just across Park Row from the foot of the Manhattan approach to the Brooklyn Bridge) now commemorates the building as the location of the founding of the American Society of Civil Engineers.

Figure 1. The Rotunda, founding site of the American Society of Civil Engineers (Haswell, 1897)

The November 5th meeting was attended by twelve engineers, with Alfred Craven presiding, in his office. Of those who explicitly signed the call, only Kirkwood did not attend. The additional seven attendees were J. W. Ayers, Thos. A. Emmet, Edw. Gardner, R. B. Gorsuch, Geo. S. Green, S. S. Post, and W. H. Talcott. (According to an anonymous letter to *Engineering News* in 1890, the magazine had been careless in its use of the term "founder member" in the obituary of the Society's third president, William James McAlpine. According to the letter writer, It was only those twelve engineers who were actually at the initial meeting in the Rotunda who were entitled to be so identified.)

At the organizational meeting, it was resolved that incorporation be sought for an American Society of Civil Engineers and Architects. The name not withstanding, membership was to be open to "Civil, Geological, Mining and Mechanical Engineers, Architects and other persons who, by profession, [were] interested in the advancement of science." Not surprisingly, perhaps, James Laurie was elected to be the first president of the new society. Also at the first meeting, a constitution was adopted, and a resolution was passed whereby prominent engineers resident outside the New York area were nominated and voted into membership, which was then offered them via correspondence from the secretary, who was Mr. Gorsuch. The notice, which was dated November 10th, gave those who received it a three-week window in which to become members by "a bare notice of their desire to become such," and by filling out accompanying forms. (Hunt, 1897)

Among the ten gentlemen answering the call to join the new society was John A. Roebling, a resident of Trenton, New Jersey, who in five years would make public a proposal for a suspension bridge between New York and the then separate city of Brooklyn. But since Roebling and others who accepted the invitation to join the new society were not then regularly in New York, they could not be counted upon to attend meetings or serve as officers. The first meeting under the new constitution was held on January 5th, 1853, with eight members in attendance. The beginnings of a library were discussed, with the result that the donation of reports, maps, plans and the like were called for. The technical program was given by President Laurie, who led a discussion on "The Relief of Broadway" by means of elevated railway tracks.

Difficult Beginnings

Though Laurie was a faithful attendee at all eight meetings held in 1853, the same could not be said for others. Some meetings were attended by as few as three members, and the average attendance was only six, even though according to the first annual report of the Society membership totaled 55. (Hunt, 1897) The leadership of the organization, being engineers, were well aware of the possibility that the endeavor could ultimately be in jeopardy, and so the Board of Direction adhered to a policy of making "no expenditures that could well be avoided, so that in case of failure the funds collected might be returned to the members." Of the $700 in receipts, after a year only $115.22 had been spent.

The fiscal conservatism of the Board meant that "no steps [were] taken towards the formation of a library, or for renting rooms for the use of the Society." (Hunt, 1897) In its first annual report, the Board, though it regretted that it could not "speak in more flattering terms of the success of the Society," still recommended that the organization be "kept up, and that renewed efforts be made to obtain additional members who are residents of the city or vicinity, and can attend the meetings." Unfortunately, conditions did not improve in 1854. Attendance at the six meetings went down rather than up. Dues were reduced, but only six new members joined. However, adherence to the fiscal policy of the Board saw the treasury increase to over $800.

An ongoing topic of discussion was the need for the Society, which had continued to meet in the Croton Aqueduct Department office in the Rotunda, to have rooms of its own in which could be established a library and museum, in which "models of patented or proposed improvements" could be exhibited. Such dedicated quarters, it was felt, were necessary to give members who did not live in New York a place to have a desk when in the city, as well as a place view and consult the reports, maps, and other items that the Society was accumulating. The 1855 report on the matter suggested trying an experiment for a year. It recommended the appointment of a committee to find a room, preferably near the office of a member willing to serve as its superintendent, and secure a desk and other furniture, "provided that the rent of the room do not exceed $250.00, and the cost of furnishing it do not exceed $150.00." (Hunt, 1897) A committee of the whole apparently discussed the recommendation, but took no action. The American Society of Civil Engineers and Architects did not meet again for over twelve years.

Failure Analysis and Ultimate Success

Although the record shows that James Laurie served as president of the Society until 1867, his major consulting jobs from 1855 on appear to have required him to spend not a little time in New England and in Nova Scotia. Among the directors during the period of inactivity was Laurie's American mentor, James P. Kirkwood, who upon Laurie's resignation on October 12, 1867 became the Society's second president and the one to whom it fell to lead the revival of the moribund organization.

In his presidential address at the first general meeting after the reorganization, Kirkwood presented what might be termed a failure analysis of the early years of the Society. He outlined what he considered to be the general causes of the failure, and asked the members to bear with him so that together they might "look our difficulties in the face and judge how we can avoid them or control them in the future." (Kirkwood, 1872)

Among the principal reasons that members had not been attracted to the Society, according to Kirkwood, was that it had no rooms of its own, even after its finances were such that it could afford them. Without a headquarters of its own, Kirkwood declared, country and city members alike had nothing tangible to point to as evidence that theirs was "something more than a parchment Society."

Kirkwood also pointed to the lack of professional communications, in the form of lectures, papers, and discussions that could give resident and non-resident members alike an "*esprit de corps.*" He considered the presentation, printing, and distribution of papers, or at least abstracts of them, to be "essential to our continuous and successful existence." Such papers did not have to be elaborate in form or time-consuming in preparation but rather could be simply "short and truthful, illustrating some point of professional difficulty or interest to which the individual's attention has been specially directed." He emphasized the benefits of learning from mistakes: "Our failures in construction, where we are at liberty to mention them, will always be of more value to others than our successes." (Kirkwood, 1872) The parallels with the failures he and his colleagues had experienced in attempting to build a new society could not have been lost on anyone in the audience.

At the meeting at which Kirkwood made his remarks, a committee on papers and printing was appointed, and it invited the membership to help achieve its goals. A note from the committee, signed by its chairman, George S. Greene, was issued in 1868 from the "Society's Rooms, 12 Chamber of Commerce Buildings, corner of William and Cedar Streets." (Greene, 1872) The Society had thus begun in earnest to follow Kirkwood's prescription for revitalization.

Unfortunately, Kirkwood resigned the presidency that same year, but fortunately William J. McAlpine was nominated to replace him. McAlpine resisted the office at first, feeling strongly that a member resident in New York should assume the all-important office. He identified himself as a non-resident engineer and let it be known that he expected to be engaged for some time in engineering projects that would keep him away from the city. The membership of the Society paid him no heed, and he was elected its third president on September 2, 1868. In his address on assuming the chair, he stressed the "benefits and advantages" of the still young Society to the American engineering profession. He also drew attention to "the effect of similar societies elsewhere, in advancing the dignity and influence of the profession," matters that occupied much of his remarks. (McAlpine, 1872)

Like Kirkwood, McAlpine looked back at failures in order to learn from them what needed to be changed if the society was to succeed. He noted that the Society could not prosper "by the exertions of any small number of its members, however zealous or influential they may be." If the Society were to succeed, he asserted, it must have the "united, cordial, earnest efforts of a considerable number for the next two or three years." It was that period of time, he estimated, that it would take the organization revived under Kirkwood's leadership to establish itself on a firm footing. He also emphasized with his own italics that the original efforts in behalf of the Society "ought not to have failed." It had "embraced nearly all of the engineers in and near the metropolis" and they had brought to the new society a "zeal and earnestness that deserved success." Yet it had not had the numbers to succeed at first, and for ten years "the Society slumbered."

McAlpine attributed the failure in part to "untoward circumstances," but he also blamed "others from the interior," like himself. McAlpine and his fellow non-resident engineers "did not perceive that we could lighten our labors, increase our knowledge, and secure better success in our various undertakings, by associating with those engaged in kindred pursuits, some of whom were working out singly the same problems." He saw the establishment of the society's rooms--the "chamber of engineering" to which engineers visiting the city could congregate to discuss professional and technical problems of common concern—as central to the ultimate success of the institution. "*Here*," he emphasized, "they can learn anything that is published in regard to anything useful in their particular pursuit, and thereby not only avoid past errors but achieve new results from advanced standpoints, which will secure to them fame and profit." The society and its members, according to McAlpine, could benefit each other.

McAlpine pointed explicitly to the "London Institution of Engineers," whose proceedings he had observed over the course of four months, as a model to which the American society might aspire. He traced the origins of the ICE to the dining club associated with Smeaton, and thought it should give "great encouragement" to American engineers. In spite of the fact that "an unfortunate personal difficulty between one of the members and Mr. Smeaton broke up the Society," McAlpine noted that it was soon reorganized and still existed, as did its "outgrowth," the thriving ICE, with a membership in 1868 of over fourteen hundred members and a treasury of enormous surplus. Once again, there was no mistaking the parallel that was to be drawn: Out of temporary setbacks can come enormous advances. This was certainly true in the physical world of engineering, and McAlpine saw it to be true also in the social world of engineering organizations.

That same year, the term "Architects" was dropped from the name of the organization, the American Institute of Architects having been founded in the meantime. Within four years (in 1872), 20 years after the Society's founding, the first volume of *Transactions* was published, with the opening pages dedicated to Kirkwood's address the note from the committee on papers and printing, with McAlpine's address following shortly thereafter. (The first volume of *Transactions* of the ICE was published in 1836, 18 years after its founding. Today, of course, both ICE and ASCE have extensive publishing programs.)

By 1871, the total membership of ASCE had grown to 259. It had survived a halting beginning, a decade-long hiatus, and near failure, but thanks to the vision of Kirkwood and McAlpine (and to their successful application of engineering failure analysis), it was revived. The revitalized organization was put on course toward becoming a successful and flourishing society of civil engineers, which in time would be the largest in the world.

Conclusion

There are striking similarities between the founding of the American Society of Civil Engineers and that of the Institution of Civil Engineers, to which ASCE can trace its intellectual roots. Both organizations were begun principally for the purpose of providing professional educational and communicational opportunities for their members, but each experienced slow growth in membership and poor attendance at meetings in its early years. However, the causes of the near failure of each of the societies to realize the vision of its founders were analyzed properly by its engineer-leaders, who evidently scrutinized their organizational difficulties as they would difficulties in construction. Being the capable engineers that they were, they took corrective measures that eventually led to the flourishing of two of the most important professional engineering societies in the world.

References

Anonymous. (1890). "Who Were the Founders of the American Society of Civil Engineers?" *Engineering News*, March 8, p. 232.

Calhoun, Daniel Hovey. (1960). *The American Civil Engineer: Origins and Conflict.* Cambridge, Mass.: Technology Press.

Greene, George S. (1872). "Note from the Committee on Papers and Printing," *Transactions of the American Society of Civil Engineers*, Vol. I, pp. 7-8.

Haswell, Chas. H. (1897). *Reminiscences of an Octogenarian of the City of New York (1816 to 1860)*. New York: Harper & Brothers.

Hunt, Charles Warren. (1897). *Historical Sketch of the American Society of Civil Engineers*. New York: ASCE.

Kirkwood, James P. (1872). "Address of the President . . . December 4, 1867," *Transactions of the American Society of Civil Engineers*. Vol. I, pp. 3-6.

McAlpine, W. J. (1872). "Address . . . on Assuming the Chair after his Election . . ., September 2, 1868," *Transactions of the American Society of Civil Engineers*. Vol. I, pp. 45-55.

Watson, Garth (1988). *The Civils: The Story of the Institution of Civil Engineers*. London: Thomas Telford.

Wisely, William H. (1974). *The American Civil Engineer, 1852-1974: The History, Traditions and Development of the American Society of Civil Engineers, Founded 1852*. New York, ASCE.

Early Development of American Civil Engineering

CIVIL ENGINEERING EDUCATION HISTORY (1741 TO 1893):
An Expanded Civil Engineering History Module

Jerry R. Rogers[1]

Abstract

In 1741, the Royal Military Academy was established at Woolwich, England for officer training for artillery and public works. Louis XV appointed Louis Perronet as chief engineer of bridges and roads in France, and he established a three- year program at the Ecole des Ponts et Chaussees, the first formal school of civil engineering. In 1802, the U.S. Congress established the U.S. Military Academy at West Point, NY, with a four year program on bridges, roads, canals and fortifications developed in 1817 by Claude Crozet. In 1835, Rensselaer School/University awarded the first U.S. civil engineering degrees to four students. The history of other early civil engineering education development in the U.S., Canada, and Mexico is summarized in an expanded Civil Engineering History Module, ending with the 1893 Chicago World Columbian Exposition (and Congress on Engineering Education) by Ira O. Baker, University of Illinois civil engineering faculty member. The 1893 Education Congress presentations led to the founding of the American Society for Engineering Education.

Introduction

In a paper (Rogers 1999: "Early Civil Engineering Education: A Selected History for Engineering Classes") (Rogers, Brenner 1999), the events and years of early civil engineering education were presented in a time-line for educators, practitioners and students for utilization in engineering and/or civil engineering classes. This international and U.S. civil engineering educational material was taken primarily from Grayson's publications (Grayson 1993: *The Making of An Engineer*, 1996) and the *American Civil Engineer* (Wisely 1974), with other submitted material (FitzSimons 1999). In a more recent paper (Jewell, Griggs, and Ressler 2001: "Early Engineering Education in the United States prior to 1850") (Rogers, Fredrich 2001), detailed information was submitted on the U.S. engineering education founded within 120 miles of four colleges: U.S. Military Academy- West Point- NY, Rensselaer Polytechnic University- Troy- NY, Norwich University- Norwich- VT, and Union College- Schenectady- NY.

---

[1]   Chair- Education Working Group- North American Alliance for Civil Engineering, Department of Civil/Environmental Engineering, University of Houston, Houston, TX 77204; phone 713-743-4276; djrogers@pdq.net

This paper expands upon the time-line (Rogers 1999) with added information (Jewell, Griggs, Ressler 2001) and with more details (Grayson 1993) (Wisely 1974) up to the 1893 founding of the Society for the Promotion of Engineering Education.

Time- Line of Early Civil Engineering Education History: A Civil Engineering History Module

The following expanded time- line (1741-1893) of civil engineering education history may be utilized in engineering courses with other topics in ethics, history and heritage, and professional development. The time-line is an example of a Civil Engineering History Module, proposed by the ASCE National History and Heritage Committee in the late 1990s and discussed with the Education Activities Committee (EDAC). In January 2002, EDAC passed a motion to submit engineering history as a part of an engineering class or classes to the ABET accreditation group. A detailed Geotechnical Engineering History Module (Exploration Milestones, Caisson/ Drilled Shaft Milestones, Pile Driving Milestones, and Lateral Support Milestones) (Parkhill 2001) "150 Years of Geotechnical Design and Construction" was published as another Civil Engineering History Module: (Rogers, Fredrich 2001).

1741- The Royal Military Academy in England was established in Woolwich to instruct officers for the artillery and engineers. With little early formal education in England, most engineers learned their profession and skills through apprenticeship with a master. In France, Louis XV appointed Jean Perronet as chief engineer of bridges and highways.

1747- Perronet established a school within the Corps des Ponts et Chaussees.

1775- The curricula became a three- year course of study and the school was renamed the Ecole des Ponts et Chaussees, the first formal school of engineering.

1778- At George Washington's request, Congress created an engineering department, although little progress was made due to a lack of volunteers. On June 9, Washington issued a formal call for a school of engineering (which was not achieved).

1794- With the recommendation of U.S. President George Washington (elected in 1789), a military school was begun at West Point in the old provost prison, destroyed by fire in 1796.

1794- Napoleon created the Ecole des Travaux Publics to train engineers for public and private service with coursework in the design/construction of bridges, roads, canals and fortifications. The school became the Ecole Polytechnique and served as a model for early U.S. engineering schools.

1795- Hinton James was cited as the first student enrolled in a state university in the U.S. (FitzSimons 1999). On February 12, 1795, James walked to the University of North Carolina in Chapel Hill, graduating on July 4, 1798 in Natural Philosophy. James became an apprentice civil engineer as an Assistant to Hamilton Fulton on public highways, canals and rivers, including the Cape Fear River.

1802- On March 19, Congress established the U.S. Military Academy at West Point, the first engineering school in America. In 1813, Alden Partridge was the first person appointed professor of engineering. In 1817, Colonel Sylvanus Thayer with Claudius Crozet, an 1809 graduate of the Ecole Polytechnic, developed a four- year program with work on the design/construction of bridges, roads, and canals, as well as fortifications. While initially intended to be part of the Corps of Engineers, the USMA was separated later from the Corps. Formal degrees were offered in 1913.

1819- Alden Partridge, formerly of the U.S.M.A., established the American Literary, Scientific and Military Academy becoming Norwich University in 1834 (with the first U.S. civilian course of civil engineering (1821)). Two Civil Engineering masters degrees were awarded in 1837.

1824- Stephen VanRensselaer and Amos Eaton established Rensselaer School with President Eliphalet Nott arriving in 1829. In 1824, subjects included land surveying, mensuration, measurements of the flow in rivers and aqueducts with the term *civil engineer* first in the 1828 catalogue. Rensselaer awarded the first U.S. civil engineering degrees to four students in 1835 and in 1849 became a polytechnic institute.

1828-1833- In Mexico, a corps of engineers included training/education in fortifications and artillery. In 1810, a Mexico college of mining was active with a name transformation in 1867 to the school of engineering (Colegio de Ingenerios Civiles de Mexico 1996).

1833-34- Universities of Virginia and City of New York began civil engineering courses. In 1835, the Virginia civil engineering school hired William Rogers, who published an 1838 book, *An Elementary Treatise on the Strength of Materials*.

1836- The College of William and Mary began a school of civil engineering with John Millington's 1839 book: *Elements of Civil Engineering*.

1837- The trustees of the University of Alabama funded a civil engineering program and University President Manly hired F.A.P. Barnard to teach engineering courses in surveying and construction techniques.

1839- Crozet was superintendent of the Virginia Military Institute, patterned after the Ecole Polytechnique. The Citadel in 1842 offered military and civil engineering.

1845- Union College (with Eliphalet Nott) was a civil engineering degree granting university. In 1841 Squire Whipple, 1830 graduate of Union College, patented the iron bowstring truss bridge. In 1847, he wrote the 1st moving load determination book: *A Work on Bridge Building*......, Utica, N.Y. with Appendix in 1869. (Griggs Jr. 1997).

1846- Harvard College and Yale College offered engineering courses (Wisely 1974).

1850- First U.S. census counted only 572 civil engineers two years before ASCE was founded in NYC in 1852 (Wisely 1974). Mechanics' institutes were growing rapidly with about 5,000 such institutes in 1850. The noted Franklin Institute was established in Philadelphia in 1825 (Grayson 1993) with a journal for mechanics and engineers.

1850- University of Michigan established a four- year engineering curriculum.

1854-1861- Civil engineering education began in Canadian colleges. In 1854, McMahon Cregan taught a civil engineering course at King's College (University of New Brunswick). In 1857, McGill University started an engineering course. In 1861, C.F.G. Robinson was the first engineering graduate in Canada.

1862- Morrill Land Grant Act provided land for agriculture and mechanic arts colleges in each state. This federal support significantly increased engineering schools to 70 total by 1872. Also, the 1862 Homestead Act and the transcontinental railroad spurred western growth. Schools offering mechanics/engineering started at Cornell (1868) and the University of California- Berkeley (1869), and later throughout the Midwest and country.

1876- The U.S. Centennial Exposition was held in Philadelphia with Alexander Holley setting up an engineering education conference between ASCE and AIME.

1882- ASCE, AIME and ASME (formed in 1880) set up a joint engineering education committee.

1893- Ira O. Baker, University of Illinois civil engineering professor and ASCE member (and department head for 39 years), organized a Congress on Engineering Education (70 attendees) at the World's Columbian Exposition in Chicago, resulting in the formation of the Society for the Promotion of Engineering Education (which became the American Society for Engineering Education (ASEE)). Ira Baker wrote several standard textbooks: *Treatise on Masonry Construction* (1889) and *A Treatise on Roads and Pavements* (1903). For teaching/research, Baker set up the Illinois cement testing lab in 1889 and later a road materials lab. John Wiley & Sons was the first publisher of engineering textbooks at the annual engineering education conference.

Summary

From the Royal Military Academy officer training for the English artillery and engineers in 1741 to early Mexico and Canada engineering- related education/training to the 1893 Engineering Education Congress at the Chicago World's Fair with 70 attendees, a Civil Engineering History Module for a "Time-Line of Early Civil Engineering Education History" was summarized. The March 19, 1802 Congressional establishment of the U.S. Military Academy at West Point was the first U.S. engineering school. Early American schools utilized examples and books from early French universities. U.S. engineering education was founded within 120 miles of the U.S. Military Academy, Norwich University, Rensselaer Polytechnic University and Union College. At about the time of the 1852 founding of ASCE in NYC, the 1850 census counted only 572 civil engineers. In 1862, the Morrill Land Grant Act provided land for agriculture and mechanic arts colleges in each state, leading to a rapid expansion of civil engineering programs. ASCE and AIME set up early engineering education committees/meetings in 1876 in Philadelphia and in 1882 with ASME. Ira O. Baker, a University of Illinois civil engineering professor organized an 1893 Congress on Engineering Education with 70 attendees, which led to the American Society for Engineering Education (ASEE).

References

Colegio de Ingenieros Civiles de Mexico (1996). La Ingenieria Civil Mexicana (ISBN 968-6272-12-7).

FitzSimons, Neil (1999). "A Sketch in the Life of Hinton James, Civil Engineer," (Rogers, Brenner 1999) *Forming Civil Engineering's Future*, ASCE.

Grayson, Lawrence P. (1993). *The Making of An Engineer*, John Wiley & Sons.

Grayson, Lawrence P. (1996). "Civil Engineering Education: A Historical Perspective," (Rogers et al, 1996) *Civil Engineering History*, ASCE.

Griggs, Jr., Francis E. (1997). "Amos Eaton Was Right!" *J. of Professional Issues in Engineering Education & Practice*, ASCE.

Jewell, Thomas, Francis Griggs Jr. and Stephen Ressler (2001). "Early Civil Engineering Education in the United States Prior to 1850," (Rogers, Fredrich 2001) *International Engineering History and Heritage*, ASCE.

Parkhill, S. Trent (2001). "150 Years of Geotechnical Design and Construction," (Rogers, Fredrich 2001) *International Engineering History and Heritage*, ASCE.

Rogers, Jerry et al (1996). *Civil Engineering History: Engineers Make History*, ASCE.

Rogers, Jerry and Brian Brenner (1999). *Forming Civil Engineering's Future*, ASCE.

Rogers, Jerry and Augustine J. Fredrich (2001). *International Engineering History and Heritage*, ASCE.

Turner, Daniel S. (1996). "The World's Oldest Civil Engineering Professor," (Rogers, Jerry et al, 1996) *Civil Engineering History: Engineers Make History*, ASCE.

Wisely, William H. (1974). *The American Civil Engineer*, ASCE.

Analytical Modeling:
Its Beginning To 1850

Francis E. Griggs, Jr. F. ASCE[1]

Abstract

Structural engineering did not move from a rule of thumb/experience-based discipline until early engineers began to develop functional relationships between some of the observed behavior of columns, beams, tension members and trusses. Men such as Archimedes, Hero of Alexandria and Vitruvius were the earliest to put into writing some of their observations around the time of Christ. They were giants, but for almost 15 centuries little progress was made in quantifying the relationships needed. DaVinci and Galileo in the late 15th century and early to mid 17th century paved the way for the development of modern Analytical Modeling. This paper primarily traces the development of Analytical Modeling in the period of the 18th and early to mid 19th century.

Introduction

It wasn't until the 19th century that engineers and builders were able to create mathematical models, confirmed by testing physical models, so they could calculate loads and stresses in actual structures or design new structures using wood, stone and metal. This paper treats the evolution of the modeling of primary structural elements: tension members, short compression members, columns, beams and trusses. Much of the material has been presented in a series of papers by the writer entitled *On the Shoulders of Giants* published in the Journal of Engineering issues and Professional Practice and an article on *S. H. Long and Squire Whipple the First American Structural Engineers* published in the Journal of Structural Engineering.

The status of Analytical Modeling was described by W. J. Macquorn Rankine, the well-known Scotsman, in his *A Manual of Applied Mechanics* first published in 1858. He wrote:

---

[1] Consulting Engineer, 30 Bradt Road, Rexford, NY 12148, fgriggs @nycap.rr.com

...the mathematical theory of a machine, - that is, the body of principles which enables the engineer to compute the arrangement and dimensions of the parts of a machine intended to perform given operations, - is divided by mathematicians, for the sake of convenience of investigation into two parts. The part first treated of, as being the more simple, relates to the motions and mutual actions of the solid pieces of a machine, and the forces exerted by and upon them, each continuous solid piece being treated as a whole, and of sensibly invariable figure. The second and more intricate part, relates to the actions of the forces tending to break or to alter the figure of each such solid piece, and the dimensions and form to be given to it in order to enable it to resist those forces: this part of the theory depends, as much as the first part, on the general laws of mechanics; and it is, as truly as the first part, a subject for the reasoning of the mathematician, and equally requisite for the completeness of the mathematical treatise, which the engineer is supposed to consult. (Rankine 1858)

In other words, the first part is statics and the second part strength of materials and design. Statics as a study can be traced back to Archimedes and even Aristotle.

The earliest Modelers

Archimedes developed the principle of the lever several centuries before the time of Christ. His problem as reported by Pappus was "to move a given weight by a given force." He correctly linked the weight to be moved and the force required to move it with the respective distances from a fulcrum to the two forces. Archimedes famous statement "give me a place to stand on, and I can move the earth" has come down through the years. It is clear that Archimedes would be using a perfectly rigid lever or one that was strong enough so as not to break. He had no idea how strong he should build his lever, as he was only discussing the statics of the problem and not the strength of the materials needed.

Hero of Alexandria who lived over a century after Archimedes in 200 BC was one of the first persons to put his ideas into print, and it is probable that he wrote at least seven books with the most important for this paper being his series called *Mechanica*. In his first book of *Mechanica* he gave the foundation of statics and dynamics as known in his time. In it he described the movement of gears, shafts, and screws leading up to a study of pulleys and other lifting devices. It has been reported that he used Archimedes' *Book of Columns* to determine the distribution of loads to columns when several columns support a continuous beam. In his second book of *Mechanica* he treated the five simple machines and in his third book he discussed lifting machines. While his work was mostly qualitative he was aware of the link between column length and the load the column could carry. He suggested wrapping wooden compression members of his lifting machines with rope, noting, "in order to strengthen a column a rope is wrapped around the column."

Vitruvius in the first century AD collected all that he could find on the subject of building and summarized it in his *Ten Books of Architecture*. He was mainly writing for architects but did include some of the machines of Hero, and Hero's mentor Ctesibius. He specified the height/ diameter ratio for stone columns but did so mainly from and aesthetic standpoint. His 10 to 1 ratio of height to diameter, however, was a good approximation to minimize buckling problems associated with long columns.

Leonardo DaVinci (1452-1519)

DaVinci was the next person to consider and place his thoughts in his notebooks in the late 15th and early 16th century. Unfortunately these notebooks were not discovered and put into print until much later. His impact on the development of Analytical Modeling, was limited but a review of his work shows he was one of the first to consider what Rankine called the second and more intricate part of Mechanics. DaVinci considered beams, columns, tension members, trusses and arches and attempted through testing and observation to understand the relationships between lengths and member dimensions. He did not fully understand the principles of elasticity, nor did he have the instrumentation to measure small deformations. He was limited to measuring what loads would cause failure or given deformations, of various structural elements and to vary lengths and dimensions of his members and determine how the failure loads or deformations varied. He then created rules, analytical models, independent of elastic properties to determine the load carrying capacities of other lengths and member dimensions. A few of his conclusions for various type members are as follows:

Beams
Ab supports 27 and is 9 beams, while cd, which is the ninth pat of it supports 3, since this is so, ef, which is the ninth part of the length of cd, will support 27, because it is 9 times as short.

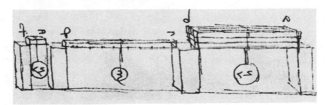

Figure 1 DaVinci's Beam Sketches

I have found that when a weight of a pound is hung from the middle of a rod of 12 ells, the rod curves by one ell, and I wish to know what weight will cause a rod 6 ells long, of the same thickness, to have the same one-ell bend...

The 6-ell rod is twice as strong in the middle as four 12-ell rods of the same thickness bound together.

Columns

If the diameter of a support is double that of another, it will support 8 times as much weight as the other, the two being of equal height.

Any support of twice the proportions (areas) of a smaller one will carry twice the load as the smaller, each of the two being in one piece and solid.

Of all supports of similar material and equal thickness, the shorter one will sustain more weight than the longer by as much as it is shorter than the longer one

Tension member

Note on the way in which you should measure the capacity of an iron wire, or what weight it can bear...Hang an iron wire two ells long or thereabouts from a strong point, then hang from it a basket or a sack or what you will, into which through a small orifice, you will pour fine sand from a hopper; and when the iron wire can sustain nor more it will break...then shorten the wire, at first by half, an see how much more weight it supports; and the make it 1/4 or its length, an so go on, making various lengths and noting the weight that breaks each one and the place in which it breaks. And you will make this test for all kinds of metals and woods, stone, ropes, and anything that will support other objects. And establish a law for each thing...(Uccelli 1938)

It is clear that DaVinci was attempting to develop rules or laws governing material behavior by experiment without having arrived at the concept of elasticity of materials. Some of his "rules" have been proven wrong based upon later studies but when his work is considered, and in the time it was considered, it is impossible not to be impressed with his ingenuity and vision.

Galileo Galilei (1564-1642)

Galileo was the first to make a detailed hypothesis of how a beam of rectangular cross section resisted a load applied transversely to its axis. Most of you know Galileo for his telescope and other work associated with the nature of the solar system. His interests, however, were much wider than this and his studies in mechanics, physics, military engineering, and model testing were significant.

In 1638 he published his work in a book entitled *DISCORSI E DIMOSTRAZIONI MATEMATICHE, intorno a due nuoue scienzse Attenenti ALLA MECHANICA & I MOVIMENTI LOCALI* or *Discourses and Mathematical Demonstrations concerning Two New Sciences pertaining to Mechanics and Local Motions*, commonly referred to as *Two New Sciences* with the first English translation being published in 1665. It is a remarkable work with amazing insights and use of geometrical principles. He presented his ideas as a dialogue between three students, Salviati, Sagreado and Simplicio, and works himself in as an *academician*. On the first day of their dialogue they discuss what happens when "a piece of wood or any other solid which coheres firmly is broken." After discussing a rope made of fibres they conclude that in "the case of a stone or metallic cylinder where the coherence seems to be still greater the cement which holds the parts together must be something other than filaments and fibres; and yet even this can be broken by a strong pull."(Galileo, 1638a)

Their conversation continued when Salviati told his friends:

...since you are waiting to hear what I think about the breaking strength of other materials which unlike rope and most woods do not show a filamentous structure. The coherence of these bodies is, in my estimation, produced by other causes which may be grouped under two heads. One is that much-talked-of repugnance which nature exhibits towards a vacuum; but this horror of a vacuum not being sufficient, it is necessary to introduce another cause in the form of a gluey or viscous substance which bind firmly together the component parts of the body. (Galileo, 1638b)

He goes on to show that a vacuum is insufficient to account for the strength of many materials and on the second day notes "whatever the nature of this resistance which solids offer to large tractive forces there can at least be no doubt of its existence; and though this resistance is very great in the case of a direct pull, it is found, as a rule, to be less in the case of bending forces." (Galileo 1683c) It is clear therefore; that at this time Galileo did not consider what behavior a gluey or viscous substance would contribute to a material other than its impact on its breaking load. He had several other experiments which mention condensation and rarefaction but never gets to the point of linking deformations and loads.

On the second day they discuss at length the beam starting with the lever of Aristotle and Archimedes. They then discuss the cantilever beam. Galileo believed that in such a beam the stress (he did not use this word) was a constant over the cross section much like it would be in a member subjected to simple tension as discussed on the first day. His classic illustration of a cantilever beam has been reproduced many times over the past 350 years. By observing how stone beams broke under load, he assumed that a fulcrum must be formed at the bottom surface of the beam at the wall. His analysis is as follows:

Proposition 1

A prism or solid cylinder of glass, steel, wood or other breakable material which is capable of sustaining a very heavy weight when applied longitudinally is, as previously remarked, easily broken by the transverse application of a weight which may be much smaller in proportion as the length of the cylinder exceeds its thickness.

Figure 2   Galileo Cantilever Beam

Let us imagine a solid prism ABCD fastened into a wall at the end AB, and supporting a weight E at the other end; understand also that the wall is vertical and that the prism or cylinder is fastened at right angles to the wall. It is clear that, if the cylinder breaks, fracture will occur at the point B where the edge of the mortise acts as a fulcrum for the lever BC, to which the force is applied; the thickness of the solid BA is the other arm of the lever along which is located the resistance. This resistance opposes the separation of the part BD, lying outside the wall, from that portion lying inside. From the proceeding it follows that the magnitude of the force applied at C bears to the magnitude of the resistance, found in the thickness of the prism, i.e., in the attachment of the base BA to its contiguous parts, the same ratio which the length CB bears to half the length BA; if now we define absolute resistance to fracture as that offered to a longitudinal pull (in which case the stretch force acts in the same direction as that through which the body is moved), then it follows that the absolute resistance of the prism BD is to the breaking load placed at the end of the lever BC in the same ratio as the length BC is to half of AB in the case of a prism, or the semi-diameter in the case of a cylinder. (Galileo 1638d)

Galileo, on the first day, had defined the "absolute resistance to fracture" as what is now called the breaking or failure load of a tension member. He determined that this resistance was a function of the cross sectional area and independent of length. Calling that force S, his equation would be:

$$S * \frac{AB}{2} = E * BC$$

In current terminology, the resisting moment was formed by resolving the uniform tensile stress distribution over the entire cross section into a force and using the half depth as the moment arm. He stated, "the fibres distributed over the(se) entire cross section(s) act as if concentrated at the center(s)." (Galileo 1638e) In other words he takes the concentrated force, applied at the half height of the rectangle, as statically equivalent to a uniform tensile stress acting over the cross section. An equation based upon his analysis would then be:

$$Moment = WL = S\frac{d}{2}$$

Where S=equivalent force=f*b*d=volume of pressure/stress diagram
b=width of beam; d=depth of beam; f=tensile strength

In modern day notation $M = WL = fb\dfrac{d^2}{2} = MZ$ where Z= section

modulus

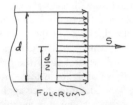

Figure 3   Assumed Stress Distribution

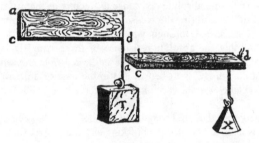

Figure 4   Galileo Proposition II Illustration

Galileo did not develop this equation but did consider a beam [ruler] supported with its long side vertical as well as its narrow side vertical, thus modifying the moment arm of his resistance. He concluded as a part of his Proposition II:

> therefore, that any given ruler, or prism, whose width exceeds its thickness, will offer greater resistance to fracture when standing on its edge than when lying flat, and this is in ratio of the width to the thickness. (Galileo 1638f)

It is clear therefore that he knew the strength of a beam was proportional to the width and the square of the depth of the cross section. His relationship was very close to the correct solution, but it is his understanding of the actual mechanics of the problem and the clarity in which he presented his ideas that is the more remarkable.

His analysis would also apply for a simply supported beam with a load at midspan as the moment is greatest at mid-span and reduces as you move toward the supports.   Reading *Two New Sciences* is a fascinating experience and at the cost of a few hours the lessons Galileo taught on those momentous first and second days can help us to understand the origins of the strength of materials.

Robert Hooke (1635-1703)

Hooke was the son of a minister on the Isle of Wright off the south coast of England.   At an early age he became interested in drawing and mechanical toys with a special interest in springs. He was interested in solving the problem of finding longitude at sea and determining the laws of gravitation especially as applied to the solar system.   He became a close associate of Robert Boyle the renowned chemist. He became Curator of Experiments to the Royal Society in London where his role was to prepare experiments and provide demonstrations of the behavior of structures for the members of the Society. He held this post from 1662 to his death in 1703.

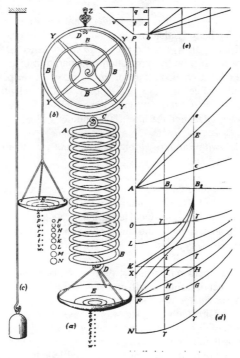

Figure 5   Hooke's Diagrams

In his book *De Potentiâ Restitutiva* published in London in 1678 Hooke first published his famous statement linking what we call stress with strain. He published an anagram in an earlier (1675) book on *A Description of helioscopes and some other instruments* in which he described his theory. The use of anagrams was common in England at this time as many researchers wanted to put their ideas into print but were afraid that others would steal them and make them their own. The anagram was a code that would be translated at a later date by the discoverer. His anagram was *"ceiiinosssttuu, id est, ut tensio sic vis..* or in English when decoded "the power of any spring is in the same proportion with the tension thereof." (Todhunter 1886a) Another theory he included in the same publication is the equally well-known relationship between the flexible chord and the rigid arch. The anagram translated into Latin was *"ut pendet continuum flexile, sic sabit contiguum rigidum inversum* or in English - "As hangs the flexible line, so but inverted will stand the rigid arch."

He notes that he had discovered these relationships 18 years prior to the publishing of his book but did not publish his discovery, as he thought he could patent the idea. He refers to his spring as any elastic body of metal or wood. His rule is contained in the statement, based upon his experiments as:

"From all which it is very evident that the Rule or a Law of nature in every springing body is, that the force or power thereof to restore itself to its natural position is always proportionate to the distance or space it is removed therefrom, whether it be by rarefaction or separation of its parts the one from the other, or by a condensation, or crowding of those parts nearer together. Nor is it observable in these bodies only, but in all other springy bodies whatsoever, whether Metal, Wood, Stones, baked Earths, Hair, Horns, Silk, Bones, Sinews, Glass and the like..."(Todhunter 1886a)

No known portrait of Hooke survives but two word description of him do survive. They are:

He is but of midling stature, something crooked, pale faced, and his face but little below, but his head is lardge, his eie full and popping, and not quick; a grey eie. He haz a delicate head of hair, browne, and of an excellent moist curle. He is and ever was temperate and moderate in dyet, etc.

As to his person he was but despicable, being very crooked, tho' I have heard from himself, and others, that he was strait till about 16 Years of Age when he first grew awry, by frequent practicing, with a Turn-Lath . . . He was always very pale and lean, and lately nothing but Skin and Bone, with a meagre aspect, his eyes grey and full, with a sharp ingenious Look whilst younger; his nose but thin, of a moderate height and length; his mouth meanly wise, and upper lip thin; his chin sharp, and Forehead large; his Head of a middle size. He wore his own hair of a dark Brown colour, very long and hanging neglected over his Face uncut and lank.... (Roberthooke.org website)

It is clear that his theory was that the force causing a rarefaction (tension) or a condensation (compression) is proportionate to the deformation (separation of its parts the one from the other or the crowding of those parts nearer together). This relationship is what is commonly called Hooke's Law. He did not know the magnitude of what we would call the constant of proportionality only that one had to exist. If he were able to conduct more exact testing he most likely would have determined the constant or modulus for various materials. This understanding of the elasticity of materials became the foundation of the study of Strength of Materials.

Leonard Euler (1707 - 1783)

Euler was born in the village of Basel in Switzerland, the son of a clergyman. He studied at the University of Basel when John Bernoulli, one of the famous Bernoulli family, was lecturing. He became one of Bernoulli's protégé's and as such was given private lessons. Later when the Russian Academy of Sciences opened, two of John Bernoulli's sons, Nicholas and Daniel, were appointed to positions. They, being aware of Euler's work, assisted him in gaining an appointment to the Academy. When Daniel Bernoulli left the Academy, Euler became head of the Department of Mathematics. One of his many publications was a Mechanics book entitled *Mechanica sive motus scientia analytice exposita* written in 1736, which dealt mostly with the use of calculus to describe the motion of a body. At about the same time he became interested in the topic of elastic curves. Like many mathematicians of his time he was looking for problems to which the new field of calculus could be applied. After moving to the Academy in Prussia at the invitation of Frederick the Great in 1744 he put his thoughts on the elastic curve into his book *Methodus inveniendi lineas curvas maxime minimeve proprietate gaudentes."* In this book, building on the suggestion of Bernoulli (Daniel), he studied the curvature that beams and columns would assume under various loadings. He prefaced his study with the statement:

Since the fabric of the universe is most perfect, and is the work of a most wise Creator, nothing whatsoever takes place in the universe in which some relation of maximum or minimum does not appear. Wherefore there is absolutely no doubt that every effect in the universe can be explained as satisfactory from final causes, by the aide of the method of maxima and minima, as it can from the effective causes themselves. (Timoshenko 1983a)

He first considered the catenary noting that for all of the possible deflected shapes of his beam (chain) under load he sought the shape that yielded a minimum potential energy, or the curve in which the deflected shape of the chain occupied the lowest center of gravity. This led to the well-known differential equation for curvature being set equal to the bending moment at a point:

$$\frac{c\,\ddot{y}}{\left(1+y'^2\right)^{\frac{3}{2}}} = -Px$$

Euler called c the absolute elasticity. He indicated that it was a function of the elastic properties of the material and for rectangular shaped members it was also a function of the width of the member and the depth squared and some constant. To determine the magnitude of c for a beam he recommended a test in which the length of the beam of rectangular cross section and its deflection under various loading be measured and c calculated using the following equation:

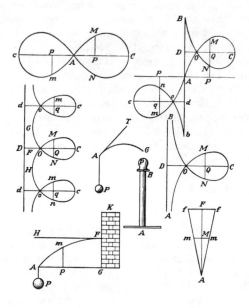

Figure 6   Euler's column and coordinate system

$$c = \frac{P l^2 (2l - 3f)}{6f}$$ where f= deflection of test beam of rectangular cross section

As we now know c is a function of the product of E, the modulus of Elasticity, and I, the moment of inertia. By running this test he did not have to determine a modulus of elasticity.

He built on the idea of Jakob Bernoulli who noted "the bending radius of an originally straight, homogeneous beam is, at any given point, inversely proportional to the moment of the bending force in relation to that point."(Timoshenko 1983b) Euler considered eight different loading conditions for a cantilever beam similar to that used by Galileo in his earlier beam study. They were primarily distinguished by the angle between the direction of the load and the tangent to the member at the point of application of the load.

One of these cases leads to the column orientation with both the load and the tangent to the member at the point of application of the load being vertical and having an angle of zero degrees between their directions. Note, Fig. 3, that this case is a column with one end fixed and one end free, sometimes called the flagpole configuration.

Euler integrated his curvature equation with y' term included thus requiring a great deal of manipulation. He then simplified his solution by assuming small deformations yielding:

$$P = \frac{c\pi^2}{4l^2}$$

which is the Euler Equation for long, homogeneous columns fixed at one end, which are loaded axially. It is clear, therefore, that the load carrying capacity of an elastic member is inversely proportional to the column length squared, whereas DaVinci had it inversely proportional to the length. Euler wrote:

Therefore, unless the load P to be borne be greater than π^2 C/ 4 l2 , there will be absolutely no fear of bending; on the other hand if the weight P be greater, the column will be unable to resist bending. Now when the elasticity of the column and likewise its thickness remain the same, the weight P which it can carry without danger will be inversely proportional to the square of the height of the column; and a column twice as high will be able to bear only one-fourth of the load. (Timoshenko 1983c)

Later in 1757 while still in Berlin, he revisited the column problem in a paper entitled *Sur La force Des colonnnes this* time neglecting the y' in the curvature equation by assuming small deflections. He arrived at the same equation for the buckling load and extended his previous discussion on the value of *c* by noting that its units must be that of a force times the square of a length. In another paper published in 1778 he replaced his *c* term with the product of the Moment of Inertia (I) and the Modulus of Elasticity (E) thus resulting in what is commonly known as the Euler Equation for a column pinned at both ends.

$$P_{crit} = \frac{\pi^2 EI}{l^2}$$

With this paper the contributions of Euler to column theory cease. Even though he considered other column problems in the future, his solutions were incorrect and had no impact on later investigators.

Many investigators realized that tests of short and medium length columns did not follow the results predicted by Euler's equation and later investigators developed relationships that could predict the load carrying capacity of shorter, more common, length columns.

Charles Augustin Coulomb (1736-1806)

Coulomb was one of the giants of early mechanics and civil engineering and one of the first to use Hooke's Law to solve a mechanics problem that had interested both DaVinci and Galileo. He was a graduate of Ecole du Genie at Mézières, at the time, the best technical school for military engineers in Europe. There he studied mechanics and materials and in addition worked on many public works projects while later serving as an engineer on the Island of Martinique.

Figure 7   Charles Augustin Coulomb

While there he wrote his famous paper that he submitted to the Académie des Sciences in 1773, shortly after returning to France. It was entitled *On an application of maxima and minima to some problems in statics, relating to architecture.* He covered many topics in this paper, but it is his solution to the beam problem that is of interest. While several French engineers studied the beam problem after Galileo, namely Mariotte and Parent, it was Coulomb who developed the analysis properly incorporating a stress distribution that could be extended to elastic bodies and incorporate Hooke's Law. Mariotte considered the beam as an elastic material but made a mistake in resolving his forces; while Parent made a correct analysis it wasn't as complete as Coulomb's. These three Frenchmen were the first to apply Hooke's law to the beam problem and thus replace Galileo's analysis that was, as noted, based upon rigid materials.

Coulomb began his paper by stating three propositions upon which the equilibrium of a co-planar force system is based. For the cantilever beam he, like Mariotte and Parent, realized that compressive stresses/strains must be present in the bottom fibers and tensile stresses/strains present in the upper fibers. He then assumed, as shown below, a general stress distribution varying from a maximum compressive stress at the bottom fiber to a maximum tensile stress on the top fiber. There had to be a location where the stress was zero leading to the concept of the neutral axis (note: Mariotte had also discussed the idea of a neutral linc). Coulomb set up his analysis for a general stress distribution that was not linear but could include a linearly elastic material. Looking at the resultants of his horizontal and shear

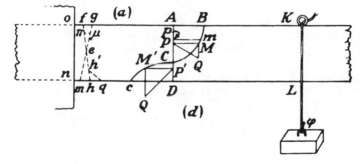

Figure 8   Coulomb's Beam and Stress Distribution over the Cross Section

forces in the tension and compression zones it can be seen that if they were added they would, combined with the applied load at the end of the cantilever equilibrate or add up to zero.

He resolved the forces in the beam fibers into horizontal (flexural) and vertical (shear) components and applied the three equations of equilibrium to the force system. The sum of the horizontal forces must equal zero so the horizontal compressive force must equal the horizontal tension force. Some researchers later wrote, erroneously, that the moments of the tensile and compression forces must be equal about the neutral axis. The sum of his two vertical forces (shear) equals the applied load and the moment of the applied load equals the magnitude of the internal couple formed by the two equal and opposite horizontal forces.

Coulomb was the first of all the investigators to consider equilibrium in its fullest sense including the vertical shear which acts across the section and he realized that this shear must exist in order to have equilibrium of the vertical forces.   Mariotte considered horizontal equilibrium by noting that the horizontal compressive force was equal to the horizontal tension force.   The other investigators prior to Coulomb were apparently only interested in rotational equilibrium.

He then took a small element near the wall, as shown on the left side of the above illustration *ofnh* and indicated how it would deform to the shape *ogmn* under load. It is clear that he was now considering an elastic material and was making the assumption that plane sections remain plane during bending or that stress is proportional to strain. He let the triangles *fgé* and *émh* represent the tension and compression forces. The forces would be located two thirds of the distance from the neutral axis to the extreme fiber and their resulting moment would be, for a beam width of b, determined by referring to the stress block of Coulomb as:

$$M = \frac{1}{2} f \frac{db}{2} \frac{2}{3} d = \frac{fbd^2}{6} = fS = f \frac{I}{c}$$

Where S=section modulus
             I= moment of inertia
             f= flexural stress on extreme fiber
             b=width of rectangular section, assumed to be 1 in Coulomb's analysis
             d=depth of rectangular section

Coulomb did not need a modulus of elasticity to solve his problem. He would need to know that stress and strain were linear over the depth of the beam, and that the member acted elastically, in order to resolve the simplest tensile and compressive stresses into forces that he could equilibrate. If he tried to compute deflections he would have to know the Modulus of Elasticity, but this factor is not a requirement to handle the first problem mentioned by Rankine, namely the statics problem with an understanding of elastic behavior.

This was the first of many breakthroughs made by Coulomb in his career, and he reluctantly submitted it for publication saying

This memoir written some years ago, was at first meant for my own individual use in the work upon which I was engaged in my profession. If I dare to present it to this Academy it is only because the feeblest endeavors are kindly welcomed by it when they have a useful objective. Besides, the Sciences are monuments consecrated to the public good. Each citizen ought to contribute to them according to his talents. While great men will be carried to the top of the edifice where they can mark out and construct the upper storeys, ordinary artisans who are scattered through the lower storeys or are hidden in the obscurity of the foundations should seek only to perfect that which cleverer hands have created. (Heyman 1972)

J. V. Poncelet (1788-1867), an École Polytechnic graduate from Metz wrote, "Coulomb's memoir contains in few pages so many things that during forty years the attention of engineers and scientists was not fixed on any of them." (Heyman 1972) Timoshenko wrote "no other scientist of the eighteenth century contributed as much as Coulomb to the science of mechanics of elastic bodies." (Timoshenko 1933d)

It remained to Navier, however, to bring Coulomb's analysis to the attention of both France and England in his *Résumé des Leçons*...In his first version he was not aware of Coulomb's effort and was one of many who equilibrated the moments of the forces about the neutral axis. In later versions, starting in 1826 or over 50 years after Coulomb's paper, this error is corrected. Navier's book included many other analyses that advanced the study of the theory of elasticity of beams, columns, plates and indeterminate structures that would be applied to engineered structures.

Thomas Young (1773-1829)

Young was one of many men of the 18th and early 19th centuries who had broad interests. He became an expert on Middle Eastern languages and was one of the first to attempt to translate Egyptian Hieroglyphs. He studied medicine and after receiving his medical degree in 1796 he attended Emmanuel College at Cambridge. Here he became interested in natural philosophy especially the nature of light and sound.

In 1799 he was elected a member of the Royal Society and was appointed to the position of Professor of Natural Philosophy by the Royal Institution. This institution

Figure 9   Thomas Young

was created for the purpose of "diffusing the knowledge and facilitating the general and speedy introduction of new and useful mechanical inventions and improvements and also teaching, by regular courses of philosophical lectures and experiments, that applications of the new discoveries in science to the improvements of art and manufactures and in facilitating the means of procuring the comforts and conveniences of life." (Timoshenko 1983e).

In 1802 he compiled his lecture notes into a book entitled *A Syllabus of Course of Lectures on Natural and Experimental Philosophy.* Section 11 dealt with the general properties of matter and Sections 19 and 20 were entitled *Of Passive Strength* and *Of Architecture.* Young was attempting to define what it was that held materials together under load. He used terms like cohesion and repulsive forces to describe tension and compression but he was not able to determine how these forces are developed in solids. He wrote, much like Hooke had earlier:

> The strength of the materials employed in mechanics depends on the cohesive and repulsive force of their particles. When a weight is suspended below a fixed point, the suspended suspending substance is stretched, and retains the form by cohesion; when the weight is supported by a block or pillar placed below it, the block is compressed, and resists primarily by a repulsive force, but secondarily by the cohesion required to prevent the particles from sliding away laterally...(Todhunter 1886b)

In 1807 he updated his lecture into *A course of Lectures on Natural Philosophy and the Mechanical Arts.* In a section entitled *Of the equilibrium and strength of elastic substances* he introduced the topic of Modulus of Elasticity or now known to some as Young's Modulus. This is the constant of proportionality that Hooke had described linking stress to strain or load to deformation. Hooke described the modulus as follows:

> The Modulus of Elasticity of any substance is a column of the same substance, capable of producing a pressure on the base which is to the weight causing a certain degree of compression as the length of the substance is to the diminution of its length. (Todhunter 1886c)

It is clear that he was describing what we would call a strain when he talked about the diminution of length and the total length. His concept of stress is a bit obtuse but the pressure on the base is what could be called a stress and he talked about the weight (force) on the base that leads to a stress. His modulus was really the product of what is now known as the modulus of elasticity and the cross sectional area of the member.

Young's description of elasticity is as follows:

The immediate resistance of a solid to extension or compression is most properly called its elasticity; although this term has sometimes been used to denote a facility of extension or compression arising from the weakness of this resistance. A practical mode of estimating the force of elasticity has already been explained, and according to the simplest statement of the nature of cohesion and repulsion, the weight of the modulus of elasticity is the measure of the actual magnitude of each of these forces; and it follows that an additional pressure, equal to that of the modulus would double the force of cohesion and require the particles to be reduced to half their distances in order that the repulsion might balance it; and in the same manner an extending force equal to the weight of half the modulus would reduce the force of cohesion, to one half and extend the substance to twice its dimension. But there is some reason to suppose the mutual repulsion of the particles of solids varies a little more rapidly than their distance, the modulus of elasticity will be a little greater than the true measure of the whole cohesive and repulsive force…(Todhunter 1886d)

Young's language is difficult to understand especially when he calls the modulus of elasticity a weight. Navier, discussed earlier, was the first to use the modulus of elasticity as the proportionality coefficient between stress and strain and published this in his 1826 book. It is of interest that Young determined the modulus of steel to be 29,000,000 pounds per square inch from tests he had run on the frequency of vibration of a tuning fork. (Timoshenko 1983f)

Young also picked up the study of columns and gave an analysis of an eccentrically loaded rectangular column that combined pure compression with the bending associated with a moment resulting from the eccentricity of load applied to the top of the column. He developed the idea of the Kern such that when the load is applied at a distance from the centroid of H/6 the neutral line is on the face of the columns opposite the eccentric load. He also treated the case of a column that was initially bent. He wrote that "considerable irregularities may be observed in all the experiments which have been made on the flexure of columns and rafters exposed to longitudinal forces; and there is no doubt but that some of them were occasioned by the difficulties of applying the force precisely at the extremities of the axis, and others by the accidental inequalities of the substances, of which the fibres must often have been in such directions as to constitute originally rather bent than straight columns." (Timoshenko 1983g)

Another of his observations concerned the load deformation relationships that exist after a member passes its elastic limit. He observed that deformations increased rapidly under steady load conditions and that this permanent deformation "limits the strength of materials with regard to practical purposes, almost as much as fracture, since in general the force which is capable of producing this effect is sufficient, with a small addition, to increase it till fracture takes place." (Timoshenko 1983g) His warning was then that any loading should be well within the elastic region.

Young's lecturing style and his writing was considered hard to understand, as he was so far advanced in his thinking that he could not explain his ideas to those who were not so gifted. As a result his ideas, far advanced beyond any one else's, did not receive the kind of dissemination they should have. Lord Rayleigh remarked that Young "from various causes did not succeed in gaining due attention from his contemporaries. Positions which he had already occupied were in more than one instance reconquered by his successors at a great expense of intellectual energy." (Timoshenko 1983h)

Eaton Hodgkinson (1789 -1861)

Hodgkinson was the son of farmer in England and was tutored by John Dalton, a well-known engineer of the time. His first publication was *On the Transverse Strain and Strength of Materials* written in 1822. He did not consider Hooke's Law and arrived at his results based upon experimental tests. His second paper, 1830, entitled *Theoretical and Experimental Researches to Ascertain the Strength and Best forms of Iron Beams* combined theory and experiment. For the beam he considered that for Cast iron the Modulus of Elasticity was different in tension than compression and that the neutral axis changes as load increases.

In 1840 he developed an equation for the strength of pillars (columns) based upon 227 experiments and published his results in a Report to the Royal Society, in the *British Association Report of 1837-1838*. For this report he was awarded a Royal Medal and elected a Fellow of the Royal Society. He tested various end conditions including rounded and flat-ended columns with cylindrical solid and hollow test specimens and varied his length/diameter ratios from 121 down to 15. His equation, for what he called breaking load, for cast iron cylindrical pillars is as follows:

$$Breaking load = A\left(\frac{h^{3.6}}{L^{1.7}}\right) tons$$

Where   h = pillar diameter in inches

L= pillar length in feet
A = a constant multiplier
= 44.3 for a hollow pillar with fixed ends
= 13.0 for a hollow pillar with round ends
= 14.9 for a solid pillar with round ends
= 44.16 for a solid pillar with fixed ends

In 1847 he became Professor of mechanical principles of engineering at University College in London. His equation was used for many years for columns in the length/diameter range he tested, and Euler's Equation was used for long slender columns.

Gordon, using Hodgkinson's data, later developed the equation for fixed end round columns used for many years as follows:

$$Breaking load = f \frac{S}{\left(1 + a\frac{L^2}{h^2}\right)} in pounds$$

Where S=cross sectional area in square inches
L=length of column
h=least external diameter (note, L and h in same units.
f=80,000 psi for cast iron and 36,000 psi for wrought iron.
a=1/400 for cast iron and 1/3000 for wrought iron.

Hodgkinson included the material strength in his A factor and multiplies his h/L factor times A to arrive at his load in tons. Gordon inputs a value for material strength with his f factor.

Summary European Efforts

By 1850 the works of Navier and Hodgkinson were known throughout Europe. Most engineers of that time frame, namely Robert Stephenson and Isambard Brunel were using the beam and column equations developed by Coulomb and Euler. Hodgkinson and Gordon had developed semi-empirical equations for the shorter more common columns. Elasticity as introduced by Hooke and extended by Young and Navier was in common use. It wasn't until the wide spread use of wrought iron and steel that even greater use was made of their work in the latter part of the 18th and early 19th century.

Developments in the United States up to 1850

Technology transfer from Europe to the United States in this time frame was very slow. The publication and translation of the French works was virtually non-existent. The dissemination of the English work, of a lesser quality than the French, did little to expand the understanding of American engineers. The USMA and Rensselaer Polytechnic Institute did use translations of French works on Civil Engineering updated by D. H. Mahan. Starting in the 1820s to the 1840s a small number of graduates of these programs were perhaps, with their professors, the only ones who knew of the French work.

The American Engineer needed to build on what they may have learned in the study of mathematics and natural philosophy and apply it to engineering practice. Other than some pamphlets that had been written by men such as S. H. Long, Ithiel Town, Herman Haupt, and a book by Thomas Pope little of value was available in the journals of the time. It was in this environment that Squire Whipple wrote his noteworthy book on Bridges in 1847. This book blazed a new trail for American Engineers to follow and it covered everything the Europeans had been working on for one hundred years or more. Whipple's work was apparently original and treated in original ways studies of beams, columns, elastic and plastic behavior, fatique and most importantly the development of techniques to analyze the truss.

Squire Whipple (1804-1886)

Whipple was born in Hardwick, Massachusetts the son of a farmer. His father, however, between 1811 and 1817 designed, built and ran a cotton-spinning mill in nearby Greenwich, Massachusetts. The young Whipple was therefore exposed to construction and materials at an early age. After selling the mill in 1817 the family moved to Otsego County just north of Cooperstown, New York to take up farming again. After receiving the best common school education available Whipple became a teacher before attending Hartwick Academy and Fairfield Academy located in central New York near his home. In 1829 he attended Union College in Schenectady, New York graduating one year

Figure 10 Squire Whipple

later in 1830. Union had a scientific curriculum, initiated in 1828, which attracted Whipple who was interested in mathematics and science. After graduating he spent the decade of the 1830s serving his apprenticeship working on the Baltimore and Ohio Railroad, the Erie Canal Enlargement, the New York and Erie Railroad, as well as several other railroads. When work was slow he designed, built and sold mathematical instruments such as transits and engineer's levels as well as drafting equipment.

In 1840 he turned his attention to the design and construction of bridges and weigh-lock scales. Having worked on the enlargement of the Erie Canal he knew that the wooden bridges that crossed the original canal were short-lived and realizing the new canal would be wider, thus requiring longer span bridges. He knew that bridges over the enlarged, modernized canal must be made out of modern materials-iron.

After some thought he designed his bowstring iron truss arch on August 22, 1840. He was issued patent No. 2,064 on April 24, 1841 for the "construction of iron truss bridges." It was obvious that Whipple knew the exact role each piece of his truss would play and how to size the member accordingly. His use of cast iron in compression and wrought iron in tension was based on economic considerations. Wrought iron was only available in small round or square bars and was very expensive. Therefore he used the cheaper cast iron for much of his bridge. That he understood the role that each member would play is shown in his discussion to the web members in his truss.

> ...diagonal ties d, d, &c. of wrought iron, or braces of cast iron, in pairs crossing one another between the vertical rods and between the arch and the thrust ties, except under the end segments, letter t, of the arch where only one tie or brace, Figs. 1, and 3, is used extending horizontally from the end of the arch to the foot of the first vertical or from the end, Fig. 5...The vertical rods may be made of round iron, and in all cases should have an aggregate strength sufficient to sustain the floor and any additional weight that may come thereon, and when the wrought iron diagonal tie is used the vertical rods should have a larger size to give them stiffness as posts...When the cast iron brace is used instead of the wrought iron diagonal tie, the vertical rod is never subjected to a thrust or negative force and may be of a smaller size and furnished only with a bolt head at one end and a screw-nut at the other, like an ordinary bolt. Otherwise, there may be one or more posts of cast or wrought iron used in conjunction with the wrought iron diagonal ties, in which case the vertical rods may be made smaller, or dispensed with entirely, the diagonal ties being enlarged so as to be adequate to sustain the whole weight. (Whipple 1847, patent application)

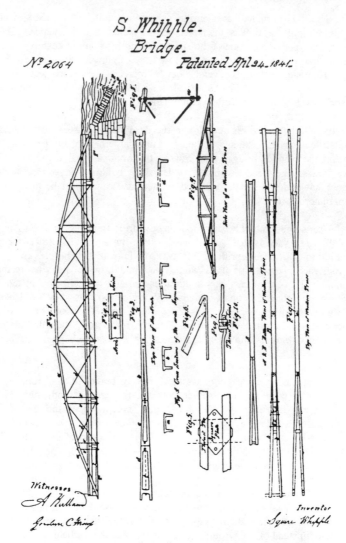

Figure 11   Whipple's Patent Application Drawing 1841

Whipple's patent, unlike any bridge patent before, shows an understanding of the structural behavior of the diagonals and verticals and the need to size the verticals to handle their loads as either a tension or compression member. He had crossing diagonals in each panel (except the end ones) in tension with posts (verticals) in compression. He never, to my knowledge, used cast iron diagonals, and thus never used his verticals in tension.

Whipple-Information Dissemination

Perhaps, his major accomplishment in the decade of the 1840s was *A Work on Bridge Building consisting of Two Essays, The One Elementary and General, the other Giving Original Plans and Practical Details for Iron and Wooden Bridges*, which he wrote and published in 1847 after having written his Essay Number One in 1846 and distributing it to some of his colleagues for their review. In his introduction to the work he wrote:

...In offering this little work to the Engineering profession, I would not be suspected of presuming that the subject has been exhausted, or that a want which has been long and deeply felt in this branch of the profession is fully satisfied. I may, however, be allowed to hope, that my labors in the field will have been the means of effecting one step of advance towards the attainment of so important a desideratum. (Whipple 1847a).

This book for the first time anywhere in the world presented the correct methods of analyzing and designing a truss. It was a giant step that later prompted several leaders of the profession to call Whipple "the retiring and modest mathematical instrument maker, who without precedent or example evolved the scientific basis of bridge building in America."(Boller 1889) In retrospect he had solved a much simpler problem that had the European engineers and it is not clear why they had not considered the problem.

What Whipple had done to analyze his trusses was to develop what is now called the method of joints. He used both trigonometric and graphical

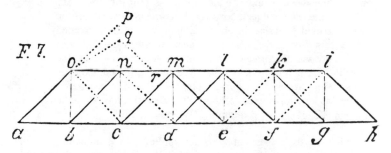

Figure 12   Whipple Graphical Analysis

means. He noted that in his work he usually used graphical means, as it was sufficiently accurate. He actually drew the force polygon for each joint right on a sketch of the truss being analyzed, thus making it easier to draw his force lines parallel to the appropriate member.

He also used to a lesser degree the method of sections, which had been considered previously by Haupt in his "Hints on Bridge Construction by an engineer." (Haupt 1841) This was the breakthrough for which Whipple is most remembered. There were, however, many other firsts in Whipple's book. The members of a truss could now be sized based upon the maximum load they would see under load. He found that the top and bottom chord should be designed for full loading across the bridge, but that the diagonals and verticals should be sized based upon a moving load placed at the point where it maximized the load in these members.

Whipple on compression member design

Whipple, working on his own, probably with little knowledge of the work of Euler and Hodgkinson, derived a column formula for more typical L/D ratios for circular compression members concluding that "the power of resistance is as the cube of the diameter directly, and as the square of the length, i.e., R is as D^3/L^2." (Whipple 1847, 53) Based upon this equation and the results of nine tests he ran on model columns, he concluded that a round compression member in cast iron with a length of 14.5 diameters should carry 46,000 [psi] and a square compression member with a length of 18 times the side dimension should carry 45,000 [psi]. He goes on to recommend that:

the hollow cylinder being evidently the form best adapted to sustain a negative strain, or a transverse strain in all directions, it would be well worth the while to have a set of thorough experiments, to determine accurately their actual strength. This has never, as far as I have learnt, been done, and therefore I shall assume the above estimate on the subject, as probably, not very far from the truth, subject, however, to correction whenever the facts and evidences shall be obtained, upon which the correction can be founded. In the mean time, since we know not the exact ratio between the greatest safe practical strength and the absolute strength of iron, and therefore should, in practice keep considerably within the limits of probable safety it becomes a matter of less importance to know the exact absolute strength, though this, of course, is desirable. (Whipple 1847b)

In other words, a large safety factor applied to the ultimate strength makes the actual value of the ultimate strength less important. His column equation for round members, he later stated, based upon several unknowns and material variations, should be modified and he recommended "as a good practical rule" to "make the power of resistance as D^3/L^2 and as D^3/L

successively, and take the mean, of the results thus obtained, as the true result." (Whipple 1847c) He thus developed an empirical relationship based upon theory and test results to give what he called the "true result." To use the formulas he tested a piece of similar cross section to arrive at its power of resistance (ultimate load) and then applied his two equations to the test sample strength to arrive at an average strength.

Whipple knowing how much American iron varied in its strength tested the actual cast iron he proposed, and used his D/L ratio to arrive at the crushing load in pounds. His equation for column strength can be expressed as:

$$P_{true} = \frac{P_{test}}{2}\left(\left(\frac{d_{act}^3}{d_{test}^3}\right)\left(\frac{L_{test}^2}{L_{act}^2} + \frac{L_{test}}{L_{act}}\right)\right)$$

Once a test was run on a column of diameter d_{test} and length L_{test} with a resulting test load of P_{test} he could determine the P_{true} for any diameter and length of column. This technique is, as noted, based upon experimental results as well as theoretical derivation, while Hodgkinson's is based solely on fitting a curve to experimental data. Whipple then recommended a column strength for ratios of L/d for the ranges of 0-2; 2-8; 8-12; 12-40; 40↑.

In the 1873 edition of his book he compared his results with those using Gordon's formula, which he acknowledged, "as the best authority upon the subject at the present day." (Whipple 1873a) His results were consistently larger than Gordon's up to a L/d ratio of 20 and smaller between L/d >20 and <50 if he compared Gordon's formula for flat ended columns. He concluded that the data compared favorably in the "range of length principally employed in bridge work."(Whipple 1873b) If he had compared his work with that of Hodgkinson, especially using his railroad data, he would have found that above L/r (slenderness ratio) of 75 he agreed almost exactly with that early British experimenter. Even using Whipple's road bridge criteria Whipple's results were only slightly above Hodgkinson's data.

It is likely that the values used by those who followed him were based, at least in part, on their lack of confidence in cast iron. Whipple firmly believed that cast iron in compression was a proper material to use in his top compression chords, given the higher cost at the time of wrought iron. With the normal L/d ratios common in trusses being built at Whipple's time his results were perfectly acceptable and safe.

Whipple on Beams

In the United States Dennis H. Mahan in his 1837 edition Course *on Civil Engineering* written for cadets at West Point clearly showed the application of Coulomb's solution. Mahan was evidently impressed by Navier stating in the introduction to his book "the European reputation of this eminent *Savant* and engineer, would render eulogium from the author more, if possible, than supererogatory. His name is connected, either as author or editor, with the ablest works on the subject under consideration, that have appeared in France within the last twenty years."(Mahan 1837) He did not, however, derive the equations in this edition. In the 1848 edition which was "mostly rewritten" he indicated, referring to the strength of a beam, that the "relations have been made the subject of mathematical investigations, founded upon data derived from experiment, which will be given in the APPENDIX..."(Mahan 1848) The copy of this edition reviewed did not have an appendix so it was not possible to determine if Mahan presented the derivation of the flexure formula. Mahan briefly discussed the results of Hodgkinson's column analysis and paper of 1840 in this edition. His 1858 edition contained a detailed derivation of the flexure formula using calculus. (Mahan 1858)

Whipple must then be given credit for first introducing equations for the design and analysis of a beam into American engineering literature. It is improbable that Whipple was familiar with Coulomb's or Navier's review of Coulomb's paper. Whipple's analysis was not very different than the method shown in current textbooks. He correctly assumed, using the Galileo cantilever illustration that "it is obvious that at every part of the cross section, the resistance to rotation is as the resistance to extension, multiplied by the distance of the part above the neutral plane. But the resistance to extension is, by the law of elasticity, as the degree or amount of extension, which is determined by the distance from the neutral plane."(Whipple 1847d)

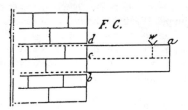

Figure 14   Whipple beam illustration

He then stated that the resistance to extension for the fibres above the neutral axis are a Sx where S is constant (equal to the slope of the stress block) and x is the distance above the axis. He then multiplied this resistance by the distance x to arrive at the resistance to rotation as Sx^2. Next he applied this stress to the differential area $t\,dx$ where t is the width of the beam. His differential resistance to rotation of all the fibres in tension above the neutral differential resistance to rotation of all the fibres in tension above the neutral axis is then $dR=s\ t\ x^2dx$. Integrating from 0 to h (the half depth of the beam) he arrived at the equation $s\ t\ h^3/3$ but letting $s\ h$ be equal to the resistance to extension at the extreme fibre and noting that $t\ h$ is the area above the neutral plane he concluded that: "the power of this part to resist rotation is equal to 1/3 the area [above the neutral plane] multiplied by the half depth of the beam, and the absolute positive strength of the material."(Whipple 1847, 62)   He continued, "now it is manifest that the part below the neutral plane exerts exactly the same amount of resistance to rotation as the part above. Therefore the total resistance to rotation about c, in other words, the resistance to rupture is equal to 1/3 the whole cross section multiplied by 1/2 the depth of the beam, and by the cohesive strength of the metal; that is equal to **1/3 C tD 1/2D**, making **D=db**, and **C**=cohesion, or positive strength of the material."(Whipple 1847, 62) Combining terms lead to, in current notation,

$$M = Ct\frac{D^2}{6} = fS$$

where S=section modulus and f=flexural strength

Whipple did not use the term section modulus but it is clear that it was included in his analysis. He pointed out that his analysis "is deduced on the supposition that the material is perfectly elastic until the strain produces actual rupture."(Whipple 1847, 63) He continued that "there are few substances, if any, and certainly wood and iron are not of the number, that fulfill this condition so nearly but that considerable discrepancies are found between the deductions of theory and the results of experiment. Indeed, in the case of cast iron, experiment shews the transverse strength to be fully twice as great as it is made to appear by the above formula."(Whipple 1847e)

Whipple on Inelastic Behavior

Whipple determined by experiment that the actual transverse strength of a cast iron beam was twice that predicted by the formula. He concluded, "I know of nothing to which to attribute this discrepancy between theory and experiment, except a want of complete elasticity in the material." (Whipple 1847f) He then proposed his hypothesis of what was happening to a beam as it was loaded to failure. His analysis is given in its entirety to show the clearness of his thinking. Keep in mind that he was discussing cast iron that had a much higher negative (compressive) strength than positive (tensile) strength.

> Cast iron, when exposed to a transverse strain, suffers extension on one side and compression on the other; and the power of resistance to both of these effects, increases very nearly as the

amount of extension or compression till they approach a certain point or maximum, and after passing this point the power diminishes. Now it is reasonable to suppose, in fact, we can hardly suppose the contrary, that for a certain interval on each side of the maximum point, the power of resistance remains nearly stationary. But this stationary interval is reached on the positive, much sooner than on the negative side, and the inevitable consequence must be that the neutral plane is transferred farther from the positive side, so as to preserve the equilibrium between the resistance to extension and the resistance to compression. Hence, the amount of resistance on the positive side is increased, both by the increased area of the cross section of extension, and increased leverage or distance from the neutral plane. Moreover, a greater portion of the fibres (so to speak) of extension, act with their full power; since, while the outside portion is passing through what we have called the stationary interval, successive portions towards the neutral plane, are reaching and approaching that interval. Hence, some considerable portion of all the fibers of extension, may act with their maximum power, whereas, if the body were perfectly elastic up to the point of actual rupture, only the outside fibers, farthest from the neutral plane, could act with absolute power, and all other parts, only in the ratio to their respective distances from said plane...I know of no more plausible manner of explaining the observed discrepancy between experiment and calculation upon this subject. (Whipple 1847g)

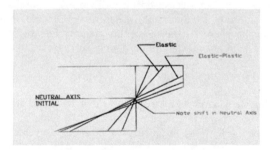

Figure 15   Whipple stress blocks elastic to plastic region

If the positive and negative strength of the material was the same, the outer fibers would both move into the plastic zone together and the neutral axis would remain fixed. In this case we would have the formation of a perfect plastic hinge upon which the theory of plastic design is based. In his 1869 appendix Whipple indicated that "it is possible, and perhaps *probable*, that a part of the discrepancy pointed out between the results of Experiment and Calculation, as to the lateral strength of Cast Iron, is owing to the assumption of too low an estimate for the Positive strength of that material; namely 18,000 lbs."(Whipple 1869)

Had anyone prior to Whipple given this matter any thought or put those thoughts into writing? Did Fairbairn and Hodgkinson discuss it in any of their writings? It appears that Young and Navier had considered this but had not given an analysis as thorough as Whipple to explain the observed behavior. It is clear that no American had published a similar analysis. St. Venant published his theory of plastico-dynamics well after Whipple put his thoughts into print.

SUMMARY

By 1850 the fundamentals of Analytical Modeling were in place for the common civil engineering applications. The analysis of beams and design of columns and tension members were well known and used by many engineering practitioners. The design of trusses, even with Whipple's work, continued for many years to be based on rules of thumb. The development of the Railroad in the United States placed more pressure on engineers to make the best use of expensive resources and this in turn required them to turn to analytical modeling efforts of the men covered in this paper.

APPENDIX, REFERENCES AND BIBLIOGRAPHY

Boller, A. P. (1889) Letter from Squire Whipple to A. P. Boller on the Development of the Iron Bridge, *Railroad Gazette*, April 19, page 253.
Galileo, Galilei, 1638a (1914) Translated H. Crew and A.DeSavlio, *Two new Sciences*, Northwestern Univ. Evanston, Ill. Page 7
Galileo, Galilei, 1638b (1914)...page 11
Galileo, Galilei, 1638c (1914)...page 109
Galileo, Galilei, 1638d (1914)...page 115
Galileo, Galilei, 1638e (1914)...page 120
Galileo, Galilei, 1638f (1914)...page 118
Griggs, Francis, E. (1996), On the Shoulders of Giants, *Journal of Professional Issues in Engineering Education and Practice*, Vol. 125, No. 4, April 1996
Griggs, Francis, E. (1996), On the Shoulders of Giants Part IV, *Journal of Professional Issues in Engineering Education and Practice*, Vol. 125, No. 4, April 1996
Griggs, Francis, E. (1999), On the Shoulders of Giants Part Three, *Journal of Professional Issues in Engineering Education and Practice*, Vol. 122, No. 2, October 1999

Griggs, Francis E, and Deluzio, A. J., (1995), Stephen H. Long and Squire Whipple, The First American Structural Engineers, *Journal of Structural Engineering*, Vol. 121, No. 9, September 1995.

Griggs, Francis E. (2002), Squire Whipple – Father of Iron Bridges, *Journal of Bridge Engineering*, Vol. 7, No. 3, May 2002.

Heyman, Jacques (1972), *Coulomb's memoir on Statics, an essay in the history of civil engineering*. Cambridge at the University Press, Cambridge UK

Heyman, Jacques (1998), *Structural Analysis - A Historical Approach* Cambridge at the University Press, Cambridge UK

Hooke R. (1678), Potentia restitutiva; or of spring explaining the power of springing bodies.

Love, A. E. H., (1944) *A Treatise on the Mathematical Theory of Elasticity*, Dover Books New York, 1944

Mahan, D. H., (1837) *An elementary course in civil engineering*, John Wiley, New York, NY

Mahan, D. H., (1848) *An elementary course in civil engineering*, John Wiley, New York, NY

Mahan, D. H., (1858) *An elementary course in civil engineering*, John Wiley, New York, NY

Navier, C. (1826) *Léçons sur l'application de la méchanique*, Paris, France.

Rankine, William John Macquorn (1901) *A Manual of Applied Mechanics*, Charles Griffin and Company, Limited, Exeter Street, Strand 1901.page 5

Timoshenko, Stephen, P. (1983a), *History of Strength of Materials*, Dover Press, McGraw-Hill Book Company, New York, NY. Page 31.

Timoshenko… (1983b), page 31

Timoshenko…(1983c), page 34

Timoshenko…(1983d), page 42

Timoshenko... (1983e), page 91

Timoshenko... (1983f), page 92

Timoshenko... (1983g), page 93

Timoshenko... (1983h), page 98

Todhunter, Isaac (1886a), *A History of The Theory of Elasticity and of Strength of Materials from Galilei to the Present Time*, University Press, Cambridge 1886. page 5

Todhunter, Isaac (1886b)…page 80

Todhunter, Isaac (1886c)…page 82

Todhunter, Isaac (1886d)…page 84

Uccelli, Arturo, (1938) The Science of Structures – contained in *Leonardo DaVinci*, Published by Istituto geografico De Agostini, Reprinted 1996. pages 264-274

Whipple, Squire (1847a) *A work on bridge building consisting of two essays, the one elementary and general, the other giving original plans and practical details for iron and wooden bridge*, Utica, N. Y., page iv.

Whipple, Squire (1847b)…page 55.

Whipple, Squire (1847c)...page 53.
Whipple, Squire (1847d)...page 62.
Whipple, Squire (1847e)...page 63.
Whipple, Squire (1847f)...page 64.
Whipple, Squire (1847g)...page 65.
Whipple, Squire (1847h)...page 200.
Whipple, Squire (1869) An Appendix to Whipple's bridge building, containing all except the original publication of the complete work Albany, N.Y. page 69.
Whipple, Squire (1873a) An elementary and practical treatise on bridge building an enlarged and improved edition of the author's original work D. VanNostrand, New York, N.Y. page 157.
Whipple, Squire (1873b)...page 158.

HISTORIC DEVELOPMENT AND USE OF TESTING/MONITORING TOOLS

By

K. Nam Shiu, FASCE[1] and Gajanan M. Sabnis Hon. MASCE[2]

ABSTRACT

This paper brings out a historical perspective of testing techniques as they were evolved. It is not implied that there were no instruments used prior to 1852, when ASCE was born; however, the emphasis is placed on the developed from that point on mainly in the United States in various aspects of testing. Testing was conducted for full scale if there was any doubt of its service life, since the structures were designed for such loads. In some cases, testing was resorted to as a means to prove the point in the court of law about the structure. This paper presents a view of the individuals who have seen development of testing and related tools both in the laboratory and in the field for the purpose of investigations. It is meant to be educational paper for the inquisitive and curious individual who would be interested in such history. List of references and in some cases websites in the end, make this paper further useful with today's technology. Some lessons learnt from the past history conclude the paper.

INTRODUCTION

This paper revisits the development and use of testing and measurement techniques in Civil Engineering. As the technology evolves in the last 150 plus years, testing and measurements have achieved many breakthroughs and many innovations have been successfully implemented. This paper attempts to touch on how testing and measurements have been used in Civil Engineering and highlight some of the new and emerging ways of applications.

DEVELOPMENT OF TESTING METHODS AND LABORATORIES

Testing of structures may be traced back to the ancient times. Perhaps, our ancestors (engineers) were not sophisticated in the mechanical testing disciplines, but they certainly were great builders. The great structures in India (Mohanjodaro), Egypt (Pyramids) and China (the Great Wall of China) are just a few of the examples. The testing was included in by Hammurabi long before mechanical testing became part of the modern codes. The idea was there, but the results were different. In a way, the mechanical testing (for failure of structure) demonstrated the quality that was expected in the work of engineers, builders and others responsible for public safety. Testing of structures in the field or in the laboratory forms important aspect of civil engineering, which allows us to learn much more about the behavior of a structure and use the information to advance our knowledge to achieve a better success of civil engineering to benefit the humanity.

----

[1]Senior Project Manager, Walker Parking Consultants, Inc., Elgin, IL 60123
[2]Professor of Civil Engineering, Howard University, Washington, DC 20059

Development of Testing in the Field and in the Laboratory

Mechanical testing was historically done in the field during the erection of a structure. Methods used for erection can be seen the ways the Greeks and Romans used the types of hoists to construct the structures. Egyptians used other approaches to erect the pyramids. Many of these methods were lost during the middle Ages and only recovered in the Renaissance. Timoshenko (1) presents an excellent treatise of structural engineering, material and structural testing and is highly recommended for all civil engineers to read. He gives interesting details of Leonardo da Vinci presentation from a note "Testing the Strength of Iron Wires of Various Lengths".

The introduction of iron and steel into structural and mechanical engineering made experimental study of the mechanical properties of those materials essential. Earlier experimenters had usually confined their attention to determining ultimate strengths, but it was later realized that properties of iron and steel depend upon the technological processes used in manufacturing. In any case, knowledge of only ultimate strength is insufficient for selecting a suitable material for a particular purpose. A more thorough investigation of materials' mechanical properties became utmost importance. Such tests required special equipment, which in turn prompted a rapid growth of material-testing laboratories, took place in many countries.

Wohler in Germany advocated the organization of government laboratories. In England, laboratories were set up by private enterprise. One of the first successful examples was that of David Kirkaldy, which was opened in 1865 in Southwark, London. Kirkaldy (2) discusses details about Napier and Sons received orders for some high-pressure boilers and marine machinery. Lightness combined with strength was critical. It was proposed to use *homogeneous-metal* for the one and *puddle-steel* for the other, instead of wrought-iron as ordinarily employed.

On the Continent, material-testing laboratories were developed original as government owned concerns. They reached a particularly high level in Germany, where they were run by the engineering schools and were used not only for industrial testing, but also for the research work of the teaching staff and for educational purposes. Such an arrangement was beneficial both to engineering education and to industry.

The first laboratory of this kind was founded in 1871 at the Polytechnical Institute of Munich. Its first director, Johann Bauschinger (1833-1893), was the professor of mechanics at the Institute. He succeeded in getting a 100-ton capacity testing machine installed. This device was designed by Ludwig Werder (1808-1885) in 1852 for *Masch,inen-baugesellschaft Nurnberg*. It proved very suitable for experiments requiring better accuracythan had been achieved before, and later most of the European laboratories installed similar machines.

Various Experimental Stations (Universities)

Grayson (3) has presented development of engineering in the US during the last two centuries and gives an excellent reading material. He discusses the beginning of the experimental stations and laboratories in various state universities along with detailed biographies of individuals responsible for this new beginning in the 1900's.

As a result of the Land Grant Act and its provisions to promote knowledge in agriculture and the mechanic arts, agricultural experiment stations and agricultural extension grew rapidly. However, similar growth in these areas did not occur in engineering. America was still an agrarian country at the turn of the twentieth century, so that research and dissemination of information in agriculture was viewed as a benefit to the majority of the people and to have direct results in terms of productivity. Productivity was an important objective since the country was on its way to becoming an urban industrialized society, and higher yields from the farms with less labor would be needed to feed the growing population. Research in engineering, however, was not considered as important, since great technological developments were made by individual entrepreneurs without the benefit of government subsidies.

Eugene Hale introduced a bill into Congress in 1896 to establish engineering experiment stations at land-grant colleges, but defeated. Although monies were not available from the federal government, individual states gradually took the initiative. The University of Illinois established the first engineering experiment station in December 1903. The following year under the newly appointed dean of engineering, Anson Marston (1864-1949), a second experiment station was established at Iowa State College of Agriculture and Mechanics Arts.

In short order, after the first engineering experiment stations were created, similar stations were organized at Pennsylvania State College, University of Missouri, University of Kansas, Kansas State Agricultural College, University of Ohio, Texas A&M College, and the University of Wisconsin. Their purpose was to assist and promote the industrial interests of the state through the publication of bulletins, research on industrial or public works problems, and to provide technological assistance to industry.

By 1924, 29 states had established engineering experimental stations within their land grant colleges. As a result, 503 engineering research bulletins had been issued, research funds amounted to $663,456, and there were 110 full-time and 358 part-time staff members. Eventually, all land-grant institutions having engineering programs followed this lead. In the area of extension services, the growth was much slower, with the Pennsylvania State College beginning a program of industrial extension in 1907.

Facilities at National Laboratories (4)

The National Bureau of Standards was established by Congress on March 3, 1901, with a charge to take custody of the standards of physical measurement in the United States and to solve "problems that arise in connection with standards." Although minor (and transient) variations occurred in the name of the institution, it was known for most of the century as NBS until Congress mandated a major name change, accompanied by new responsibilities, in 1988. Thus the "Bureau" completed its first century as the National Institute of Standards and Technology, or NIST. This volume commemorates the centennial by presenting brief accounts of selected classic publications of NBS/ NIST which illustrate at the same time the rich history of its scientific and technical accomplishments and the broad scope of its contributions to the Nation. If asked to select one word that best describes the work of the institution, most people familiar with NBS/ NIST would choose "measurement." Indeed, the theme of precise, accurate measurements runs through its first century's history.

With so many diverse audiences to serve, communication of the results of its work has been a major concern since the founding of NBS. Rosa (5), one of the 11 initial NBS staff members (the only one with the title of "Physicist") outlined his vision of the future of the Bureau. In addition to its mission to maintain the standards of measurements in the United States and to advance the art of precise measurements, he added a third mandate, "To distribute information regarding instruments and standards to manufacturers, state and city sealers of weights and measures, scientific and technical laboratories, and to any and every one applying for such information" - a daunting challenge which the Bureau has taken very seriously. Rosa went on to stress the interdependence of the three functions and to explain that the distribution of information would be "accomplished through correspondence, circulars and bulletins issued by the bureau," thereby setting the stage for the broad publication program.

Development of Testing Standards (ANSI and ASTM)

The American National Standards Institute (ANSI) (6) has served in its capacity as administrator and coordinator of the United States private sector voluntary standardization system for more than 80 years. Founded in 1918 by five engineering societies and three government agencies, the Institute remains a private, nonprofit membership organization supported by a diverse constituency of private and public sector organizations. Throughout its history, the ANSI Federation has maintained its primary goal of enhancing global competitiveness of U.S. business and the American quality of life by promoting the integrity of and facilitating voluntary consensus standards and conformity assessment systems. The Institute represents the interests of nearly 1,000 company organizations, government agencies, institutional and international members through its office in New York City, and its headquarters in Washington, D.C.

National Standardization

ANSI does not itself develop American National Standards (ANS); rather it facilitates development by establishing consensus among qualified groups. The Institute ensures that its guiding principles -- consensus, due process and openness -- are followed by the more than 175 distinct entities currently accredited under one of the Federation's three methods of accreditation (organization, committee or canvass). ANSI-accredited developers are committed to supporting the development of national and, in many cases international standards, addressing the critical trends of technological innovation, marketplace globalization and regulatory reform.

International Standardization (ISO)

ANSI promotes the use of U.S. standards internationally, advocates U.S. policy and technical positions in international and regional standards organizations, and encourages the adoption of international standards as national standards to meet the needs of the users. ANSI is the sole U.S. representative and dues-paying member of the two major non-treaty international standards organizations, the International Organization for Standardization (ISO), and, via the U.S. National Committee (USNC), the International Electrotechnical Commission (IEC). ANSI was a founding member of the ISO and plays an active role in its governance. ANSI is one of five permanent members to the governing ISO Council, and one of four permanent members of ISO's Technical Management Board. U.S. participation, through the U.S. National Committee, is equally strong in the IEC. The USNC is one of 12 members on the IEC's governing Committee of Action and the current president of the IEC is from the United States.

Through ANSI, the United States has immediate access to the ISO and IEC standards development processes. ANSI participates in almost the entire technical program of both the ISO (78% of all ISO technical committees) and the IEC (91% of all IEC technical committees) and administers many key committees and subgroups (16% in the ISO; 17% in the IEC). As part of its responsibilities as the U.S. member body to the ISO and the IEC, ANSI accredits U.S. Technical Advisory Groups (U.S. TAG's) or USNC Technical Advisors (TAs). The U.S. TAG's (or TA's) primary purpose is to develop and transmit, via ANSI, U.S. positions on activities and ballots of the international technical committee.

In many instances, U.S. standards are taken forward, through ANSI or USNC, to the ISO or IEC where they are adopted in whole or in part as international standards. Since the work of international technical committees is carried out by volunteers from public and private sector organizations, not ANSI staff, the success of these efforts often is dependent upon the willingness of U.S. industry and the U.S. government to commit the resources required insuring strong U.S. technical participation in the international standards process.

American Society for Testing and Standards (ASTM)

Organized in 1898, ASTM International (7) is one of the largest voluntary standards development organizations in the world. ASTM is a not-for-profit organization that provides a forum for the development and publication of voluntary consensus standards for materials, products, systems and services. More than 32,000 members representing producers, users, ultimate consumers, and representatives of government and academia from over 100 countries develop documents that serve as a basis for manufacturing, procurement, and regulatory activities.

ASTM develops standard test methods, specifications, practices, guides, classifications, and terminology in 130 areas covering subjects such as metals, paints, plastics, textiles, petroleum, construction, energy, the environment, consumer products, medical services and devices, computerized systems, electronics, and many others. ASTM Headquarters has no technical research or testing facilities; such work is done voluntarily by the ASTM members located throughout the world. More than 11,000 ASTM standards are published each year in the 73 volumes of the Annual Book of ASTM Standards. These standards and related technical information are sold throughout the world.

Standards development work begins when a need is identified by members of the committee or when other interested parties approach the committee. Task group members prepare a draft document/standard, which is reviewed by its parent subcommittee through a series of letter ballots. After the subcommittee approves the document, it is submitted concurrently to the main committee and the Society. All members are provided an opportunity to vote on each standard. All negative votes cast during any portion of the entire balloting process must include a written explanation of the voter's objections. Each negative vote or comment must be fully considered before the document can be submitted to the next level in the process. Final approval of a standard depends on concurrence by the ASTM Committee on Standards that assures that proper procedures were followed and due process was achieved. Only then is the ASTM standard published.

ASTM's Web site provides members flexibility in their standards development efforts. One of the most significant benefits of the Web site is the Standards Development **Forums** that provide committees with the ability to draft standards on-line, accelerating the review and ultimately the publication process. These forums also provide increased access to the process to international members and those who may have had limited participation due to travel constraints.

DEVELOPMENT OF MONITORING TOOLS

Historically, civil engineering is closely tied with testing and measurements. Some of the representative tools are discussed as follows:

Transit and Level

Even in the earliest civil engineering project, testing and measurements have been always an integral part of the construction. For example, to make sure vertical surfaces are plumb, our forefathers made use of the simple yet effective plumb bulbs. Of course, later on the transit and level have been used extensively in surveying and even symbolize civil engineering as a whole.

Mechanical Strain Gages (8)

As the structures and engineering projects become more complicated, the demand for stronger and more durable materials increases. To fully exploit the building materials, mechanical properties of commonly used materials (wood, steel) such as Modulus of Elasticity, and Poisson's ratio were measured. Dial gages, mechanical stain gages, load cells and compression machines were used. Measured data were incorporated into design manuals. Even today, we still use some of these design charts and measured materials properties.

The mechanical strain gage in its basic form uses mechanical systems such as levers, gears, or similar means for magnification of the strain measurement. Although this appears simple, the magnification from one gear and/or lever to another causes mechanical interaction, such as friction, lost motion, inertia, and flexibility of the parts, and if not overcome, some of these shortcomings reduce in measurement accuracy. In most instances, mechanical strain gages are limited to static stain measurements. With their size and inherent structure, it rules out any dynamic applications. In spite of some of the disadvantages of mechanical gages, they are used often, primarily because they are self-contained and easy to use. The strains to be measured are shown on scales or dials, and no additional equipment is required for readouts. They are reusable, which makes them economical.

The Whittemore gage, manufactured by Baldwin Locomotive Works, has been in use for many years. It consists of two frame members connected together by two elastic hinges, which provides a parallel frictionless motion. Conical points are attached to the frame legs, which are inserted into the attachment holes on the structure; these points define the gage length. The strains are measured with an integral dial indicator.

The main disadvantage of this gage is the potential error induced when the gage is repositioned on the structure for each strain reading. This error is minimized by having the operator to develop a consistent technique and also by having the same operator to read the same gage throughout a given test. Nevertheless, the gage is extremely useful for long-term (creep) measurements on concrete members, for measuring distortion in shear panels, and in other similar applications where measurement over a relatively long gage length is permissible.

Electrical Resistance Strain Gages

Electrical strain gages use the principle of change in the electrical characteristic in the gage material caused by strain. The important advantage of electrical strain gages is the ease to work with the electrical output, such as recording, and display. Of the various kinds of electrical strain gages, the resistance type is the most commonly used because of its many advantages. The resistance type of strain gage functions as a resistance element in an electric circuit. Strain produces a change in the magnitude of the electrical resistance associated with the gage. This change is recorded and related to strain during calibration, after the gage is bonded to the structure.

Although the principle of metallic resistance was observed by Kelvin in 1856, the first use of this principle related to strain measurement was by Carlson and Eaton in 1930. This first gage developed was unbonded, of the metallic type, with a single wire wound over two pins that were embedded in the structure. This gage had appreciable mass, size, and gage length, and consequently it was not used very much as a practical strain gage.

Simmons and Ruge in 1938 independently conceived the idea of bonding the wire either directly to the test specimen or to a thin paper backing which was in turn bonded to the specimen. Since then, a considerable amount of development has taken place in bonded electrical resistance strain gages, using etched foil circuits as well as wire circuits. Today strain gages are available in a large variety of shapes and sizes, and gage lengths as small as 1/64 in. These resistance-type gages, such as the SR-4® strain gage (initials of Simmons and Ruge), are the best tools for strain measurements in all types of structures and structural elements.

Because of the relatively low cost and lightness, a large number of these gages can be used to evaluate the strain behavior of a structure under given loading. Electric resistance strain gages have achieved the widest applications for strain measurements, including use in airplane parts, boat hulls, structures, buildings, bridges, and machine parts.

Analog Measurements

In the early years, testing was quite labor intensive. Readings and measurements would need to be made manually. The measurement process not only takes a lot of time, but manual monitoring becomes inadequate, particularly in vibration and dynamic situations.

The Tacoma Bridge collapse ushered us into the structural dynamic age. With the need to understand structural dynamics, innovation follows closely behind. In the 1960s, analog recording of data on magnetic tapes becomes feasible. In addition, accelerometers and other electronic stain gages were invented. Recorded data were plotted on analog paper.

Instead of 4 to 10 inches long mechanical gages, electric stain gages can be as small as half to quarter of an inch. With that, we can collect a lot more useful information during testing. This made testing of full sized structures more meaningful. We can then command a better understanding of complicated structures.

At the same era, computers began to come into the picture. Analytical tools such as Finite Difference, and Finite Element Analyses FEA, provide engineers with valuable tools to evaluate complicated structures. This together with the ease of measurements extended the confidence of building complicated structural elements.

Long Term Measurements for Concrete Structures

Measurements in concrete pose a special problem. Since concrete is a non-homogeneous material, displacements or strains cannot be easily converted to stresses through the Modulus of Elasticity as for elastic material such as steel. Time-dependent properties such as creep and shrinkage make the relationship between stress and strain highly non-linear. Therefore, long term monitoring since casting of concrete is necessary before stresses in concrete structure can be evaluated.

However, this poses a special problem for measurements in concrete structures. Almost all electric stain gages measure strains by monitoring the changes in electrical resistance. To make the measurements, a standard voltage or an electric current is applied to the strain gage. However, it is almost impossible to keep the initial readings from the strain gage constant because of electric drift. As a result, a special gage called Carlson strain meters were invented and were used successfully to monitor long term behavior of concrete structures.

Data Acquisition Systems

Recording of data is one of the crucial parts of testing, and must be carefully planned and checked out well in advance. Modern data acquisition equipment ranges from a simple, manually operated strain indicator box to automatic, sophisticated systems that record data continuously on magnetic tape or on paper charts. The range of satisfactory equipment for a static test is very broad because recording time is usually not crucial except when failure is approached and strains and displacements are changing rapidly. On the other hand, dynamic tests require a data acquisition system that is capable of monitoring and recording many channels in a fraction of a second. Information from equipment manufacturers is of crucial importance in selecting a new data acquisition system.

Reduction of data must be considered in selecting data acquisition systems. Numbers (strains, displacement, etc.) that are written down by hand must later be reduced by hand or after putting the data manually into computer storage; both processes are prone to human error and should be avoided whenever possible. Fortunately, the electronics revolution has made it possible for even a modest-budget laboratory to have a data acquisition system tied to a desktop calculator, to a microcomputer, or to a small minicomputer. These systems not only make data acquisition itself much easier, but perhaps even more importantly, the data is stored permanently on magnetic tape, ready for reduction and conversion into stresses, stress resultants, and plots and tables that can be used directly in reports on the experiments. Data acquisition systems are classified as intermittent, semi-continuous and Continuous.

Harris and Sabnis (8) present detailed discussion on the above.

Various Sites providing Information on Related Testing Accessories:

With the access of Internet, various commercial entities provide details of many products in their websites. Some of these are given here as reference material for the reader to pursue.

www.load-cells.org
www.dataacquisitionsystems.com
www.pressure-gauges.com
www.thermocoupleassemblies.com
www.leak-detectors.net

History of Some Manufacturers of Testing Equipments and Testing Tools

Baldwin Lima and Hamilton (BLH) (9) incorporate deep roots in tradition and American Industrial History. The 'B' in BLH actually stands for Baldwin; the original Baldwin Locomotive Company. In 1883, commissioned by the Philadelphia Museum, Baldwin produced a miniature locomotive for an in-house exhibition. Eventually, Baldwin built more steam locomotives than any other manufacturer in the world. Since the beginning of trade, some kind of measure of weight had to be established. Not only did this measure have to be uniform, it also had to be honest. In order to weigh or measure anything, there has to be a standard for comparison. The equal arm balance scale or the unequal arm beam scale has been used for thousands of years as the standard for comparison. It is still, by far, the most commonly used technique in the world for determination of weight. However, as discussed earlier, strain gage was invented approximately sixty years ago to make electronic weight measurements reliable and economically practical. With the strain gage as his technology base, BLH quickly produced and patented the first load (load cell) pressure, and torque transducers.

BLH is a technology leader in developing precision transducers and instrumentation uniquely tailored to the process weighing and web tension measurement industries. Cylindrical KIS® Beams (load cells) reject side, torsion, and torque forces to provide the "cleanest", most accurate weight measurement signal possible. HTU Transducers measure both horizontal (X) and vertical (Y) force components simultaneously to precisely calculate resultant web tension force.

In 1880, Tinius Olsen (10) armed with new patents reflecting his revolutionary ideas and designs, started producing testing machines. Primarily because of his efforts, an industry of relatively standard products emerged from what had been, until then, one-of-a-kind "engineering specials". The Tinius Olsen Testing Machine Company, Inc. is privately owned, and currently managed by the fourth generation of the Olsen family, and continues to offer a comprehensive line of advanced testing machines.

Testing equipment is designed and manufactured to comply with ASTM standards. In most case, machines and instrumentation also meets comparable ISO and other standards in use around the world. These instruments, engineered with the latest technology and software application are produced by employees with the same commitment to quality and workmanship that has been an integral part of our design and manufacturing traditions from the very beginning.

Tinius Olsen materials testing systems are ideal for testing a wide range of material, products and structures in tension, flexure, shear and compression: metals, construction materials, plastics, paper, textiles, non-wovens, composites, rubber, foods, and full sized products for quality assessment purposes

HISTORICAL APPLICATIONS IN CIVIL ENGINEERING

The following is a brief discussion of some of the application of testing in Civil Engineering.

Testing to certify new structures

Throughout the development of the building codes, testing the actual structure under the full design loads has been always a way to bypass any strength requirements in building or bridge codes. The reason is simple. If one can prove that the as-built structure can actually resist the fully factored design loads, there is no reason why the structure is deemed unsafe even though the structural calculations do not work out to meet the code specific limits.

This philosophy has carried on even as of to day. A good example is outlined in Chapter 20 – Evaluation of Existing Structures in American Concrete Institute (ACI) 318 Building Code.

Research and Development

If any one has been to a structural laboratory, one will find one thing in common and that is the testing facilities and the instrumentation hardware. Testing and monitoring is almost an inseparable part of a full size structural laboratory.

Research, as its name implies, venture into new types of structure or building technologies. For example, new technology such as development of seismic resisting structures, investigation of wind loading, external post-tensioning, plastic design of steel structures, etc. In those studies, scaled models are built and tested to failure in order to study their ultimate behavior (8). Together with the detailed modeling, the new concepts and technologies are developed and verified.

In addition, structural laboratories have also been used to evaluate the existing structures. Full scaled I-section girder has been extracted from highway bridges and tested for strengths and prestress loses. Other incidence include testing full size guide way girders to prove the sufficiency of torsion capacity by using open sections of double-tees instead of close box-girder sections

Monitoring to confirm behavior
As new technologies are imported from overseas, engineering applications of those technologies have to be closely scrutinized. Two of the outstanding examples are application of prestressing technology and the use of concrete segmental method in North America.

Prestressing Construction:
Prestressing construction has widely used in Europe before North America started using this structural system. Experimental testing was conducted in the Laboratory and followed by full size monitoring of several actual structures. All these tests have been fully documented in different technical journals.

Concrete Segmental Technology
About twenty years ago, segmental concrete technology was imported from Europe, an extensive instrumentation program was undertaken by Construction Technology Laboratories (CTL) and funded by Federal Highway Administration (FHWA) to monitor the long term behavior of the bridges designed by using the French code and French design programs. In the meantime, the time dependent properties used in the European code were closely compared with the generally accepted American ACI Committee 209 recommendations. The testing and monitoring confirmed that the technology is good and viable. No significant difference can be found between using the European code versus the American code. Nowadays, this concrete segmental technology is widely accepted and used in many Civil Engineering applications.

USE OF COMPUTERS IN FIELD TESTING AND MEASUREMENTS

Field testing and measurements on structures are the oldest and yet one of the most effective means to evaluate structural behavior. Measurements include strain, temperature, deflection, and dynamic response. However, detailed field measurements are rarely used during and after construction except in cases of reported abnormality in structural behavior or as a result of construction problems. Until recently, the advantages of field measurements have been overshadowed by the highly labor-intensive operation and the inherent high cost. With the advent of microcomputers, monitoring of large and complex structures during construction is now feasible at a fraction of the cost of the former manual systems. Manual readings are replaced by automatic data acquisition, controlled by a microcomputer.

The microcomputer and associated digital technology have changed the way we used to work both in the structural laboratory and in the field. The impact of microcomputers on the science of field measurement is mainly with regard to cost and time. The many benefits of field monitoring of structures are now available at an acceptable cost. Cost is reduced due to automatic recording compared to manual methods. The reduced cost of determining the behavior of buildings and bridges is not the only benefit of these three new measuring systems. Data returned for analysis are in a form that can be quickly reduced and evaluated by computer. A short turn-around time means that the behavior data are available when needed. Weinmann, Shiu and Hanson (11) present the benefits of field monitoring during construction and the life of the structure.

Actual and anticipated behavior of a structure can be different from the design. Differences may result from the many assumptions used in design, construction tolerances, the ever-changing environment at the job site, and less than ideal construction practices. As a result, field measurements are needed to provide important feedback to designers, contractors, and owners alike during construction to assure dependable structural performance. Moreover, field measurements can be used to evaluate effects of abnormal loading and to determine strength sufficiency of structures during and after construction. Detailed field measurements are now becoming recognized as a "tool" for inspection programs especially for highway bridges. For structures with sensors installed before construction, time-dependent behavior of the structure can also be evaluated.

The microcomputer system for monitoring during construction has several functions. The primary function is to control the data collection process of the data acquisition system. The microcomputer will control the sequence of time and date for scanning. Secondary functions are to display, reduce, and/or store data obtained on site. The data display can be done on a video screen such as a CRT, some kind of printer, or some other peripheral device. Data can also be reduced from measured signals to engineering units such as temperatures, strains, loads, displacements, or rotations. If necessary, noise filtering, signal enhancement, or averaging of readings can also be performed by the microcomputer. Recorded data are stored for later use in either internal random access memory (RAM) or external magnetic storage mediums such as cassette tapes or diskettes.

Another feature of the microcomputer is the internal real-time clocks. These timers allow the user to preprogram the computer for anticipated events with unattended data gathering. Microcomputers can also be self-starter to gather data upon given conditions or circumstances. This enables monitoring of the construction during critical phases. Specified sensors can also forewarn dangerous conditions.

With the rapid development of computer hardware and increasing capability of computer programming, the impact of microcomputer use in the field has been significant. Realization of microcomputer application to its fullest potential for on-site construction monitoring is just beginning. Application of field monitoring during construction is changing from a luxury to the necessity as structures become more and more complex. This high-tech advancement will enable construction to reach another unprecedented level. Further development can only be limited by our own imagination, creativity, ingenuity, and openness to change.

ADVENT OF ELECTRONIC AGE

The late 1900's was an era of fast pace development in electronics. The advent of electronics drastically revolutionized the testing and monitoring industries. Testing and behavior monitoring have the image and reputation of expensive, tedious, and labor intensive. However, electronic DDAS, new monitoring sensors such as tilt meters, and vibrating wire gages change the outlook of testing and monitoring. With automation, the costs have come down significantly. About 20 years ago, it would cost about $1,000 per sensor for field testing including installation. Nowadays, the costs could be easily cut in half or more.

The major changes in technologies include:

• Digital versus analog measurements
• Remote sensing and monitoring

Digital Technology

With the faster and faster CPUs and digital technologies, analog measurements are replaced with digital measurements at reduced costs. Large magnetic recorders are no longer required. Instead, small solid state computers and DDAS can be installed easily in the field and in the laboratory. The fast scanning rate can generate so much data points that it practically can provide similar data base as analog recording. The large amount of digital data can now be processed quickly with over 2.0 gig CPU speed and stored in large hard drive or CD disks.

Remote sensing and monitoring

The digital technology also allows remote monitoring with collected data transferred via Internet. The wide spread use of cell phone also allow structures in remote site be monitored by an on-site computer and collected data passed back to the control center via wireless phone connections. Specific applications include monitoring behavior of moving auger of the treatment plant or the reason why rotary blades cracked in a high-speed engineer. All these would be impossible without the remote sensing and the highly sophisticated computer technologies.

EMERGING APPLICATIONS

Of course, the availability of new technology can only be taken advantage of the ingenuity of Civil Engineers. Specifically, two applications of monitoring have been implemented.

- Asset Management of Bridges
- Smart Structures

Asset Management of Bridges

The aging infrastructure in North America has been well publicized. A large number of bridges are in need of repair and maintenance. In the past, bridges especially in the remote areas have not been regularly inspected or maintained. As these structures age, remedial actions are badly needed in order to keep up the service levels. As a result, many State Highway agencies have launched state-wide bridge management system, using computer data base and inspection rating system.

In some cases such as in Florida Department of Transportation (FDOT), bridges are load-tested to verify the strength sufficiency. With the advance in remote sensing, critical bridge elements can be monitored easily and their performance recorded on a regular basis. The centralized management system allows global monitoring of the critical infrastructures.

Smart Structures

More and more buildings are instrumented with sensors everywhere. Sensing includes regulation of building temperatures for heating and cooling. Some bridges have been instrumented using the weight-in-motion technology to monitor truck traffic in interstate highways.

Critical structures such as long span bridges can be monitored to forewarn dangerous condition. For examples, vibration monitoring in long span bridges can forewarn motorists of excessive highway sway.

Many accelerometers have been installed in numerous strategic buildings in high seismic areas to collect information of how earthquake affects buildings. This has been implemented for over 15 to 20 years ago and has since yielded valuable data to understand how buildings can be design to be earthquake-resistant.

CONCLUSIONS: LESSONS LEARNED ON TESTING AND MONITORING

Surely, testing and monitoring have been instrumental in the development of Civil Engineering. Unfortunately, most of our civil engineering students have no exposure to this important technology during their college years. As we look back in the past, we have definitely learned some lessons regarding use of testing and monitoring.

1. Determine what to measure

Ease of instrumentation and the relative low costs have allowed more sensors to be installed. However, more sensors do not necessarily mean more meaningful information. One has to maintain the discipline of choosing the necessary number of sensors and placing them at the strategic locations to collect the information that we want. Getting a large amount of information does not always be helpful. One should select discreetly what to measure before launching any instrumentation program.

2. Understand what we measure

A good understanding of how sensors works is important. This will allow us to correctly interpret what the data means. Factors that could affect the readings should be carefully studied and identified. Especially for field measurements, readings can be affected by many environmental changes and those factors have to be isolated before we can get hold of the meaningful data.

3. Refine what we know

Collecting all these information is great. But this is only part of the bigger picture. Making good use of the collected data is the key for taking full advantage of the testing and monitoring program. As we have seen in the past, a better understanding of how structures behave eventually will allow us engineers to push the envelope further and improve the process of designing and exploiting the emerging or new structural materials.

REFERENCES

1. Timoshenko, S.P., "*History of Strength of Materials*", McGraw Hill, 1953

2. Kirkaldy, I D., "Results of an Experimental Inquiry into the Tensile Strength and Other Properties of Various Kinds of Wrought-iron and Steel," Glasgow, 1862

3. Grayson, L.F., "*The Making of an Engineer*", John Wiley, 1993

4. http://nvl.nist.gov/pub/nistpubs/sp958-lide/html/002-intro.html

5. Rosa, E.B., "The National Bureau of Standards and its Relation to Scientific and Technical Laboratories", *Science* **21**, 1905

6. http://www.ansi.org/public/about.html

7. http://www.astm.org/FAQ/1.html

8. Harris, H.G. and Sabnis, G.M., "*Experimental Techniques and Structural Modeling*", CRC Publications, Boca Raton, FL, 1999

9. http://www.blh.com

10. http://www.tiniusolsen.com/tinius.html

11. Weinmann, T.L., Shiu, K.N. and Hanson, N.W., "Monitoring with Microcomputers as You Build", Proceeding of ACI Special Publication: "*Computer Applications in Concrete Technology*", SP-98, ACI, Farmington Hills, MI, 1987

Historic Development of U.S. Transportation Systems

George Washington, the Potomac Canal and the Beginning of American Civil Engineering: Engineering Problems and Solutions

Robert J. Kapsch, Ph.D., M.ASCE[1]

Abstract

The American Society of Civil Engineers was organized in 1852. But American civil engineering began much earlier. To understand the beginning of American civil engineering, it is useful to look at America when civil engineering and American civil engineers did not exist but were coming into existence. The history of the Potomac Company, established in 1785, provides us with a vehicle for examining America without civil engineers and civil engineering. As the company's first president, it was George Washington who first tried to hire American civil engineers to undertake the planning, design and construction of the Potomac Canal – but there were none. England and France had engineers but the company was not willing to pay foreign engineers to come to America. The company did occasionally use English engineers already in America as consultants. It was Washington and his Board of Directors who made the engineering decisions for the canal. The principle engineering decision was the decision to use sluice navigation as a means of improving the Potomac River navigation. In engineering terms, as shown later by early American engineers Thomas Moore and Isaac Briggs, this decision was not only wrong but counter-productive—sluice navigation made the river more dangerous and difficult to navigate. By 1823 the Potomac Canal was widely viewed as a major failure, not only financially (which it was) but also from an engineering perspective. The failure of the Potomac Company was directly related to the lack of engineering expertise used in its planning, design and construction. It was decided that the Potomac Company had to be abandoned and replaced by a new stillwater canal, the Chesapeake and Ohio Canal. The new canal was to be planned and designed by an army of civil engineers from the Erie Canal and elsewhere, under the direction of Benjamin Wright, the father of American civil engineering. These men employed totally different methods than were used on the Potomac Canal. Detailed plans were prepared. Levels and other precise survey instruments were used. Work was broken into individual contracts monitored by a large corps of engineers. Between the beginning of the Potomac Canal (1785) and its demise (1828), civil engineering had come to America and Americans had become civil engineers.

[1] Senior Scholar in Historic Architecture and Historic Engineering, National Park Service, 1100 Ohio Drive, S.W., Washington, D.C. 20242; telephone 202-619-6370; Robert_Kapsch@nps.gov.

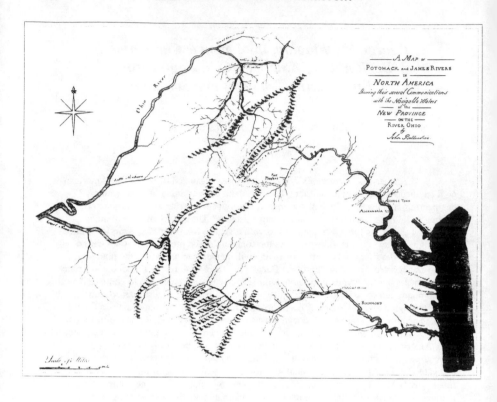

Figure 1. John Ballendine's map, "A Map of Potomack and James River in North America Showing their several Communications with the New Province on the River Ohio." This undated map is believed to have been executed in 1773 or 1774 in London for John Ballendine to promote the development of the navigation schemes of the Potomac and James Rivers to the Ohio Territory. As such it is one of the earliest maps developed for American canal development. Ballendine had been sent to England to study the canals of that country, especially the Duke of Bridgewater's two canals built by James Brindley near Manchester. This map was part of a detailed plan to develop the Potomac River navigation through a series of bypass canals to be constructed around the principal obstacles to navigation in the Potomac River. Navigation between these bypass canals was to be river navigation with adjoining horse path. The Potomac Company, organized in 1785, was influenced by Ballendine's plan but did not follow it with respect to bypass canals. Use of image courtesy of the Library of Congress.

I. Introduction

In the year 2002 we celebrate the 150th anniversary of the establishment of the American Society of Civil Engineers, begun in 1852. But the establishment of the Society was not the beginning of American civil engineering. Rather, it was an act that formally recognized that there was a profession of civil engineering in America. American civil engineering began much earlier. To investigate when American civil engineering actually started, it is useful to look at America when civil engineering and American civil engineers did not exist. The history of the Potomac Company (1785-1828) provides a vehicle for examining the early beginnings of American civil engineering and American civil engineers. Engineering decisions for the Potomac Company were made by George Washington, the first president of the company, and his board of directors. These decisions were augmented with occasional consultation from English engineers, such as James Brindley, the nephew of the famous English canal builder; William Weston, the English canal builder who worked on the Schuylkill & Susquehanna Navigation and sometimes called the father of American canals; Nicholas King, principally known for his surveying work for the City of Washington; and Christopher Myers, who was fired by the company for misuse of company funds and labor. In addition, Benjamin Henry Latrobe designed an extension to the Potomac Canal that if it had been constructed would have extended the Potomac Canal from its terminus at Little Falls through the City of Washington to the Navy Yard on the Anacostia River.

The Potomac Canal was a financial failure and in 1828 it was replaced by the Chesapeake and Ohio Canal (1828-1924). Unlike the design and construction of the Potomac Canal, the Chesapeake and Ohio Canal was begun by an army of engineers whom Benjamin Wright, called the Father of American Civil Engineering, brought with him from the recently completed Erie Canal and elsewhere. These men employed totally different methods than were used on the Potomac Canal. Detailed plans were prepared. Levels and other precise survey instruments were used. Work was broken into individual contracts monitored by a large corps of engineers. Between the beginning of the Potomac Canal (1785) and its demise (1828), civil engineering had come to America and Americans had begun civil engineers.

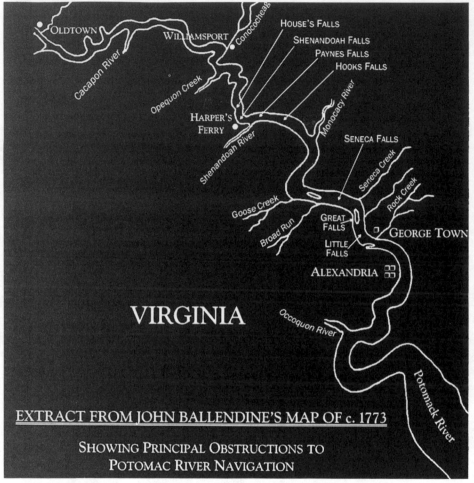

EXTRACT FROM JOHN BALLENDINE'S MAP OF c. 1773

SHOWING PRINCIPAL OBSTRUCTIONS TO
POTOMAC RIVER NAVIGATION

Figure 2. There were five principal obstacles to improving he navigation from Georgetown (on the tidal waters of the Potomac River) to Cumberland, Maryland, which Ballendine addressed in his plan: Little Falls; Great Falls; Seneca Falls; Payne's Falls, the Spout and Shenandoah Falls; and House's Falls. Ballendine planned for bypass canals at the first four of these obstacles. Once above House's Falls (located above Harper's Ferry, West Virginia) navigation was relatively unimpaired to Cumberland, Maryland. Ballendine died in 1781 and never saw his plan implemented. When the Potomac Company was organized in 1785, George Washington, its first president, and the company's Board of Directors, decided only to build bypass canals at Little Falls and Great Falls. At the remaining locations sluice navigation would be used. A major factor in this decision was a lack of engineers who could plan, design and execute the bypass canals proposed by Ballendine – American civil engineers did not exist and the company was not willing to pay the money to bring over from Europe an English or French engineer.

II. The Potomac River

The Potomac River drains a basin of approximately 14,670 square miles, including the District of Columbia and parts of Maryland, Pennsylvania, Virginia and West Virginia. It is approximately 400 miles long. At its mouth, at the Chesapeake Bay, the Potomac is eleven miles wide. Further upstream, at Harper's Ferry for example, the Potomac River averages 1300 feet in width. Up to Georgetown and Alexandria, the Potomac is part of the Chesapeake Bay tidal estuary. The region that the Potomac River drains receives 35 to 45 inches of rain per year (60 inches in the higher elevations). Flow rates vary from 559,000 gallons per minute (gpm) at Cumberland, Maryland to 4.2 million gpm at Point of Rocks, Maryland and 5.1 million gpm at Washington, D.C.[1]

In the eighteenth and early nineteenth century, shipping by water was much less expensive than shipping by land – perhaps only one twentieth the cost of land transportation.[2] The tidal Potomac provided water transport as far as Georgetown and Alexandria – ports established for the tobacco trade and extensively used in the eighteenth and nineteenth centuries for grain export. Above Harper's Ferry (located approximately 60 river miles above Georgetown) the Potomac River was navigable by narrow river craft as far as Cumberland (located approximately 185 river miles above Georgetown). Once at Cumberland it was possible to use wagons to cross the Appalachians to the Ohio River valley. If the obstacles between Georgetown and Harper's Ferry could be overcome, the Potomac River could be used as a highway to link tidal Maryland and Virginia to the Appalachian region and beyond.

There were five obstacles to river commerce between Georgetown and Harper's Ferry:

1. Little Falls (3 river miles above Georgetown). Here the Potomac River drops thirty-seven feet. Downstream navigation through the Little Falls is not possible and upstream navigation difficult and dangerous. If the Potomac River was to be used as a highway, Little Falls would have to be bypassed by a canal.

2. Great Falls (14 river miles above Georgetown). The next obstacle upstream was the Great Falls of the Potomac, where the river falls seventy-six feet. Even more so than Little Falls, the Great Falls was impassable to river traffic. Like Little Falls, the Great Falls would have to be bypassed by a canal.

3. Seneca Falls (22 river miles above Georgetown). With a drop of only seven feet, it was possible for downriver traffic to navigate Seneca Falls, but only at times of high water and only with great difficulty and danger. Upstream traffic was also very difficult and dangerous.

4. Paynes Falls, the Spout, and Shenandoah Falls (55 to 62 river miles above Georgetown). Proceeding upriver from Seneca Falls the next obstacle was a

series of rapids near the town of Berlin (now called Brunswick), Maryland. The first of these was Payne's Falls, where the Potomac River crossed the Catoctin Mountains. Continuing upriver, next was, "The Spout," and "Bull Ring Falls," both immediately below the town of Harper's Ferry and the mouth of the Shenandoah River. Above Harper's Ferry was Shenandoah Falls[3] where the Potomac fell fifteen feet in a mile. These rapids were also difficult and dangerous but could be navigated downstream at a time of high water.[4]

5. House's Falls (65 river miles above Georgetown). Four miles beyond the head of Shenandoah Falls was House's Falls, where the Potomac River fell three feet. Above House's Falls was relatively clear navigation to Cumberland, Maryland.

The proponents of developing the Potomac River for navigation believed that if these five obstacles could be overcome then the wealth of the upper Potomac River could be easily shipped to the port cities of Georgetown and Alexandria. Perhaps even the wealth of the Ohio River valley beyond.

III. Early Proposals to Open the Potomac River to Navigation

A. Military Transport for General Braddock (1755).

One of the first proposals to use the Potomac River as a water highway came as a result of General Braddock's advance on the French and their Indian allies as part of the French and Indian War. It was proposed to use the Potomac for military transport. In 1755 Maryland Governor Horatio Sharpe and Sir John St. Clair traveled the Potomac River to determine if the river could be used. They determined that it could not.[5] Instead of utilizing the Potomac, Braddock improved existing trails and built Braddock's Road from Cumberland toward Fort Duquesne (Pittsburgh). He and his men were ambushed on this road and defeated on July 9, 1755.

B. Proposal of 1762.

Shortly before the treaty of Paris was signed ending the Seven Year War between England and France (1763), interest in developing the Potomac River as an avenue of commerce was rekindled. A proposal to open the Potomac River for navigation from Cumberland, Maryland to just above Great Falls was made.[6] The men involved in this proposal had intense speculative interest in this area. For example, Colonel George Mercer, the son of John Mercer, secretary and legal counsel to the Ohio Company, was named by this announcement as one of the Treasurers. Thomas Cresap, the Ohio Company's field manager, was named as one of the individuals authorized to take subscriptions.[7] This also proposal suggested that the Great Falls of the Potomac could be somehow opened or bypassed and that "Skillful Gentlemen have agreed to view the Great Falls..." and to report if this is practicable.[8] There is no reference to whom these "Skillful Gentlemen" might be or if they even undertook their study of the Great Falls. At this time civil engineering had not yet been invented but the reference to "Skilful Gentlemen" is clearly a reference to a man or men who had the skills of what would become a civil engineer. No work was undertaken on the Potomac River navigation under this proposal.

C. John Semple's Proposal of 1769.

Perhaps the first serious, detailed proposal for improving Potomac River for navigation was made by John Semple, an immigrant from Scotland who first settled in Prince Charles County, Maryland, but after relocated to Virginia and who bought and developed the Keep Trieste Furnace above Harper's Ferry.[9] In his proposal Semple first reviewed previous proposals for Potomac River navigation improvements. He reviewed the reason for the failure of the proposal to use the Potomac for carrying military cargo for General Braddock's advance and also the reason for the failure of the 1762 proposal. Semple explained that the failure of the proposal to use the Potomac for military transportation was because of the construction of a road over the mountains in Pennsylvania by General Forbes. This road, "put an entire Stop to all further proceedings."[10] The proposal announced in the advertisement of 1762 was stopped because it was, "...thought by many to be too heavy for private persons to accomplish."[11] Semple explained that some individuals proposed taking this effort to the State Legislature for assistance but, "some of that respectable body being consulted, they advised the postponing of it to a more favorable opportunity. The people at that time being heavily burthened with taxes occasioned by the late war..."[12] Semple thought that now, 1769, was an appropriate time to open the Potomac to navigation.

Semple's proposal included the following elements:

1. Little Falls. Semple, like the 1762 proposal, proposed to begin his river navigation improvements above Great Falls, thus avoiding the obstacles of both Little Falls and Great Falls. His plan began river navigation improvement at the Widow Brouster's, two miles above Great Falls,[13] an established landing for Potomac River boatmen.

2. Great Falls. Like Little Falls, Semple envisioned no work to be undertaken at this location.

3. Seneca Falls. Semple believed that there was sufficient water during the winter and spring months and sometimes as late as June or July for boats properly constructed to pass from the upper end of Seneca Falls to Payne's Falls, some forty miles above Georgetown.[14] Passage at Seneca Falls would be via "gates" – either locks or flash locks – installed in each of the two dams that were to be constructed here. Each of these dams were to be five feet high.

4. Paynes Falls, the Spout and Shenandoah Falls. Once at Payne's Falls, there are three separate obstacles to be overcome: Payne's Falls itself, the Spout and Shenandoah Falls. Semple proposed that Payne's Falls be passed by means of a sluice or channel.[15] This improvement would also cost £250.[16] Above Payne's Falls is the Spout. Semple's plan called for this rapid to be passed by another channel.[17] Above the Spout is Shenandoah Falls, beginning at Harper's Ferry. Semple recommended that a stillwater canal, with locks, be constructed around these rapids. This bypass canal would be a half mile long.[18]

5. House's Falls. And at the head of Shenandoah Falls, above Harper's Ferry, Semple proposed that the natural channel, which someone had previously tried to improve, be made into another stillwater bypass canal.[19] This improvement would cost, according to Semple, £800. Semple stated that from here, above House's Falls, one can continue 130 miles upstream to Cumberland.[20]

Semple also addressed how other areas of the river might be cleared and improved. He suggested appointing overseers who would clear out navigable channels in the riverbed for year-round river travel -- a harbinger of the sluice navigation work that the Potomac Company was to later undertake.[21]

Semple's proposal became the basis for the proposal introduced in the House of Burgesses in the same year, 1769[22] by Richard Henry Lee and George Washington, "for clearing and making navigable The river Potowmack, from the great Falls of the said River up to Fort Cumberland."[23] No action was taken on the bill.

It was not only Virginians who were interested in opening the Potomac River to navigation. In his proposal, Semple makes reference to others who were developing plans to make the Potomac River navigable:

> ...some Gentlemen have lately carefully viewed the river and computed the Expence of removing the different obstacles, that obstruct and make the passage of Vessells Difficult...[24]

Probably Semple is referring to Marylanders interested in opening up the Potomac River to navigation, especially Thomas Johnson.

D. Thomas Johnson's Sluice Navigation Proposal of 1770.

Thomas Johnson, the man who nominated George Washington to be the Commander in Chief of American forces during the Revolutionary War and who later became the first Governor of the State of Maryland, was also interested in improving navigation of the Potomac River. Johnson's letters to Washington on this subject have been preserved, along with his drawings of lock gates and a route for a canal around Great Falls.[25] His drawing of the lock gates foretold the design to be used by both the Potomac Company and the Chesapeake and Ohio Canal Company while his route for the bypass canal around Great Falls was approximately the route that was used almost twenty-five years later by the Potomac Company.

Where Semple favored locks to overcome the falls and rapids of the Potomac, Thomas Johnson did not. In a letter to George Washington, dated June 18, 1770, Johnson discussed what would come to be called "sluice navigation." This involved blowing channels out of the bedrock of the river bottom to develop chutes or sluices down which the narrow riverboats could pass.[26]

Johnson wrote to Washington why he preferred sluice navigation to the use of locks in overcoming falls such as the Shenandoah Falls – that sluice navigation appeared a feasible means to Johnson to navigate the rapids above Harper's Ferry and if locks were constructed, they would be vulnerable to damage by floods and ice storms.[27]

In an undated Maryland Subscription Paper thought to have been the subscription paper referred to by Johnson in his June 18, 1770 letter to Washington, Johnson again discusses what would come to be called sluice navigation at the Shennandoah Falls above Harper's Ferry. He describes how sluice navigation could be combined with two small locks that would be filled in the fall with stones to prevent flood and ice damage. These locks would be emptied of those stones in spring allowing river commerce would resume.[28]

Fifteen years later Washington, as President of the Potomac Company, would decide on sluice navigation over the construction of locks in these areas -- perhaps influenced by Thomas Johnson's earlier arguments. The sluice around Shenandoah Falls, called the Long Canal, would be constructed complete with a towing path, but without the locks, much as had been recommended by Johnson.

In his letter of June 18, 1770 to George Washington, Johnson appeared confident that the Maryland subscription for the proposed Potomac River navigation improvements would be widely supported. He encouraged Washington to urge Virginians to support the project. On July 20, 1770, Washington wrote to Thomas Johnson on the plan for improving the navigation of the Potomac River. Washington wrote of the general strategy that would be needed to establish the Potomac Company – a combination of legislative authority from both the Maryland and Virginia state legislatures, and a combination of public and private capital.[29] The Maryland subscription wasn't widely supported and this initiative failed.

E. John Ballendine's Plan for Bypass Canals and River Navigation of 1774.

While these plans were being put forth for navigation on the Potomac River, James Brindley was constructing the first true canals in England, for the Duke of Bridgewater. The first of these two canals was constructed 1760-1763 and extended west from Manchester seven miles to the Duke of Bridgewater's mine at Worsley. The second, constructed 1762-1773, extended west from the first canal to the River Mersey. Traffic from Manchester could then descend the River Mersey to Liverpool.[30] These canals were immensely successful, in lowering the cost of coal delivered to Manchester from the mines at Worsley, in providing an ocean outlet for goods produced in Manchester and in improving the general quality of transport in England. The success of the first and second Duke of Bridgewater's canals sparked the canal revolution in England. Those men planning the improvement of the navigation of the Potomac River were aware of these developments in the mother country. They decided to send a representative to England to study Brindley's canal works so that similar improvements could be made upon the Potomac.

John Ballendine, John Semple's half brother, was selected to go to England to study these early canals. Despite a somewhat dubious reputation,[31] he had the support of George Washington who thought he had a natural genius for this sort of thing and thought he should be encouraged.[32] On May 8, 1772 Ballendine secured support from a number of prominent men and it was that support that enabled him to travel to England to examine canals and locks in order to develop a better plan for improving the Potomac River for navigation.[33] While in England he visited the Duke of Bridgewaters' canals, and others, in England.[34] While in London in 1773, Ballendine had printed a proposal for improving the Potomac River navigation to secure English support.[35] He was not successful in gaining that support.

Ballendine returned home to America. On September 8, 1774, he announced in the *Maryland Gazette* that he was back from Europe with a number of engineers and artificers to open the Potomac to navigation at and above Little Falls and proposed a meeting of principal subscribers for this work to be held September 26, 1774.[36] The meeting was held October 10th and the subscribers met and endorsed his plan. But implementation depended on adequate funds being received. He added a note to his announcement stating that a minimum of £30.000 Pennsylvania currency needed to be subscribed to achieve success.[37]

Although most of the numerous newspaper announcements taken out by Ballendine during this period have been published,[38] Ballendine's detailed plan (a copy of which is in the National Archives) has not. It was an ambitious plan and one that was advanced over the previous plans set forth for the improvement of the navigation of the Potomac. The plan also influenced George Washington and the

Board of the Potomac Company as they developed their plans, in 1785, to build the Potomac Canal.

Unlike previous proposals for improving the Potomac River for navigation, Ballendine began his plan at tidewater and not above Great Falls. His first recommendation dealt with a bypass canal around Little Falls:

1. Little Falls. At Little Falls Ballendine proposed a three mile bypass canal equipped with four lift locks to overcome the thirty-seven foot drop in elevation.[39] Ballendine went on to explain that the three mile long canal bypassing or skirting the Little Falls should be thirty feet wide and four feet deep.[40] The dimensions of the locks would be adequate to admit barges one hundred feet along, fifteen feet wide and four feet deep – large boats compared to the river vessels of the day. He also recommended, at a cost of £1200, that a dam be constructed at the upper end to divert water from the Potomac River into the Little Falls bypass canal. His plans for the Little Falls bypass canal were very close to what was constructed twenty years later by the Potomac Company. There was six miles between the head of Little Falls and the base of Great Falls. Ballendine was aware that the river in this area was full of rocks and rapids, such as Stubblefield Falls, and that work would also be required in this stretch of the river. His plan included clearing rocks in the river and building a horse tracking path along the river bank.[41] Ballendine wrote that the horse path, "...must be laid with stones secure from freshes...." By planning that these paths, "...would be secure...," he is suggesting that the paths would naturally be in the flood plain, would be periodically flooded but without damage and would be reused after the flood resides.

2. Great Falls. After Little Falls, the second major obstacle to upstream river navigation was at Great Falls where a drop of seventy-six feet must be overcome. As at Little Falls, Ballendine proposed that this obstacle be overcome by a bypass canal, with locks. He was aware that a portion of the excavation would have to be undertaken through solid rock and he allowed extra expense for that excavation. As at the Little Falls bypass canal, he specified unusually large locks for the time – one hundred feet long by sixteen feet wide and four feet deep. He also specified a small dam at the head of the bypass canal to divert water from the river into the canal.[42] Ballendine did not specify whether this bypass canal was to be built on the Virginia side, where is was to be built twenty years later by the Potomac Company, or the Maryland side, where the Chesapeake and Ohio Canal was to be constructed over fifty years later. Work between Great Falls and Seneca Falls would have been similar to work between Little Falls and Great Falls: rocks were removed from the river and a horse path built alongside the river.[43]

3. Seneca Falls. The third obstacle to Potomac River navigation, after Little Falls and Great Falls, was Seneca Falls. Ballendine proposed a thirty-foot wide canal, four feet deep and one and an half miles in length on the Virginia side of the river, with a single lock to overcome the seven-foot drop of the Potomac at Seneca Falls. This canal would begin at "Iron Landing" where a regulating lock would be constructed to bring water into the canal from the Potomac. The canal would terminate at its lower end at Mill Branch.[44] Semple had recommended the construction of two dams here (and the use of the existing third dam, having been built by Ballendine). The Potomac Company would not follow these recommendations but instead constructed a sluice navigation through Seneca Falls. From Seneca Falls to Payne's Falls Ballendine proposed work similar to the work he proposed between Little Falls and Great Falls, or the work proposed between Great Falls and Seneca Falls, Ballendine recommended removal of rocks in the river and the construction of a horse path between the upper end of Seneca Falls and the beginning of Shenandoah Falls.[45]

4. Payne's Falls, The Spout and Shenandoah Falls. Ballendine's plan groups Payne's Falls, The Spout and Shenandoah Falls under the heading of Shenandoah Falls. Here he recommended a three mile long bypass canal equipped with two locks.[46]

5. House Falls and above. To the above improvements for Potomac River navigation, Ballendine added improvements in the river from Shenandoah Falls to Cumberland, 120 miles, and locks at House's Falls and at Anita (i.e. Antietam) Creek.[47]

Ballendine's proposal would have cost £ 45,000 (about $150,000) as compared to Semple's proposal of £ 5,000 (about $17,000).[48] Raising the amount of money required by Ballendine's proposal must have seemed daunting. Nonetheless, Ballendine and his supporters seemed confident that such an amount could be raised.

Ballendine, in the same proposal, estimated that the tolls needed to be collected to pay for these large costs these improvements would probably generate £ 11,875 per year[49] – about $40,000. In other words, the tolls from the increased river traffic would more then pay for the expenditures within four years. In developing this estimate of future revenue, Ballendine estimated yearly tolls on the following downriver commodities: tobacco (£ 750); wheat and other grain (£1667); Indian corn (£833); flour (£1875); pig and bar-iron (£1500); timber, plank, stave (£2000); lime, coals and freestone (£500) and one-fourth of the downriver total for shipping upriver salt, rum, sugar, fish, dry goods, to provide the total annual estimated yield of £11,875.[50]

On October 25, 1774, Ballendine advertised a meeting to be held on November 12, 1774 to elect the Trustees of this enterprise and to announce that he had begun work on the canal at Little Fall.[51]

Whether Ballendine ever started building his canal at Little Falls has been debated. Many plans had been put forth over the years by various prospective builders for improving the Potomac River for navigation, virtually none were carried out.[52] The evidence indicates that, although Ballendine never completed his canal, he did at least initiate construction.[53] Shortly after beginning construction of the bypass canal at Little Falls, Ballendine transferred his operations from the Potomac River to the James River.[54] He announced that the reason for this was the lack of cooperation by the Maryland Assembly on the Potomac River proposal.[55]

Ballendine's 1774 proposal for navigation along the Potomac River was only one aspect of a three part plan led by George Washington, Thomas Johnson and John Ballendine – although it was the most prominent of the three parts. After the war, Washington, in a letter to Thomas Jefferson dated March 29, 1784, spelled out the three parts of the plan that was initiated by he, Johnson and Ballendine a decade earlier – the first part was the public plan advertised by Ballendine, the second part was the legislation that Washington was to have the Virginia Assembly enact and the third part was similar legislation to be enacted by the Maryland Assembly upon the action of Thomas Johnson.[56] Several parts of the plan had been put in place by 1775: the public part by Ballendine and Virginia's part by the Acts of 1772 and 1775 in the Virginia House of Burgesses. But the American Revolution had begun. All plans to improve navigation of the Potomac River were placed on hold.

Like most Americans, Ballendine's attention swung to the war. In conjunction with a new partner, John Reveley, Ballendine proposed to the Virginia Convention in May 1776 that they advance him and his partner funds for the establishment for a furnace and foundry so as to cast cannon and ball for the use of the Continental Army. The convention agreed to furnish £5000 in return for a mortgage on Ballendine's and Reveley's property and for iron to be furnished. The State was to build and operate the foundry and use water from Ballendine's canal, then being dug. Problems arose. Ballendine could not supply the iron ore contracted for, from Buckingham County. Shortly thereafter, on January 1, 1781, British vessels sailed up the James, entered Richmond on January 5, 1781, and destroyed Ballendine's and Reveley's property, including the foundry and its shops and their dwellings. Ballendine never got over this setback and died the same year, on October 14, 1781.[57]

IV. The Establishment of the Potomac Company

A. Enactment of Potomac Company Legislation (1784)

Following the successful prosecution of the war, Washington saw that now, 1784, was the opportune time to restart the plan that Ballendine, Johnson, he and others had initiated a decade earlier.[58]

To Richard Claiborne he wrote of the three steps that he saw as necessary to undertake the improvement of the Potomac River for navigation: 1. acquire enabling legislation for the company from both Maryland and Virginia; 2. organize the company, and 3. undertake an actual survey of the river by an engineer. On the last of the three steps, Washington would continue to emphasize the need for an engineer to guide the construction of the Potomac Company up until his death in 1799.[59] This engineering survey of the river was never undertaken by the Potomac Company.

In the previous year, 1783, the Maryland General Assembly passed a resolution for Charles Beatty and Normand Bruce to survey the Potomac, "...as laying the foundation for opening the navigation of Potomack."[60] It is not clear if this survey was ever undertaken. But others were thinking about using the Potomac River for navigation.

Late in 1784, bills for the Potomac River navigation improvement by the new Potomac Company had passed both the Maryland and Virginia Assemblies.[61] The differences between the two were resolved by a committee which met for that purpose, in Annapolis on December 22, 1784.[62] Washington was notified on January 16, 1785, that the Virginia legislature had passed the revised Act, as had the Maryland Legislature. The books for the new company were opened February 8, 1795.[63]

Washington was buoyed by the quickness in which the stocks in the new company were sold. In a letter to the Marquis de Lafayette he wrote:

> Of the £50,000 Sterlg. Required for the Potomac navigation, upwards of £40,000, was subscribed before the middle of May and encreasing fast ...[64]

The company was capitalized, therefore, at approximately the same level as estimated by John Ballendine some ten years before.

B. Mount Vernon Compact

Although the new company had authorizing legislation, from both the Virginia and Maryland legislatures, it also needed legal protection from any future disputes between the two states over the Potomac River. In 1777, both the states of Maryland and Virginia appointed Commissioners, "...to consider the most proper means to adjust and confirm the rights of each to the use and navigation of and jurisdiction over the Bay of Chesapeake and the Rivers Potomack and Pocomoke...." The war prevented an agreement from being reached at that time but after the war Washington and others moved the legislature of Virginia to appoint commissioners to meet with commissioners from Maryland on this subject. On January 16, 1785, the General Assembly of Maryland also appointed commissioners. The commissioners met at Alexandria on March 21, 1785 and at Mount Vernon on March 28, 1785 where they drew up the Compact of 1785, also called the Mount Vernon Compact.[65] This Compact stated:

> Sixth. The River Potomack shall be considered as a common highway, for the purpose of navigation and commerce to the citizens of Virginia, Maryland and of the United States, and to all other persons in amity with Aid states, trading to or from Virginia and Maryland.
>
> Seventh. The citizen of each state respectively shall have full property in the shores of Potomack river adjoining their lands, with all emoluments and advantages thereunto belonging, and to the privilege of making and carrying out wharves and other improvements, so as not to obstruct or injure the navigation of the river; but the right of fishing in the river shall be common to, and equally enjoyed by the citizens of both states.
>
> *Provided* That such common right be not exercised by the citizens of the one state, to the hindrance of the fisheries on the shores of the other state; and that the citizens of neither state shall have a right to fish with nets or seins on the shores of the other.[66]

The Compact was ratified by the Maryland General Assembly on November 21, 1785 and by the Virginia legislature on January 3, 1786.[67] With the Compact in place, Washington and the supporters moved to form the first Board of Directors of the new company. George Washington was elected President and the former Governors of Maryland Thomas Johnson and Thomas Sim Lee, and Colonel Fitzgerald and Lee of Virginia were elected Directors.

In a letter to the Marquis de Lafayette, dated July 25, 1785, Washington explained to Lafayette the basic development plan that he and the company's board were pursuing. First they would pursue the relatively easy sections of the river, primarily Seneca Falls and Shenandoah Falls. Second, they planned to hire a chief engineer to develop the more difficult portions of the navigation, the bypass canals of Great Falls and Little Falls.[68]   Thirty years after he had originally proposed the improvement of the Potomac River for navigation purposes, Washington led the company to actually undertake the work.

C. Organization of the Company

On Monday, May 30, 1785, the Board of Directors of the newly formed Potowmack Company held their first meeting, in Alexandria. After George Gilpin administered the Oath of Office prescribed by the Acts of Assembly enacted by Virginia and Maryland to Company President George Washington, Directors John Fitzgerald, Thomas Sim Lee, and Thomas Johnson; and after John Fitzgerald administered the same Oath to George Gilpin, the Board set down to business.[69]

After appointing a Treasurer for the new company, William Harthorne of Alexandria, the President and Board addressed the subject of improving navigation on the Potomac River. Their first decision was to employ two sets of hands, each consisting of 50 people and under the direction of a, "skilful person." One set of hands would be set to work improving the Potomac between Great Falls and Payne's Falls (i.e. primarily Seneca Falls). The other group of men were to work on the Potomac from the upper part of the Shennandoah Falls to, "...the highest place practicable on the North Branch."[70]

These were the least difficult tasks. The more difficult tasks, the bypass canals at Little Falls and Great Falls, and improving river navigation down from Harper's Ferry through the Spout and to the foot of Payne's Falls, would be addressed later.

From the beginning, Washington knew that he and the Board of Directors would need the advice and supervision of an experienced engineer for the construction of the bypass canals at Little Falls and Great Falls. In February of 1785 he wrote to the Marquis de Lafayette, "...that one thing ... is certain, namely ... a skilful Engineer, or rather a person of practical knowledge will be wanted to direct and superintend (the work)."[71] He wrote that such a person may have to be obtained from Europe and that he would prefer a French engineer, although other members of the Board might prefer one from England due to the language. Later he wrote to George William Fairfax that the Company needs a, "...skilful Engineer, a man of practical knowledge to conduct the business...," but that, "...where to find him we know not at present." In the meantime, Washington wrote Fairfax that, "...the less difficult parts of the river will be attempted, that not time may be lost effecting so important and salutary an undertaking."[72]

At the first meeting Washington and the Potomac Company Board of Directors began deviating from Ballendine's plan, which called for locks at Seneca Falls and Shenandoah Falls, in favor of what would later be called sluice navigation.

Washington and the Board of Directors had a choice of three types of canal: (1) a sluice navigation, (2) a system of dams and locks, and (3) an independent canal.

The first alternative, sluice navigation, was the least expensive. It amounted to clearing the river of rocks and building sluices – underwater ramps in the river – down which the boats could descend and up which the boats could be pulled. Sluice navigation could be dangerous as the boats could be easily upset in their downward descent of the sluice and they, their cargo and their crew could be lost. Upriver navigation was usually very laborious as the boats had to be manually hauled up the sluice against the current. Frequently water flow was not adequate.

The second alternative, dam and lock construction, was more expensive, usually much more. This alternative involved constructing dams across the river, usually made of wood cribs with rocks in their interior. These dams turned the river into a series of ponds which facilitated navigation. Navigation between ponds was by canal locks built into the dams. Not as dangerous or difficult as sluice navigation, this alternative was susceptible to damage by floods and winter ice.

The third alternative, an independent canal, sometimes called stillwater navigation, was also expensive but was the preferred alternative of the late eighteen and early nineteen centuries. The independent canal would usually be dug next to a river and would be provided with canal locks to overcome changes of elevation. A towpath would be constructed adjacent to the canal. The Duke of Bridgeport's canals in England, the model of the day, were stillwater canals.

Ballendine's plan was to build bypass or skirting canals to get around the falls and rapids. Washington and the first Board of Directors of the Potomac Company would choose in-river sluice navigation for Shenandoah Falls, Payne's Falls and Seneca Falls to minimize costs.

Washington and the Board of Directors had to decide on a maximum boat size. This would be used to size the width of the locks and determine the width and depth of the canal. There is no recorded discussion in the records of the Potomac Company on the ideal size of boat but Washington's view, expressed in 1783, was that the canal should accommodate vessels between 6 and 10 tons and larger vessels drawing three feet of water and which are no more than 60 feet long.[73]

On Tuesday, May 31, 1785, Washington and the Board met again and decided to advertise for 100 men to undertake this work.[74] At the same meeting, the Board authorized the construction of four work boats, two each for the two work parties. These boats were to be strong boats, " ... each to be thirty five feet long, Eight feet wide or upwards and not less than twenty Inches deep in the common manner of the Flats used at the Ferries on Potowmack above Tide Water."[75] Two were to be contracted for by Captain Abraham Sheppard of Sheppardstown and the other two by Col. Josias Clapham. Work began.

V. Engineers and Engineering Decisions of the Potomac Company

A. James Brindley

Washington strongly believed that he needed a competent engineer to oversee the planning, design and construction of the Potomac Canal. Such a man was James Brindley, who claimed to be the nephew of James Brindley, the canal builder of the Duke of Bridgewater's canals outside of Manchester. He also claimed to have had extensive experience on English canals under the supervision of his Uncle. If so, Brindley would have been the most competent canal engineer in the United States at that time. Washington certainly thought so:

> Mr. Brindley, nephew to the celebrated person of that name who conducted the work of the Duke of Bridgewater & planned many others in England, possesses, I presume, more *practical* knowledge of Cuts & Locks for the improvement of inland navigation, than any man among us, as he was executive officer (he says) many years under his uncle in this particular business ...[76]

And in a letter to John Fitzgerald and George Gilpin, dated March 31, 1786, Washington again expressed the same thought:

> ...as it is *said* no person in this Country has *more practical* knowledge than Mr. Brindley ...[77]

In the same letter, Washington outlined why he believes the Potomac Company could not bring an English engineer over from England to supervise the works of the Potomac Company (a tactic which Robert Morris used in 1792 to bring William Weston over to America to become the chief engineer of the Schuylkill & Susquehanna Navigation Company of Pennsylvania) -- it was too expensive. He proposed the idea that several American canal companies unite so as to share the cost of an European engineer and also to share his services.[78]

Washington also discussed the possible use of English and French engineers and that he has made inquiries into this matter. He quoted Lafayette as saying that what is needed is not a military engineer but what the French call, "Ingènieurs des ponts & chausses" -- the equivalent of a civil engineer.[79]

George Washington and the Potomac Company never brought over an engineer from either England or France. Instead, Washington would propose to use Brindley, then working for the Susquehanna Canal Company, as an engineering consultant to the company.[80] In so doing, Washington was particularly concerned about the siting and design of the locks on the Great Falls bypass canal and suggested that Brindley visit Great Falls and review the matter.[81]

In explaining why this was necessary, Washington wrote what could be considered the universal justification for the employment of a civil engineer:

... Taking Mr. Brindley to the works *now*, may, ultimately, save expence; at the sametime, having a plan before us, it would enable us at all convenient times, to be providing materials for its execution ...[82]

Despite the exhortations by Washington there is no evidence that the Potomac Company ever hired Brindley as a consulting engineer -- perhaps they tried but his obligations to the Susquehanna Company probably prohibited him from spending much time at the Potomac Company's works. However, Brindley did visit Great Falls and did provide some recommendations as to the work being undertaken.[83] He visited with James Rumsey, the company's superintendent.

B. James Rumsey

When the Board met on July 1, 1785, a number of persons applied to work as the Directors' Assistants. None of the applicants were found, "...Equal to the Superintendance of the Business," and none had sufficient, "Credentials of their Abilities and Integrity to be employed in the Subordinate Departments..."[84] It was at this time that George Washington recommended James Rumsey to the Board as Superintendent.[85] Rumsey was then building a house for Washington in Bath, Virginia.[86] On July 2, 1785, George Washington sent a letter to James Rumsey suggesting that he consider applying for the position of Superintendent and encouraging him to meet Colonel George Gilpin who was traveling to Seneca Falls and Shenandoah Falls for the purpose of hiring men for the two work gangs that the Company had authorized.[87] Rumsey followed Washington's advice.

Rumsey was hired on July 14, 1785, by the Company as Principal Superintendent at a salary (including expenses) of £200 per year Virginia Currency (approximately $666/year) – or about four times more than the Company paid its laborers.[88] At the same meeting, the Directors hired W. Richardson Stewart, of Baltimore, as one of two Assistants under Rumsey for a salary and expenses of £125/year Virginia Currency (approximately $416).[89] Since they could not find another suitable candidate for Assistant Superintendent, only Stewart was hired at this time. It is clear that the Board did not expect Stewart to have engineering skills for a year later Rumsey made a series of ten charges against Stewart, several of which related to Stewart's lack of knowledge on how to construct hoisting machines. The Board dismissed all charges indicating that they were pleased with Stewart.[90]

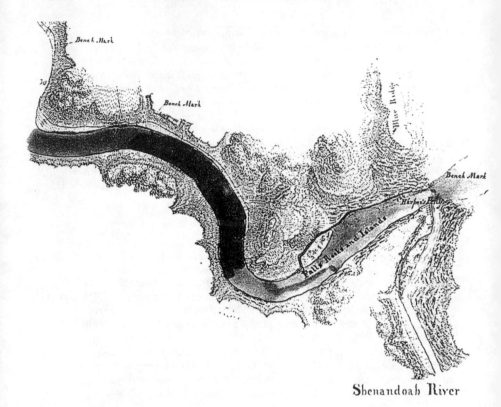

Shenandoah River

Figure 3. Harper's Ferry and the Potomac Company's Long Canal as mapped by
the U.S. Topological Bureau in 1825, Lt. Col. J. J. Abert assisted by Lts. W.H. Swift,
J. Macomb, J.K. Findlay, N.B. Bennett, H. A. Wilson (drawn by Lieut. Bennett).
This map was prepared as a reconnaissance for the Chesapeake and Ohio Canal, the
stillwater navigation intended to take the place of the Potomac Canal. At the
confluence of the Potomac River and Shenandoah River (right) the downstream
entrance of the Long Canal could be seen. At the middle of this image can be seen
the entrance to the Long Canal, as discussed by George Washington. Most of the
Long Canal was a sluice navigation along the Maryland (upper) shore of the
Potomac River. Where the drawing is labeled "Falls Rocks and Islands" were
Shenandoah Falls, later renamed "Harper's Ferry Falls" probably so as to avoid
confusion with the rapids on the Shenandoah River. The Long Canal was one of the
first works undertaken by the Potomac Canal, in 1785. It was obliterated by the
construction of the Chesapeake and Ohio Canal in 1830. Image courtesy of the
National Archives, Record Group 77.

C. Early Engineering Decisions in Favor of Sluice Navigation

The Board of the Potomac Company could not find an American civil engineer to supervise the works and were not willing to pay to bring over from Europe an English or French engineer. They therefore hired Rumsey – not an engineer but a qualified mechanic. The principal early engineering decisions were to be made by Washington and the Board of Directors and implemented by Rumsey.

The first, and principal engineering decision to be made was by Washington and the Board of Directors to either build a series of stillwater bypass canals (as proposed by Ballendine in his plan of 1774) or to build a sluice navigation with stillwater bypass canals only at Great Falls and Little Falls (as suggested by Thomas Johnson in his plan of 1770).

Washington and his companions were very much aware of John Ballendine's plans for bypassing Seneca Falls by the use of locks and a bypass canal. But they were also concerned about the cost of Ballendine's plan. On August 2, 1785, Washington and the Board began a trip to Seneca Falls and the Shenandoah Falls to investigate conditions for improving river navigation in these two areas. Their first stop was Mr. Goldsborough's house, at the head of Seneca Falls, approximately twenty miles above Georgetown. On August 3, 1785, Washington and the members of the Board examined Seneca Falls from canoes provided by James Rumsey.[91] Washington and the Board decided to use sluice navigation since they found the water to be of sufficient depth and felt that if the rocks were taken from the river in a straight path, than this navigation would be adequate. Locks and tracking paths would therefore not be necessary.[92] Sluice navigation through Seneca Falls would suffice.

From Seneca Falls Washington and his party traveled upriver to Keep Trieste furnace, about one mile above Harper's Ferry and at the head of Shenandoah Falls. Washington and his party traveled through the Shenandoah Falls to Harper's Ferry at the mouth of the Shenandoah River. In so doing, they decided that a sluice navigation, later to be called the Long Canal, would be dug out of the bed of the Potomac River along the Maryland shore. This sluice navigation would have a tracking path along the shore to allow horses to pull boats upriver through the rapids. As at Seneca Falls, at Shenandoah Falls Washington and his Board decided not to follow Ballendine's recommendation to build a bypass canal with locks.

Next, Washington and the Board members traveled down the rapids immediately below Harper's Ferry – sometimes called Bull Ring Falls. After Bull Ring Falls was the Spout and then Payne's Falls. Again, Washington decided that sluice navigation constructed in the riverbed adjacent to the Maryland shore was better than building a bypass canal.[93]

Washington and his party then traversed Payne's Falls, the last difficult river rapids between them and the head of Seneca Falls.[94] Once their party has cleared Payne's Falls, they met on the bank of the Potomac River and reaffirmed their decision made at the first meeting of May 30, 1785, to improve the navigation of the Potomac River here through sluice navigation and without the use of the tracking paths, bypass canals and locks as recommended by Ballendine.[95] This decision was formalized and entered into the proceedings of the President and Directors on the following day, Monday, August 8, 1785.[96] What remained was to instruct Rumsey to use the workers that he had been employing in the Shenandoah Falls for this purpose.[97] The Board also gave Rumsey a letter of direction allowing him a great deal of latitude in executing what would come to be known as the Long Canal and the Seneca Canal.[98]

In writing to William Grayson some two weeks later, Washington summarized the trip to Seneca Falls and Shenandoah Falls and the decision to use sluice navigation:

We have got the Potomac navigation in hand: workmen are employ'd under the best manager and assistants we could obtain, at the Falls of Shenandoah and Seneca; and I am happy to inform you that, upon a critical examination of them by the Directors, the manager and myself, we are unanimously of opinion that the difficulties at these two places, do not exceed the expectations we had formed of them; and that the navigation thro' them, might be effected without the aid of Locks: how far we may have been deceived with respect to the first (as the water, tho' low may yet fall) I shall not decide; but we are not mistaken I think in our conjectures of the other.
...[99]

Bypass canals would require locks. And, at this time, no lock had been yet constructed in the United States. Washington was concerned about the cost of such locks and that they might be constructed in the wrong place. He shared his concerns with lock construction with John Fitzgerald and George Gilpin.[100]

By these decisions, the Potomac Company had committed to a river navigation system largely based on sluice navigation in the riverbed of the Potomac. Tracking paths along the river banks would not be constructed. Bypass canals would still be used at Great Falls and Little Falls but the remainder of the river navigation system would be dependent on sluice navigation and river travel. These decisions were made by George Washington and the Board of Directors of the Potomac Company with little or no engineering input. Work progressed on the in-river sluices. By 1791 the company began work on the Little Falls bypass canal.

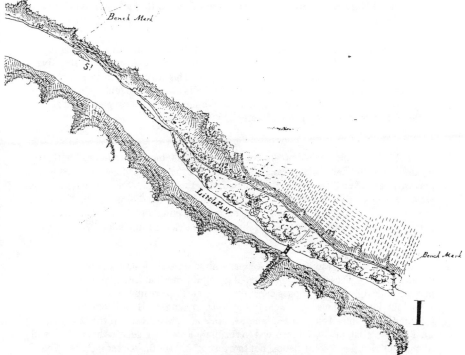

Figure 4. The Little Falls bypass canal as mapped by the U.S. Topological Bureau in 1825, Lt. Col. J. J. Abert assisted by Lts. W.H. Swift, J. Macomb, J.K. Findlay, N.B. Bennett, H. a. Wilson (drawn by Lieut. Bennett) . This map was prepared as a reconnaissance for the Chesapeake and Ohio Canal, the stillwater navigation intended to take the place of the Potomac Canal. Since the Potomac Canal used few engineers, few engineering drawings exist of their navigations. At the right of his bypass canal is Woodward's or Lock Cove where the masonry locks were constructed in 1818 to replace the wooden locks, also built here, by Leonard Harbaugh. At the middle is chain bridge, the fourth bridge to be constructed at this location. At the left of this canal is its upriver entrance and a small dam extending out into the Potomac. This was the first major bypass canal built by the Potomac Company and the wooden locks constructed by Harbaugh in 1795 are some of the earliest locks constructed in the United States. It was later obliterated by the construction of the Chesapeake and Ohio Canal in 1828. Image courtesy of the National Archive, Record Group 77.

D. Leonard Harbaugh and the Construction of the Little Falls Bypass Canal (1791-1795)

Work began on the Little Falls bypass canal at the beginning of 1791. The bypass canal was to be 3,814 yards long (2.16 miles), six feet deep and twenty-five feet wide at the top and twenty feet at the bottom. This was larger in section than the sluice canals excavated at Seneca Falls, House Falls and Shenandoah Falls, which were sixteen to twenty feet wide and four to five feet deep. The bypass canal at Little Falls was to be equipped with three[101] wooden locks.[102]

In 1792 the Potomac Company employed over 100 men under James Smith, in digging the Little Falls bypass canal. The route for the canal had been cleared the previous year. Ten black slaves were included in this work force.[103] The remaining were free labor. By November 5, 1792, the Board authorized hiring up to 200 slaves at $60/year[104] to facilitate this work. In the same meeting, the Board authorized the purchase of the wood needed to build the wooden locks at Little Falls.[105]

In January 1794, the Potomac Company hired over sixty slaves to work in the coming year.[106] These slaves continued digging the canal at Little Falls and that portion of the Great Falls bypass canal that extended upstream from the basin. By 1794 John Fitzgerald was President of the company and the Board consisted of George Gilpin, John Templeman, James Keith and Tobias Lear. At the December 27, 1794 meeting, the Board resolved to retain their two overseers, Abner Meek and Michael O'Heara and some hands, not to exceed 30, over the winter months. They appointed a committee, consisting of George Gilpin, John Templeman and Tobias Lear, to meet with Leonard Harbaugh and urge him to finish the locks at Little Falls immediately and, if they approved of his work to date, to enter into a contract with Harbaugh for finishing these wooden locks within a given time.[107] This Harbaugh would accomplish.

Leonard Harbaugh[108] was an experienced builder. He would complete both the Little Falls and Great Falls bypass canals for the Potomac Company. His son, Thomas[109] would begin work for the Company in 1803. Harbaugh was a builder from Baltimore. He was attracted to Washington, D.C. by the first Federal building program, 1791-1800. His first project was a masonry arch bridge over the Rock Creek at K Street, constructed for $8668. Unfortunately the arch collapsed shortly after construction but at no fault to Harbaugh's reputation as he was contracted to rebuild it. A respected builder in the new Federal city, Harbaugh was the President of the Architect's and Builder's Society of that city – a very early example of such an organization.[110]

But Leonard Harbaugh was a builder and not an engineer. Who designed the locks constructed at Little Falls – some of the first locks constructed in the United

States? There is the suggestion that an European engineer designed the wooden locks at Little Falls for Harbaugh to construct.[111] Nicholas King who designed the locks at Great Falls for the Potomac Company would not arrive in Washington City until May or June of 1796 and therefore after the locks at Little Falls had been constructed.[112] Therefore they were not designed by King. But at that time there were a number of European engineers and architects in Washington City related to the construction of the public buildings of the new Federal city. Several could have designed these locks.

By 1794 it became apparent that Harbaugh would be able to complete the Little Falls bypass canal in the near future. The company's attention turned to the completing the Great Falls bypass canal -- the last uncompleted link in the Potomac River navigation..

E. William Weston and the Construction of the Great Falls Bypass Canal (1791-1802)

With completion of the Little Falls bypass canal soon to be accomplished, the main challenge facing the Potomac Company was the construction of a skirting or bypass canal around the seventy-six foot drop of the Potomac River at Great Falls. This canal was to be 1200 yards long (0.68 miles) six feet deep, twenty-five feet wide at the top and twenty feet wide at the bottom – to be excavated mostly from solid rock. The excavation of the upper end of the canal, from the upstream end to the basin at the middle, began in 1791. The lower portion, from the basin to the downstream end, would contain the locks, either four or five. By 1795 lock seats for five locks were being excavated and masonry locks were planned to be constructed. A mistake in the siting of this canal or in the construction details would be very expensive and otherwise harmful to the company.

It was in the summer of 1794 that the Potomac Company first invited the famous English engineer, William Weston, to visit the works at Great Falls and to give his opinion of that construction.[113] Whereas the Potomac Company, in 1785, had declined to bring from Europe an English or French engineer to supervise its operations, Robert Morris, President of the Schuylkill & Susquehanna Navigation Company of Pennsylvania, had no such qualms. In November 1792 he brought over from England William Weston to supervise his design and construction operations. Weston was the son of Samuel Weston, an engineer who had worked for James Brindley. Before he left for America, William Weston had been involved in a secondary role in the construction of the Oxford Canal and as the principal engineer and managing contractor for the design and construction of the Trent Bridge at Gainsborough (1787-1791). Although not particularly prominent in England, in America, because there were so few engineers, he assumed dominance of his profession.[114]

President George Washington, who had previously urged the use of English engineer James Brindley to consult on the construction of the Great Falls and Little Falls bypass canals in 1786,[115] now encouraged the Potomac Company to hire William Weston as a consultant. On December 12, 1794 he wrote Potomac Company President Tobias Lear urging him to bring Weston to the Potomac Canal:

> ...I will not neglect any *fair* opportunity of facilitating a visit from Mr. Weston to that quarter (i.e. the Potomac Canal)... [116]

On December 21, 1794 he again wrote Lear urging him to bring Weston to the Potomac Canal.

> ...A good opportunity presenting itself on Thursday last, I embraced it, to enquire of Mr. Morris, if the Directors of that company might entertain any hope of deriving aid from Mr. Weston's opinion, respecting the Lock seats at the Great fall of that river; his answer was; 'Mr. Weston, from some peculiar circumstances attending their own concerns, had been prevented from visiting that spot, as was intended, but that he was now expected to be in this City in a few days (as I understood) when he wd. propose, and urge his going thither.[117]

Inclined planes, in lieu of locks, were considered for use on the Potomac Canal. Washington wrote to Lear about the possibility of the use of inclined planes, in lieu of locks, at Great Falls,[118] and why Washington did not favor their use – because their wasn't machinery equal to the task and because Washington did not see such inclined planes in general use. But perhaps the major reason was that Weston had told him that no method of raising and lowering boats was superior to lift locks. The Potomac Company pressed on with their plans to use locks at Great Falls.

On January 12, 1795, Washington again continued to press Lear about bringing Weston to the Potomac Canal.:

> ...the Canal Company of this State (has given permission) for Mr. Weston to visit the falls of Potomack, and that he might be expected at the federal city about the first of next month...[119]

And again on March 5, 1795, President Washington wrote Lear about Weston's visit:

From what you have written and from what I have heard from others, I hope Mr. Weston is on the Potomack 'ere this and that much benefit may be expected from his Visit. He is certainly a judicious man. with both theory and practice united. I am pleased to hear that the Locks which have been erected at the little falls have stood the test of a first trial so well; and this pleasure will be increased if Mr. Weston should make a favorable report of them.[120]

Weston visited the Potomac Canal in March 1795. The President and Directors reported on that visit in their annual report on August 6, 1795. Weston had approved of the locks constructed at Little Falls but recommended that the lock seats at Great Falls, then under construction, be relocated. Instead of turning toward the river at the basin (located at approximately mid-point of the Great Falls bypass canal), Weston recommended that the canal be continued in a straight line to the Potomac River. Although this change meant abandoning much work undertaken on the lock seats between the basin and the river, the Company enthusiastically accepted it.[121]

For this consulting work Weston was paid £370.[122] This was an enormous sum of money at a time when laborers were usually paid less than £2 per month for sixty hours per week of labor, sunrise to sunset. Because of these and similar payments to him for his services, Weston was able to return to England in 1801 and retire, although he was only 38.[123]

F. Isaac Roberdeau and William Weston

Although Weston was viewed as a competent engineer, he was also viewed as not having contributed to the development of American civil engineering and American civil engineers. Benjamin Henry Latrobe's view of Weston may be thought of as typical:

Mr. Weston, brought over to execute the Delaware and Schuylkill Canal, and afterwards engaged in the Western navigation of New-York, has returned, leaving no pupils, and without having completed his operations.[124]

This is not quite a fair assessment as Weston did leave a few, albeit not many, American civil engineering pupils behind. One of which was Isaac Roberdeau. Roberdeau was born in America but studied engineering in England. While the Potomac Company was beginning the Little Falls bypass canal, Roberdeau was working for Pierre L'Enfant in laying out the new plan for the new Federal City of Washington. It was Roberdeau that led the work party that demolished Carroll's house, under orders from L'Enfant – the immediate reason that L'Enfant, and

Roberdeau, were dismissed by the Commissioners of Public Buildings of the new Federal City.[125] Roberdeau accompanied L'Enfant to lay out the new industrial city of Patterson, New Jersey but in 1792 began working for Weston on the Schuylkill & Susquehanna Navigation.[126]

It was Roberdeau that wrote the first treatise on canal construction written by an American in the United States, in 1796.[127] This manuscript was never published but displayed a comprehensive knowledge of the art of the design and construction of canals and canal structures by an American at about the time that Harbaugh was constructing the locks at Little Falls.

Roberdeau never worked directly on the Potomac Canal but in 1818 he became Chief of the new U.S. Army Topological Bureau of the War Department and in this capacity would have known Thomas Moore and Isaac Briggs and perhaps had some influence on their recommendations on the Potomac Canal.

G. Captain Christopher Myers and the Construction of the Great Falls Bypass Canal, 1795-1796.

By June 12, 1795, the Board had advertised for contractors to undertake the excavation of the new lock seats at Great Falls at the location recommended by William Weston. No one applied. They therefore decided to undertake the first three locks from the basin, including the walls in-between, using the company's work force. They also decided to continue to advertise for a contractor to undertake the other three lock seats, i.e. those that would have to be hewn out of rock.[128] In addition, they decided to advertise for bricks to build the locks.[129] On June 22, 1795, the announced that they had agreed with Edward Noss to make 500,000 bricks and to lay them in the lock seats at Great Falls.[130]

On December 22, 1795, the Board authorized that the Company to advertise for building stone and lime needed for the new locks at Great Falls.[131]

It was at the same meeting that the Board ordered a quantity of three inch diameter rope, to be mounted in the existing ringbolts installed in the rock below Great Falls to assist the boats in hauling themselves up the river.[132] Many of these ringbolts can still be seen today by canoeists.[133]

On December 22, 1795, the Board asked Company President Tobias Lear to invite Captain Christopher Myers to meet with them in Georgetown and to review his proposal for taking over the position of Chief Engineer of the company.[134] They had previous conversations with Myers and President Washington, always interested in the involvement of engineers in the affairs of the Potomac Company, had written them to encourage them to move with haste with any plans they may have to hire the English engineer.[135]

Washington went further. He had informed Captain Myers that should Myers choose to visit the Directors, that Washington would provide him with a letter of recommendation. On December 25, 1795, he informed Lear that he had written the letter of recommendation but that Lear should himself critically examine Myers' qualifications.[136]

Captain Myers was hired by the Potomac Company on January 4, 1796 at a salary of $1600 per year.[137] With the hiring of Myers, the Lear and the Board of Directors undertook a new push to complete the Potomac Canal. At the same meeting where Myers was hired, the Board authorized the hiring of 100 slaves for the coming construction year, 1796.[138] The Company, however, had trouble finding this many slave laborers. In addition to the slaves, because of the rock excavation at Great Falls, the Company intended to recruit several miners acquainted with, "the best manner of blasting & Quarrying Rock etc."[139]

In February 1796, the Board began to make preparations for the oncoming construction season at Great Falls. They authorized, for example, that a dormitory for the workers be constructed.[140]

Having contracted with Marks and Nicholas Noss in June 1795 for 500,000 bricks for the locks, in March 1796 the Board found out from its engineer, Captain Myers, that the bricks made by the contractors were not adequate. Nicholas Noss made inquiries whether it was possible to drop the contract, agreeable to both parties.[141] This was done.

The bypass canal at Great Falls was uppermost in the minds of Potomac Company President Lear and his Board. However they were also concerned about other portions of the Potomac River navigation. In the summer of 1796, the Board asked Captain Myers to examine the Potomac River navigation at Payne's Falls, the Spout and Shannandoah Falls and to report back what actions were necessary to place the river navigation in the best possible condition.[142] These were the locations of the in river sluice navigations. These sluices tended to fill with river deposited gravel and sand and had to be periodically cleaned.

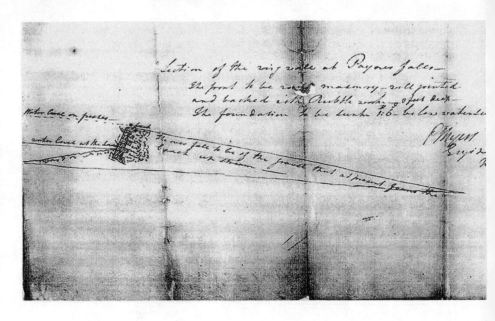

Figure 5. "Section of the wing wall at Paynes falls," by C. Meyers, Engineer, November 30, 1796. This is one of the few engineering drawings to survive of the works of the Potomac Company and the only known surviving drawing executed by Captain Christopher Myers during the year he was engineer of the Potomac Company. The drawing shows the batter, height and construction detail of a wing wall presumably constructed at Payne's Falls. The height of expected flood and the normal level of water are shown by dotted lines. Wing walls such as this were used by the Potomac Company to "swell" (i.e. increase) the water at certain locations to facilitate navigation. Thomas Moore and Isaac Briggs were critics of construction such as this as the masonry walls frequently injured boats, cargoes and crew under some river conditions. National Archives, Record Group 79, Entry 162.

In March 1796, the Board disapproved a plan for Captain Myers' house at Great Falls as being too large, and, instead, dictated that the house be twenty or twenty five feet wide by thirty five feet deep and two stories tall.[143]

On September 13, 1796, the Board ordered that the locks at Great Falls be extended to fourteen feet in width, instead of the twelve feet initially ordered. At the same meeting, the Board ordered Captain Myers to purchase up to thirty more indentured servants in Baltimore to work at Great Falls.[144]

Throughout 1796, the Potomac Company Board of Directors showed their determination to open the Great Falls bypass canal. At the same meeting, September 13, 1796, the Board approved a performance oriented award system for Captain Myers that would award him up to $5000 if he would finish the locks at Great Falls so that loaded boats could pass the Great Falls within a year or $2500 if he would finish the locks at Great Falls so that loaded boats would pass within fifteen months.[145] This was an innovative approach for completing the Great Falls bypass canal and the $5,000 bonus represented an enormous sum of money for the times. The pace of construction only somewhat quickened.

To complete the Great Falls bypass canal meant that the Great Falls locks had to be completed. To complete these locks meant that cut stone had to be cut and delivered to the site. On October 1, 1796, the Company contracted with John Henry to quarry and cut, "…Eighteen thousand feet of stone for the purpose of building Six Locks at the Great Falls…" for £95 Virginia currency.[146] The stone was to be quarried adjacent to Seneca Creek, Maryland, upriver and across the Potomac River from the upper end of Great Falls. The Company provided transportation of the stone across the river, including loading and unloading. Captain Myers put the freed slave, George Pointer, in charge of the company's boats for this operation.[147] The number of boats to be loaded in any given day was to not to exceed four.[148] Stone for these locks may have also come from the, "White Quarries," also from Seneca.[149] Quarrymen would be required at the Seneca quarries and advertisements for quarrymen began to appear in the newspapers.[150]

Despite this flurry of activity and offered bonus, the locks were not to be finished soon. Captain Myers would not get the bonus and it would take the canal company another seven years to complete the Great Falls bypass canal.

Not all work was at the Great Falls. The in river sluices continued to demand the Company's money and attention. By January 1797 the Potomac Company had a contractor and a crew working to clean the river channel at Payne's Falls, under the supervision of Captain Henry.[151]

On November 2, 1796, Captain Meyers proposed to the Board that his salary be raised to $4000 per year, a proposal that was rejected by the Board.[152] It was at this time that dissatisfaction with Company engineer Myers began to grow amongst Board members. At the January 6, 1797 meeting the Board expressed their dissatisfaction with their engineer and, in particular, his request for an absence of three to four weeks from the works.[153] The Board called a meeting to be held at Great Falls six days later in order to examine the works and to insure that no, "... extravagant and unnecessary Expenditure of money has taken place here during the past year ---"[154] The Board compared the Company's belongings to inventories and examined the accounts. By the next day the Directors had suspended the Company Clerk, Mr. Goulding.[155] By the following day, Saturday, January 14, 1797, the Board had decided that "...measures be taken to carry on the work during the present year upon a different arrangement from the last..." They decided to hold a meeting to discuss this new approach for completing the Potomac Canal at Great Falls by holding of the Board of Directors and Stockholders four days hence, on January 18, 1797, at the Union Tavern in Georgetown.[156]

The Directors had become very unhappy with Captain Myers and the progress that he and his men had made at Great Falls during 1796. The Board became particularly concerned on what Myers had expended Company resources for. They began an investigation. They found that Captain Myers had expended £11,724 19s 11d -- £7,266 5s 9d at Great Falls.[157]

This was a huge amount of money for the Company – one fourth of its initial capitalization, or $32,500. What was worse was that much, much more work needed to be accomplished before the Great Falls bypass canal would be open. As the Directors reported:

> How far the Excavation and other work done on the Lock seats correspond with the labour expended at that place, it is impossible for the Directors to judge. A Considerable Excavation has been made between the Bason and the River, in the Course of the Lock seats; but much more still remains to be done.[158]

During their investigation, the Directors discovered that Captain Meyers had company workmen, using company supplies, building houses on his property.[159] The Directors also found that although there were other buildings at Great Falls including various huts for as many as 150 workers, the 79 feet long by 18 feet wide by 7 feet high dormitory that the Board directed Captain Myers to construct in February 1796 was still not constructed.[160]

The Directors also found that Captain Myers may have not followed the Board's order to construct the Great Falls buildings in the cheapest manner, as instructed in February 1796, although it was difficult for them to tell if this the case.[161]

Company President Tobias Lear and Directors John Templeman and John Mason finished the report of their investigation with a rebuke to Captain Myers:

> The Engineer will settle with the Company for the Materials and labour belonging to them, which have been used in his service, also to settle respecting the House built on his Lot.[162]

Myers did not refute that Potomac Company men and supplies were used on his house. He informed Lear and the Directors that the total cost of these men and materials, from July 1, 1796 to December 31, 1796, was only £ 44 4 4-3/4, Virginia Currency.[163] In defense of his actions in using Company workers to build his house, Captain Myers stated that he could not hire any other workers at Great Falls and therefore had to use Company workers.[164]

The Board of the Potomac Company was extremely displeased with the performance of Captain Myers. Their visit and inspection of the works at Great Falls on January 12-14, 1797, set the stage for a larger review to be held at the Union Tavern in Georgetown, to include the stockholders, scheduled for Tuesday, January 17, 1797. The purpose of this meeting was to review Captain Myers performance and to discuss the findings of the Board's investigation. Captain Myers did not appear at this meeting.[165] Myers failure to appear sealed his fate with the Potomac Company.

On Tuesday, May 2, 1797, the Board fired Captain Myers. Resentment against Myers had been welling up for some time and the expression of that resentment against Myers took the form of a lengthy statement copied into the Proceedings of the President and Board of Directors.[166]

The Board of Directors concluded these charges by stating that Captain Myers actions resulted in, "…a great and unnecessary Expenditure of Money – the work retarded and probably, in many Instances, improperly executed…"[167]

Shortly thereafter, Captain Myers took out an advertisement requesting the public, "…to suspend any opinion, that may arise from certain CALUMNIES…"[168] Myers was to die a short time after.[169] His widow was to establish an inn at Great Falls.

Figure 6. "View of the mouth of George Washington's 'Potowmack' Canal at the Great Falls," 1943, photographed by Theodor Horyduzak. The Great Falls bypass canal was the greatest achievement of the Potomac Company and the greatest achievement of the Great Falls bypass canal was Lock 5 (here shown filled with debris as seen from the downriver side). Lock 5 was largely blown or otherwise hewn out of solid rock by Leonard Harbaugh and his workforce of slaves and free men. The downstream gate was eighteen feet in height – more than double the usual height of canal gates. As impressive as this accomplishment was, this photograph shows a longstanding problem of the Potomac Company – low water. The man in the center of the photograph is pointing to normal river levels. Critics of the Potomac Company argued that the Potomac Company only opened the river to navigation for 33 to45 days per year (the navigation was closed during the winter months). Photograph courtesy of the Historic American Engineering Record, Library of Congress.

H. Leonard Harbaugh, Nicholas King and the Completion of the Great Falls Bypass Canal

On June 6, 1797, the Board moved to replace Captain Myers by entering into an agreement with Leonard Harbaugh, who had built the locks at the Little Falls bypass canal, to superintend the construction of the locks at the Great Falls.[170]

While the dispute with Myers was continuing, Company President Tobias Lear moved to gain a replacement source for the engineering services that were to have been provided by Myers. He entered into an agreement with Nicholas King, a young English mapmaker/surveyor, to prepare drawings for the canal route to be constructed at Great Falls. King had immigrated to the United States, arriving in New York City in January, 1794. By 1797 he was in Washington, D.C.[171] By May of 1797 King had transmitted to the Potomac Company a sketch of the course of the canal with differences of level indicated but not accurate horizontal distances and bearings.[172] For this work he was paid £13 – 13s.[173] Three years later, in 1800, Nicholas King designed the locks at Great Falls that were to be built by Leonard Harbaugh.[174]

Leonard Harbaugh was in overall charge of construction at Great Falls. Maurice Delany was hired as the lock builder; Richard Elliott as the excavator and B. J. Machall as overseer. By September 1797, over 100 workers were employed by Harbaugh.[175] Like much of the work done throughout the construction of the Potomac Canal, it was a mixed work force consisting of free whites working along side black slaves.[176] Harbaugh himself was a slave owner, with slave Harry, a carpenter, working on the locks.[177] Harbaugh also employed his fourth son, Joseph Harbaugh, then eighteen, at the works as a carpenter.[178] As before, masonry was shipped across the river from Peters Quarry at Seneca.[179] The stone was hauled from the quarry to the river's edge by Matthew Kennedy and his workers.[180] Abner Meeks and his crew hauled the stone across the Potomac to Great Falls.[181] Richard Gridley, the blacksmith, provided the hardware.[182]

Money was now the largest obstacle to the Potomac Company's attempts to complete the canal at Great Falls. Myers extravagance in 1796 had hurt the company. On August 7, 1797, action was taken to authorize the Board to mortgage its future tolls, not to exceed $16,000.[183] Mortgaging future income is a serious matter and the Board felt that they needed to show that they were serious about completing the Great Falls canal and therefore the Potomac River navigation. To the authorization to mortgage future income, the Board appended the statement:

...and that no money be applied to any other object until the Navigation be made practicable at the Great Falls, unless in those Instances where obstructions prevent the navigation in some other parts.[184]

"Obstructions (to) prevent the navigation in some other parts...," did happen. On October 4, 1797, the Board had to take action to clear the channel at the head of Payne's Falls from an accumulation of sand and gravel and to repair a wall along the Long Canal at Shenandoah falls. In authorizing this expenditure, the Board noted that the work would not exceed $250.[185]

The year 1798 began with only eleven workmen, mostly slaves, at work at Great Falls under John Panton.[186] By the following month, February 1798, this work force was increased to twenty-four men, mostly slaves owned by John Templeman.[187] Slaves worked not only as laborers but also as blacksmiths, referred to in the payrolls as "Doctor Stuart's Julious," and, ""Ditto his Nelson."[188] By March this work force was further increased to thirty-two men.[189] The Company could no longer afford to sustain as large a work force and subsequently reduced this number. The traditional summer high construction months of 1798 saw only six to twelve workers employed at Great Falls.[190] Work on the Great Falls locks was accordingly slowed.

On June 5, 1798, Company President Tobias Lear announced to the Board of the financially strapped organization that he had the loan of "...the Sum of two thousand six hundred and odd dollars of Six percent Stock of the United States," from Daniel Carroll of Duddington, "...and from General Washington the Sum of three Thousand four hundred and ninety eight dolls. Of like Stock, which have been put into the hands of the Treasurer..."[191]

It was not enough. On the following day, June 6, 1798, the Company released Leonard Harbaugh as Superintendent of the Works, along with Mr. Elliott, the Overseer, as:

...the funds of the Company would not admit of engaging hands to go on with work, and their Services being therefore no longer necessary.[192]

Leonard Harbaugh was hired back on September 4, 1798, to report on the flood damage to the canal at Little Falls, Great Falls and Seneca and to report on the obstructions near Shenandoah Falls.[193] The hands that were working at Great Falls were instructed to be employed at :

Banking and other necessary repairs until a Trunk with a flue Gate can be fixed in the waste Dam near the Guard Gate to carry off the mud – when they are to be set to Loosening and Dragging the mud in the Beason and Canal until they are sufficiently cleaned –[194]

On November 6, 1798, the Board had authorized that a petition be presented to the Assembly of Maryland asking them to take another 100 shares of the Company's stock. The Company's treasury must have been extremely low for at the same meeting the Board directed that the Company's horses, "...be sent down to Alexandria immediately and there sold for the best price the(y) will command."[195]

But help was to be forthcoming from the Maryland Assembly. On January 20, 1800, the Board authorized that a contract be let for completing the five Locks needed at Great Falls:

Contract for Potomack Navigation

At the Great Falls of Potomack there are five Locks to be constructed. The seat for one of these Locks is already excavated. The seats of the other three are Still to be excavated. The Presd. And Directors are anxious to contract for the completion of the said Locks. Or they will contract separately for the Excavation of the Lock or any of them or any other part of the Work...[196]

At the annual meeting of the Stockholders in August 1800, they announced to the Stockholders why they were able to move ahead with this contract:

... we have Granted Bond in the name of the Company, to the persons Who Subscribed the Bond to the State of Maryland dated the 10th of December last (i.e. 1799) in the penalty of 150,000 Dollars ...[197]

The Company was again not successful in obtaining a contractor to undertake the needed work. Therefore they carried out the work using hired labor under the Company's direction. They entered into an agreement with Leonard Harbaugh to supervise the work.[198] They began in April 1800 with fifteen workers. The Company was able to increase this to thirty-five, although the average number of workers for the 1800 construction season was around twenty four.[199]

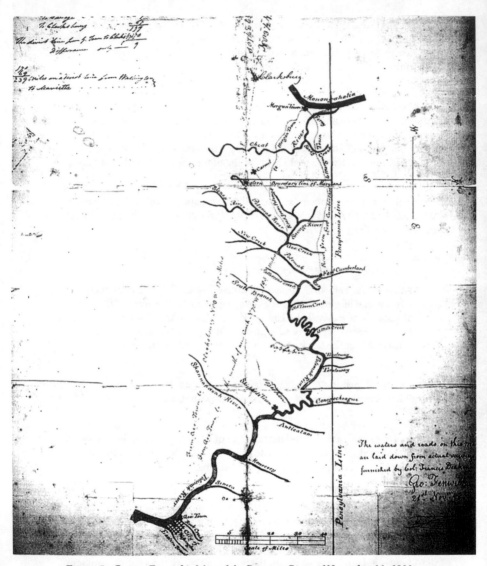

Figure 7. George Fenwick's Map of the Potomac River of November 21, 1800.
Financial help for the struggling Potomac Company came at the beginning of 1800.
Completion of the Great Falls bypass canal was assured. The Company began
planning for lateral canals that would feed downriver commerce to the main stem
navigation. This map was probably used by the company to plan their work on these
new lateral canals, as well as to direct the work crews charged with the continuing
task of keep the main stem open for navigation. Contrast this map with the detailed
maps developed for the new Chesapeake and Ohio Canal shown in Figures 1 and 2.
Use of image courtesy of the National Park Service.

At the annual meeting of the Stockholders in August 1800, the Board announced to the Stockholders why they were able to move ahead with this contract:

> ... we have Granted Bond in the name of the Company, to the persons Who Subscribed the Bond to the State of Maryland dated the 10th of December last (i.e. 1799) in the penalty of 150,000 Dollars ...[200]

The Company was again not successful in obtaining a contractor to undertake the needed work. Therefore they carried out the work using hired labor under the Company's direction. They entered into an agreement with Leonard Harbaugh to supervise the work.[201] They began in April 1800 with fifteen workers. The Company was able to increase this to thirty-five, although the average number of workers for the 1800 construction season was around twenty-four.[202]

Because of their problem in getting adequate numbers of free workers, the Company once again turned to hiring slaves. On December 17, 1800 the Board authorized the hiring of up to 60 negroes at from sixty to seventy dollars per year.[203]

The construction of the locks at Great Falls posed other problems to the company. To get through the rock at Great Falls meant blasting with black powder – a lot of black powder. Eventually a worker had to be injured because of this heavy reliance on black powder for blasting. On August 3, 1801, the President and Board of Directors reported one such incident to the stockholders, a laborer by the name of Robert Wisely damaged his eyes from an explosion and was left blind.

> The Directors suppose it may be proper to mention to the Stockholders that in the prosecution of the work at the Great Falls this Summer an unfortunate accident happened to a Labourer named Robert Wisely then in the employ of the Company.[204]

More building stone was ordered January 13, 1801, again from the Seneca Quarries across the river at Seneca.[205] On May 4, 1801, the Board directed that the dimensions of the locks be changed, most made higher.[206]

On January 6, 1802, the Potomac Company was able to announce that the Locks at the Great Falls were now completed.[207] The Great Fall bypass canal was open for commerce in February 1802. These locks were the pride of the Potomac Canal. Several had been blown out of solid rock. Thomas Harbaugh, Leonard's son, furnished us with a description of the lock and canal system built at the Great Falls by his father.[208]

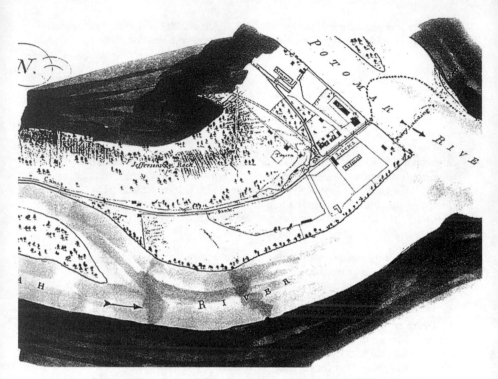

Figure 8. Nicholas King and Leonard Harbaugh, delineators. Detail, "Plan and Section of a Canal at the lower Falls of the Shenandoah, and its junction with the public Canal at Harper's Ferry and the Necessary Locks," February 1803. The most important of the lateral canals was the navigation improvements planed for the Shenandoah River. The Potomac Company hired surveyor Nicholas King to work with Leonard Harbaugh to develop a proposal to extend this new lateral canal through Harper's Ferry. They proposed a canal that would link with the U.S. Government Arsenal power canal which would provide the company an improved navigation around Shenandoah Falls (later renamed Harper's Ferry Falls)and thus provide an alternative to the sluice navigation of the Long Canal. Approval for construction was required from the War Department. This approval was never forthcoming and the company had to build a less ambitious navigation system further up the Shenandoah River. Use of image courtesy of he National Archives, Record Group 79.

On the eve of opening the bypass canal at Great Falls, in December 1801, the Company ecstatically reported to the Governor of Virginia that the canal would be opened soon and, "...on in opening the Locks at the Great Falls will throw considerable funds into the hands of the Company..."[209] This wasn't to happen. Despite the success of constructing the Great Falls bypass canal, the company never achieved financial success.

I. Leonard Harbaugh, Nicholas King and the Completion of the Feeder Canals

While Harbaugh was cutting the final locks at Great Falls, the Potomac Company began to plan for the expansion of the Potomac River navigation system. In particular, the Board of Directors began planning the construction of lateral canals that would feed commerce to the main Potomac River navigation.

The most important navigation system of the lateral canals was that along the Shenandoah River. In 1802 and 1803 the Virginia Legislature granted the charter to open the Shenandoah River to the Potomac Company. The company began planning the Shenandoah River navigation. In addition to the Shenandoah River work, the Board, on May 11, 1802, ordered several new projects. They ordered that the condition of the wooden locks at the Little Falls (of the Potomac) be examined, that twenty men be ordered to remove the obstructions on the river between the Little Falls and the Great Falls, and that an examination of the Mill dams on Conococheague Creek and the Monocacy River be inspected by Leonard Harbaugh with a view of improving navigation on those rivers, and other work.[210]

These projects displayed the new financial optimism of the Board of Directors. But this new optimism was not to last. By August 2, 1802, the President and Directors were reporting to the stockholders that due to unusually dry weather, the increase in tolls from the navigation were not as large as had been expected.[211]

After the Great Falls bypass canal opened tolls would increase but never would meet expectations. In their annual report to the stockholders in the following year, 1803, the President and Board would again complain about low water and a scarcity of boats but would express the, "...opinion of the President and Directors, good reason to believe that the Stock of the Company cannot fail rising in the public estimation "[212] The Company continued with its plans to improve the Potomac. They reported on the use of two teams of workmen to improve river navigation: one team beginning at tide water and proceeding upriver to the head of Seneca Falls, the other beginning at Seneca Falls and continuing upriver to the mouth of the Shenandoah. The Company continued its interest in developing lateral canals. In particular, they reported on the company projects in developing navigation along the Shenandoah, the Conococheague and the Monocacy Rivers.[213]

With regard to the Shenandoah navigation, the Potomac Company President James Keith presented to the Stockholders a report prepared by Leonard Harbaugh and Nicholas King on the construction of a canal along the Shenandoah and particularly through Harper's Ferry. Harbaugh and King had developed a proposal which would construct the Shenandoah River lateral canal through the middle of Harper's Ferry. This new canal would link with the existing U.S. Arsenal power canal and would therefore offer an alternative passage around the Shenandoah Falls of the Potomac River to the difficult sluice navigation of the Long Canal along the Maryland shore. The annual report stated, "...there seems to exist but little difficulty as to Bringing the arrangement (with the War Department, who controlled Harper's Ferry due to the armory) in Such manner as would be thought reasonable by the Board..."[214] Keith was very optimistic but the King/Harbaugh proposal for constructing a canal through Harper's Ferry was never approved and never constructed.

The Company did construct a series of five bypass canals or river sluices on the lower Shenandoah River. Starting with the furthest river improvement upriver on the Shenandoah from Harper's Ferry and the mouth of the Shenandoah, Littles Falls, these were:

1. Littles Falls[215] (10-3/4 feet drop) eight miles south of Harper's Ferry. The company built one lift lock and a basin. The lock was 100 feet long, 12 feet wide and had a lift of 8 feet. The lock was built of granite and freestone.

2. Wilson Upper Falls[216] (12-1/2 feet drop). The company built one lift lock. The lock was 100 feet long, 12 feet wide and had a lift of 12 feet. The lock was described as having been built of granite and freestone.

3. Bulls Falls[217] (4 feet drop). The company built a sluice.

4. Wilsons Lower Falls (6-1/2 feet drop). The company built one lift lock. The lock was also 100 feet long, 12 feet wide and had a lift of 6 feet. The lock was also constructed of granite and freestone.

5. Saw Mill Falls (17 feet drop) A quarter mile south (upriver) of Harper's Ferry. Here the company built two lift locks, both 100 feet in length, 12 feet in width. One had a lift of 9 feet and the other of 8 feet. They were also constructed of granite and freestone.[218]

Although these works were constructed and opened for river traffic they were little used. The Potomac Company did not take any other action to open the Shenandoah River for navigation. Because of this, the Potomac Company could not meet their charter for opening the Shenandoah River for navigation and in 1815 asked the Virginia legislature for a five year extension to do so. There was dissatisfaction with the Potomac Company and in 1818 the company was forced to transfer their the locks and other property to the Shenandoah Company for $15,000.[219]

The Potomac Company investigated other feeder canals to improve river commerce on its main stem navigation. In the annual report of 1803, the Stockholders were informed that $!500 had been received for the purpose of improving the Monocacy River and that a "chute" (i.e. sluice) had been added at Swingleys Mill on the Conococheague Creek.[220]

In Potomac Company President John Mason's letter to Secretary of the Treasury Albert Gallatin, January 20, 1808, Mason mentions the Company's attempts to improve feeder canals on the Shenandoah and to develop feeder canals on streams such as the Conococheague, the Monocacy, Pattersons Creek, the Opequon and the Cacapon. He did not mention the company's efforts to open Antietam Creek for navigation.

The citizens of Maryland raised $20,000 to improve Antietam Creek from the Pennsylvania line to the Potomac River and the Company accepted it.[221] On April 7, 1812, Thomas Harbaugh was appointed Superintendent of the works on Antietam Creek.[222] He initiated a survey of Antietam Creek and found twenty dams along its the length. He estimated that twenty-one locks and one sluice gate along the thirty-eight and a half miles of Antietam Creek would be required to make it navigable so as to overcome its 164-1/2 foot fall in that distance. His estimate for that work was $86,942.45.[223] In his Journal he would round that number up to $100,000[224] – the same amount that Mason had earlier written Gallatin would be required to finish the entire Potomac River navigation.

This amount, $100,000, was a huge sum for the financially troubled company, but some work was undertaken on Antietam Creek. Whatever work had begun was suspended by the Board on February 3, 1814.[225] The Company did not have sufficient funds to pursue this work.

By 1814, except for maintenance and repair of existing canal structures, all additional construction of the Potomac Canal and its feeders had been halted.

J. Benjamin Henry Latrobe and the Potomac Canal Extension (1802)

One additional Englishman became involved in the planning and design of the Potomac Canal, Benjamin Henry Latrobe. Latrobe developed plans, never constructed, to extend the Potomac Canal from its terminus at Woodwards Cove just below Little Falls to the Navy Yard on the Anacostia River. In 1802 Latrobe prepared four drawings detailing the construction this extension into Georgetown and from Georgetown to the Navy Yard (the drawings of the proposed canal from Georgetown to the Navy Yard were never prepared). One of the purposes of this proposed canal was to provide an adequate supply of water to the new dry dock to be constructed at the Navy Yard. Congress did not provide the funding for this project and it was never constructed.[226]

K. Josias Thompson and the Masonry Locks at Little Falls (1810-1816)

In 1810, the Potomac Company offered the position of Chief Engineer to Thomas Moore of Sandy Springs, Maryland.[227] Moore had constructed a causeway to Mason's island in 1805 and 1806 and therefore gained the attention of John Mason, a director of the Potomac Company and later to become president of that company.[228] Moore worked for the company on improving wing walls and similar river constructions. He was appointed by Jefferson as one of the three commissioners for the siting of the National Road and was generally well regarded. Moore declined the position and instead recommended Josias Thompson, "...in such strong terms as to induce the Company to appoint him without further enquiry..."[229]

Thompson worked for the Potomac Company from June 23, 1810 through June 23, 1816.[230] Although originally a carpenter and surveyor,[231] he functioned as an engineer for the company. He was responsible, for example, for rebuilding the wooden locks at Little Falls in masonry. Although he did not finish the task by the time he left in 1816, most of the work had been accomplished. The old wooden locks built by Harbaugh in 1795 had given way in 1815. The new masonry locks, four in number and comprising a total lift of thirty-seven feet, were built adjacent to the old wooden locks at Woodward's Cove and were finally completed in 1818.[232]

Thompson undertook other engineering tasks for the company, such as another study of possible navigation improvements to Antietam Creek.[233] Thompson was well regarded by the company's Board. One Board member wrote:

> The qualities I ever admired in Mr. Thompson were his peculiar and happy manner of managing the several Labourers, Freemen and Slaves, employed in the Company's service, the facility with which he conducted his many

operations, and his address in securing the interest, confidence and friendship of persons with whom the company had to negociate.[234]

Thompson left the company in 1816 with the masonry locks at Little Falls only partially completed. At this point, the Potomac Company no longer had sufficient funds to pay a chief engineer.

L. Thomas Moore and Isaac Briggs and the Demise of the Potomac Company

Thomas Moore and his brother-in-law, Isaac Briggs, played an instrumental role in the demise of the Potomac Company. In 1805/1806, Moore gained the attention of John Mason by constructing a causeway from Virginia to Mason's Island (now named Theodore Roosevelt Island), opposite Georgetown. Mason's Island was the summer home of John Mason. He was the fourth son of George Mason and a wealthy merchant and landowner in Georgetown. Later, John Mason would become a director and finally President of the Potomac Company. Moore would become one of the three Commissioners appointed to the National Road by Thomas Jefferson and would develop expertise in the behavior of rivers by direct study. In 1818 John Mason would recommend Moore to become Engineer to the Virginia Board of Public Works. This position, along with the position of Chief of the Department of War's Topographical Bureau (then held by Isaac Roberdeau) was one of the most important American positions in the then emerging fields of civil engineering.[235]

Moore was chosen as Engineer to the Virginia Board of Public Works and was given a salary of $3,500 per year plus expenses.[236] In this capacity, Moore was responsible for drafting the first of three reports which would determine the future of the Potomac Company.

On August 2, 1819, the Potomac Company asked the State of Virginia for a study of the Potomac Canal to be conducted by the Engineer for the Virginia Board of Public Works, Thomas Moore. No doubt Mason thought that his friend Thomas Moore would give a favorable report that would facilitate the funding of the financially troubled Potomac Company by the State of Virginia.[237]

In 1820 Thomas Moore issued his report of the Potomac Canal. He reported that only $18,000 to $20,000 would be needed by the Potomac Company to complete the Potomac River navigation and that this amount would be much less than the estimated $1,114,300 that would be required for a stillwater canal from Georgetown to Cumberland. But the Moore Report of 1820 is also mildly critical of the Potomac Company pointing out that it had frequently not made wise use of its previous funding.[238]

The second report that would influence the fate of the Potomac Company was the 1822 Report, *The District of Columbia Report*, of the District of Columbia Committee of the U. S. House of Representatives. This report recommended that the Potomac Company be the recipient of massive amounts of new money -- $2.5 million. This money would be used to pay off the old debts of the Potomac Company and the Potomac Company would use the remainder to build a stillwater canal, from Georgetown to Cumberland, Maryland.[239]

But many disagreed with the recommendations of the 1822 *District of Columbia Report*. To them, the Potomac Company was not only a financial failure but also an engineering failure. They believed that a stillwater canal should be constructed to Cumberland, but not by the Potomac Company. This reaction to *The District of Columbia Report* led to a third report, *The Commissioners Report of 1823*. The states of Maryland and Virginia appointed commissioners to study the Potomac Canal. The study was initially led by Thomas Moore. Moore died, on October 3, 1822, shortly after the group finished their trip down the Potomac but before the report had been prepared.[240]

Moore was replaced by his brother-in-law, Isaac Briggs. Briggs was the son of Samuel Briggs who established a cut nail factory in Rock Creek, Washington, in 1791. Briggs became a surveyor and assisted Andrew Ellicott in surveying and laying out Washington. He was later appointed Surveyor General on the Louisiana Purchase and, still later, became a contractor on the Erie Canal under Benjamin Wright.[241]

Under Briggs, *The Commissioners Report of 1823* became a general indictment of the mismanagement of the Potomac River navigation under the Potomac Company. The Commissioners found that the Potomac Company was a financial failure case which had only issued one dividend, that the Potomac Company had not met the legal terms of its charter, that the Potomac River navigation could only be used 33 to 45 days a year due to low water, that the Potomac River navigation was dangerous and that it placed a tremendous burden on merchants and farmers.[242]

But the most vehement argument put forth by the Commissioners in *The Commissioners Report of 1823* was that the Potomac Canal was an engineering failure. The core argument by the Commissioners was that sluice navigation, adopted by George Washington and the first Board of Director of the Potomac Company, not only did not assist river navigation but that it was counter productive to river navigation. The sluices cut into in the river bed of the Potomac River developed areas where the velocity of water tended to be increased. This increase in water velocity made downriver transit more dangerous and upriver navigation more difficult. In addition, the increase in water velocity tended to decrease the level of available water further making navigation more difficult or even impossible. The Potomac Company became aware that this was occurring and installed masonry walls in the river bed to "swell" the water at the upstream side of these sluices. But these walls themselves became dangerous to river navigation. Also, the Commissioners continued, because of the lack of knowledge of river behavior, many of these sluices tended to fill with rock and gravel and had to be continually dug out.[243]

The Commissioners Report of 1823 was, therefore, essentially an engineering indictment of the Potomac Company and the Potomac Canal. Not only had large sums of money been spent on the Potomac Canal but those sums of money had been spent by men who did not understand the ramifications of river navigation. Instead of improving river navigation on the Potomac River, much of the work undertaken by the Potomac Company had actually denigrated river navigation.

After the issuance of *The Commissioners Report of 1823*, there was no longer any question that the Potomac Company would not be entrusted with additional funds for further river improvements along the Potomac River. The Company's demise seemed certain. This would occur in 1827.

In 1827 a new canal company was formed, the Chesapeake and Ohio Canal Company. This new company was formed to construct a stillwater canal between Georgetown and Cumberland, Maryland. On October 1, 1827, the Potomac Company transferred its property to the Chesapeake and Ohio Canal Company[244]

In stark contrast to the development of the Potomac Company, the Chesapeake and Ohio Canal had almost an army of engineers to plan, design and build it. Chesapeake and Ohio Canal President Charles Mercer had hired Benjamin Wright as its Chief Engineer. Benjamin Wright would later be named the "Father of American Civil Engineering," for his role in constructing the first Erie Canal (1817-1825) and, perhaps more importantly, for his role in educating a whole generation of American civil engineers.

The first spadeful of earth was turned for the Chesapeake and Ohio Canal on July 4, 1828. Only shortly after, on August 23, 1828, the Board of Directors of the new canal established a Board of Engineers.[245] Benjamin Wright was selected as the Chief Engineer of the Board and the Office of Director was offered to Nathan S. Roberts (assistant to Benjamin Wright on the first Erie Canal and later Chief Engineer of he Erie Canal Enlargement between Rochester and Buffalo, as well as many other engineering assignments)[246] and John Martineau (also Wright's assistant from the Erie Canal).

By November 22, 1828, the Board of Directors of the Chesapeake and Ohio Canal had established a Corps of Engineers for supervising the construction of the Eastern Division. This Corps of Engineers was organized into five residencies extending from Georgetown to Pont of Rocks (approximately the first 50 miles of the new canal). These engineers included:

First Residency: Thomas F. Purcell, resident engineer; Charles D. Ward, assistant.

Second Residency: Daniel Van Slyke, resident engineer; Herman Böye, assistant.

Third Residency: Wilson M. C. Fairfax, resident engineer; William Beckwith, assistant.

Fourth Residency: Erastus Hurd, resident engineer; Charles B. Fisk, assistant.

Fifth Residency: Alfred Cruger, resident engineer; Charles Ellet, assistant.[247]

The methods of the Chesapeake and Ohio Canal Company also differed greatly from the Potomac Company. Where the Potomac Company hired slaves, indentured workers and free laborers and placed them under the supervision of overseers, the Chesapeake and Ohio Canal Company surveyed work to be undertaken, engaged in quantity takeoffs, developed profiles, let contracts in half-mile sections and monitored the contractor's work so that payment could be made based on a quantities of material excavated or filled.

The canals and sluices of the Potomac Company were either abandoned or were abolished through the construction activity of the Chesapeake and Ohio Canal Company. The Shenandoah River improvements that were transferred to the Shenandoah Company proved no more successful under the new company as under the Potomac Company. They were eventually buried in mud. The lower portion of the Potomac Company's Little Falls bypass canal was enlarged in 1828 to become the prism of the new Chesapeake and Ohio Canal. The upper portion became the feeder canal for the new canal. The Potomac Company's long canal, opposite Harper's Ferry, was obliterated by the construction of the new canal in the same location in 1830. The Seneca Falls sluice was abandoned and is used by canoeists today. The Potomac Company's greatest accomplishment, the bypass canal at Great Falls, was kept open by the new company until 1830 when it was abandoned in place.

VI. Summary and Conclusions

At the time that the Potomac Company built their canal system, civil engineers were extremely rare in the United States. In lieu of a supervising engineer, the Potomac Company made their own engineering decisions. George Washington very much desired to hire an engineer to oversee the company's plans and operations. But in 1785 there were no American civil engineers to be hired and the company was not willing to bring an engineer from England or France for this purpose. In the absence of an engineer, Washington and the company's Board of Directors made the engineering decisions themselves. The most far reaching decision of George Washington and the Board of Directors was not to follow John Ballendine's recommendation for bypass canals but, rather, of to rely extensively on sluice navigation (except, of course, at Great Falls and Little Falls).[248] Ballendine's plan to build tracking paths on one of the banks of the Potomac River, thereby facilitating upriver traffic, were also not implemented. The Potomac Company did use some English engineers, but usually for only a short period of time (Brindley, Weston, King) or with unfortunate results (Myers).

The decisions of the Potomac Company were clearly made to save money and were made without benefit of engineering advice. These decisions were later greatly criticized by early American civil engineers Thomas Moore and his brother-in-law Isaac Briggs, on behalf of the Virginia Board of Public Works. Moore and Briggs reported that the improvements made by the Potomac Company not only did not greatly assist navigation of the Potomac River but also proved counter-productive to that purpose. They reported that the sluice navigations built by the Potomac Company increased river velocity within the sluices and decreased water levels. Further, the sluices tended to fill with river-borne debris. Walls constructed by the company to "swell" the river waters proved dangerous to river commerce. Moore and Briggs reported that the navigation system of the Potomac Company could only be used during a short period of the year. They recommended that a stillwater canal be constructed to Cumberland, Maryland, but not by the Potomac Company.

In 1828 the Chesapeake and Ohio Canal Company began construction of the stillwater canal to Cumberland. Unlike the Potomac Company, they began their design and construction operations with an army of engineers led by Benjamin Wright, later called the father of American civil engineering. These engineers were largely trained on the Erie Canal. The procedures used by the engineers of the Chesapeake and Ohio Canal Company would be familiar to most practicing American civil engineers today: accurate surveys undertaken, work designed before construction begun, contracts let to qualified contractors; and supervision of the contractors work undertaken by resident engineers.

Between the beginning of the Potomac Canal (1785) and its demise (1828), civil engineering had come to America and Americans had become civil engineers.

[1] U.S. Geological Survey, "River Basins of the United States: The Potomac," (Washington, D.C.: U.S. Government Printing Office, 1991).

[2] Commissioners of Maryland and Virginia, "Report of the Commissioners of Maryland and Virginia to Survey the River Potomac, Communicated to the Senate January 27, 1823," U.S. Congress, 17th Cong., 2d Sess., No. 535, published in American State Papers, Miscellaneous, 988-1007.

[3] Although located by eighteenth and nineteen century writers above (i.e. upriver) Harper's Ferry and the mouth of the Shenandoah River, several twentieth century writers incorrectly locate Shenandoah Falls as below Harper's Ferry. When the surveyors came through this area in the 1820s to locate the new Chesapeake and Ohio Canal, these rapids were renamed "Harper's Ferry Falls," presumably to avoid confusion with the rapids upriver of Harper's Ferry on the Shenandoah River.

[4] George Washington in 1754, at the age of twenty-two, was one of those who safely navigated this portion of the Potomac River. It may have been this trip that influenced Washington in his views of sluice navigation versus bypass canals. For Washington's description of this trip, see Robert J. Kapsch, "The Potomac Canal: A Construction History," Canal History and Technology Proceedings, Volume XXI, (Easton, PA: Canal History and Technology Press, 2001), 148-150.

[5] Sharpe writes to Braddock, "...we set off to explore Potowmack River which proved from the number of shoals & falls to be of no Service in transporting either Artillery or other Baggage in our passage down...," William Hand Browne, Archives of Maryland: Correspondence of Governor Horatio Sharpe, Vol. I, 1753-1757, (Baltimore: Maryland Historical Society, 1888), 167.

[6] The announcement in the February 11, 1762 issue of the *Maryland Gazette* (Annapolis) read: "The opening of the River Potowmack and making it passable for Small Craft, from Fort Cumberland ...to the Great-Falls, will be of the greatest Advantage to *Virginia* and *Maryland* , by facilitating Commerce with the Back Inhabitants, who will not then have more than 20 Miles Land-Carriage to a Harbour where Ships of Great Burden load annually; whereas at present many have 150; and what will perhaps be considered of still greatest Importance, is, the easy Communication it will afford the Inhabitants of these Colonies with the Waters of the Ohio."

[7] Douglas R. Littlefield, "Eighteenth-Century Plans to Clear the Potomac River," The Virginia Magazine of History and Biography, Vol. 93, No. 3 (July 1985), 298. Douglas R. Littlefield, op. cit., 298.

[8] The announcement read: "Some Skillful Gentlemen have agreed to view the *Great Falls* in the Spring and if they should Report the opening or passing of them practicable (which is now generally believed) it is proposed that whatever Balance remains in the Treasurers Hands after compleating the first Design, shall be appropriated to that Purpose...," Douglas R. Littlefield, op. cit., 298.

[9] The dating of this proposal, which is undated, at 1769 is based on Grace L. Nute, "Washington and the Potomac: Manuscripts of the Minnesota Historical Society 1764-1796," American Historical Review, XXVIII (1923), 497-519; 705-721, footnote 14, 502.

[10] Grace L. Nute, op. cit., 500.

[11] Ibid.

[12] Ibid.

[13] Widow Brouster's is believed to have been on the Virginia side of the river above Great Falls..

[14] Ibid., p. 501.

[15] Semple wrote: "Paynes falls consistith of a narrow Rift of Rocks, extending across the River which may be passed through a naturall channel in land that may be improv'd so as to admit the passage of Vessels...," Grace L. Nute, op. cit., 501.

[16] Currency shown is Virginia currency (1 Virginia £ = 3.33 U.S. $) unless otherwise indicated.

[17] Semple wrote: "...from thence [i.e. Payne's Falls] it is about two miles and a half to that remarkable fall called the Spout att present the most difficult and dangerous above the great falls which ariseth not so much from its Height, as from the water of almost the whole River being confined and forced through a narrow Rocky Passage that makes it rapid, and subjects Vessells to the danger of filling as they pass, but notwithstanding the apparent difficulty a safe and easie passage may be had by a channel dug within land having locks placed in it by computation at the Expence of £800," Grace L. Nute, op. cit., 501.

[18] Semple wrote: "The next obstacle is above Harpers ferry about a mile above the Spout, this tho in appearance not so formidable as the last, will be found on tryuall by much the most expensive, requiring a channel Dug and wall'd along the Side of the River for at least half a mile with locks placed in it, at proper distances, to execute which in an effectual manner will require the sum of two thousand Pounds," Grace L. Nute, op. cit., 501.

[19] Semple wrote: "The next obstacle and last of any consequence, is at the head or beginning of what is called Shanadoah Falls, where there is already a naturall channel, formed between the main and an Island which channel was formerly begun to be improved and partly dug but never completed and now much choaked and filled up, which being cleared, the channel dug deeper and enlarged, and Dams and locks placed in it Vessells will pass through with readiness and safety To the Levell water above the Falls," Grace L. Nute, op. cit., 501.

[20] Semple wrote: "There is no very materiall obstruction but want of water over Shallow places where the River is low in autumn and the latter end of Summer, which in many places nay almost every part may be much improved at the expence of about £900 and following the same plan as is afterwards proposed, for clearing the Gravelly Shoals below Shanandoah Falls...," Ibid.

[21] Semple wrote: "From Shanadoah downwards, such shallow places when the river is low being all gravelly Shoals may be improved and made passable in any time of the year at a small expence and trouble, by adopting the plan by which severall rivers to the northward have been improved Viz. appointing overseers on the Shoals in the same manner as on public Roads, and allotting to them the Taxables contiguous and convenient to their respective Shoals to clear and make channels through them. The greater Stones being removed, the River would in these places soon naturally wash itself into Such channels as woud thereafter require little or no assistance; This plan being followed and Fish and other Dams removed and every thing presented for the future, that would any ways hinder or prejudice the passage of Vessells up and down the River, which Dams and obstructions, our neighbouring Colony has wisely prohibited, will render it in a very short time readily passable att all times be the river high or low," Ibid.

[22] Fairfax Harrison, op. cit., 540.

[23] J.P. Kennedy (ed.), Journals of the House of Burgesses, 1766-1769, (Richmond, 1906).

[24] Grace L. Nute, op. cit., 500.

[25] Ibid. Johnson's sketch of a proposed lock gate, enclosed in his November 4, 1785 letter to George Washington, is published in Douglas R. Littlefield's. "Eighteenth-Century Plans to Clear the Potomac River," The Virginian Magazine of History and Biography, Vol. 93, No. 3 (July 1985), while his plan for a bypass canal around Great Falls is published as Exhibit VII to Ricardo Torres-Reyes, "Potowmack Company Canal and Locks: Historic Structures Report, Great Falls, Virginia, (Washington, D.C.: Division of History, Office of Archeology and Historic Preservation, National Park Service, May 1, 1970). At the time Johnson drew this sketch, no such lock had been constructed in what was to become the United States.

[26] Writing of the Shenandoah Falls Johnson stated: "...from Catons Gutt (i.e. the top of Shenandoah Falls) to Paynes Falls about 5 Miles Distance will we think in prudence our present Object and 2500£ Pennsylva. Currency it is thought by an Englishman in whom I have very great Confidence and a German who has been long employed in blowing Rocks will reduce Shannandore to allow a tolerable passage and make a towing Path ...," Grace L. Nute, op. cit., 506.

[27] Johnson wrote: "We choose to blow a Passage rather than attempt Navign. Through Locks because the Falls no where appear too steep for Vessells to come down if they had but Room enough and this plan is the more eligible as it avoids a very strong Objection to Locks from the Freshes Ice etc...," Ibid.

[28] Johnson wrote: "...suppose an Entrance 2 – ½ feet deep is made into the Gut it takes off something in the most necessary part. Why might not 3 or 4 feet be again taken off by a low River Look the sides made of [rough?] Loggs well pinned together or bolted with Iron Bolts filled in with large Stones of which there's such plenty at Hand in the same Manner as Stone Dams. When there's any Danger from Ice or Trees the Water would gen'lly if not always be over the Lock and Ice and Trees would probably pass over without touching. If it's feared they would not to prevent their hanging the Lock might in the Fall be filled with Stones and no great Labour to clear it out in the Spring. Two such Locks 4 Feet each would certainly do the Business or perhaps of 3 feet or perhaps one might do. If a Stone Towing path was built which might be done for 3 or 400 £ and large Iron pins with broad

Heads fixt in the Rocks at proper Distances around, which occasionally to take a turn of a long strong pointer by which one Hand might hold on what three of the other Hands gained in pulling leaving one Hand in the Boat to keep her from the Sides with a light Setting-pole it seems likely five Hands would carry such a Boat through any water in which she could swim ...," Grace L. Nute, op. cit., 512.

[29] Washington wrote: " ... I conceive, that if the Subscribers were vested by the two Legislatures with a kind of property in the Navigation, under certain restrictions and limitations, and to be reimbursed their first advances with a high Interest thereon by a certain easy Tolls on all Craft proportionate to their respective Burthens, in the manner that I am told works of this sort are effected in the Inland parts of England or, upon the Plan of Turnpike Roads; you woud add thereby a third set of Men to the two I have mentioned and gain considerable strength by it: I mean the monied Gentry; who tempted by lucrative views woud advance largely on Acct. of the high Interest. This I am Inclind to think is the only method by which this desirable work will ever be accomplished in the manner it ought to be ...," Letter, George Washington to Thomas Johnson, July 20, 1770. John C. Fitzpatrick (ed.), The Writings..., op. cit., Volume 3, 18-19.

[30] A. W. Skempton, et. al., A Biographical Dictionary of Civil Engineers in Great Britain and Ireland, (London: Thomas Telford Publishing, Institution of Civil Engineers, 2002), 79.

[31] Skaggs wrote: "Ballendine slowly gained a reputation for business dishonesty. He shorted the weight of iron shipments to customers. Severe reprimands by the county court apparently had little effect upon his conduct. When in 1760 George Washington found a shortage in a shipment from Occoquan, he asked his attorney George Johnson to secure an indictment against the ironmaster...," David Curtis Skaggs, op. cit., 289.

[32] Washington wrote to Thomas Johnson, about Ballendine: " ... I acknowledge that, Mr. Ballendine has a natural genius to thing's of this sort, which if properly encouraged may lend much to publick utility, I cannot help adding, that, his Principles have been loose; whether from a natural depravity, or distress'd circumstances, I shall not undertake to determine; how far therefore a Man of this cast is entitled to encouragement every one must judge for themselves; for my part I think, if he applies the Money Subscribed, to the end proposed, the Publick will derive great advantages of it; on this acct. it is, alone, I wish to see him encouraged, and on this principle it is, I have taken the liberty of mentioning him to Govr. Eden, Colo. Sharpe, Majr. Jenifer and yourself; because I think the opening of the Potomack will at once fix the Trade of the Western Country...," Letter, George Washington to Thomas Johnson, May 5, 1772, John C. Fitzpatrick, (ed.), The Writings... op. cit., Volume 3, 83.

[33] Grace L. Nute, op. cit., 516.

[34] Letter, George Washington to Thomas Johnson, May 5, 1772, John C. Fitzpatrick, (ed.), The Writings... op. cit., Volume 3, 82.

[35] A copy of the proposal is published in George Armroyd, A Connected View of the Whole Internal Navigation of the United States, (Philadelphia, 1830), 208-214, and its economic arguments summarized in defense of the proposed Chesapeake and Ohio Canal in U.S. Congress, "Chesapeake and Ohio Canal," 20th Cong. 1st sess., House Report No. 47, January 22, 1828, 44-45.

[36] Corra Bacon-Foster, Early Chapters in the Development of the Patomac Route to the West, (New York: Burt Frankin, 1971), 26. This is a reprint of, "Early Chapters in the Development of the Potomac Route to the West," Records of the Columbia Historical Society, Washington, D.C., Volume 15 (1912), 96-322. Pagination cited is from the reprint.

[37] Ballendine wrote: "N.B. As nothing effectual can be properly done for less than £30,000, this subscription is not binding unless the value of £30.000 Pennsylvania currency, be subscribed," Ibid.

[38] The October 25, 1775 notice of Ballendine moving to the James River is reproduced in, Randolph W. Church, "John Ballendine: Unsuccessful Entrepreneur of the Eighteenth Century," Virginia Cavalcade, VIII (1959), 39-46, 42; his announcement of the meeting of September 26, 1774, the announcement of the appointment of Trustees of October 1774; the announcement to hire 50 slaves of December 22, 1774 and the October 28, 1775 notice are published in Corra Bacon-Foster, op. cit., 26-29.

[39] Ballendine's plan stated: "From the tide-water in the cove at Woodward's to the little island above the lower falls is three miles; in this distance there is a fall of thirty-six feet, which will require four locks, estimated at £ 1000 each...," NA-RG79, Entry 179, Miscellaneous Accounts, John Ballendine, of the Expence in Removing Obstructions in Potowmack River From Tide-Water to Fort Cumberland, being about two hundred miles, (1774). "The cove at Woodward's ..." would be the cove at what is

now called Fletcher's Boat House on the Chesapeake and Ohio Canal, located at milepost 3.1 miles above Georgetown and adjacent to the Abner Cloud House, not yet built at the time of Ballendine's plan. This cove has also been called Garrison Cove, Lock Cove or Lock Harbour. NPS, "Copy of Condemnation papers for land around Little Falls to be used by Potomac Co.," July 30, 1793, Records of the Potomac Company, Chesapeake and Ohio Canal National Historical Park, Acc. #2498.

[40] Ballendine's plan stated: "As the falls from tide water to the upper end of the island aforesaid, are too rapid for the river navigation, a level cut must be made along the Maryland side of the river, from the cove at Woodward's, to the upper end of the island aforesaid, three miles in length, thirty feet wide, and four feet deep, estimated at £1250 per mile...," Ibid.

[41] Ballendine's plan stated: "From the regulating lock at the island, it is about six miles to the Great Falls, and the river wide, so that removing the single stones in the river for boats to go loaded with safety, may cost £600. The track for the horses by the river side, up to the Great Falls being six miles, and in most places full of rocks, it must be laid with stones secure from freshes, and may be done for £ 400 per mile...," Ibid.

[42] Ballendine's plan stated: "At the Great Falls there is a fall of about seventy-six feet in the distance of one mile and a quarter, which will require eight locks at £ 1000 each. ... Extra expense in the first and upper locks, for sinking and extending parts of the lock through a solid rock, so as to admit barges or rafts one hundred feet long, sixteen feet wide, and four feet deep   800. ... Extra expense in a small dam, for raising and deepening the water 100 ... Extra work in clearing and turning the water, so as to raise the vessels in the upper lock, and secure from freshest. ...," Ibid.

[43] Ballendine's plan stated: "From these Falls to Seneca Falls is about six miles, many rocks to remove and clear in the river, also some shoals to deepen   1600
A track for the horses the same distance, must be laid with stones along the river side, at £ 250 per mile   3500."
Ibid.

[44] Ballendine's plan stated: "As the fall at this place is too rapid for river navigation, a level cut must be made on the Virginia side, from the Mill Branch to the Iron Landing at the head of the falls, one mile and an half in length, thirty feet wide and four feet deep, estimated at £ 1250 per mile   1875.
For spreading the earth, etc. on the lower side of the canal, or a road, etc. to track up with horses, £ 25 per mile,   37 10 0
Extra expense for a regulating lock at the Iron Landing   400."
Ibid.

[45] Ballendine's plan stated: "From Seneca to Shenandoah Falls is about sixty miles; removing the loose stones in the river and other obstructions, estimated at £ 25 per mile,   1500
The track for horses the same distance along the river side, being chiefly clay, and large trees to cut and clear away, may be done for   1500."

Ibid.

[46] Ballendine's plan stated: "The Shenandoah Falls are [indistinct] three miles in length; in that distance there is a fall of [indistinct] but being so gradual no more than [indistinct- 2 locks?] will be required at £ 40 each ...   800
A track for horses the same distance with large stones and a secure from freshest, at £ 500 per mile,
                                                                                           1500

Clearing and removing rocks in the river – the same distance   800."
Ibid.

[47] Ballendine's plan stated: "From the Shenandoah Falls to Fort Cumberland, is about one hundred and twenty miles by water, in this distance there are a few sudden or rapid falls, but the removing the stones and some rocks for a clear passage in so many places, may be estimated at   3000"
Extra expense in swelling the river, and deepening the shoals in many places, may amount to
                                                                                           2000

Extra expense in two river locks at Houses and Anita [i.e. Antietam],   800

A track for the horses, one hundred and twenty miles to be made along the bank, large trees to be cut and clear away, £ 25 per mile, from Shenandoah to Fort Cumberland,   3000."

Ibid.

[48] A cost comparison between the two proposals:

		Ballendine	Semple
Bypass canal at Little Falls		£ 9425	n/a
Little Falls to Great Falls		£ 3000	n/a
Bypass canal at Great Falls		£ 9900	n/a
Great Falls to Seneca Falls		£ 3100	0
Bypass canal at Seneca Falls		£ 4562-10	£ 250
Seneca Falls to Shenandoah Falls		£ 3000	0
Shenandoah Falls (incl. Payne's Falls)		£ 3100	£ 3250
Shenandoah Falls to Cumberland		£ 8000	£ 900
Houses Falls and Antietam		£ 800	£ 800
	Total	£ 44,887	£ 5,000
(Virginia Currency)		(about $150,000)	(about $17,000)

[49] NA-RG79, Entry 179, "Miscellaneous Accounts, John Ballendine, of the Expence in Removing Obstructions in Potowmack River From Tide-Water to Fort Cumberland, being about two hundred miles," (1774).

[50] Ibid.

[51] The announcement read: "This meeting is judged to be the more necessary as the subscriber is now at work on the locks the lower Falls on the Maryland side of the river with what hands he has," Corra Bacon-Foster, op. cit., 28.

[52] Several plans are referred to. The details of Thomas Johnson's plan of 1770 are, for example, not known except as they are mentioned in Washington-Johnson correspondence. Corra Bacon-Foster mentions Charles Beatty, of Montgomery County, and Norman Bruce, of Frederick County, who at the direction of the Maryland legislature, were to make a survey of the Potomac River and to submit an estimate of how much it would cost to make it navigable. Washington had access to this report. Corra Bacon-Foster, op. cit., 124. Washington was also aware of the plans of Stephen Sayre to, "... (develop a) navigation thro the great & Little Falls, without a *single* Lock...," Letter, Stephen Sayre to George Washington, October 15, 1784. W.W. Abbott and Dorothy Twohig (eds.), The Papers of George Washington, Volume 2, op. cit., 99. No doubt there were other plans.

[53] See Robert J. Kapsch, , "The Potomac Canal: A Construction History, " op. cit., 161-162.

[54] Randolph W. Church, op. cit., 42.

[55] The *Maryland Gazette* read: "Falls of the James River, Oct. 25, 1775. At the earnest solicitation of many gentlemen on Potowmack and influenced by my own interest on that river I have been endeavoring to open its navigation from tide water upwards, and have been at considerable expense in preparation etc. To forward that useful work, but the necessity of a Maryland Act of Assembly co-operating with one passed in Virginia and which I have not been able to obtain has obliged me to decline it for the present. ...," Corra Bacon-Foster, op. cit., 29.

[56] Washington wrote: " More than ten years ago I was struck with the importance of it [i.e. the plan for improving navigation of the Potomac River]; & despairing of any aid from the public, I became a principal mover of a Bill to empower a number of subscribers to undertake, at their own expense, (upon conditions which were expressed) the extension of the Navigation from tide water to Wills's Creek (abot 150 Miles) ... I was obliged even upon *that ground* to comprehend James River, in order to remove the jealousies which arose from the attempt to extend the Navigation of the Potomack. The plan however, was in a tolerable train when I set out for Cambridge in 1775, and would have been an excellent way had it not been for the difficulties which were met with in the Maryland Assembly, from the opposition which was given (according to report) by the Baltimore Merchants, who were alarmed, and perhaps not without cause, at the consequence of water transportation to George Town

of the produce, which usually came to their market.. ...," Letter, George Washington to Thomas Jefferson, March 29, 1784, W.W. Abbot and Dorothy Twohig (eds.), The Papers of George Washington: Confederation Series, Volume 1, January 1784 – July 1784, (Charlottesville, Va.: University Press of Virginia, 1994), 238.
[57] Randolph W. Church, op. cit., 43-44, 46.
[58] Washington wrote: "The local interest of that place (Baltimore) joined to the shortsighted politics, or contracted views of another part of the Assembly, gave Mr. Thomas Johnson who was a warm promoter of the Scheme, on the No. side of the River, a great deal of trouble. In this situation things were when I took command of the Army -- The War afterwards called Mens attention to different objects -- and all the money they could or would raise, was applied to other purposes; but with you, I am satisfied that not a moment ought to be lost in re-commencing this business...," Letter, George Washington to Thomas Jefferson, March 29, 1784. W.W. Abbot and Dorothy Twohig (eds.), The Papers..., op. cit., Volume 1, 238.
[59] Washington wrote: "... First, because Acts of the Assemblies of Virginia & Maryland, must be obtained to incorporate private adventurers to undertake the business -- 2d. the Company must be formed before anything can be done -- 3d. an actual survey of the waters, by skilful Engineers, (or persons in that line) must take place & be approved before the points at which the navigation on different waters can be ascertained....," Letter, George Washington to Richard Claiborne, December 15, 1784, W.W. Abbot and Dorothy Twohig (eds.), The Papers..., op. cit., Volume 2, 184-185.
[60] Letter, William Paca to Charles Beatty, Normand Bruce, August 1, 1783, Maryland Archives, Maryland State Papers, Series A. Washington had access to this report. His comments are reproduced in W.W. Abbot and Dorothy Twohig (eds.), The Papers ..., op. cit., Volume 2, 131-133.
[61] The bill passed the Virginia Assembly in October and the Maryland Assembly in November, 1784. Corra Bacon-Foster, op. cit. 44.
[62] Ibid., 45.
[63] Letter, George Washington to Thomas Johnson, January 17, 1785, John C. Fitzpatrick, (ed.), The Writings..., op. cit., Volume 28, 31.
[64] Letter, George Washington to the Marquis de Lafayette, July 25, 1785, Ibid., p. 207.
[65] Edward B. Mathews and Wilbur A. Nelson, Report On the Location of the Boundary Line Along the Potomac River Between Virginia and Maryland In Accordance with The Award of 1877, (Baltimore, 1928), 3-4.
[66] Ibid., 18.
[67] Ibid., 4.
[68] Washington wrote: "The first dividend of the money was paid in on the 15th of this month; and the work is to be begun the first of next [i.e. August 1785], in those parts which require least skill; leaving the more difficult 'till an Engineer of abilities and practical knowledge can be obtained ... ," Letter, George Washington to the Marquis de Lafayette, July 25, 1785, John C. Fitzpatrick, (ed.), The Writings..., op. cit., Volume 28, 207.
[69] NA-RG79, Entry 160, Proceedings of the President and Directors, Potomac Company, 1785-1828, May 30, 1785, 1.
[70] Ibid.
[71] Letter, George Washington to the Marquis de Lafayette, February 15, 1785, John C. Fitzpatrick, (ed.), The Writings..., op. cit., Volume 28, 73.
[72] Letter, George Washington to George William Fairfax, June 30, 1775, John C. Fitzpatrick, (ed.), The Writings..., op. cit., Volume 28, 184.
[73] Washington wrote the following note: "Vessels of 6, & not exceeding 10 Tons burthen, are recommended tho' the Canals should be made to suit larger Vessels & Rafts drawing 3 feet water & 60 feet long ... ," W.W. Abbot and Dorothy Twohig (eds.), The Papers ..., op. cit., Volume 2, 133.
[74] The advertisement read: "Ordered that Advertisements be inserted in the Alexandria, Baltimore and some one of the Philadelphia Papers giving Notice that this Board will meet at Alexandria on the first Day of July next to agree with a Skilful Person to conduct the opening and improving the Navigation of the Potowmack river from the Great Falls to Payne's and from the upper part of the Shanadoah to the highest place practicable on the North Branch, and also to agree with two Assistants and Overseers – Also that liberal wages will be given to any Number not exceeding one hundred good Hands with Provisions and a reasonable Quantity of Spirits; that a further Encouragement will be

given to such as are dextrious in boring and blowing Rocks in which Service a proportion of the Men will be employed and that the Conductor of the Work or some other Person authorized will attend at Seneca on the third Day of July next and at Shanadoah on the 14th Day of July to contract with the men who may offer for this Service ... ," NA-RG79, Entry 160, Proceedings ..., May 31, 1785. There is also some evidence that some preliminary work may have been undertaken in the autumn of 1784. George Gilpin wrote George Washington, on July 10, 1785: "I then crossed the river just above the falls [i.e. Seneca Falls] to the Maryland side and went down to where the huts was in which the people lived last fall and then to a Mr. Goldboroughs at whose house Johnson and Clapham lodged When they attended the works ... ," Grace L. Nute, op. cit., 713.

[75] Ibid.

[76] Letter, George Washington to William Moultrie, May 25, 1786, W.W. Abbot and Dorothy Twohig (eds.), The Papers ..., op. cit., Volume 4, 73.

[77] Letter, George Washington to John Fitzgerald and George Gilpin, March 31, 1786, W.W. Abbot and Dorothy Twohig (eds.), The Papers ..., op. cit., Volume 3, 615.

[78] Washington wrote: "It appears to me therefore, that of the cost of bringing from Europe a professional man of tried & acknowledged abilities is too heavy for one work; it might be good policy for several Companies to unite in it; contributing in proportion to the estimates & capital sums established by the several Acts. ... ,"Letter, George Washington to William Moultrie, May 25, 1786, W.W. Abbot and Dorothy Twohig (eds.), The Papers ..., op. cit., Volume 4, 74.

[79] Washington wrote: "... [I] had written both to England & France, to know on what terms a person of competent skill could be obtained--& have received the following answer from my friend the Marqs de la Fayette; 'There is no doubt but what a good Engineer may be found in this country to conduct the work. France in this point exceeds England; & will have I think every advantage but that of the language ... an intimation that you set a value on that measure, will ensure to us the choice of a good Engineer. They are different from the military ones, and are called Ingènieurs des ponts & chaussées ... ,'" Letter, George Washington to William Moultrie, May 25, 1786, W.W. Abbot and Dorothy Twohig (eds.), The Papers ..., op. cit., Volume 4, 74.

[80] Washington wrote: "I have engaged him [Brindley] to call upon Colo. Gilpin on his rout back. Mr. Brindley and Mr. Harris[80] took the Great Falls in their way down, and both approve of the present line for our Canal. The first very much, conceiving that 9/10th of the expence which must have been incurred in the one first proposed, will be saved I the second—the work be altogether s secure—and the discharge into the river below, by no means unfavorable. He thinks however, that a good deal of attention and judgment is requisite to fix the Locks there; the height of which, he observes, must be adapted to the ground, there being no precise rule for their construction; Locks running, frequently, from 4 to 18 feet -- & sometimes as high as 24 – The nature, & declension of the ground, according to him, is alone to be consulted, and where these will admit of it, he thinks the larger the locks are, the better, because more convenient," Letter, George Washington to John Fitzgerald and George Gilpin, March 31, 1786, W.W. Abbot and Dorothy Twohig (eds.), The Papers ..., op. cit., Volume 3, 615.

[81] Washington wrote: "With respect to this part of the business [the siting, design and construction of the locks at Great Falls], I feel, and always have professed, an incompetency of judgment; nor do I think that theoretical knowledge alone, is adequate to the undertaking. Locks upon the *best digested plan* will certainly be expensive--& not properly constructed, & judiciously placed, may be altogether useless. It is for these reasons I have frequently suggested, though no decision has been had, the propriety of employing a professional man. Whether the expence of importing one has been deemed altogether unnecessary; or, that the advantages resulting therefrom are considered as unequal to the cost, I know not; but, as it is *said* no person in this Country has *more practical* knowledge than Mr. Brindley, I submit it for consideration, whether it is not advisable to engage him to take the Falls on his way home—examine—level--& digest a plan for locks at that place. ... ," Letter, George Washington to John Fitzgerald and George Gilpin, March 31, 1786, W.W. Abbot and Dorothy Twohig (eds.), The Papers ..., op. cit., Volume 3, 615.

[82] Letter, George Washington to John Fitzgerald and George Gilpin, March 31, 1786, W.W. Abbot and Dorothy Twohig (eds.), The Papers ..., op. cit., Volume 3, 616.

[83] Rumsey wrote Washington: "This will be handed to you by Mr. Brindley, we have had the pleasure of his, and Mr. Harris's Company Since yesterday, and they Boath approve of what is Done and

proposed Here ... ," Letter, James Rumsey to George Washington, March 29, 1786, W.W. Abbot and Dorothy Twohig (eds.), The Papers..., op. cit., Volume 3, 611.
[84] Ibid.
[85] Washington wrote: "As I have imbibed a very favorable opinion of your mechanical abilities, and have no reason to distrust your fitness in other respects; I took the liberty of mentioning your name to the Directors, and I dare say if you are disposed to offer your services, they would be attended to under favorable circumstances... ," Letter, George Washington to James Rumsey, July 2, 1785, John C. Fitzpatrick, (ed.), The Writings... op. cit., Volume 28, 189.
[86] Letter, George Washington to James Rumsey, June 5, 1785. John C. Fitzpatrick, (ed.), The Writings..., op. cit., Volume 28, 159. Fitzpatrick identifies Bath as being the present-day Warm Springs, Virginia, but this is incorrect. Bath would have been the present-day Berkeley Springs, West Virginia.
[87] Letter, George Washington to James Rumsey, July 2, 1785, John C. Fitzpatrick, (ed.), The Writings..., op. cit., Volume 28, 189-190.
[88] NA-RG79, Entry 160, Proceedings ..., July 14, 1785, 5.
[89] Ibid.
[90] Robert J. Kapsch, "The Potomac Canal: A Construction History," op. cit., footnotes 184 and 185, 227-228.
[91] John C. Fitzpatrick (ed.), The Diaries ..., op. cit., Volume II, (Boston and New York: Houghton Mifflin Company, 1925), August 3, 1785, 395.
[92] Washington wrote: "The Water through these Falls is of sufficient depth for good navigation; and as formidable as I had conceived them to be; but by no means impracticable. The principal difficulties lye in rocks which occasion a crooked passage. These once removed, renders the passage safe without the aid of Locks and may be effected for the Sum mentioned in Mr. Jno. Ballendine's estimate (the largest extant), but in a different Manner than that proposed by him. It appearing to me, and was so unanimously determined by the Board of Directors, that a channel through the bed of the River in a strait direction, and as much in the course of the current as may be, without a grt. increase of labour and expence, would be preferable to that through the Gut, which was the choice of Mr. Ballendine for a Canal with Locks. The last of which we thought unnecessary, and the first more expensive in the first instance, besides being liable to many inconveniences which the other is not; as it would probably be frequently choaked with drift wood, Ice, and other rubbish, which woud be thrown thereby through the several inlets already made by the rapidity of the currts. In freshes and others which probably would be made thereby; whereas a navigation through the bed of the River when once made will, in all probability, remain for ever, as the currt. here will rather clear, than contribute to choak the passage... ," Ibid., 396.
[93] Washington wrote: "Here we breakfasted (i.e. at Harper's Ferry); after which we set out to explore the Falls below; and having but one Canoe, Colo. Gilpin, Mr. Rumsey (who joined us according to appointment last Night) and myself, embarked in it, with intention to pass thro' what is called the Spout (less than half a Mile below the ferry). But when we came to it, the Company on the Shore, on acct. of the smallness, and low sides of the Vessel, dissuaded us from the attempt, least the roughness of the Water, occasioned by the Rocky bottom, should fill and involve us in danger. To avoid the danger therefore we passed through a narrow channel on the left, near the Maryland shore and continued in the Canoe to the lower end of Pain's fall distant, according to estimation, 3 Miles." John C. Fitzpatrick (ed.), The Diaries..., op. cit., Volume II, August 7, 1785, 400.
[94] Washington wrote: "These falls [i.e. Payne's Falls] may be described as follow [i.e. beginning at Harper's Ferry and traversing the Spout]): From the Ferry for about 3 hundred yards, or more, the Water is deep with rocks here and there, near the Surface; then a Ripple; the Water betwn. which and the Spout, as before. The Spout takes its name from the rapidity of the Water, and its dashings, occasioned by a gradual, but pretty considerable fall, over a Rocky bottom which makes an uneven surface and considerable swell. The Water however, is of sufficient depth through it; but the Channel not being perfectly straight, skilful hands are necessary to navigate and conduct Vessels through this rapid. From hence, there is pretty smooth and even Water with loose stone, and some rocks, for the best part of a Mile; to a ridge of rocks which cross the river with Intervals, thro' which the Water passes in crooked directions; but the passage which seemed most likely to answer our purpose of Navigation was on the Maryland side, being freest from rocks but shallow. From hence to what are

called Pain's falls the Water is tolerably smooth, with Rocks here and there. These are best passed on the Maryland side. They are pretty swift, shallow, and foul at bottom, but the difficulties may be removed. From the bottom of these Falls, leaving an Island on the right, and the Maryland Shore on the left, the easy and good Navigation below is entered ... ," Ibid., 400-401.

[95] Washington wrote: "At the foot of these falls (i.e. Payne's Falls) the Directors and myself (Govr. Lee having joined us the Evening before) held a meeting. At which it was determined, as we conceived the Navigation could be made through these (commonly called the Shannondoah) Falls without the aid of Locks, and by opening them would give eclat to the undertaking and great ease to the upper Inhabitants, as Water transportation would be immediately had to the Great Falls from Fort Cumberland, to employ the upper hands in this work instead of removing the obstructions above, and gave Mr. Rumsey directions to do so accordingly, with general Instructions for his Governmt. ... ," Ibid., 401.

[96] The company Proceedings stated: "The President and all the Directors having yesterday viewed and examined the Shanadoah Falls from the flat water above to that below were unanimously of Opinion that the Navigation may be carried through the falls without a Lock and that the purpose of the Incorporation would be best promoted by the speedied removal of the Obstructions within the above described space. ... ," NA-RG79, Entry 160, Proceedings ..., August 8, 1785, 8.

[97] The company Proceedings stated: "It is therefore ordered, that the Party directed by the former Order to be employed above the Shanadoah Falls be immediately employed in clearing and improving the River for Navigation from Payne's upwards through the Shanadoah Falls. ... ," Ibid.

[98] The Long Canal, on the Maryland shore opposite Harper's Ferry, was obliterated by Chesapeake and Ohio Canal construction in 1832-1833. The new Chesapeake and Ohio Canal was constructed in the sluice of the Patowmack Canal for nearly two miles. Letter, Thomas F. Purcell, Civil Engineer, to Edward Colston, January 27, 1835, Document No.1, Virginia Senate Journal, 1834-1835, 1.

[99] Letter, George Washington to William Grayson, August 22, 1785, John C. Fitzpatrick, (ed.), The Writings..., op. cit., Volume 28, 234.

[100] Washington wrote: "With respect to this part of the business [i.e. concerning locks and bypass canals] I feel, and always have confessed an entire incompetency: nor do I conceive that theoretical knowledge alone is adequate to the undertaking. Locks, upon the most judicious plan, will certainly be expensive; and if not properly constructed and judiciously placed, may be altogether useless. It is for these reasons therefore that I have frequently suggested (though no decision has been had) the propriety of employing a professional man. ... ,"Letter, George Washington to John Fitzgerald and George Gilpin, March 31, 1786, John C. Fitzpatrick, (ed.), The Writings..., op. cit., Volume 28, 397-398.

[101] Some sources say four locks.

[102] Replaced by masonry locks in 1818.

[103] NA-RG79, Entry 179, Payroll for April 28 to July 28, 1792, Little Falls.

[104] The resolution stated: "Resolved that a number not exceeding two hundred Negro slaves be hired at a price not exceeding Sixty dollars per year The time to commence the first day of January Next, the hire to be paid half yearly.—The owners of the Slaves to provide good cloathing and a good Blanket for each. In case of Sickness or Elopement, Such time to be discounted out of the hire or made good by labour- ... ," NA-RG79, Entry 160, Proceedings ..., November 5, 1792, 42.

[105] Ibid., p. 43.

[106] NA-RG79, Entry 179, "Number of Yearly Negroes in the Potowmac Companys Employ: their Owners Names and the Time they Commenced," October 1794.

[107] NA-RG79, Entry 160, Proceedings..., December 27, 1794, 49.

[108] Leonard Harbaugh was born in 1749 in Pennsylvania and died 1822 in Washington, D.C. He moved to Baltimore sometime before 1773 and again to Washington, D.C. in the early 1790s. Leonard Harbaugh, A Journal, Of Accounts, Etc. Thomas Harbaugh, with the Potomac Company and Others, from 1803 to 1833, Unpublished manuscript, copy at Western Maryland Room Hagerstown (Maryland) Library, 71.

[109] Thomas Harbaugh was born in 1777 in Baltimore and died 1857 in Baltimore. He was the third son of 12 sons and 1 daughter of Leonard Harbaugh and Rebecca Rinehart. Ibid.

[110] Robert J. Kapsch, The Labor History of the Construction and Reconstruction of the White House, 1793-1817. (Unpublished Ph.D. dissertation, University of Maryland, 1983), 381.

[111] Board of Directors member Doddridge wrote" " I am informed that the Potomac Company had formerly employed an European Engineer to lay out and construct their works. These were found insufficient and perishable, and they are now in the course of being replaced, with the happiest prospects by others of Mr. Thompson's construction [Thompson, at this time, had been rebuilding the Little Falls locks in masonry]. ... ," Letter, P. Doddridge to W.C. Nicholas, Board of Public Works, Virginia, April 16, 1816. Annual Report of the President and Directors of the Board of Public Works to the Legislature of Virginia, December 1816,21.

[112] Ralph E. Ehrenberg, "Nicholas King: First Surveyor of the City of Washington, 1803-1812," Records of the Columbia Historical Society of Washington, D.C. 1969-1970, Francis Coleman Rosenberger (ed.), Volumes 69-70, (Washington, D.C.: Columbia Historical Society, 1971), 39.

[113] NA-RG79, Entry 160, Proceedings ..., July 14, 1794, 48.

[114] A.W. Skempton, et. al., A Biographical Dictionary of Civil Engineers in Great Britain and Ireland, Volume 1: 1500-1830, (London: Thomas Telford for the Institution of Civil Engineers, 2002), 773.

[115] Letter, George Washington to Thomas Johnson, December 28, 1786, W.W. Abbot and Dorothy Twohig (eds.), The Papers..., op. cit., Volume 4, 487.

[116] Letter, George Washington to Tobias Lear, December 12, 1794, John C. Fitzpatrick, (ed.), The Writings..., op. cit., Volume 34, 54.

[117] Letter, George Washington to Tobias Lear, December 21, 1794, John C. Fitzpatrick, (ed.), The Writings..., op. cit., Volume 34, 66.

[118] Washington wrote: "The plan of Mr. Claiborne Engineer, as far as I understand it, is to avoid locks altogether. The vessels are received into a basket or cradle, and let down by means of a laver and pullies; and raised again by weights at the hinder extremity of the laver, which works on an axis at the top of a substantial post fixed about the centre of the laver... ," Ibid, 66-67. This is an early reference to the use of inclined planes and boat cradles for moving boats over land elevations, such as was later used on the Morris Canal and the Pennsylvania Main Line. An inclined plane was built at Great Falls but this was a ramp for rolling barrels down to waiting vessels below Great Falls. Claiborn is Richard Claiborne who was awarded 3,555 acres of bounty land in Ohio for his service during the Revolutionary War. He was closely associated with James Rumsey in Rumsey's efforts to develop steam propulsion for vessels and was interested in portage sites between the Potomac and Ohio basins. See footnotes, Letter, George Washington to Richard Claiborne, December 15, 1784, Abbot and Dorothy Twohig (eds.), The Papers..., op. cit., Volume 2, 185. "Claiborn's engineer," is not identified.

[119] Letter, George Washington to Tobias Lear, January 12, 1795, John C. Fitzpatrick, (ed.), The Writings..., op. cit., Volume 34, 85.

[120] Letter, George Washington to Tobias Lear, March 5, 1795, John C. Fitzpatrick, (ed.), The Writings..., op. cit., Volume 34, 132.

[121] Tobias Lear wrote: "At the last annual meeting the Company directed that Mr. Weston, the person who has the direction of the Susquehannah & Schuylkill Canal should be requested to take a View of the Great Falls and give his opinion as to the most eligible way of conducting those Locks from the foot of the Canal to the River – the application to him was immediately made but his presence could not be procured till the middle of March, during which time every kind of work at the Great Falls was necessarily Suspended. Upon examination of the Ground & River Mr. Weston recommended it to the President & Directors to relinquish the place where considerable progress had been made in sinking some of the Lock seats & to conduct the Locks to that part of the river which had been originally marked out for that purpose but given up upon the Representation of Mr. Smith the Conductor of the work at that time, from the view of Saving the expence of one of the Locks – at the same time Mr. Weston took a view of the several pieces of work done upon the river from the Shanandoah Falls to tide water especially the Locks at the Little Falls all of which as far as executed met with his warmest approbation – he also took a view of those parts of the River Shanandoah where the principal impediment to that navigation are to be encountered. His observations upon different parts were put into writing which are ready to be produced to the Company – No time was lost after Mr. Weston's View of the Great Falls in taking the necessary measures to carry his recommendations into effect... ," NA-RG79, Entry 160, Proceedings..., August 6, 1795, 60-61. Weston's report has not been located and has apparently not survived.

[122] Corra Bacon-Foster, op. cit., 90.
[123] A.W. Skempton, et. al., A Biographical Dictionary of Civil Engineers in Great Britain and Ireland, Volume 1: 1500-1830, (London: Thomas Telford for the Institution of Civil Engineers, 2002), 774.
[124] Letter, Benjamin Henry Latrobe to John Spear Smith, April 25, 1816, as published in Annual Report of the President and Directors of the Board of Public Works to the Legislature of Virginia, December, 1816, 26.
[125] Robert J. Kapsch, The Labor History of the Construction and Reconstruction of the White House, 1793-1817, op. cit., 104.
[126] Committee on History and Heritage, American Society of Civil Engineers, A Biographical Dictionary of American Civil Engineers, (New York: American Society of Civil Engineers, 1972), Volume 1, 101. Don Postle, "An Early American Civil Engineer: Isaac Roberdeau, Canal History and Technology Proceedings, (Easton, PA: Canal History and Technology Press, 1999), Volume XVIII, 173-200. Richard Shelton Kirby, "William Weston and His Contribution to Early American Engineering, Transactions, The Newcomen Society, Volume XVI 1935-1936, 1-17.
[127] Isaac Roberdeau, "Mathematics and Treatise on Canals," manuscript, 1796. Manuscripts Division, Library of Congress.
[128] NA-RG79, Entry 160, Proceedings..., June 12, 1795, 52.
[129] Ibid., 52.
[130] Ibid., 54.
[131] The order read: "Ordered that Advertizement be published in the Alexandria, George Town, Shepherd Town & Frederick Papers that the Directors will purchase Three thousand Cubic feet of Lime stone to be delivered at the lower end of the Canal at the Great Falls before the 15th Mar, 1st April & 15th May next and each piece of Stone to be not less than two feet by four, not less than four inches thick and a Quantity not less than Six thousand Bushels of good unslacked Lime to be delivered at the same place... ."NA-RG79, Entry 160, Proceedings..., December 22, 1795, 65.
[132] The record stated: "That a Quantity of 3 Inch Rope not less than two Coils, be procured and fixed in the Ringbolts below the Great Falls, for the purpose of hauling Boats up the River... ," NA-RG79, Entry 160, Proceedings..., February 5, 1796, 68.
[133] Osgood R. Smith, "Present Reminders of Early Commerce on the Potomac River above Washington," Pamphlet, 1983, 8 pages.
[134] NA-RG79, Entry 160, Proceedings..., December 22, 1795, 65.
[135] Washington wrote: "If the directors are in want of such a character, as the enclosed letter describes, it may be well to intimate it as soon as possible; as it is not likely that Mr. Myers will remain long unemployed, as lock navigation is contemplated in many parts of this country. I have not seen the Gentleman myself, but understand from others that his testimonials are full and ample; and that he is a stout, healthy man. ... ,"Letter, George Washington to Tobias Lear, November 30, 1795, John C. Fitzpatrick, (ed.), The Writings..., op. cit., Volume 34, 381.
[136] Washington wrote to Lear: "I mention this because the letter of Mr. Myers seems to imply more, and as much depends upon the skill, industry and other qualifications of an Engineer, or Person employed in such a work, that you may examine him critically yourselves; for it is proper I should observe that I have no other knowledge of Mr. Myerss fitness than is derived from his own Acct, and some papers which he has shewn, but which I had not leisure to examine correctly. ... ,"Letter, George Washington to Tobias Lear, December 25, 1795, John C. Fitzpatrick, (ed.), The Writings..., op. cit., Volume 34, 411.
[137] NA-RG79, Entry 160, Proceedings..., January 4, 1796, 67.
[138] Ibid, 66.
[139] Ibid., February 1, 1796, 71.
[140] The resolution stated: "Resolved, that measures be taken immediately to complete the House already begun on the lot belonging to the Potomak Company at the Great Falls and to erect such other works as may be necessary for the accommodation of those people who may be employed by the Potomak Company at that place in the cheapest manner that will answer the purpose: And that to effect these objects, twelve thousand feet of Inch Plank be purchased and sent to Great Falls by the best and most expeditious means. The dimensions of the Building contemplated for the accommodation of the people, are 79 ft long, 18 Ft wide and 7 ft high in the Clears to be covered with Plank. ... ," Ibid., February 5, 1796, 68.

[141] NA-RG79, Entry 160, Proceedings..., March 21,1796, 73.

[142] NA-RG79, Entry 160, Proceedings..., 1796.

[143] Ibid., 72.

[144] The Board's order read: "Ordered that the Engineer Capt. Myers do go to Baltimore and endeavour to buy any number of Irish Labourers not exceeding Thirty as he may approve of at a credit of Sixty or ninety days and convey them to the Great Falls. .. ," Ibid., September 13, 1796, 83.

[145] The order read: "Ordered that Capt. Myers the Engineer be allowed Twenty Four hundred Dollars p ann. for this year and the next and the further sum of Five Thousand dollars if he compleats the Locks at the Great Falls so that loaded boats can pass in twelve months or Two Thousand five hundred if in Fifteen Months or in Proportion for any time above twelve and under Fifteen Months, from the first Day of the Present month and that the same be communicated to him by letter... ," Ibid.

[146] NA-RG79, Entry 164, "Contract With John Henry," October 1, 1796.

[147] NA-RG79, Entry 190, "The Petition of George Pointer to the President and Directors of the Chesapeake and Ohio Canal," September 5, 1829, published in Appendix, to Robert J. Kapsch, "The Potomac Canal: A Construction History, op. cit., 216-219.

[148] NA-RG79, Entry 164, "Contract With John Henry," October 1, 1796.

[149] NA-RG79, Entry 159, Proceedings..., January 13, 1801.

[150] One advertisement read: "LIBERAL WAGES
        WILL be given to a few men who are acquainted with the quarrying of freestone. Apply to
        Thomas Paxton, or John Delahunty, at the Seneca quarries; within ten miles of the Great
        Falls. ... " Washington Gazette, June 22, 1796, 3.

[151] Ibid., January 6, 1797, 90.

[152] NA-RG79, Entry 160, Proceedings..., November 2, 1796, 87.

[153] NA-RG79, Entry 160, January 6, 1797, 90.

[154] Ibid., January 12, 1797, 91.

[155] Ibid., January 13, 1797, 92.

[156] Ibid., January 14, 1797, 92.

[157] Ibid., January 14, 1797, 92-93.

[158] Ibid., January 14, 1797, 93.

[159] The record stated: "Two other Houses under one Roof similar to those described, excepting the ground story have been built on a lot belonging to Capt Myers, one occupied by the low gate maker the other now used as an Office for the Clerks. ... ," Ibid.

[160] The record stated: "The House intended to be built for the Potomak Company Lay (on) the Ground story and part of the first Story run up. The joists of the lower Floor are laid. Flooring and other Plank are on the spot for it but the Board have ordered that nothing more be done on it at present. ..." Ibid.

[161] The record stated: "The Engineer having ordered to have the Accomodations at the Great Falls built in the cheapest manner that the nature of the Thing would admit of, the Measurement and Value of the Work is the only means now left to show if this order has been complied with or not. ... ," Ibid.

[162] Ibid.

[163] NA-RG79, Entry 179, "Account of the Several artificers and Labourers, in the Potomac Company Service, employed by ther Engineer at his house... 1st of July to the 31st December 1796."

[164] Myers wrote: "The reason why I employed for my own use any of the Companys hands, was my not having been able to hire any servants in the whole country: chiefly owing to the advanced period of the season when I settled at the Great Falls. It therefore became a matter of necessity and not of choice. I have now provided myself with people to do all my own work, independent of any assistance from the companys people. ... ," Ibid., Letter, To the Board of Directors, From Capt. Christopher Myers, January 21, 1797.

[165] NA-RG79, Entry 160, Proceedings..., January 17, 1797, 95.

[166] The record stated: "Resolved that the Engineer Capt. Christopher Myers be immediately dismissed from the service of the Potomack Company for the following reasons.
        1st. Because he refuses proper and reasonable Communications with the Directors, and to
        deliver plans which are already made, as well as to furnish others called for.

2d. Because he so often absents himself and gives so little of his time to the works of the Company that they have suffered very Materially in consequence.

3d. Because he has not furnished, when applied to, proper working plans to the different Artificers, and thence they have been either impeded in the progress of their business, or done it in such a manner as to make it necessary to pull it to pieces again, to the great Loss and Detriment of the Company.

4th. Because, he has not furnished, when called on by the Directors, Bills of Materials proper to be employed.

5th. Because he has neglected to inspect, or suffered to be passed, materials prepared for the Potomack Company's use by Contract, until the Directors driven to the necessity, or withdrawing their Confidence, employed an Inspector, when the several months after such materials were begun to be prepared and a proportion was condemned by the Inspector appointed, he at length makes a Report and acknowledges that a part of them are unfit for service, -- Instance the Cutt Stone. ... ," NA-RG79, Entry 160, Proceedings..., May 2, 1797, 100-101.

[167] Ibid., May 2, 1797, 101.

[168] The advertisement read: "C A U T I ON

The public is requested to suspend any opinion, that may arise from certain C A L U M N I E S in circulation, tending to injure my character; made on the force of a difference that has taken place between the directors of the Potomac Company and myself:

WE HAVING MUTUALLY AGREED

TO LEAVE THE BUSINESS TO ARBITRATION.

Therefore invidious and improbably tales, I have the confidence to think, will not obtain any degree of credit in the public mind; to whom, in a little time, I shall present a statement of the several causes that have induced me to resign the situation of Engineer to the Potomac Company.

Matildaville   \

Great Falls of Potomac] C. MYERS

May 24, 1797"   /

Alexandria Advertiser, August 4, 1797, 1.

[169] NA-RG79, Entry 190, "Petition of Captain George Pointer to the President and Directors of the Chesapeake and Ohio Canal," September 5, 1829, published in Appendix, to Robert J. Kapsch, "The Potomac Canal: A Construction History, op. cit., 216-219.

[170] Ibid., June 6, 1797, 105.

[171] Ralph E. Ehrenberg, "Nicholas King: First Surveyor of the City of Washington, 1803-1812," Francis Coleman Rosenberger (ed.), Records of the Columbia Historical Society, 1969-1970, (Washington, D.C.: Waverly Press for the Columbia Historical Society, 1971), 31-65. Silvio A. Bedini, "The Kings of Washington, D.C. (1796-1818): Cartographers and Surveyors," With Compass and Chain: Early American Surveyors and Their Instruments, (Frederick, Maryland: Professional Surveyors Publishing Company, Inc., 2001), 571-579.

[172] NPS, Letter, Nicholas King to Mr. Lear and Templeman, dated May 12, 1797, Records of the Potomac Company, Chesapeake and Ohio Canal National Historical Park, Acc. # 2498 (86); [269].

[173] Footnote 30, Ralph E. Ehrenberg, op. cit. Ehrenberg incorrectly lists King's payment as $13.13. King was paid £13 –13s, Virginia Currency, approximately three times $13.13. NA-RG79, Entry 160, Potomac Company "Waste Book, 1785-1800," 326.

[174] Ralph E. Ehrenberg, op. cit., Footnote 49.

[175] NA-RG79, Entry 179, "Note of New employes in the Service of the Potomac Company, Great Falls from an estimate made Sept. 5th 1797," Records of the Potomac Company. Included were 4 masons; 8 stone cutters; 8 carpenters and sawyers; 14 white laborers with 2 white servants; 4 blacksmiths; 16 black laborers; 2 female cooks; 2 female Company servants; 3 children; and a number of other men, probably also laborers.

[176] There were some free blacks working as laborers. For example, "Negro Dick Free," NA-RG79, Entry 167, Potomac Company, Wastebook: 1785-1800, 321.

[177] NA-RG79, Entry 179, "Names of the People in the employ of the Potomack Company, Great Falls, Oct. 3d '97."

[178] Ibid.

[179] The Potomac Company paid Peters £100 per year Virginia currency (£80 Maryland currency) for the use of his quarry. NA-RG79, Entry 167, 323.

[180] Ibid., June 8, 1787, 325.

[181] Ibid., June 7, 1787, 325.

[182] NA-RG79, Entry 168, Ledger (1796), 1. Gridley was also working for the Commissioners of Public Buildings and would, a year later, have his contract abruptly terminated by the Commissioners on suspicion that he was "too Saucy..." and the author of a pamphlet which ridiculed the Commissioners. See Robert J. Kapsch, The Labor History of the Construction and Reconstruction of the White House, 1793-1817, op. cit., 285-292.

[183] NA-RG79, Entry 160, Proceedings..., August 7, 1797, 113.

[184] Ibid.

[185] Ibid., October 4, 1797, 123.

[186] NPS, Monthly payroll, January 1798, Records of the Potomac Company, Chesapeake and Ohio Canal National Historical Park, Acc. 2498, Number 292.

[187] NPS, Monthly payroll, February 1798, Records of the Potomac Company, Chesapeake and Ohio Canal National Historical Park, Acc. 2498, Number 293.

[188] Ibid.

[189] NPS, Monthly payroll, March 1798, Records of the Potomac Company, Chesapeake and Ohio Canal National Historical Park, Acc. 2498, Number 294.

[190] NPS, Monthly payrolls April, June, July, August, September, November, 1798, Records of the Potomac Company, Chesapeake and Ohio Canal National Historical Park, Acc. 2498, Numbers 297, 299, 300, 301, 302, 304. In July 1798, for example, workers included: Edward Sweeney, Stone Mason; Dominick Burns, Alexander Wallace, Charles Gough, Enoch Lovely and Patrick O'Hara, laborers. Some of these employees were long time Company employees. O'Hara, for example, had worked for the Company since 1786.

[191] NA-RG79, Entry 160, Proceedings..., June 5, 1798, 133-134.

[192] Ibid., 134-135.

[193] Ibid., 148.

[194] Ibid.

[195] Ibid., 153-154

[196] NA-RG79, Entry 160, Proceedings..., January 20, 1800, 181.

[197] Ibid., August 4, 1800, 200.

[198] Ibid., August 3, 1801, 219.

[199] Ibid., August 4, 1800, 201.

[200] Ibid., August 4, 1800, 200.

[201] Ibid., August 3, 1801, 219.

[202] Ibid., August 4, 1800, 201.

[203] Ibid.

[204] Robert Wisely was injured: "... by the explosion of a rock charged with powder he had his face much wounded, and his Eyes so injured that he has not since recovered his sight. Altho from the best information obtained on the Subject it appears that the accident happened thro' the man's own imprudence by meddling with the blasting business not committed to him, and with which he was unacquainted. Yet as the intention was probably good and as the poor man's Situation was a deplorable one – The Directors thought it but right to have the necessary care taken of him until except as to his eyesight he was restored to health; and since to contribute so far to his support as to

allow him a ration per day estimated at fifteen pence Maryland Currency. ... ,"NA-RG79, Entry 160, Proceedings..., August 3, 1801, 222-223. The last payment was made to Robert Wiley in October 1804. Wiley sent a series of letters to Company Paymaster Joseph Carleton asking for further financial support. NPS, Records of the Potomac Company, Chesapeake and Ohio Canal National Historical Park: Acc. #2498 (81) [264]; #2498 (52) [235]; and # 2498 (53) [236], July 30, 1805, January 11, 1806, January 23, 1806.

[205] The record stated: "Agreed with George Jacobs to quarry at Senica and deliver along side the boats 3000 feet of Freestone of the White kind... ," Ibid., January 12, 1801, 209.

[206] The record stated: "Resolved, that the following alterations be made, viz., Lock No. 1 instead of rising eighteen feet shall rise Twenty one feet. No. 2, instead of rising fifteen feet shall rise eighteen feet. Lock No. 3, instead of rising Twelve feet shall rise only Ten feet eight inches and that the width of said Lock No. 3 be so increased as when filled to admit two Boats of the usual size. Lock No. 4, shall be (indistinct) as is Lock but be completed as is correct to communicate between Locks No. 3 and 5. Lock No. 5 instead of rising twelve feet shall rise sixteen feet so that with the rise of eleven feet (indistinct) by the locks already finished the whole rise of seventy six feet eight inches shall be absorbed by the said Locks and that said Lock No. 5 if necessary be enlarged in its Contents by adding either in the length or Breadth ... ," Ibid., May 4, 1801, 212.

[207] Ibid., January 6, 1802, 224.

[208] Harbaugh wrote: "The Locks at the Great Falls of Potomac are Six in number, situated 14 miles above Georgetown, a greater part of them are blown out of the solid rock, a work of much labour & expence. A Further Discription of the size etc. of the Locks are— From Gate No. 1 of Lock No. 1, at the Bason—to gate No. 2 is 390 feet in length 25 feet wide and 6 feet deep – from Gate No. 2 to Gate No. 3 is 112 feet in length 15 feet wide and 14 feet deep, called Lock No. 2, thence from gate No. 3 to gate No. 4 is a small Bason 400 (feet) in length 30 feet wide and 5 feet Deep, thence from gate No. 4 to gate No. 5 is Lock No. 3 is 97 feet long 12 feet wide and 19 feet deep walls made of red free stone, cut thence from Gate No. 5 to a large apron is 97 feet in length 20 feet wide and 4 feet deep, thence from the apron to gate No. 6 is Lock No. 4 – 100 feet in length 12 feet wide and 16 feet deep, one side and the bottom in solid rock, thence from Gate No. 6 to Gate No. 7 is Lock No. 5 – 95 feet Long 12 feet wide and cut down in the rock about 25 feet in Depth – making both sides and Bottom all solid rock, thence from gate No. 7 to Gate No. 8 – is Lock No. 6 – is 84 feet long 12 feet wide and 24 feet Deep with 1-1/2 Sides and all the Bottom solid rock --- part of this Lock has been cut down through the rock 43 feet in Depth – the River rises in the Lock in high water 53 feet in height. The whole length from upper to lower gate is 1375 feet.

> Fall of Water
>
> From Lock No. 1 into Lock No. 2 is 9 feet
>
> From Lock No. 2 into Lock No. 3 is 15 feet
>
> From Lock No. 3 into Lock No. 4 is 11 feet
>
> From Lock No. 4 into Lock No. 5 is 18 feet
>
> From Lock No. 5 into Lock No. 6 is 20 feet

Total Fall is   73 feet ... ," Ibid. Harbaugh's terminology is a bit skewed. What he refers to as Lock Number One was the 390 foot long canal entrance to what is usually known as Lock Number One. Hence there are five locks at Great Falls and not six. Harbaugh was not the only one to count the locks as six in number. Manaseh Cutler visited the Great Falls locks on January 30, 1802 and reported six locks. Corra Bacon-Foster, op. cit., 104.

[209] NA-RG79, Entry 160. Proceedings..., January 6, 1802, 232.

[210] NA-RG79, Entry 160, 235-238.

[211] The Report to the Stockholders read: "Upon this event (opening the locks at Great Falls) it was generally expected that the Stock of the Company would immediately become productive to the holders and we had no doubt of being able to lay before you at this meeting such a State of the Tolls as would afford a handsome dividend – We are however sorry to say that notwithstanding in common years the River is now navigable from George's Creek to tide-water without interruption during a

considerable part of the year – and that there was certainly large quantities of Flour and other produce prepared and intended to have been sent down the River to market on the opening of the Navigation at the Great Falls, such has been the remarkable low state of the waters hitherto since the Locks were finished in consequence of the total want of Snow last winter in the upper Country, and the rains then and since proving only very moderate and partial, that the River could only be used for transporting produce at short intervals after some of these partial Rains… ," NA-RG79, Entry 160, "Report of the President and Directors of the Potomack Company to the Stockholders," Proceedings…, August 2, 1802, 246-247.

[212] Ibid., 268-270.

[213] Ibid., 269-271.

[214] Ibid., 270.

[215] Located east of Bloomery, West Virginia, where the present State Highway 9 crosses the Shenandoah River. Littles Falls is at the bend of the Shenandoah where the river turns north. This area was also known as Hopewell Mills. James Herron, in his 1832 map for the New Shenandoah Company, identifies this as Newcomers Mill.

[216] This is probably Millville Mill, also known as Snyder's Mill and so described on James Herron's 1832 map prepared for the New Shenandoah Company.

[217] Millville Station, the present Millville.

[218] NA-RG79, Entry 162, Records…, Letter, Potomac Canal President John Mason to Secretary of the Treasury Albert Gallatin, January 20, 1808. These are the locks that Thomas Harbaugh describes building. Thomas Harbaugh, A Journal,… op. cit.

[219] Corra Bacon-Foster, op. cit., 211.

[220] NA-RG79, Entry 160, "Report of the President and Directors of the Potomac Company to the Stockholders at their annual meeting at George Town on the first day of August 1803," August 1, 1803, 271.

[221] Corra Bacon-Foster, op. cit., 111.

[222] Thomas Harbaugh, op. cit..

[223] NA-RG79, Entry 162, Letter, Thomas Harbaugh to John Mason, President of the Potomac Company, "Memorandum of the Antietam Creek," no date. This memorandum was probably written shortly after Harbaugh was assigned to Antietam Creek as Superintendent, April 7, 1812.

[224] Thomas Harbaugh, op. cit.

[225] Ibid.

[226] Darwin H. Stapleton, The Engineering Drawings of Benjamin Henry Latrobe, (New Haven and London: Yale University Press for the Maryland Historical Society, 1980), 117.

[227] Letter, William Stewart to Charles F. Mercer, June 22, 1816, Annual Report of the President and Directors of the Board of Public Works to the Legislature of Virginia, December 1816, 19.

[228] Letter, John Mason to Philip Norborne Nicholas, May 27, 1818, recommending Moore to the post of engineer for the Virginia Board of Public Works.

[229] Letter, William Stewart to Charles F. Mercer, June 22, 1816, Annual Report of the President and Directors of the Board of Public Works to the Legislature of Virginia, December 1816, 19.

[230] Ibid.

[231] Latrobe wrote: " Mr. Thompson, a very respectable carpenter of this city was employed as surveyor on the Cumberland Road, and also in the improvement of the Powtomac navigation; that alone, is sufficient to provide his competency… ," Letter, Benjamin Henry Latrobe to the President of the United States (James Madison), April 8, 1816, Annual Report of the President and Directors of the Board of Public Works to the Legislature of Virginia, December 1816, 19.

[232] Letter, to the Virginia Board of Public Works from the President and Directors, Potomac Company, December 9, 1817, Annual Report of the President and Directors of the Board of Public Works (Virginia), 1816/1817, 20.

[233] The Petition of Captain George Pointer to the President and Directors of the Chesapeake and Ohio Canal, September 1829, National Archives, Record Group 79, Entry 190, as reproduced as "Appendix," to Robert J. Kapsch, "The Potomac Canal: A Construction History, op. cit., 217.

[234] Letter, William Stewart to Charles F. Mercer, June 22, 1816, Annual Report of the President and Directors of the Board of Public Works to the Legislature of Virginia, December 1816, 19.

[235] Letter, John Mason to Philip Norbone Nicholas, May 27, 1818, Manuscripts Division, The Library of Virginia.

[236] Robert J. Kapsch, "The Potomac Canal: A Construction History," op. cit., 210.

[237] Ibid.

[238] Ibid.

[239] Ibid., 210-211.

[240] Ibid. 211.

[241] Ibid., footnote 362, 234.

[242] Ibid., 212-214.

[243] Ibid., 214-215.

[244] NA-RG79, Entry 162, Correspondence, October 1, 1827.

[245] NA-RG79, Entry 184, Proceedings of the President and Directors, Chesapeake and Ohio Canal Company, August 23, 1828, 46-47.

[246] Committee on History and Heritage, American Society of Civil Engineers, Francis E. Griggs (ed.) and Karen O'Conner (res. asst.), A Biographical Dictionary of American Civil Engineers, (New York: American Society of Civil Engineers, 1991), Volume 2, 97.

[247] Orders of the President and Directors of the Chesapeake and Ohio Canal Company, in relation to the distribution of the Corps of Engineers, November 22, 1828, published in Appendix to the Report of the General Committee of Stockholders (of the Chesapeake and Ohio Canal Company), June 6, 1829, li.

[248] Washington believed that sluice navigation of the Little Falls was possible and therefore a bypass canal would not be needed at this site, either. He wrote: "The little fall, if a Rock or two was removed, might be passed without any hazard, more especially if some of the Rocks which lye deep and which occasion a dashing surface could be removed. ... ," John C. Fitzpatrick (ed.), The Diaries ..., op. cit., Volume II, September 22, 1785, 416-417. Washington's under emphasis of the degree of difficulty of passing through Little Falls was commented on by the editor of Washington's Diaries, Fitzpatrick. Fitzpatrick writes: "There was, evidently, a greater volume of water in 1785 than there is to-day, as it is impossible to conceive of any boat of size passing down Little Falls as it is now. It is a perilous and almost impossible passage even for canoes and rowboats.: Ibid.

NA = National Archives.

RG = Record Group.

SI Conversions. This paper was written in the original English dimensions used by the eighteenth and nineteenth century engineers and builders. SI conversions are as follows:

1 Mile = 1.6093 kilometers

1 Square Mile = 2.590 square kilometers

1 Foot = 0.3048 Meters

1 Inch = 2.540 centimeters

"To Make the Crooked Ways Straight and the Rough Ways Smooth:"
Surveying and Building America's First Interstate Highway
Dr. Billy Joe Peyton

The history and prosperity of the United States has been greatly impacted by the transportation systems that have evolved over time. Early public leaders recognized the importance of internal improvements and how the development of an integrated national transportation network must necessarily include navigable waterways and land routes connecting the eastern seaboard with the "western waters." It was within the framework of a national "internal improvements" movement that the federal government built the nation's first interstate highway, a formidable undertaking in the context of its time. The project represents the young nation's earliest attempt to minimize sectional differences through internal improvements and cement a tangible bond between transportation systems and settlements on opposite sides of the Appalachian Mountains. Alternately referred to as the National Road, Cumberland Road, Great Western Road, Uncle Sam's Road, or simply The Road, it was the most ambitious--and costly--public works project of its day. Occupying a unique place in our nation's transportation heritage, no other route has played a larger role in the growth and development of the United States than this landmark highway. (Jordan, *National Road*, foreword)

Early Transportation in the United States
The Appalachian Mountains formed an insurmountable barrier to the North American interior for the first two centuries after European colonization. Little had changed by the end of the colonial era, when a dearth of reliable all-weather roads still kept most settlers clinging to within 150 miles of the coast. Few thoroughfares led to the interior at this time, and those that did proved to be little more than enlarged animal or American Indian trails. This meager highway system remained in a deplorable state at the dawn of the 19th century, as the latest in modern construction techniques had not yet been introduced into the United States from Europe.

The need for an integrated national improvements plan became increasingly important as thousands of intrepid souls made their way over the Central Appalachians after the U.S. Army defeated a large American Indian force at the Battle of Fallen Timbers (near Toledo, Ohio) in 1794. With the opening of the trans-montane region, settlers quickly populated the rich agricultural lands of the Old Northwest Territory that would later become the states of Ohio, Indiana, and Illinois.

Most decision-makers agreed that any viable transportation nexus to the nation's interior must logically include navigable waterways and feature one or more overland routes to connect the eastern seaboard with the "western waters." George Washington and other early visionaries considered several geographic alternatives for breaching the mountains, but few surpassed the advantages of the Potomac-Monongahela corridor that penetrated the Appalachian front just west of Cumberland, Maryland. (Bacon-Foster, *Potomac Route*, 147-49)

One of the earliest organized plans to develop the regional transportation potential of the Potomac-Monongahela corridor dates from 1784, when commissioners representing Maryland and Virginia met to organize the Potomac Company. (Bacon-Foster, *Potomac Route*, 147-49) However, the first improved road came 30 years earlier when British Major General Edward Braddock commanded an ill-fated military expedition on a westward trek from Baltimore to the Forks of the Ohio in 1755.   History records Braddock's mission to seize Fort Duquesne (at modern-day Pittsburgh) as a failed attempt to avenge George Washington's defeat at Fort Necessity and oust France from the Ohio Valley during the French and Indian War. This disastrous affair culminated in the total defeat of Braddock's army and cost the general his life.   En route to their fateful encounter on the banks of the Monongahela River, Braddock's troops hacked their road through the mountain fastness so they could take their supply wagons over Nemacolin's Path, a winding foot trail named for the Delaware Indian scout who had blazed it.   Using only hand-held axes, mattocks, and other simple digging, felling, and grubbing tools, the soldiers chopped and cleared a twelve-foot wide swath through the rugged terrain that became Braddock's Road.

At the conclusion of the Revolutionary War, expanding national business and trade brought a corresponding increase in country-to-city traffic in the newly-created United States--especially near the nation's largest cities. Only after the nation sorted out its post-Revolutionary economic problems could it begin to deal with internal improvements. Before that time, Western settlers had few overland alternatives for breaching the central Appalachians. Their choices included the Wilderness Road (shown on the map as the Cumberland Gap Road), a popular passage used by Daniel Boone to reach Kentucky and Tennessee; the Pennsylvania Road, which ran from Philadelphia to the forks of the Ohio (also called Forbes Road after British General John Forbes widened it during the French and Indian War) and connected with an important post road to New York; and Braddock's Road [Figure 1].

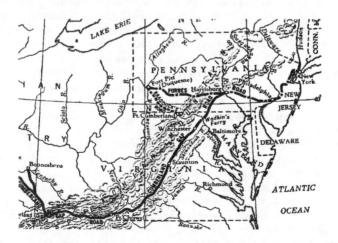

[Figure 1]   Map showing the routes of three important colonial roads that crossed the central Appalachian Mountains. (from *America: Its History and People*, New York: Harper & Row, 1934)

Most travelers in 1800 opted against taking an overland route to the country's interior because most existing roads were little more than cleared dirt paths that deteriorated with each passing season, despite the growing national interest in developing a safe and reliable transportation nexus. Responsibility for road construction and repairs usually fell on local governments with little money or interest, a situation that often resulted in feeble or token maintenance efforts. Most states had neither begun to build stone-paved and finished roads nor carried out regular repairs on existing routes in any systematic fashion, and so even the most important westward passages were being reclaimed by nature from lack of maintenance. Indicative of the situation nationwide, by the dawn of the 19th century even the popular Braddock's Road had devolved back into a mere thread comparable to the old buffalo trail from which Nemacolin had cut his original path.

Planning a National Road

Despite deplorable road conditions, westward migration soared between 1790 and 1810. The nation's population nearly doubled from 3.9 million to 7.2 million during this period, as did territorial possessions after acquisition of the Louisiana Territory in 1803. (*U.S. Census*) With physical expansion came opportunities to extend commercial markets in the west, as well as increased pressure to settle and develop the rich lands beyond the Ohio River. Rapid population growth led to Ohio statehood in 1803, a seminal event that presented an irresistible opportunity for the federal government to become actively involved in the increasingly popular internal improvements movement. In 1802, Congress had passed a key act creating a "2 per cent fund," setting aside two percent of the proceeds from the sale of public lands in Ohio were set aside for building roads "to and through" the state. Establishment of this fund made it possible for the federal government to begin finalizing its plan for building a national road linking the eastern seaboard with the Ohio Valley. (*Public Statutes at Large*, vol. 2, 226) In response to this growing federal interest in internal improvements, Albert Gallatin researched and wrote a landmark study for the U.S. Congress in 1808 known as the *Report on the Subject of Public Roads and Canals*. (*New American State Papers*, Transportation, vol. 1) This comprehensive document presented internal improvements in the context of a boon to transportation, communication, and the economy, and it received wide praise for being far-reaching and national in its scope.

Constructing a substantial roadway over the rugged central Appalachians presented an enormous and difficult task, but it made perfect sense to business and professional leaders, and the American public. Where would such a road be built? The U.S. Senate Committee on Internal Improvements debated this point before finally settling on a preferred corridor by process of elimination All prospective routes north of Philadelphia and south of Richmond were immediately excluded from consideration because the law specified that the highway must strike the Ohio River at some point contiguous to the state of Ohio. Any thoughts of an all-Virginia route extending up the James and down the Kanawha rivers proved inappropriate because it would enter Ohio in the sparsely populated southern part of the state, a prospect that did not meet the necessary expectations for economic expansion. (*American State Papers*,

Transportation, vol. 2, 97)

Given the fact that the citizens of Ohio had the most commercial intercourse with Baltimore and Philadelphia made these cities the logical favorites for an eastern connector. Since Pennsylvania already had the "spirit and perseverance" to build a road from Philadelphia to the western waters and Maryland was engaged in building a road westward from Baltimore, therefore the committee felt that intervention in either state's efforts would "produce mischief instead of benefit." Since Maryland had no vested interest in building a road beyond the mountains, the Committee on Internal Improvements deemed it expedient to recommend:

> laying out and making a road from Cumberland, on the northerly bank of the Potomac...to the Ohio river, at the most convenient place on the easterly bank of said river, opposite to Steubenville, and the mouth of Grave Creek, which empties into said river, Ohio, a little below Wheeling in Virginia. This route...will cross the Monongahela at or near Brownsville, sometimes called Redstone, where the advantage of boating can be taken. (*American State Papers*, Transportation, vol. 2, 97)

With a general blueprint for its location, Congress passed an act to regulate the layout and construction of a "national" road. The act set broad parameters for route placement, but it left "the manner of making said road, in every other particular" up to the president. Specifically, it authorized President Jefferson to determine the exact route by appointing "three discreet and disinterested citizens of the United States" to lay out the road and spend an appropriated sum of $30,000 to defray the expenses of planning said route. (*American State Papers*, Transportation, vol. 2, 97)

Construction from Cumberland to Wheeling

The president held ultimate jurisdiction over the Cumberland Road. His authority included the right to appoint a board of road commissioners and to accept or reject their recommendations, adopt measures to secure the consent of states through which the road passed, and exercise judgement and control over construction, administration, and finances. Later authority granted to the president gave him the right to appoint superintendents who would oversee construction and repairs.

Thomas Jefferson ultimately chose three expertly qualified commissioners to lay out the route—Joseph Kerr, Thomas Moore, and Elie Williams. Kerr was a professional surveyor and U.S. senator from Ohio, while Moore, who hailed from Maryland, later served as chief engineer of the Virginia Board of Public Works and also helped to plan the Chesapeake & Ohio Canal. Chief commissioner Elie Williams lived much of his life in Hagerstown, Maryland, and had seen service as a colonel in the Revolutionary War and as commissary to General Light Horse Harry Lee during the Whiskey Rebellion. His extensive involvement with internal improvements began in 1797 when he served on a planning committee for the Baltimore Turnpike. After completing his work on the Cumberland Road, Congress appointed Williams as a surveyor on a proposed Potomac River canal which later became the C & O Canal.

(Jordan, *National Road*, 78; Ierley, *Traveling the Road*, 229)

It is doubtful whether any of the commissioners had formal training in engineering or surveying; rather they learned their skills in apprenticeships carried out under practicing professionals. Nevertheless, all three had extensive surveying experience and a strong sense of purpose, and in the final analysis they deserve high marks for their outstanding contributions to the project. The team's mission was to lay out a road from Cumberland to the east bank of the Ohio River, opposite the northern boundary of Steubenville, Ohio, and adjacent to the mouth of Grave Creek (near Wheeling), West Virginia. After laying out the road the commissioners presented their findings to the president in a plan detailing proposed distance, elevations, topography, markers and monuments, and estimated expense.

As specified in the March 29, 1806 act to regulate the layout and construction of the Cumberland Road the commissioners hired one surveyor, two chainmen, and one marker to assist on the project. Wages for the surveyor totaled three dollars per day plus expenses, while the chainman and marker each earned one dollar per day plus expenses. A vaneman and a packhorse man (with horse) were then added to the team in order to expedite the assigned tasks. (*Message from the President Transmitting Report*, 6)

Members of the surveying team began their duties on or about September 1, 1806, while the commissioners met in Cumberland on September 3 to begin daily forays in search of an optimal route. Making their way westward toward Wheeling, they stayed in hostelries or in private homes with area residents, often accompanied in their explorations by knowledgeable locals who knew the area and offered suggestions on the best possible ground.

Successful layout of a feasible route depended on the expertise of chief surveyor Josias Thompson, who not only had to master the skillful use of his surveying instruments but who had to maintain meticulous field notes. In addition, his job required a great deal of physical stamina to carry cumbersome equipment through the mountainous terrain. Thompson used a magnetic compass and 66-foot long surveyor's chain to complete his measurements, and he left wooden stakes and iron markers adorned with red flannel flags to mark proposed grades and alignments. Thompson's progress was slow at the outset, and it soon became apparent that he would require help to accomplish his work. Arthur Rider, who had been hired as the vaneman on the expedition, assumed the role of second surveyor on September 22, a position he held until December 1, when he became Thompson's assistant involved in copying field notes and drafting a preliminary report. (Searight, *Old Pike*, 30)

It took about a month for the commissioners to reach Fayette County, Pennsylvania, which they finally did in the first week of October to begin scouting in the general vicinity of Washington and Brownsville. They spent the remainder of October exploring possible routes through Western Pennsylvania and much of the first two weeks of November in and around Wheeling in search of the best terminal point on the Ohio River. (*General Journal Kept During Examination of Route*, 10 October-13 November 1806)

A harbinger of what lay ahead, the season's first snowfall came during the third week of November. More inclement weather followed, until expedition members decided to suspend work for the season after a second measurable snowfall on December 6 made it impossible to continue. On mutual agreement, the commissioners retired to Cumberland with every intention of resuming work the following spring. While in Cumberland, they agreed to submit an interim report to the president. After instructing Josias Thompson "to prepare with all convenient dispatch a compleat & comprehensive map of all their work," the commissioners left Cumberland with plans to "meet in Washington as soon as their journal & report could be made up to be presented with their plat to the President of the United States." (Ierley, *Traveling the Road*, 39) The commissioners submitted their interim report on December 30, 1806. In it they revealed their insights on a number of topics, such as the inaccurate maps at their disposal and the amusing and somewhat vexing problem of inhabitants in the surveyed area who "conceived their grounds entitled to a preference." They also candidly admitted: "the duties imposed by the law became a work of greater magnitude, and a task much more arduous than was conceived before entering upon it." (Report of the Commissioners, *Message from the President*, 5, 15; Searight, *Old Pike*, 29)

Most importantly, the interim report announced that Cumberland, Maryland, would be the eastern terminus of America's first interstate highway. This came about without controversy, partly out of propriety and partly out of geographic necessity resulting from its advantageous location nestled at the eastern slope of the Appalachian Mountains near the Eastern Continental Divide. Here, Wills Creek slices a deep water gap into Wills Mountain at "The Narrows," a fortuitous erosional accident that firmly established Cumberland 's reputation as a strategic frontier outpost and gateway to the West, as well as a most attractive place to begin constructing a national road.

After a careful study of the ground west of Cumberland, the commissioners fixed their attention on determining the most desirable path to the Ohio River. Four basic criteria guided their decision. They sought: (1) the shortest distance between navigable points on the eastern and western waters; (2) to cross the Monongahela River at a point that maximized the potential of portage to the country within reach of it; (3) a point on the Ohio River most capable of combining river navigation with road transportation and which considered the potential growth of the lands north and south of that point, and; (4) the shortest road with the most benefits. Methods and materials were important factors for consideration, too. As the commissioners affirmed, "nothing short of a firm, substantial, well-formed, stone-capped road can remove the causes which led to the measure of improvement." (Report of the Commissioners, *Message from the President*, 7-8)

The selected route originated at Lot Number 1 in the heart of downtown Cumberland. According to the commissioners, the formidable topography would not allow traditional cut-and-fill construction techniques to reduce the grade to an acceptable five degrees over the mountains. Consequently, they suggested crossing the mountains and hollows obliquely, which would require a great deal of hillside digging but preclude the need for costly cuts and fills. As far as utilizing Braddock's Road, its indirect course and frequent elevation changes exceeded limits of the law and consequently "forbid the use of it, in any one part for more than half a mile, or more than two or three miles in the whole." (Report of Commissioners, *The New American State Papers*, Transportation, vol. 1, 111) Thus, the commissioners recommended an entirely new route, one that would cost considerably more than a thoroughfare incorporating parts of the old road.

After submitting their interim report in late December of 1806, expedition members returned to their homes in anticipation of renewing their efforts the following spring. As it turned out, they had to wait to resume work until the Pennsylvania legislature granted permission for the road to pass through its borders. Following much deliberation, state politicians granted their consent only if the route passed through both Uniontown and Washington, Pennsylvania. However, approval came too late for team members to return to the field in the spring. In fact, they did not resume their efforts until the fall of 1807. They subsequently continued their explorations until December when winter weather halted their progress. The abbreviated season's activities concluded with the submittal of a second interim report to the president on January 15, 1808.

The third and final season of fieldwork began in the spring of 1808 and continued throughout most of that summer. Thomas Moore and Elie Williams collaborated to write the final project report, which they submitted to the president on August 30, 1808. (*American State Papers*, Miscellaneous, vol. 1, 940-41) The two commissioners wrote the document without the assistance of Joseph Kerr, who left the expedition at the end of 1807 due to domestic concerns and did not return. By the time Kerr departed, the surveying team had finished locating, grading, and marking the route from Cumberland to the Ohio River. Shortly thereafter, the government let contracts for clearing portions of the right-of-way.

Moore and Williams called for the new road through Maryland, Pennsylvania, and (West) Virginia in their report, to be constructed "in the style of a stone-covered turnpike." (*American State Papers*, Miscellaneous, vol. 1, 940-41) Owing to the ruggedness of the landscape and requirement by law for a 66-foot right-of-way, it should be "nothing short of well constructed" with "completely finished conduits" to render it passable at every change of season, after rainstorms and snowfalls. The commissioners went on to estimate project costs at $6,000 per mile, excluding the expense of building bridges over principal streams. (*American State Papers*, Miscellaneous, vol. 1, 940-41)

In the final analysis, all members of the Cumberland Road expedition deserve to be commended for their contributions to our nation's transportation heritage. For three exhaustive seasons between 1806 and 1808 the group ran comprehensive surveys over 131 miles of mountain wilderness. During that time they carried out extensive field explorations, met with local residents and community leaders, spent countless hours in careful deliberations, and submitted to the president three carefully studied and detailed reports which fully documented their efforts. In the process, they received very few complaints or criticisms from the president, members of Congress, or the American people. They completed their duties with great vigor, professionalism, and meticulous attention to detail--no small accomplishment considering the rather meager compensation and imprecise orders under which they operated. Indeed, their dedication to purpose formed the very foundation upon which our nation's first interstate highway, the "Road from Cumberland, in the State of Maryland, to the State of Ohio," became a reality.

Building a National Road

The selected route took in some of the most ruggedly spectacular mountain landscape in the eastern United States. Connecting with the privately funded and built Baltimore Pike east of Cumberland, the National Road snaked its way in a more or less northwesterly direction over some of the most formidable peaks in the central Appalachians. Elevations in Western Maryland reached nearly 3,000 feet above sea level as the road crossed over Big and Little Savage, Little Meadow, and Negro mountains. After crossing into Pennsylvania it climbed Chestnut Ridge and Laurel Hill on the general alignment of Braddock's Road to a point where the former route veered toward Pittsburgh (at the summit of Laurel Hill) just east of Uniontown. Continuing westward, the route passed over less rugged hill terrain between Brownsville and Washington, where it took "as straight a course as the country will admit to the Ohio, at a point between the mouth of Wheeling creek and the lower point of Wheeling Island [Figure 2]." From Wheeling the road connected with Zane's Trace, an important existing post road running from the west bank of the Ohio River through Zanesville, Ohio, to a terminal point at Limestone (now Maysville), Kentucky. A portion of this key route later became incorporated into the western extension of the National Road that eventually ran through Ohio, Indiana, and Illinois.

[Figure 2] Original route of the 131-mile National Road from Cumberland, Maryland, to Wheeling, (West) Virginia. East of Cumberland the highway connected with the Baltimore Pike, a privately-funded turnpike. (drawing by Billy Joe Peyton)

Following approval of their proposed general alignment, the commissioners eagerly anticipated moving the project beyond the planning stage. In fact, contracts had already been let for the partial clearing of timber and brush before they even completed their final report, with all the specified work to be completed by March of 1808. However, this early flurry of progress proved to be premature, as momentum quickly ceased due to congressional squabbles over the project's constitutionality. It ultimately took three more years before actual road construction began.

Aside from the initial $30,000 allotted in 1806 for project planning and surveying, Congress approved meager annual appropriations averaging just $48,000 in each of the first three years between 1810 and 1812. Despite a chronic scarcity of funds that slowed the overall rate of progress, construction commenced on the initial ten miles of highway west of Cumberland in the spring of 1811. Most of this work was done by the late fall of 1812. Progress on the next eleven miles began in the summer of 1812 and did not finish until early 1815. By the fall of 1815, construction contracts were let on the first forty-five miles of highway from Cumberland to a point six miles west of Smithfield, Pennsylvania. (Searight, *Old Pike*, 319-20)

Work continued swiftly after 1816. All contracts were let in Pennsylvania to the (West) Virginia line by the spring of 1817, with the exception of a segment from Washington to Brownsville, Pennsylvania. U.S. mail service between Wheeling and Washington, D.C. began in 1818, although parts of the road around Wheeling still lacked a finished topcoat of stone. The final contract in (West) Virginia was let in 1820, with all construction subsequently completed to the Ohio River the next year. (Searight, *Old Pike*, 319-20) In total, the 131-mile road took slightly over 10 years to build at a cost of $1.7 million, an average of around $13,000 per mile--more than double the original estimate. (Searight, *Old Pike*, 319-20)

General financial and contractual management for the project rested with the Treasury Department under its secretary, Albert Gallatin. From his office in Washington, Gallatin maintained close contact with the on-site construction superintendent responsible for letting contracts, disbursing project funds, and providing on-site engineering and construction supervision. Meanwhile, from his base of operations at Cumberland, the superintendent carried out his assigned duties and responsibilities that included originating work orders, arranging for the payment of drafts and vouchers, and maintaining a steady correspondence with the treasury secretary on all matters of importance. He let contracts, supervised construction, and insured the successful completion of all contractual obligations, requisitioned supplies and outfitted workers with tools and materials as needed. He also responded to a variety of unforeseen circumstances, such as disgruntled workers who grumbled about their pay, spring freshets that swept away survey markers, and contractors that complained of work being too difficult to complete

Albert Gallatin chose David L. Shriver, Jr. of Cumberland as first superintendent of construction. An unheralded figure that is scarcely mentioned in most published works on the Cumberland Road, Shriver deserves to be remembered for his dedication to the task at hand. Moreover, he merits a rightful place alongside Thomas Jefferson, Albert Gallatin, and Henry Clay as one of the foremost individuals responsible for the project's overall success. Shriver began his professional career in partnership with his brother to improve their family property. He first served as a member of the Maryland House of Delegates from Frederick County before giving up his political aspirations to become superintendent of the Reistertown Turnpike. After successfully completing that project, Shriver received his appointment as superintendent of the Cumberland Road, a post that he held until being selected in 1820 as one of three commissioners to lay out the western extension of the road beyond the Ohio River. (Shriver, *History of Shriver Family*, 115)

Shriver served as the lone superintendent on the National Road until 1816, when it became necessary to divide the project administratively into an eastern and western division with the boundary line near the Monongahela River just east of Brownsville, Pennsylvania. (Searight, *Old Pike*, 319-20) Josias Thompson, a Virginian and the surveyor on the original expedition to lay out the route, became Shriver's counterpart in the west.

Competition among prospective bidders who wished to build a section of road proved to be intense. Would-be contractors submitted bids for work advertised in local papers and on posters displayed in public places. A typical bid included government-mandated specifications for the prospective work, material samples (such as the types of stone and mortar to be used), and total price based on a "per perch" construction rate, with one perch equaling 16½ feet in length. (Morse & Green, *Searight's "Old Pike*," 66) Once awarded a job, contractors often followed their own plan for completing the work, a situation which sometimes led to quarrels with surveyors over acceptable angles and grades. Since most contractors were builders and not engineers, they cared little for the potential effect of a one or two degree difference in grade. But, such a deceivingly slight change could have a huge impact on a team of horses straining to pull a heavy load over the mountains. Other potential problems could arise when contractors decided to sublet their work to subcontractors or even to sub-subcontractors, thus creating a potential recipe for disaster that could lead to major problems of accountability.

Burly axmen began the construction process by felling all the trees within the federally mandated 66-foot right-of-way. Next came the choppers, grubbers, and burners whose job it was to remove and dispose of trunks, snags, and debris. Roots were grubbed out by hand, while horses and oxen strained against heavy chains in the attempt to pull stumps out of the ground. Such work might take weeks to complete in areas heavily forested with thick primeval growth. After clearing and grubbing out the right-of-way, the roadbed had to be prepared by a pick-and-shovel wielding army who cut down small hills, filled low-lying hollows, and hauled away excess earth and rock. (Jordan, *National Road*, 84) With all advance preparation complete, the graders, stone crushers, and pavers descended on the area to lay a bed for America's road.

Frenzied construction activity dominated the local landscape, as area residents often turned out in large numbers to view the work for themselves. Because a number of contracts were simultaneously in progress at any one time, it is difficult to determine exact completion dates for individual sections of roadway. To the casual observer construction probably appeared to progress haphazardly, and yet contracts were let systematically for specific "sections" of roadway averaging a couple of miles or more in length. One Pennsylvania farmer observed an army of workers descending "a thousand strong, with their carts, wheel-barrows, picks, shovels, and blasting tools, grading those commons, and climbing the mountainside, leaving behind them a roadway good enough for an emperor to travel over." (Ierley, *Traveling the Road*, 48)

Work orders emanated from the division office under the direction of the superintendent, who was aided by as many as three assistants, one inspector with an assistant, a superintendent of masonry, and several principal overseers. Meanwhile, two office clerks handled paperwork and filled out all necessary forms and requisitions. (*U.S. Senate Documents*, 173) Overseers kept records of hours worked by laborers and owners of teams, while contractors and workers received payment monthly for work completed. (*U.S. House of Representatives Documents*, 24th Congress, 27) Government reimbursements for miscellaneous purchases were made upon presentation of a bill of sale, while reconciliation of accounts and disbursements were prepared monthly. The superintendent kept meticulous records of expenditures for inclusion in his annual report to Congress, as well as to provide a check to minimize the potential for fraud or embezzlement.

Most of the Cumberland Road was built under the contract system, despite the fact that Superintendent David Shriver cited inadequate pay to laborers and the potential ruin of the road as reasons to abandon it in favor of paying workers a daily rate. An 1813 labor shortage even brought about a plan for using slaves, which never came to pass. (Jordan, *National Road*, 87-88) Labor crews consisted almost entirely of Irish and English immigrants, many of which settled in the northern panhandle of (West) Virginia and western Pennsylvania. Construction camps consisting of roadside shacks and tents echoed at night with the sounds of traditional British ballads. Little is known of the plight of individual workers, but one observer near Wheeling noted that the immigrant workers seemed "well fed, well clothed and comfortable," adding that "the Irish here have not lost in our esteem; two or three times we have been beholden to individuals of that nation for good-natured little services....I heartily return them the good wishes they so frequently expressed as we passed them." (Ierley, *Traveling the Road*, 54-55)

In its authorizing language for the Cumberland Road, the U.S. Congress specified a roadway of raised stone, earth, or gravel and sand measuring four rods (66 feet) of cleared width, with a maximum grade of five degrees (8.75 per cent). An 1811 contract specified that all stumps should be grubbed out and the roadbed leveled to 30 feet in width. Fill, where needed, would include the proper allowance for settling--no stumps, logs, or wood could be used. Ditches or "water courses" were to be constructed, if not already present. (*American State Papers*, 175-77)

The 66-foot wide roadway set a standard that is surprisingly close to modern specifications of 72 feet for an urban four-lane controlled-access freeway. Also, the 8.75 percent maximum grade was amazingly stringent through such steep grades and narrow valleys, especially since modern construction standards allow for maximum grades of up to eight percent on certain rural primary routes in hilly or mountainous terrain. (Merritt, *Standard Handbook*, 16 14, 16-15) An 1818 survey disclosed only one location that failed to meet the accepted standard, that being along a stretch of highway in the mountainous regions between Cumberland and Uniontown that measured 9.2 percent, or 5.25 degrees. (Ierley, *Traveling the Road*, 41)

While the surveyors, engineers, and laborers who laid out and built the road certainly deserve credit for their outstanding contributions, the roadbed itself did not exactly embody any cutting-edge technology. Caspar Wever, superintendent of the

National Road's western extension, described the original roadway east of the Ohio River as being little more than a prepared "bed, or channel" of the prescribed width about one foot deep, inside of which was a hand-placed base of stone about one foot in height. He observed that another layer of stone laid above the first formed a fairly even surface, and that the large base stones were set edgewise and appeared to be laid "promiscuously," with attention paid only to balancing the thickness of the pavement and firmness of the surface. Furthermore, he noted that the substrata was covered by a six-inch layer of stone, broken by hammer-wielding laborers and sized by hand to pass through a three-inch ring. (*U.S. House of Representatives Documents*, 20th Congress, 2-3) Treasury Secretary Albert Gallatin confirmed this construction method in a letter in which he describes the road as "covered with a stratum of stones twelve inches thick, all the stones to pass through a three-inch ring." (Searight, *Old Pike*, 374)

According to 19th century road builders, two different iron hammers should be employed for breaking stones into their correct sizes, with the smaller stones for the upper strata made to pass through a 3" iron ring [Figure 4]. Workers would then use a pronged fork "instead of a shovel for taking up the stones to throw upon the road. The advantages attending its use are, that a man can take up the stones much quicker and easier than with a shovel, and free from all dirt and extraneous matter, which, in the case of broken angular stones, is of importance. (Law, *Common Roads*, 118-20)

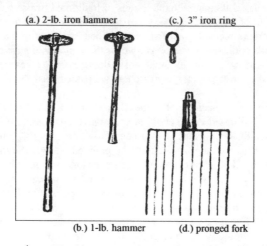

(a.) 2-lb. iron hammer   (c.) 3" iron ring

(b.) 1-lb. hammer   (d.) pronged fork

[Figure 3] Typical 19th century road-building tools: (a.) 2-lb. iron hammer for breaking large stones for the bottom strata of the roadbed, (b. smaller 1-lb. iron hammer for breaking smaller stones for the upper strata of the roadbed, (c.) 3" iron ring to size broken stones for upper strata, (3) pronged fork to spread stones on road. (drawn by Billy Joe Peyton from sketch in Henry Law, *Rudimentary Art of Constructing and Repairing Common Roads*, 119)

While the U.S. Treasury Department did a commendable job of handling general project administration, the agency as a whole had no demonstrated expertise in the technical aspects of highway engineering or construction. Within the ranks of the

Treasury Department, only David Shriver had first-hand knowledge of modern road and bridge building techniques. Given the multitude of responsibilities that fell on his shoulders, he no doubt had precious little time to devote to engineering problems. Perhaps this explains some of the problems with the roadbed as observed by Caspar Wever. More importantly, it gives rise to speculation as to why the federal government did not provide more professional engineering expertise at the outset. A related question is why did the U.S. Army Corps of Engineers, the nation's congressionally-mandated engineering agency, not participate in either planning or building the National Road? The answer is quite simple. During the initial construction (1811-1821), the Corps of Engineers had no federal authority to participate in civil works projects of any type. Limited strictly to military engineering projects, the agency did not gain the authority to work in the civil sector until congressional passage of the General Survey Act of April 30, 1824. The act authorized the president to use Army engineers to survey road and canal routes "of national importance, in a commercial or military point of view," and employ engineers "in the public service which he deemed proper." (USACE, *History of Corps of Engineers*, 37, 39)

Once granted approval to perform civilian work, the Corps of Engineers wasted little time getting involved. After assuming full construction management of the project in 1825, the agency directed the government's extension of the route west of the Ohio River and coordinated a complete overhaul of the original roadbed in Maryland, Pennsylvania, and (West) Virginia prior to turning ownership of the route over to those states. (USACE, *History of Corps of Engineers*, 37)

What Style Is It?

Two commonly accepted construction technologies could have been used to build the Cumberland Road. The first was developed in the 1750s by French engineer Pierre-Marie Jerome Tresaguet, who believed that good drainage and proper selection of materials to be essential ingredients for success. His specifications called for a crowned subsurface and stone pavement of a uniform depth throughout, with a bottom layer of stone seven inches thick laid on end and hammered in place. On top of this he placed a layer of hammered-in smaller stones; the surface, which in profile had a central crest with each side sloping down to the edges, consisted of small hard stones. Finally, he separated the substrata from the wearing surface, as opposed to forming one uniform foundation. Tresaguet's system eventually spread to Central Europe, Switzerland, and Sweden after 1760. (Singer, *History of Technology*, 527-28)

The second system in common use at the time was named for English engineer Thomas Telford, a former stone mason who developed his method around 1802. He advocated a level subsurface set edgewise and fit tightly by hand, with irregularities broken off by a hammer and the interstices filled with stone chips. Smaller stones were added on top to a depth of seven inches at center and sloping to three inches at the sides. Horses compacted the small stones placed at the center, and then one-and-a-half inches of clean gravel covered the whole road. (Kirby, *Engineering in History*, 200-04)

Tresaguet influenced Telford's work to a great extent, but their systems differed considerably in cost. Telfordization typically cost more because of its heavy foundation, extra expense of placing several stone layers, and the fact that top layers of pavement wore out quickly if not maintained in good order. As a result, Telford's techniques gained popularity in Britain, but they were not economically feasible in the United States where distances were greater and money and labor more scarce.

Aside from these two popular construction styles, a third system that developed in the early 1800s would eventually overshadow both Tresaguet and Telford. This technique was contrived by James Loudoun MacAdam, a Scottish engineer and road surveyor. Inspired by Tresaguet, MacAdam first improved on the Telford system by using small broken granite pebbles as his primary material. MacAdam firmly believed that a stone surface did not function to support vehicular weight, but rather formed a covering over the natural soil to support the road's weight. He deemed a heavy metal (broken rock) course as unnecessary, and the porosity of large stones proved unacceptable. MacAdam felt thickness of the road to be no consequence, but "if water passes through a road and fills the native soil, the road, whatever its thickness, loses its support and goes to pieces." (Singer, *History of Technology*, 534) He also believed in a flat foundation, which contrasted with the crowned surface of the Tresaguet system. Because MacAdam did not have rigid rules for construction, he varied the number of necessary layers. An imperviousness to water was paramount, inasmuch as moisture destroyed a road's weight-carrying capability if allowed to penetrate to the soil beneath. (Singer, *History of Technology*, 534)

MacAdam enforced the use of a two-inch ring for sizing stones that he claimed should not exceed six ounces in weight. To him, every stone over an inch in size was "mischievous" because it increased permeability and might be easily dislodged by wheels. He did not roll his roads, but preferred applying stones in thin layers and allowing traffic to compact each successive application. He also sought to keep the road as level as possible to avoid sharp ascents and descents detrimental to proper drainage. Unfortunately for builders of the Cumberland Road, MacAdam's techniques gained recognition as a durable and inexpensive road surface only after the 1816 publication of his seminal work titled *Remarks on the Present System of Road-Making*. (Turner & Goulden, *Great Engineers & Pioneers*, 329-30)

A commonly held myth is that the Cumberland Road originally had a macadamized surface--it did not. In truth, MacAdam's principles did not spread to North America until 1823 when the Boonsboro Pike, a feeder highway to the Cumberland Road in Maryland, had repairs done on the macadam plan. (Rose, *Historic American Roads*, 36) Although the National Road eventually did receive a macadam surface as a result of government-funded repairs in the late 1820s, the original construction techniques appear to be something of a regional variation on the Tresaguet system, except in certain places containing a single layer of broken stone pavement [Figure 4]. (*U.S. House of Representatives Documents*, 20th Congress, 2; Highway Administration, *America's Highways*, 20) While this system looked acceptable at the surface, it proved to be little more than a ditch filled with rocks of varying sizes that wore out quickly under heavy traffic.

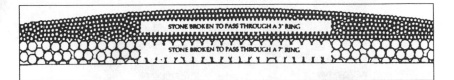

[Figure 4] Cross-section of the original Cumberland Road. Built as a hybrid of sorts, with a 7" base layer and a 3" top layer that was cambered to allow for drainage, it most closely resembled a variation of the Tresaguet plan. (from a drawing by John Hriblan, WVU IHTIA)

An Abundance of Superb Stone Bridges

As construction moved westward through the Maryland countryside, the issue of carrying the route over rivers and streams became vitally important. The first major watercourse to be bridged was the Little Youghiogheny (or Casselman) River near Grantsville, Maryland. In response to government-advertised bids, Abraham Kerns and John Bryson collaborated in August of 1813 to submit the winning proposal for "a stone Bridge across the Little Youghagany (sic) River, on the United States Western Road for the sum of nine thousand dollars...we will build the Bridge in a completed and workmanlike manner so that it will be substantial, permanent and neat in all its parts." Under their articles of agreement, Kerns and Bryson would build their bridge for $9,000, of which $4,000 would be paid when the span was half-finished. The contractors promised to "find at our own expence (sic) all materials lime excepted, and dig the foundations, so that the total cost to the United States, will not exceed the sum above mentioned." (Correspondence, Kerns & Bryson to Shriver, 2 August 1813)

Construction at Little Crossing began without delay in 1813, but incessant rains the next summer led to extended work stoppages and prolonged problems with the bridge foundation. Already behind schedule, the inclement weather worsened the contractors' shaky financial situation and left them short of much-needed cash with which to buy materials and pay their laborers. Kerns & Bryson eventually overcame their financial woes and completed their contract in June of 1815. All work on the bridge, including the rubble fill, was completed in 1816.

On November 16, 1814, the contractors confidently reported their success "in turning the largest and we think the most permanent stone arch in the United States." (Correspondence, Kerns & Bryson to Shriver, 17 November 1814) Theirs was no idle boast, for they had erected the largest masonry arch in the nation. Semi-circular in design, measuring 80 feet in diameter and three feet thick, it was the most prominent feature of Little Crossings Bridge--a marvelous structure over 300 feet long, 50 feet high, backed with 100-foot wing walls [Figure 5 & Figure 6]. Replete with its delicate arch, massive cut stone exterior, prominent wing walls, and locally quarried rubble fill, the bridge still stands as a monument to its builders. Located north of I-68 and east of Grantsville, Maryland, it is both a National ASCE Landmark and National Historic Landmark. (IHTIA & NPS, *HABS/HAER Documentation Casselman Bridge*)

[Figure 5 & Figure 6]   Two views of Little Crossings Bridge over the Little Youghiogheny (Casselman) River in Western Maryland. The prominent 80-foot stone arch was the largest ever constructed in the U.S. at the time. (photos by John T. Nicely, WVU IHTIA)

About 12 miles west of Little Crossings the Cumberland Road enters Pennsylvania in Somerset County, east of the village of Addison. In this vicinity the highway generally followed existing ridges without the need for great cuts or fills. West of Addison the road left the ridges and wound its way down to the Big Youghiogheny River, site of the greatest stone arch bridge on the road.   When Superintendent David Shriver viewed the crossing site from the banks of the river in June of 1815, he paid careful attention to the stream width and closely examined the ground before determining possible bridge pier locations. Based on his observations he proposed two bridge plans: one for a span finished in hewn stone with three arches measuring 90, 75, and 60 feet, and another for a four-arch configuration of 80, 66, 53, and 41 feet, respectively. (Correspondence, Shriver to Dallas, 30 June 1815)

Officials in Washington questioned several points about Shriver's proposal, expressing concern for the stability of the three-arch plan and particularly questioning whether an arch of 90 feet would be prone to settle, or worse, total collapse. These fears were not altogether unfounded, as no U.S. engineer had ever attempted to build a masonry arch of that size.   In fact, the largest standing arch spanning 80 feet in length had just been erected over the Little Youghiogheny River, and skeptics feared that it might someday collapse under its own weight. "After maturely considering" all possibilities, Shriver ultimately made the final decision to erect a three-span stone arch bridge at Great Crossings. (Correspondence, Dallas to Shriver, 11 July 1815)

Among the contractors interested in building the Great Crossings bridge was the firm of James Kinkead, James Beck, and Evan Evans, who joined forces in a three-man partnership for the express purpose of bidding on the project. Both Beck and Evans were stonemasons of some repute, while Kinkead had considerable bridge building experience. Their bid turned out to be the winner and they were promptly awarded a contract to build the specified structure of sandstone. Under the articles of agreement, which they signed on September 5, 1815, the contractors specified their intention to erect a masonry bridge with three spans measuring 90, 75, and 66 feet in the clear for the fixed price of $40,000. They also agreed to supply all construction materials at their own expense, except for lime to be furnished by the U.S. government. (Articles of Agreement, 15 September 1815) Payment would be in four equal installments of $10,000, due on the following schedule: (1) when President James Madison approved the contract; (2) completion of work to the height of the spring of the arches, with centers completely raised and work approved; (3) completion of the whole of the arches, with the centers slackened and work approved; (4) upon completion of the bridge and final approval. (Articles of Agreement, 15 September 1815)

Work at Great Crossings began in September of 1815 and continued steadily until the following summer, when wet weather retarded the firm's progress. Despite some minor setbacks, the contractors continued to provide all necessary materials for erection of the west abutment and piers. Kinkead, Beck, and Evans ultimately completed their master work in 1817, at which point the government let a new contract for filling the bridge with rubble stone. When it opened for traffic, the Great Crossings bridge at Somerfield, Pennsylvania, became the largest masonry arch span in the United States. (Searight, *Old Pike* 219) Adorned with handsome detailed parapets reminiscent of a medieval fortress, it embodied state-of-the-art bridge-building technology and displayed the finest in masonry construction [Figure 7 & Figure 8]. And yet, these magnificent bridges contradict the fact that the roadbed was of inferior construction. It remains a mystery as to why the government elected to spend top dollar for its bridges while at the same time cut corners on the quality of the roadbed itself. Whatever the motive, the superb bridges were showpieces that rivaled the finest structures being built in Europe at the time.

Nineteenth century travelers often stopped to admire the impressive structures, especially the fine specimens over the Youghiogheny and Little Youghiogheny rivers. One typical observer was Uria Brown, a surveyor and conveyancer from Baltimore who in 1816 wrote that "the Bridges & Culverts actually do Credit to the Executors of the same, the Bridge over the Little Crossings of the Little Youghegany [sic] river is positively a Superb Bridge." As he reached the Great Crossings he astutely observed that "they have Commenced the erection of the Bridge over this River, [and] no doubt from the specimen of the work already on the road, but this Bridge will be a superb & Magnificent Building." (Ierley, *Traveling the Road*, 49-50)

[Figure 7 & Figure 8] Two views of Great Crossings Bridge before its inundation by the waters of Youghiogheny Reservoir in 1943. (photos from Charles Morse Stoltz, *The Architectural Heritage of Early Western Pennsylvania: A Record of Building Before 1860*, Pittsburgh: University of Pittsburgh Press, 1936)

It is not surprising that Great Crossings Bridge inspired awe in travelers, for it was one of the finest masonry arch bridges of the early 19th century. For nearly 125 years it faithfully carried the National Road up and over the river at Somerfield, Pennsylvania, until the construction of nearby Youghiogheny Dam in 1943 resulted in its closure and the subsequent destruction of the town.   While all vestiges of Somerfield's buildings are long gone, the bridge still survives under the deep waters of Youghiogheny Reservoir, a flood-control impoundment created when the U.S. Army Corps of Engineers built the nearby dam.   Six decades of submersion has certainly taken its toll on the structure, but it remains surprisingly intact and discernible.   On rare occasions when exceptionally dry conditions result in extremely low water levels (typically in the fall), it emerges like a creature of the deep from beneath the waters of the lake [Figures 9, 10, 11].   This rare and unusual spectacle has occurred twice in recent memory--in 1991 and again in 1998. (Photodocumentation by IHTIA)

[Figures 9, 10, 11] Great Crossings Bridge as it appeared in the fall of 1998 after a severe summer drought left the reservoir at its third-lowest recorded level. (photos by John T. Nicely, WVU IHTIA)

In addition to the two exemplars at Big and Little Crossings, contractors built numerous other exceptional stone arch bridges along the Cumberland Road. Notable survivors include the triple-arch stone structure over Wheeling Creek at Elm Grove, West Virginia, and the unusual "S-Bridge" at Taylorstown, Pennsylvania. In addition, numerous smaller single-span arch bridges and culverts that once carried the Cumberland Road to its terminus on the Ohio River still exist, either intact or in ruin [Figure 12].

[Figure 12] Still in use, this small single-span masonry culvert is located east of Little Crossings Bridge near Grantsville, Maryland. (photo by John T. Nicely, WVU IHTIA)

Rediscovering the Road

All across the United States, highways slowly deteriorated into a deplorable condition by the late 1800s, escorted into their Dark Ages by the railroads. Likewise, the National Road had fallen into a sorry state by then, too, with most of its traffic gone and grass growing up through its pavement. In 1879, *Harper's Weekly* reported that "the old iron gates have been despoiled, but the uniform toll-houses, the splendid bridges, and the iron distance posts show how ample the equipment was." (*Harper's Weekly*, 816) The route languished in relative obscurity for a half-century.

The first call to repair our nation's thoroughfares came in the 1880s through the efforts of grassroots organizations such as the League of American Wheelmen, a dedicated group of bicyclists that allied itself with farmers who needed decent roads to carry their produce from farm to market. From this fledgling effort a national Good Roads Movement formed with the mission to "pull America out of the mud." This movement helped to galvanize popular support in the cities, primarily through civic leaders who realized that good roads were only possible with adequate funds raised through the taxation of urban and rural property. (Rose, *Historic American Roads*, 73; Highway Administration, *America's Highways*, 41)

In 1911, U.S. Congressman Albert Douglas decided to drive his automobile from his office in Washington, D.C. all the way home to Chillicothe, Ohio. Along the National Road he experienced an excruciating four-day journey in his Ford that sputtered up and over the mountains of Maryland, Pennsylvania, and West Virginia. On day four he crossed the Wheeling Suspension Bridge and entered "the promised land" of Ohio. Commenting on the voyage after reaching his destination, Douglas announced that highway conditions ranged from good to horrible. (Jordan, *National Road*, 385) His intrepid adventure not only brought increased attention to the need for uniform and reliable long distance roads in the United States, it demonstrated the growing preference among Americans for driving automobiles instead of taking trains.

Echoing public sentiment of a century earlier, citizens began to demand federal aid for better roads. A Post Office Appropriation Bill of 1912 made it possible for federal funding to be used for "post-road" improvements. (Rose, *Historic American Roads*, 87) After World War I, highway travel in the United States increased rapidly and the federal government found it necessary to adopt a national transportation policy. The result was passage of the Federal Highway Act of 1921, intended to encourage states to build connector highways that were "interstate in character." National Road revitalization became a reality under this legislation.

Given this blossoming love affair with the auto, interest in a series of transcontinental highways quickly developed. First in 1913 came the Lincoln Highway (U.S. Route 30), followed by several other "national trail" associations that incorporated parts of historic highways into their alignments. Among those being promoted or in existence by 1920 were the Dixie Highway, Pikes Peak Ocean-to-Ocean Highway, Yellowstone Trail, Midland Trail, and the National Old Trails Road – a transcontinental corridor that included the old National Road right-of-way. (Rose, *Historic American Roads*, 88; Highway Administration, *America's Highways*, 109)

A National Old Trails Association had already formed when Indiana Congressman Henry A. Barnhart introduced a bill to continue the transcontinental highway through Ohio, Indiana, Illinois, and Missouri. Reminiscent of the original 1806 act to lay out the Cumberland Road, the bill authorized three national highway commissioners to be appointed by the president. (Jordan, *National Road*, 386) Beginning in the early 1920s, the old National Road underwent a complete upgrade. From Cumberland to St. Louis, the old stone pavement began to be replaced by a ribbon of concrete nine inches thick and eighteen feet wide. Brick paving also became popular for a time, and in some places it replaced concrete or stone surfacing. (Reedy interview) Meanwhile, Congress authorized the continuation of the National Old Trails Road to the Pacific Ocean.

A uniform highway numbering system introduced in the 1930s ultimately led to its designation as U.S. Route 40. Erection of standardized U.S. highway shields shortly thereafter made it official--the Cumberland Road had come full circle as part of a new national highway stretching from Atlantic City, New Jersey, to San Francisco, California. Route 40 served as one of the nation's arterial corridors for nearly four decades until it, too, was bypassed by a newer, faster, and safer route. It is no coincidence that its successor, Interstate 70, shadows the footprint of the original Cumberland Road for hundreds of miles on its way west. In this modern world of high-tech solutions to many engineering questions it is gratifying to know that much of the original alignment, laid out by Elie Williams, Thomas Moore, and Joseph Kerr and built by English and Irish laborers almost 200 years ago, remains the route of choice today.

References

American State Papers (1834). Gales and Seaton, Washington, D.C.

Bacon-Foster, Cora (1912). "Early Chapters in the Development of the Potomac Route to the West." Proceedings of the Columbia Historical Society, Washington, D.C.

"Cumberland Road" (19 December 1805). 9th Congress, 1st session, *American State Papers*, Transportation, volume 2, Wilmington, Delaware: Scholarly Resources, Inc., 1972.

"Cumberland Road" (30 August 1808). *American State Papers*, Miscellaneous, volume I, 10th Congress, 2nd session, No. 258.

Gallatin, Albert. (4 April 1808). "Report on Roads and Canals," 10th Congress, 1st session, *The New American State Papers*, Transportation, volume 1, General, Scholarly Resources, Inc., Wilmington, Delaware, 1972.

Harper's Weekly Magazine (November 1879).

Ierley, Merritt (1990). *Traveling the National Road: Across the Centuries on America's First Highway*, Overlook Press, Woodstock, New York.

Jordan, Philip D. (1948). *The National Road*, The Bobbs-Merrill Company, Indianapolis.

Kirby, Richard S., et al. (1956). *Engineering in History*, McGraw-Hill, New York.

Law, Henry (1850). *Rudimentary Art of Constructing and Repairing Common Roads*, John Weale, London.

Merritt, Frederick S. (1983). *Standard Handbook for Civil Engineers*, McGraw-Hill, New York.

Message from the President of the United States Transmitting A Report of the Commissioners (31 January 1807). A. & G. Way, Printers, City of Washington.

Morse, Joseph E. & H. Duff Green (1971). *Thomas B. Searight's "The Old Pike,"* Green Tree Press, Orange, Virginia.

National Archives and Records Administration (1806). *General Journal Kept During Examination of Route for Cumberland Road by the Commissioners in 1806* (10 October-13 November 1806), Record Group 77, entry 179, box 5, Washington, D.C.

National Archives and Records Administration (1813). Correspondence, Abraham Kerns & John Bryson to David Shriver (2 August 1813). RG 77, entry 179, box 1, Washington, D.C.

National Archives and Records Administration (1814). Correspondence, Abraham Kerns & John Bryson to David Shriver (17 November 1814). RG 77, entry 179, box 1, Washington, D.C.

National Archives and Records Administration (1815). Correspondence, David Shriver to A.J. Dallas (30 June 1815). RG 77, entry 179, box 1, Washington, D.C.

National Archives and Records Administration (1815). Correspondence, A.J. Dallas to David Shriver (11 July 1815). RG 77, entry 179, box 1, Washington, D.C.

National Archives and Records Administration (1815). Articles of Agreement for Bridge at Great Crossings (15 September 1815). RG 77, entry 179, box 2, Washington, D.C.

Public Statutes at Large of the United States of America (1846). volume 2, Little & Brown, Boston.

Reedy, Ralph, former road worker, personal interview (1993) by Dan Reedy, Livingston, Illinois.

Rose, Albert C. (1976). *Historic American Roads: From Frontier Trails to Superhighways*, Crown Publishers, New York.

Searight, Thomas (1894) *The Old Pike: A History of the National Road*, by the author, Uniontown, Pennsylvania.

Shriver, Samuel S. (1889). *History of the Shriver Family and Their Connections, 1684-1888*, Press of Guggenheim, Weil & Company, Baltimore.

Singer, Charles, et al. (1958). *A History of Technology*, Volume 4, *The Industrial Revolution*, Oxford University Press, New York.

Turner, Roland T. and Steven L. Goulden, editors (1981). *Great Engineers and Pioneers in Technology*, St. Martin's Press, New York.

U.S. Army Corps of Engineers (1986). *The History of the U.S. Army Corps of Engineers*, U.S. Government Printing Office, Washington, D.C.

U.S. Census Bureau (1790-1810). *U.S. Census Statistics*.

U.S. Congress (1808). *American State Papers*, 10th Congress, 2nd session.

U.S. Department of Transportation, Federal Highway Administration (1976). *America's Highways, 1776-1976: A History of the Federal Aid Program*, U.S. Government Printing Office, Washington, D.C.

U.S. House of Representatives (1816). 20th Congress, 2nd session, *U.S. House of Representatives Documents*, No. 185.

U.S. House of Representatives, (1824). 24th Congress, 2nd session, *U.S. House of Representatives Documents*, No. 302.

U.S. Senate (1822), 23rd Congress, 2nd session, *U.S. Senate Documents*, No. 266.

West Virginia University Institute for the History of Technology & Industrial Archaeology and National Park Service Historic American Engineering Record (1994). *HABS/HAER Documentation of the Casselman Bridge*, Morgantown, West Virginia, IHTIA.

West Virginia University Institute for the History of Technology & Industrial Archaeology (1991 & 1998). Photodocumentation of Exposed Great Crossings Bridge, WVU IHTIA, 1535 Mileground Road, Morgantown, W.Va., 26506.

Williams, Elie, et al., "Cumberland Road, Report of the Commissioners" (15 January 1808). *The New American State Papers*, Transportation, volume 1, General, Scholarly Resources, Inc., Wilmington, Delaware, 1972.

The Baltimore and Ohio Railroad and the
Origins of American Civil Engineering

John P. Hankey

Abstract

The Baltimore and Ohio Railroad has long been regarded as the first "modern" railroad in America and an important work of civil engineering, but its true significance has been clouded by 150 years of myth, assertion, and unreliable history. A careful review of the evidence reaffirms the engineering significance of the B&O and its stature as one of America's most complex and innovative antebellum projects. The B&O Railroad demonstrates the profound effect civil engineering has had on our landscape, culture, and development. Assessing the mythology surrounding the company offers a model for how works of civil engineering acquire mythic or iconic status.

Introduction

The American Society of Civil Engineers and the Baltimore and Ohio Railroad share anniversaries in 2002. A century and a half previous, a small group of men formed the ASCE in New York to bring a higher degree of order and professionalism to the burgeoning field. Also in 1852, the B&O Railroad completed its "Main Stem" between tidewater at Baltimore, and the Ohio River at Wheeling in what was then Virginia. Coincidentally, 2002 marks the 175th anniversary of the founding of the B&O as well.

The links between the B&O and the ASCE are significant. The B&O was one of the largest, most complex, and most innovative civil engineering projects of the early nineteenth century. It helped shape the field of American civil engineering during its formative period, and many of its officers and engineers played critical roles in the development of the profession. Three men who established their careers on the B&O and had significant parts in its construction–Albert Fink, Mendes Cohen, and Francis L. Stuart–later served terms as president of the ASCE.

The author can be reached at johnphankey@aol.com.

Occasions like this encourage society at large to reflect on the events which brought it to this point. For historians, they are invitations to revisit old conclusions and apply new perspectives. Unlike engineering, which has at least a few dependable natural laws and scientific principles, the humanities exist in a constant state of flux and reinterpretation. Historians regard the idea of "historical truth" or immutable "fact" with great suspicion. History is not reproducible, verifiable, reducible to formula, or even tangible (although its artifacts may be). It is a discipline rooted firmly in human nature and perception, and for that reason is subject to continual revision.

Anniversaries also are one way we have of keeping score. Surviving a century and a half of change and turmoil is irrefutable evidence that an organization has been successful in the past, and that it is agile enough to have a future. The same is true for civil engineering works. By attaining significant anniversaries, they acquire the kind of status and cultural importance which enhances the odds for their survival. Perhaps that is the point: Anniversaries are cultural events which add layers of meaning to human endeavors such as engineering, and the physical artifacts they produce.

More broadly, revisiting the first half-century of the B&O's existence offers the opportunity for deeper understanding of the place civil engineering occupies in American culture. The issues range from the obvious–the role of this one railroad in the development of antebellum engineering–to more nuanced, and perhaps more universal, insights into the relationship between technology and American life. Anniversaries for their own sake have little utility, but those which prompt thoughtful reflection and new insights repay the effort many times over.

This paper's goals are modest. It derives in part from a larger work documenting the early history of American railroading, and the history of the Baltimore and Ohio in particular. A project such as this must be cross-disciplinary and confront the sheer mass and inertia of two centuries of work of wildly varying quality. It is not unlike trying to refurbish the Brooklyn Bridge and recover its appearance when new. The genuine article is in there somewhere, but it never will be possible to know it completely.

I am aware of the dangers of partisanship, the limitations of the data available, and the latitude of legitimate interpretations. Whatever conclusions ultimately result from this project will themselves be qualified, nuanced, and subject to revision. That is the nature of this kind of history, and why it is so useful to bring these ideas and arguments before the civil engineering community now, while there is ample time for exchange and discussion. Please regard this paper as the beginning of a conversation. I look forward to a collegial exchange of views and constructive critique.

My methodology is straightforward. I will frame a series of questions which, taken together, review the antebellum B&O in reality, the B&O of legend and myth, and the company's roles in the early development of civil engineering. Just as no single question can frame the issues, it will take a series of answers and examples to form a reliable conclusion. A new assessment of this company's relative importance to America culture and civil engineering will emerge step by step, point by point.

Part I: The Questions

Like so much in the past, the seemingly straightforward history of the B&O Railroad is not quite what it appears to be. There are "big answers" to the "big questions"–but they have to be qualified, which complicates the whole process. Was the B&O America's first railroad? Yes, with a few major stipulations. Is it one of America's most important civil engineering projects? Absolutely, but not for the reasons usually cited. Does the company deserve the privileged status it enjoys in American culture? It certainly does, but only when you separate truth from flackery.

Then there are the second order questions, which often lead in unexpected directions. If the B&O came to be regarded as the "most historic" railroad in the country after the Civil War, how did people think of it before that war? How aware was the railroad itself of its role in evolving American culture? Beyond the goal of creating a transport link, what attitudes motivated these men, and complicated their work?

There are questions of sources and methods. Despite adequate primary sources (such as corporate minute books and diaries), documentary evidence is thin. Some conclusions will have to be read from non-traditional kinds of evidence. Attempting a synthesis is further complicated by the fact that the railroad was new to most people. We must avoid imposing our modern ways of thinking on the antebellum period. People of that time understood their past and future very differently than we, and their present was as discontinuous, fast moving, and disorienting as ours seems today.

Finally, there is opportunity. Residing in the more modest history of the B&O are broader insights and useful lessons about the emergence of engineering in a distinctively American format. Rather than avoid them, it seems more honest and useful to make a few broad conclusions about the place of technology in antebellum America, with the hope that it sparks further discussion and perhaps new conclusions.

Defining the Parameters. Any coherent assessment of the B&O has to engage business history, economics, historical geography, political and social history, and a half-dozen other fields. Because of our tendency towards segregation and focus, it would be tempting to restrict this analysis to a narrow conceptualization of civil engineering: The permanent way, bridges and earthwork, tunnels, water systems, and other structures. That would also miss the point completely.

For our purposes, the B&O Railroad was a single, integrated work of American civil engineering. It was, after all, a solution to what was essentially a civil engineering issue–the lack of reliable transportation infrastructure. We will not quibble as to which part of the project belongs to someone else, especially because antebellum engineers themselves had not quite figured out what their discipline comprised.

The B&O Railroad also is a cultural artifact. That is true of engineering in the same way as for fine art, language, and any other product of human endeavor. Too often technical disciplines lose sight of the fact that they arise from, and are shaped by, highly contingent cultural factors. Civil engineering is important in the United States because we, as a culture, embrace technology in particular ways and have certain expectations for its results. Those attitudes and values came from a specific time (the Enlightenment) and place (Europe), and do not prevail in most of the world.

Culture is a self-reflexive process, of course. The works of engineering and technology that we, as a society, create subsequently reshape our culture. Developing innovative works of engineering or new technology requires a convergence of need, willingness, ability, means, and risk taking. Once proven successful, they have the power to alter society and culture, which then further adapts its technology, and so on.

Because the B&O and other early railroads had their origins in the flux of antebellum America, that kind of give-and-take is especially apparent. These railroads arose from a distinct set of needs and conditions informed by the existing American culture. The engineering used to create them, and the physical form that they took, were deeply influenced by the attitudes and expectations and shared beliefs of that time.

Simultaneously, the effects of the railroad–rapid and networked transportation, greater efficiency, time discipline and grand physical structures–worked directly and indirectly to change the culture. At no other point in our history has technology had such a rapid, profound, and lasting effect as in the first six decades of the nineteenth century. In 1800 the US was local, agrarian, and marginally commercial. By 1860 it was becoming industrial, urban, interconnected, and recognizably modern. That, in part, is why the Civil War happened when it did. That war was as much a response to the stresses of industrialization and modernization as to other root causes.

What follows is an engineering history of a American cultural artifact. It also is a cultural history of an important American engineering project. However one defines it, a careful look back suggests that the B&O Railroad played a significant role in America's embrace of technology and our fondness for large scale engineering projects. And that separating engineering from culture is even more difficult than it seems.

Assessing the Antebellum B&O. For nearly a century and a half, the B&O has been represented, and generally accepted, as "America's First Railroad." It nurtured an image as a pioneer, technological wonder, and courageous work of American progress. Throughout its existence, the B&O consistently made history part of its core identity. The company embraced its past and the critical roles it played in American history. As the machinery of cultural production and corporate public relations ground away, a body of legend and myth came to surround the railroad's first four decades.

To a degree almost unique among American corporations, the B&O employed what historian David Houndshell termed an "industrial creation myth"–a set of stories which describe the company in heroic terms, and which represent it as more than just an engineering project. This mythical B&O developed in parallel with the actual B&O, and over time acquired almost independent existence. It was a product of myth-making by the railroad itself, and of cultural production in American society at large.

People with little knowledge of American history or technology came to know of this railroad, if only from the stories reproduced in grade school texts and magazine articles. Generations of children learned about the B&O's first locomotive, the so-called "Tom Thumb," and how it raced a horse and lost. Railroad travelers regarded the B&O as the "historic" route, possessing some intangible advantage over faster or better-located rivals. In all sorts of places–on postage stamps, in advertising, in news photos and stories--the B&O turned up disproportionately.

Two qualities make the B&O's case different from other railroads, and from similar instances in American popular culture. First, there was substantial truth to its claims. The B&O was a complex enterprise cast into the thick of unfolding American history. Second, through a variety of mechanisms, the mythos of the B&O was embellished, refined, legitimated, and made part of the American cultural canon.

Often, B&O history was presented as fable; at other times, the company simply made assertions of priority or importance. Throughout its existence, the B&O freely appropriated all sorts of symbols and meanings. For its emblem, it used the United States Capitol. It claimed connections with George Washington and status as an agent of modern civilization. At world's fairs and public extravaganzas staged over the course of nine decades, millions viewed pageantry and exhibitry portraying the history of the B&O in the same light as the founding of the Republic. Americans understood the company as somehow special or important, even if they couldn't say why. In the language of the historian, the B&O acquired privileged status in American culture.

Figure 1. This woodcut from an 1855 *Ballou's Pictorial Drawing Room Companion* (published weekly in Boston) shows the B&O's first bridge–the Carrollton Viaduct–rendered in the "romantic" or "picturesque" style.

Thus the first, and most basic, questions posed by this anniversary are quite straightforward: Does the B&O Railroad deserve that privileged status? Was it really the first, oldest, most historic, or most significant American railroad? Is the mythology surrounding the B&O merely the product of skillful corporate flackery and a desire to understand American progress in a linear, easily assimilated form? Or is the antebellum B&O in fact a great work of civil engineering, with the added burden of accreted myth?

It would be satisfying, and perhaps even useful, to answer those questions convincingly. Rigorously addressing the mythology would clarify the origins of both the ASCE and the B&O Railroad. If the company was truly noteworthy and established a model for American railroad development, then it deserves greater recognition as a seminal work. But answering the first set begs a more convoluted second set of questions. Where did that inherited tradition come from? Who created it and for what reasons? Why should this matter, especially as the railroad no longer even exists?

We still are left with the task of judging the relative importance of the B&O Railroad in the antebellum period, and what insight it offers into the emergence of civil engineering as a professional discipline. Only now the process is complicated by the need to engage the notoriously disorderly cultural realm, the presence of previously unconsidered actors and agencies, and the opportunity to find larger meaning as a case study. It is not sufficient simply to ask *was* the B&O important, and *why*. It is equally necessary to ask *how*, and *for what reasons* it became legendary and mythic.

Separating the real antebellum B&O from the mythical B&O becomes the initial challenge of any analysis. The first exists in the realm of reality, the second in the realm of perception, myth, and culture. Both had profound impacts on the origins of American civil engineering. The "real" B&O helped establish engineering methods, principles, and professional identity. The other, "mythic" B&O helped form domestic attitudes toward technology and the manipulation of nature, and aided the adoption of an engineering ethos for American development.

Two Railroads. The first of these "two B&Os" obviously is the railroad itself, and the sum of the parts and actions which comprised its transportation functions. We know surprisingly little about how actually it was built and functioned, or the gritty texture of railroading before the Civil War. Still, it is possible to describe the company in terms of its engineering and economic rationale, construction sequence and operations, and then to make tentative conclusions as to what its place in engineering history might be.

The creators of the B&O were supremely practical men who clearly understood this as a massive, speculative construction project. They had to push an undeveloped technology 300 miles through forests, over rivers, and across formidable ridges and valleys. There were precedents to draw upon, but those works were materially different in their challenges and assumptions. This B&O Railroad is a work of engineering, and must be judged on that basis. Did it fulfil its design criteria? Was it successful on its own terms? Could it economically and reliably move goods and passengers?

The second, or "mythic" B&O Railroad derives from the physical reality of the first, but exists in the realm of perception, myth, and culture. This B&O exists independently of time, physical reality, or normal causality. It is what people *thought* the B&O Railroad was, or more accurately, what *meanings, values, and symbolism* they attached to the project. Like any legitimate and effective body of understanding, it had consistency, persistence, and qualities which made it self-replicating. What in one realm were structures and civil engineering works, in the other became cultural monuments, historic places, and artifacts of American history.

This mythic B&O did not serve a transportation function, but it had utility nonetheless. The stories deeply informed the organizational culture of the company and manifested themselves in ways as diverse as employee loyalty and the naming of specific locations along the railroad. The mythic B&O influenced decision making by external audiences such as politicians, bankers, and railroad customers.

The real B&O was subject to continual improvement, reconstruction, and reinvention as the technology changed and the functions of the railroad broadened. For different, but equally compelling, reasons, the mythical B&O likewise evolved and grew. Some of that change was strategic and intentional, some wholly unanticipated. Each successive generation had to be educated as to this mythical B&O, but on a property ripe with history and heritage, there were ample opportunities to do so.

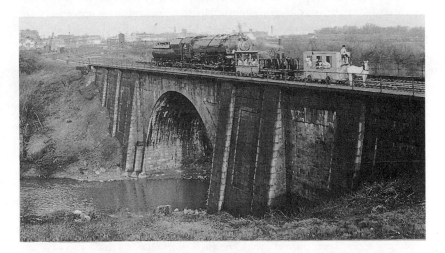

Figure 2. Literally from its opening in 1829, the Carrollton Viaduct served as a prop for special events. For the viaduct's centennial in 1929, the B&O posed a "replica" horse car, the "operating replica" of the "Tom Thumb," and its largest passenger locomotive, 4-8-2 No.5500, the "Lord Baltimore."

Part II: The Antebellum B&O In Reality

To assess whether the B&O was significant, and perhaps even monumental, requires a few parameters. There are no hard definitions or strict standards, although the ASCE National Historic Civil Engineering Landmark designation criteria are useful. At minimum, a landmark project should be original, substantial, paradigm shifting, and–in a seeming contradiction–both exceptional in its execution and typical in the utility or universality of its function. It should have novel or aesthetically pleasing aspects. A project which truly synthesizes a new system or style of engineering is especially worthy. There should be a quality of size and impressive scale about it because monumental works, by definition, need a commemorative purpose or great scale. General recognition as an "important project" is excellent evidence.

Some level of institutional self-awareness is a prerequisite for true significance. It provides a basis for a larger role in society, and helps in taking a project beyond the state of the art. Finally, the most fundamental attribute of a great work is how well it performed its intended function. If a railroad was neither economically viable nor an effective transport system, then no matter how impressive its vision or how sizeable its plant, it is hard to argue for its historical significance.

On those terms, the antebellum B&O Railroad was a pivotal and significant project. It deserves standing with other great works, such as the Brooklyn Bridge or the taming of the Mississippi. This kind of evidence is cumulative. There is no single instance which permits us to conclude that the B&O was any more important than a half-dozen other antebellum railroads. It is the weight and volume of evidence, however, which argues strongly that the company may even be more significant than the inherited mythology suggests.

For our purposes, there are two useful ways to frame the engineering significance of the "real B&O" in the antebellum period: 1) What was the railroad's time, place, and context? 2) What was the physical nature of the original B&O, and was it significant? Reviewing the actual form of the railroad reveals anew just how large, complex, and innovative Baltimore's project really was.

The Time, Place, and Context of the Antebellum B&O

The political and economic character of America was unsettled throughout the antebellum period. Transportation was a pressing issue, and the country intensely debated the responsibility for "internal improvements" (as roads, canals, maintenance of navigation, and later, railroads, were known). The result was a patchwork of federal initiatives, state projects, and private entrepreneurship. No clear national policy guided the constant competition for scarce resources–a legacy which persists to this day.

Through the 1820s, the United States was a sparsely populated, fundamentally pre-industrial, culturally backward nation. We may have had the most modern and progressive system of government in the world, but we were largely a nation of rubes. Americans looked to Europe for most of their manufactured goods and luxury items. We imported our culture and standards of taste, and modes of higher education, science, and technology. Transportation remained primitive, and congress shied away from creating the kind of national network proposed by Albert Gallatin in 1808.

That narrow view of transportation seriously disadvantaged the major east coast cities, which since Colonial times had been both domestic central places and entrepots for coastal and international trade. With the opening of the Mississippi/Missouri/Ohio watersheds to effective navigation after the War of 1812, the large and growing "western trade" (especially commerce between the US and Europe) found easier access to the American interior via New Orleans than through the old Atlantic ports, all of which were blocked from the west by the Appalachian Mountains.

One of the earliest engineering projects to breach the transportation frontier was the 1806-1818 National Road. This political experiment and engineering compromise provided a federally supported, hard-surfaced road connecting the seaboard with the interior. The route more-or-less equally benefitted Pennsylvania, Maryland, and Virginia. More so than early canals, the National Road was a tentative expression of technology as an agent national policy and valid public investment. It was successful, but the inherent expense of land carriage discouraged turnpikes elsewhere.

New York, which had the most to lose from diversion of trade and lack of growth, responded with a canal. Nature provided the splendid, fjord-like Hudson River through the Northern Appalachians as far as Troy, and then a gentle and low Mohawk River Valley to the west. No comparable east-west corridor existed for 700 miles to the south. The Erie Canal was a $7 million investment, 40 feet wide, 4 feet deep, and 363 miles long. It opened with much fanfare in 1825, and became an immediate success.

Throughout this period, the tramway lurked as another possible answer to the American transportation dilemma. Both tramways and railroads are fixed guideway systems; think of a tramway as a proto-railroad. By 1800, Great Britain had well over 1000 miles of horse-powered, low speed tramway primarily for mineral carriage. Many used iron rails, and a team of animals might haul a dozen cars coupled together. Stationary steam engines pulled trains up hills. As early as 1799, British railway promoters proposed a cross-country tramway linking London with Portsmouth, and by 1801 James Anderson had described the basic principles of a modern railroad system for England. British mechanics built dozens of crude steam locomotives before 1820.

Thus the eastern cities thus had three models to breach the Appalachians. First was the Erie Canal, based on the British system of canals. A canal, however, was ill-suited to mountain crossings and required a reliable source of water. Second was the National Road or other "improved" roads, which offered a hard surface for wagons and stages. Their disadvantage was low efficiency and high maintenance cost. Third was the "rail-road," then regarded as an elaboration of a common road. Because only tramways existed, it was difficult for Americans to make the conceptual leap to this next iteration.

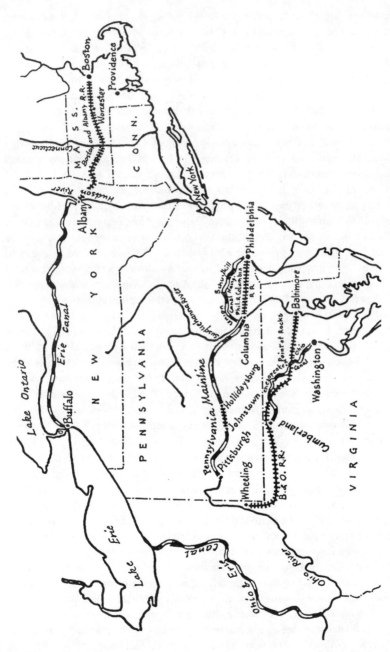

Figure 3. In a detailed analysis appearing as volume 57 of the *Transactions of the American Philosophical Society* (1961), Julius Rubin sketched out the primary northern trans-Appalachian canal and railroad projects of the 1820s and 30s.

In response to the particular competitive threat posed by the Erie Canal, coastal states and their proxy cities selected from these alternatives according to geography, hydrology, and a willingness to embrace new technology. Boston was stymied by long winters, a location far from the interior, and the formidable Berkshire range of the Appalachians. Philadelphia had the great advantage of Pennsylvania bounding tidewater on the east and the Ohio River on the west, and the great disadvantage of having the entire breadth of the Allegheny Mountains in between.

Baltimore was triply disadvantaged. Every route west encountered ridge and valley domain not far from Baltimore, and the distance to the Ohio Valley was great. Any link to the interior required passage through Pennsylvania or Virginia, which were both hostile to a Maryland project. (West Virginia did not become a state until 1864.) The Washington, DC, ports (Georgetown and Alexandria) lay at tidewater on the Potomac River, and shared a formidable Allegheny crossing. Richmond, and more southerly ports such as Charleston and Savannah, faced substantially the same obstacles, although their severity lessened to the south.

The traditional view was that progressive cities and states chose railroads. Less astute or forward-thinking cities either hesitated, or chose canals or roads to penetrate the interior. A more fair and correct interpretation is that intelligent men, with little engineering advice, made the best decisions they could based on their understanding of the economic, political, geographic, and technological realities obtaining in their particular place. Local culture also had a great deal to do with these decisions.

Within the space of five years or so, Boston, Baltimore, and Charleston decided to build railroads. Alexandria and Richmond chose canals, while Philadelphia began a hybrid canal/portage railroad system which incorporated the disadvantages of each and the strengths of neither. By 1840, the U.S. had lost interest in canals and adopted the railroad as its primary means of connection. The B&O was neither the first railroad chartered in the United States, nor the first to turn a wheel. It was, however, the earliest and most successful of these "railroad" solutions.

The Railroad Anomaly. The context of Baltimore's choice was desperation. Through the 1820s, the city faced dramatic decline as an Atlantic port. It seems far fetched today that so important an entrepot could slip into obscurity. But Baltimore's memory of the ports it vanquished was vivid as the city contemplated a similar future. Desperation as a motive is simplistic, but it helps make sense of the anomaly Julius Rubin pointed out forty years ago. The question is not why Baltimore was the first to adopt the modern railroad, but why cities better suited and wealthier had not done so a decade earlier.

By 1810, a few small mineral tramways were operating near Philadelphia. Americans had access to a growing body of British railway literature and European engineering talent. No less a visionary then Benjamin Henry Latrobe reported to congress on the feasibility of railroads in 1808. As early as 1812, Revolutionary War veteran John Stevens published a credible proposal for a steam-powered railroad as an alternative to the Erie Canal. Oliver Evans designed a crude steam powered railway, and Stevens himself received a New Jersey charter for a railroad from New York harbor to the Delaware River. Even as the successful Granite Railway (a heavy duty mineral tramway) opened near Boston in 1826, Pennsylvania ignored the recommendations of its own engineer, William Strickland, and built a canal to the west rather than a railroad.

By that time, the few American engineers interested in railroads could have consulted five editions of Thomas Gray's *Observations on a General Iron Rail-way*, Nicholas Wood's *A Practical Treatise on Rail-roads*, or Thomas Tredgold's *A Practical Treatise on Rail-roads and Carriages*, the latter published in New York. There were reports available on British tramway projects, and the press covered the topic fully. Why, then, was there so little interest in tramways or railroads in the United States?

The adoption of new technology always has a lumpy quality to it. Thomas Kuhn's classic model of "shifting paradigms" for the progress of science works well for older technology. There usually is a "technological salient" or leading edge, followed by general adoption if the technology proves useful and effective. One would expect the railroad to have followed the same pattern as textile technology (brought to the US at about the same time). With such good examples so plainly in use in Great Britain, there should have been the gradual and incremental adoption of tramways throughout the Early Federal and National Periods (roughly 1790 through 1820).

Instead, there was resistence to railroads until the late1820s. Places like Boston, New York and Philadelphia, with commercial and scientific connections to Europe and progressive attitudes regarding technology, could have at least tinkered with tramways, but did not. That makes the turnaround in attitudes between 1828 and 1832 even more striking. Seemingly overnight, the railroad became the preferred transportation option. Partly that was a response to the 1830 opening of the Liverpool and Manchester Railway in Great Britain, but the shift has not been satisfactorily documented or explained. It remains one of the most perplexing aspects of early American engineering.

This also may be an example of how the B&O Railroad found myth to be useful. The "official" explanation (meaning, sanctioned by the B&O and largely unchallenged) of the origin of the railroad in the United States was that a chap named Evan Thomas, while visiting England in 1826, stumbled upon the "new" British invention called the "rail-road." Thomas immediately wrote back to his brother Philip Thomas, a prominent Baltimore banker destined to become one of the founders and first president of the B&O. With that information, Baltimore immediately adopted the railroad. The day was saved–Baltimore's terrible fate (to be relegated to backwater status by the Erie Canal) was averted by the availability of this wonderful new technology. The implication was, of course, that the US had little prior knowledge of tramways or railroads.

Possibly the B&O found it simpler to adopt a straightforward (and heroic) story of luck and quick thinking, rather than deal with the complexities of American politics and engineering history. That myth distorted the history of American railroading for over a century. After it became apparent that no single railroad legitimately could claim priority, each company adopted its own creation myth shaped to fit the convenient facts. Is there a way around this anomaly? In advance of more serious scholarship, is there a plausible explanation for the sudden popularity of the railroad in the late 1820s?

A decade ago, historical geographer James Vance Jr. argued that American railroad development was more indigenous and historically contingent than the standard academic model of technology transfer. Vance's thesis, which is gaining acceptance, assigns significance to a convergence of need and means in the critical decade between 1825 and 1835. He emphasizes the different physical challenges and operating environment in North America, coupled with scarce resources. The result was a distinct style of engineering and railroad development right from the beginning. That thesis works exceedingly well in explaining the sudden prominence of the B&O.

In the explanation I am proposing, the B&O remains the first coherent, large scale expression of the principles of the modern railroad (as distinguished from a tramway) proposed and operated successfully. It retains the distinction of unprecedented reach, vision, complexity, and thus some status as the pivot point in American railroad development. It was the engineering project which most directly ignited the railroad boom and which established a different paradigm for land transportation.

But now the context of its creation is iterative and opportunistic, rather than heroic and lucky. Prior projects–some roads, some canals, some railroads–contributed important precedents or principles. From the Erie Canal came the realization that a financial investment of that magnitude in an engineering work of that scale could make a profit and change the nature of basic commerce. From the Granite Railway came American experience with heavy British tramways. The promoters of the B&O did indeed scour the literature for useful information, and there was enough of an industrial base in Baltimore to make construction of a railroad feasible.

Even the opportunism of the project was incremental, and carefully considered. Baltimore had a decade to assess how it would respond to the challenge presented by both New Orleans and the Erie Canal. The city examined canal options, tried to broker inclusion into the C&O Canal project, and explored upgrading its connections with the National Road. The several dozen men who risked their reputations and personal fortunes to back the B&O railroad were undeniably brave and incontestably visionary. Only now, while their desperation seems no less real and compelling, their gamble appears much more calculated, rational, and astute.

When first proposed, the B&O itself was an anomaly as one of America's largest and most complex engineering efforts. It was also a political project, and an economic development gamble, and a grand experiment in business development and the mechanisms of capital formation and administration. But in its essence, the B&O was a continuous and elaborate work of integrated civil and mechanical engineering built on a scale not previously conceived by a single American company. Within a few years, it became the pattern for dozens of equally audacious railroad projects, including the mid-century proposals for a transcontinental railroad.

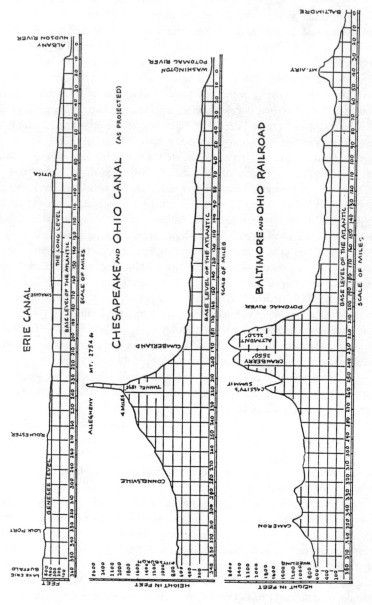

Figure 4. These profiles, adapted by Julius Rubin from J.L. Ringwalt's *Development of Transportation Systems in the United States* (1888), show the routes of the Erie Canal and B&O Railroad, and the proposed route of the C&O Canal west of its terminus in Cumberland, MD.

The Nature of the Original B&O

We must distinguish between what I term the "original B&O," and the postbellum railroad that continued to grow beyond its founders' wildest vision. The original B&O was the realization of the railroad as conceived in 1826-27. Two branches brought the antebellum operating route to a little over 500 miles. Railroad operations remained essentially the same from 1857 until 1867, when the company leased a small connecting railroad. From that point, the B&O grew to its maximum length of 6200 route miles during the middle of the 20th century.

The Washington Branch extended 30 miles to the capital from a point on the Main Stem west of Baltimore. Construction began in 1833, and the line opened two years later. Its most dramatic feature was a spectacular stone viaduct patterned after the Sankey Viaduct in Great Britain. Its designer was Benjamin H. Latrobe, son of Benjamin Henry Latrobe, one of America's most noted engineer/architects.

The second important addition was the Northwestern Virginia Railroad, conceived in the 1840s as an alternate route to the Ohio River. Construction began in 1852, as men and material became available with the imminent completion of the Main Stem to Wheeling. The 104 mile "branch" (which, like the Washington Branch, in reality was a mainline railroad) left the Main Stem at Grafton, in what is now north central West Virginia. Its terminus was Parkersburg, on the Ohio River ninety miles downstream from Wheeling. Parkersburg had long been the company's preferred Ohio River terminus, but in that era politics, and not engineering or economics, often determined railway location. That line opened in 1857 as an integral part of the B&O. Throughout this period, the railroad built and expanded its Baltimore terminal facilities, especially after reaching Cumberland in 1842. It established working connections with a half-dozen other railroads and participated in both through and intermodal traffic.

This original B&O is bounded in time by the Civil War, in space by the states of Maryland and Virginia, and in attitude by the concept that a railroad was a discrete project. For most of this period, the railroad adhered to its stated goals: To create an all-weather land link between Baltimore and the Ohio River, and serve as a conduit intercepting as much western trade as it could for the benefit of Maryland interests.

During its construction, the B&O passed through three distinct organizational phases, each lasting about ten years. As a shorthand, I describe them as the "Pioneer Phase," the "Redesign Phase," and the "Completion Phase." These do not necessarily correlate with construction or with the tenure of officers, but they do correspond to the evolution of the field of railroad location and construction, and a deepening sophistication on the part of the railroad's management. One might describe these phases in terms of "design ethos"–the sum of various philosophies, technical standards, engineering approaches, and construction practices. To the B&O's management, it was an experiment in what we would describe as "design-build." They had to make it up as they went along.

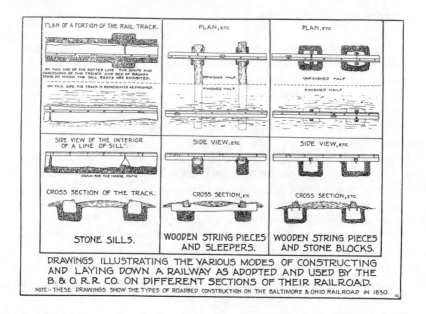

Figure 5. On the first eighty miles of track, the B&O tried stone rails, wood stringers on stone blocks, and wood stringers on wood cross ties, all capped with 2" wide and 3/8" to 1/2" thick iron strap rail. None worked well.

The Pioneer Phase. The "Pioneer Phase" took the railroad from Baltimore Harbor to the Potomac River at Harpers Ferry, and lasted from 1828 to1838. In this period, the first generation of the company's engineers built just over eighty miles of main line and the first of four Potomac River crossings. Among the issues which stalled the B&O at the water's edge were design changes, the Panic of 1837, location disputes with the C&O Canal, the retirement of the railroad's first president, and lack of money. No one had anticipated how complex building even a simple tramway-style railroad would be.

By the early 1830s, the company's engineers contemplated building a dozen major bridges, over a hundred minor spans and culverts, two long tunnels, thousands of perches of masonry, and the placement or removal of hundreds of thousands of cubic yards of earth and rock. Needed support facilities included substantial machine shop and repair facilities, at least fifty depots, dozens of water stations, warehouses at major traffic nodes, hotels and eating houses, pier and wharf facilities, and command and control systems. This scale and complexity alone would have been remarkable. But the B&O also had to surmount the Eastern Continental Divide at 2600 feet from a standing start at Tidewater in Baltimore, and do it all with only black powder and hand tools.

The B&O initiated construction with only superficial understanding of existing British tramways, at a time when methods, engineering standards, and basic principles of railroad construction had yet to be fixed. During a board meeting in 1836, one of the directors characterized the early stages of the project as "new, crude, and doubtful." The company estimated that it would take five years and cost $5 million to complete the original project, or a little more than $88 million in 2002 dollars. This was at a time when few private entrepreneurial projects exceeded $1 million in capitalization.

None of America's engineering disciplines had become professionalized, or even conceptualized as distinct fields. Few men had the training and experience required of a project such as this, so the B&O solicited help from the Army, which assigned varying numbers of Topographical Engineers (West Point trained officers) to assist with surveying and route location. Jonathan Knight was lured away from the National Road to be the company's first Chief Engineer at a salary of $3000 per year.

The Redesign Phase. The second phase of the B&O's construction was the price it paid for being a pioneer. It represented a slow, but complete, turnover in management and a similar top-to-bottom rethinking as to route, financing, engineering, and construction. Track was extended only a hundred miles to Cumberland, Md. The company had to pause, modernize its thinking, consolidate its finances and talent, and then poise itself to move forward on a different basis. The goals remained the same, but the entire concept of a railroad–and the B&O in particular--changed dramatically.

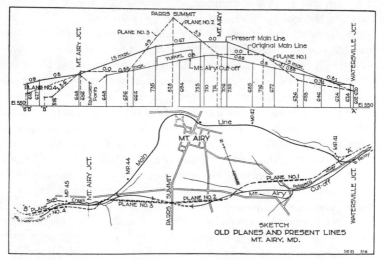

Figure 6. At Mt. Airy, the headwaters of the Patapsco River 43 miles out, the railroad hit Parr's Ridge. The first crossing was by British-style inclined plane. By 1839, the B&O had built a 5.5 mile bypass with 1.5% grades.

The challenges facing the B&O were financial, political, technological, and also a kind of spiritual malaise. This was a major transitional period, when the design ethos moves from an elaboration of British tramway designs to something more closely resembling modern American practice. This also is the interval in which the railroad began to act as a real transportation system hauling substantial tonnages, as opposed to poaching traffic from existing turnpikes and operating at a low level of intensity.

Throughout the "Redesign Phase," the railroad continued the work of almost completely replacing its original 1820s-1830s stone track. It resurveyed and realigned tight curves, renewed bridges, improved earthwork and grades, built two major "cut-offs" to straighten kinks and avoid hills, and generally went back over the existing physical plant with what amounted to partial reconstruction of the original Main Stem to Harpers Ferry. The company also expanded its workshops, increased its locomotive fleet, and regularized its system of train operations. The B&O worked out many of the principles of railroad administration and operations management which would inform the industry until after the Second World War.

One of the first tasks in this phase was to attain Cumberland, a growing city deep in the Allegheny Mountains of Western Maryland. Surveying for the route began in 1838, and construction commenced the next year. Unlike most of the previous track built, the extension from Harpers Ferry employed a track structure more closely resembling modern track. Rolled iron rails with an inverted "U" section rested on lengths of wood, which in turn bore on wood cross ties set in rock ballast. The railroad reached the city in late 1842 and immediately assumed the role of the eastward extension of the National Road, which terminated at Cumberland.

At that point, Cumberland was the westernmost point reached by a railroad from Tidewater. More importantly, it was the collection point for a number of bituminous coal mines which previously had lacked efficient transportation. The traffic grew slowly, but throughout the decade increasing tonnages of coal went east via the B&O to Baltimore, Washington, and aboard increasing numbers of steam vessels. The net effects were far reaching and beyond the scope of this analysis. Suffice to say that the availability of substantial quantities of inexpensive mineral fuel changed and accelerated the character of industrialization up and down the East Coast.

Early in the phase, Knight and Latrobe hired a young Austrian architect/engineer named John Rudolph Niernsee to be head draughtsman and office engineer. He had trained in Prague and Vienna and had an excellent education in engineering mathematics. Latrobe was a fair mathematician, as was Knight, but both essentially were self-taught. Niernsee brought the latest theory and methods from the mainstream of Continental practice, and it is possible to discern Niernsee's influence in the increasing sophistication of the company's bridges, engineering reports, and experiments. After Latrobe became Chief Engineer in 1842, he pursued an expanded program of bridge design exercises, culminating in the ca. 1850 Bollman Truss design—one of the earliest, and most successful, all-metal railroad bridge patterns.

Figure 7. This 1855 cut shows the form of Bollman's ca. 1850 suspension truss design, developed with B.H. Latrobe and others on the engineering staff. The only surviving Bollman Truss bridge became ASCE's first NHCE Landmark.

The Completion Phase. The third phase opened in 1848 with the election of Thomas Swann as president of the B&O. General financial conditions were favorable in the US, making it feasible to raise the capital needed to finish the railroad. This "Completion Phase" was one of intense, superbly directed activity resulting in the opening of the railroad to two termini on the Ohio, establishment of a through route to Cincinnati and St. Louis, and refinements which made the B&O a fully functioning, modern railroad.

Swann was a lawyer/politician who had the good sense to stay out of Latrobe's way. When he took office, there was a bit of a race on to see which railroad would connect tidewater with the Ohio first. Realizing that they had one opportunity to move the project forward, Swann and the directors decided to make a bold, final push to Wheeling despite that city's interference with line location and financing. Latrobe thought it could be done in three years for about $7.5 million. It took a little longer and cost just over $8 million. His long tenure with the company had stood him well.

This final leg–201 miles–traversed some of the most beautiful, and harshest, country this side of the Rockies. The best route Latrobe and his location engineers found followed a series of watercourses and had a maximum grade of 2.2%. This was on the edge of feasibility for an adhesion railroad, and grades of that magnitude faced traffic in both directions. But Latrobe considered, in minute detail, the entire process of transportation. At the same time he was settling on stiff ruling grades, he was ordering locomotives capable of besting them, cars capable of withstanding them, and repair shops to keep the whole system fluid. It was a closely calculated solution.

In 1849 Latrobe hired Albert Fink, who had recently earned an engineering degree at the Darmstadt Polyteknikum before being forced to emigrate due to his reformist liberal political views. He was one of many technically trained "48ers" who fled the German states during a wave of Prussian-inspired political repression. He brought with him even more formidable skills than Niernsee ten years earlier, and he became (at the age of 23) construction engineer for the entire extension to Wheeling.

Latrobe planned the project meticulously using the latest instruments, detailed maps, and the best men on his staff. The work itself was a model of logistics and administration, with construction of the most difficult and critical sections begun first. Essentially all of the line was under contract by 1849. At one point in 1851, nearly 4800 men and a thousand horses were at work, and Latrobe was spending $200,000 per month to get the line built. These were extraordinary figures for an antebellum project.

There were two long tunnels on this segment, and Latrobe took personal charge of their design and execution. Despite labor troubles, material shortages, continual meddling by the Commonwealth of Virginia, the need to avoid Pennsylvania (the line skirted the southwest corner by a half-mile), and the myriad woes which still plague large projects, a rough Irish crew closed the track at a large glacial boulder on Christmas Eve, 1852. Everything about this final phase (including the subsequent construction of the Northwestern Virginia Railroad) was well planned, crisply executed, and successfully concluded, in greatest measure due to Latrobe's abilities.

Figure 8. The Kingwood Tunnel was the longest bore on the Main Stem. Its neoclassical detailing is typical of Latrobe's work.

The Original B&O. As with many new technologies and works of innovative engineering, the final statistics of the B&O's "Main Stem" were far greater than at first anticipated. It took a quarter of a century to reach the Ohio River, due partly to politics and financing issues and partly because the company had to invent so much as it went along–and reconstruct much of what it built originally. The route stretched nearly 380 miles between water's edge in Baltimore and the steamboat landing in Wheeling, or ninety miles longer than envisioned. And the cost trebled, to just over $15 million (excluding the branches), at a time when monetary inflation was not a factor.

As a completed work of engineering, the original was impressive for its statistics alone. On the eve of the Civil War, the B&O comprised 519 route miles and roughly 825 track miles. The extensive marine terminus at Locust Point in Baltimore handled enormous quantities of coal, wheat, tobacco, and other commodities. Nearly 5000 employees produced an average of 270,000,000 ton miles per year before the Civil War wrecked the railroad. The ton mile is a measure of railroad productivity (obviously, one ton of freight carried one mile) devised by Jonathan Knight. It is still used today.

The B&O laid rail of up to 85 lbs per yard, exceedingly heavy at a time when most railroads used rail of half that weight. The company ran large, idiosyncratic "Camel" locomotives as the backbone of its fleet and built a substantial proportion of its own motive power. There were no less than 57 engine houses located around the railroad with an aggregate capacity of 230 stalls. The roster itself totaled roughly 250 locomotives, 125 passenger cars, and 3700 freight and work cars. The railroad claimed to have the largest complement of rolling stock on the continent.

The line of road and fixed plant was no less impressive. On the Main Stem alone were 186 bridges totaling 15,088 feet, divided roughly equally between stone, wood, and composite iron spans. Fourteen tunnels carried the Main Stem through obstructions, with a total length of 12,694 feet. These included the 4100 foot Kingwood Tunnel and the 2360 foot Board Tree Tunnel, which were some of the most impressive in the country when opened. Many of the bridges were an Albert Fink variant of the Bollman Truss, while some were impressive composite stone and iron viaducts.

Steam locomotives use a great deal of water, and the original B&O had 94 water stations along the line to replenish them. The company had 12 substantial machine and repair shops and another 33 minor repair facilities. Eighty five depots and 30 telegraph stations stretched out along the Main Stem and branches, ranging from the substantial stone and brick buildings at Martinsburg, Virginia to small, cramped wood shanties in the middle of nowhere. The B&O had use of a telegraph line along most of its route.

The railroad was successful. In the three years between its completion to Parkersburg and the onset of the Civil War, it had gross revenues of between $4 and $4.5 million, with net revenues well in excess of $1 million per year. The company was considered one of the most efficiently run railroads in the world and was making the transition from mixed public/private ownership to purely private capital status.

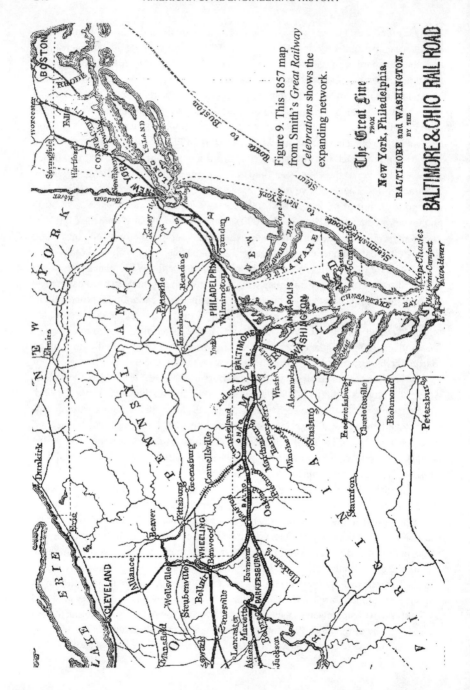

Figure 9. This 1857 map from Smith's *Great Railway Celebrations* shows the expanding network.

Part III: The B&O of Legend and Myth

A great body of myth and legend grew to surround the building of the B&O Railroad. It was an uneven, non-linear, and at times, contradictory process. These myths, stories, and legends should be treated with respect and understanding, for they were not merely the response of an innocent or unknowing people to mysterious forces. While they come from the same human impulse to weave the new and strange into a continuous narrative of lived experience, these railroad stories do not arise from fear, bewilderment, awe, or ignorance. Their sources are mostly European, and their expression distinctively and pragmatically American.

The mythology of the B&O Railroad arises from practicality, through a process akin to technology transfer. Americans inherited a rich and diverse mythical tradition, encompassing everything from classical Greek and Roman mythology, Norse, Germanic, and Celtic myths, to African, Native American, and even Asian mythical culture. Ordinary people used mythology, along with the Bible, for instruction, guidance, politics, explanation, evocation, and as a framework for everyday life.

Americans borrowed the structure and utility of its inherited mythologies, then filled in the details according to what was useful here. In principle it was the same as Samuel Slater memorizing textile technology in his native Britain, then replicating it in New England under a different set of circumstances. The spinning and weaving process was essentially the same. But we supplied our own raw material, embellished the product with our aesthetic sensibility, and used the cloth in particular ways.

Mythology, legend, and memory had important functions in an antebellum America facing daunting waves of change. Americans contended with bitter sectional rivalries, which ultimately rent the Union. The promise of unlimited westward expansion produced its own opportunities and stresses. Following the War of 1812, immigration boomed, reaching a peak after the failure of European reform movements and the Irish potato crop in the 1840s. Finally, a bewildering array of new, powerful, and unpredictable technology exploded upon a largely agrarian, craft-based society.

Individuals and institutions fit the new and unfolding reality into their established, and still-serviceable, cultural framework. For a company such as the B&O on the leading edge of much of this change, the natural response would have been to drape the project in the familiar, and accepted, garments of popular mythology. For a society grappling with social dislocation, revolutionary economic modernization, and profound changes in concepts such as space, time, and intensity, the natural response would have been to place the new B&O in the existing framework of mythology.

Like so much else I have described, this was an iterative, evolving, bi-directional process. It was neither controlled by the B&O, nor was it unconscious on the part of society. No one necessarily decided to "use" myth and legend explicitly. Antebellum Americans simply proceeded according to the culture framework of their time, using myth and legend constructively, idiomatically, and according to the prevailing logic.

Figure 10. In a bit of early landmarking, the railroad chose a large glacial boulder near Moundsville, Va. (now WV) as the spot to "close" the tracks, completing the B&O to Wheeling. Roseby Carr was construction foreman.

Engaging the Mythos.

Throughout the nineteenth century, the B&O, and to lesser degrees, other railroads, actively used mythology and the mechanisms of cultural production. That is common sense, and many engineering projects directly engaged the process. Any human endeavor will be influenced by the habits of mind and personal biases of its designers and managers. In a society which used myth and allusion as an ordinary mode of discourse, one expected railroads to have their share of mythological influence.

My contention is that the B&O used different forms of culture and myth more explicitly, more consistently, and over a longer duration than other American companies, and that the consequences of that engagement with American culture have yet to be fully discerned, much less analyzed and conceptualized. Because the mythic B&O helped articulate and transmit what Americans understood as their shared values, the company came to serve a cultural purpose well beyond mere transportation.

This process is important to the history of engineering, for it illustrates specific mechanisms Americans used to incorporate technological progress into their culture. Conversely, the B&O example reveals how new technologies change existing perceptions, expectations, and behavior. The mere fact of great engineering works or new technology affects little and means less. What ultimately matters is how individuals, and society as a whole, respond to the work of science and engineering. Myth and legend are effective teachers and mediators. An example will suffice.

The *Niles Weekly Register* for October 25, 1835 commented that "four engines were built by Phineas Davis, who, from his first effort in constructing the York, to the full attainment of the Herculean powers of the Arabian and Mercury, has made rapid advances in perfecting these machines . . ." Readers understood the allusion to Hercules, the Greek mythic figure. They grasped the intent in naming a locomotive for a strong and accomplished breed of horse (the Arabian). Mercury was the Roman god of speed and commerce; naming a locomotive after him made perfect sense. Phineas Davis himself had a heroic Old Testament first name. This one sentence fragment ranges over a goodly span of Western Civilization and packs a lot of meaning.

Other B&O locomotives bore names such as Atlas, Vesta, Juno, and Pegasus. Every president of the US was thus honored, as well as animals with locomotive-like qualities (Stag, Reindeer, Elephant, Elk). Cars had names of a different character, such as Maryland, Ohio, and Carroll. The B&O revealed an acute awareness of the cultural framework of the project, and a willingness to engage the machinery of cultural production. Naming equipment was not whimsy. It was astute judgement.

The leaders of the B&O were mindful that they were proposing, designing, and executing a massive and novel engineering project in a period of great change. They did not have the benefit of wide acceptance of the technology, an orderly capital market, or even a reasonable certainty that the railroad could be built at all. But they did have many ways to engage myth and legend, which fall into three broad categories.

Cultural Production. This is an obvious, but useful, term describing a range of acts and effects which somehow add to, or manipulate, the body of understandings which define a people's culture. In this case, it represents the output of formal and informal initiatives by the railroad in "the manufacture of public opinion," in the words of a B&O subsidiary company executive in 1850. There was nothing cynical about this.

Cultural production included creating spectacle (parades, equipment demonstrations), performance (the ritual ceremony placing the First Stone), visual elements (art, lithographs, or three dimensional representations), rhetoric (verbal or written representations), and monuments (purpose-built memorials or structures with monumental attributes). The key aspects were intent and agency, or more bluntly, an ulterior motive. When the railroad treated newspaper editors west of the Ohio River to a free excursion to Washington in 1860, it was cultural production *and* savvy politics.

Cultural Co-option. The B&O quite cheerfully adopted and used whatever emotions, beliefs, attitudes, or values it thought useful from American culture. The company reached out and borrowed a mythical framework for the project, in precisely the same ways that the patriots during the Revolution selected aspects of European civilization and culture with which to frame the fight for "liberty." It is not at all unusual that the B&O would have done so. What is surprising is the sophistication with which so many elements of American culture were made useful to the project.

This mode of engagement with culture differs from, but underlies, the various acts of cultural production. Naming locomotives "Gladiator" or "George Washington" is a mild example. Engaging Daniel Webster as company counsel is another, very subtle, form of cultural cooption. Think of it as the reasoning behind the visible or tangible manifestations. It is a way to create meaning, associate with existing mythology and culture, and invite people to understand the project in specific ways. If Webster (a cultural figure even when alive) took the B&O as a client, then the B&O must have been worthy of the great man's effort. Cultural cooption was a little easier then.

Cultural Accretion. This is the flip side of cultural cooption. The term describes the layers of meaning which a project acquires through the actions of broader and more general cultural mechanisms. Cultural accretion is unpredictable and erratic, and may even result in unfavorable or undesirable connotations. In many ways it is the result of cultural production and cooption–the reflections of meanings as processed and cast back upon the project. Society at large also may see in an engineering project some trait, metaphor, or meaning which resonates with its mood or sensibilities.

Skillful managers of evolving culture are careful to use cultural production and cooption in ways which encourage desirable forms of accreted meaning. A company's reputation in the community is an example–it can be seeded and maintained, but it has to be created externally and accrues over time. Just after the line opened to Wheeling, the B&O suffered a tragic derailment due to the roadbed settling unevenly. The company moved fast–and quietly–to settle claims lest it get saddled with a reputation as unsafe or unsound. When most historians or cultural observers treat the relationship between engineering and culture at all, this is the aspect most likely to be examined.

Figure 11. In 1891, the B&O commissioned Francis B. Mayer to paint "The Builders of the B&O"–a large canvas in the prevailing heroic allegorical style.

The Process of Myth Making.

The evidence of how the B&O intentionally engaged American culture has to be distilled from unconventional evidence and informed conclusions. That might be one reason why this kind of mythic discourse in engineering has been largely overlooked. Nowhere in the primary sources or ordinary records will there be a file for "myth making" or "corporate policy with regards to cultural manipulation."

Just a few antebellum examples of myths with their origins in the B&O's early years will illustrate the nature and extent of these cultural engagements. At this preliminary stage of research and analysis, many questions remain as to fact and interpretation. Still, it is possible to see a variety of patterns and a consistent corporate awareness of the usefulness of myth and legend. One of the most straightforward examples unfolded at the beginning of construction in 1828.

Establishing a Mythical Association. The men who conceived the B&O as a railroad were of the first generation born after the War for Independence. They came of age at a time when the Founders and their direct political heirs were crafting the machinery of government and the great national myths. Some of the B&O's early promoters were themselves Revolutionary War veterans and original patriots. The others knew firsthand the words, voices, and ideas of Jefferson, Madison, Adams, and the rest. They were close to the sources of the country's founding myths and legends, which was the basis for a formal ideology of expansion and development.

By training and inclination "the Founders" were Enlightenment citizens. They shared the belief that the United States was the legitimate inheritor of the progressive civilization of Ancient Greece and Rome. These men (women were largely absent from public life and nowhere to be found in technology) were familiar with science and the idea of progress, and they believed that the destiny of the American nation was to subdue the continent. The promoters of the B&O understood civil engineering–building a railroad like no other–as the means to apply that ideology to their economic futures.

The B&O was a railroad project. But it was informed by, and is an expression of, strong ideological and patriotic beliefs. That was not true of most later railroad projects, and is rarely true of ordinary engineering works. One of the most effective ways to express ideology and reinforce its place in culture is through the creation and elaboration of related myth and legend. That was the purpose of the B&O's commencement celebration in Baltimore. With carefully staged public spectacle (a massive all-day parade, ceremony, and party), the railroad borrowed the ideology of progress and associated itself with the great myths and legends of the founding of the nation. This was an obvious case of cultural production and cultural cooption.

The company chose July 4, 1828, as the day to begin the project officially. It staged a massive march through the streets of Baltimore modeled after the patriotic parades held in connection with Independence. The 5,000 or so participants in the parade included political entities, trades and crafts, citizen's groups, and the military. Virtually every segment of society was explicitly represented. (Missing from the parade, of course, were unfree labor, women, and competing cities.)

The symbolism was explicit, direct, and effective. Floats and moving tableaux addressed themes such as "The Union" and progress and westward expansion. There was oratory, patriotic music, allusions to the coming prosperity, and entertainment as the parade took a full three hours to pass. Over 60,000 people lined the route that day, making it the largest parade held in the United States to that point. Those with a good line of sight might have caught a glimpse one of the most venerated men in the country: Charles Carroll of Carrollton, nearly ninety years old and the sole surviving signer of the Declaration of Independence.

The terminus of the parade was the spot chosen by the railroad for turning the first symbolic spade of earth, at which the "First Stone" was placed with great fanfare and full Masonic ceremony. Charles Carroll officiated, making the formal public utterance that he "considered this act second only in importance to signing the Declaration, if even second to that." At that time, the Catholic Church was in a bitter, contentious fight with Freemasonry, which Rome regarded as not merely heretical but also possibly satanic. The fact that Carroll, a devout Roman Catholic, would take part in a high Masonic ritual (complete with pagan and pre-Christian allusions) is an indication of how important both the Grand Lodge of Maryland and Carroll himself thought this occasion was. To people in 1828, it must have seemed a bit bizarre.

Figure 12. The company routinely created images to "fill in the gaps" where none existed. This 1927 Stanley Arthurs painting depicts Charles Carroll at the First Stone ceremony in a manner intended to evoke the nation's founding.

The site itself was at the Baltimore City line a few yards from Gwynn's Falls, which would be bridged the next year by a magnificent two track granite viaduct patterned after the 1727 Causey Arch tramway viaduct in Great Britain. The new "Carrollton Viaduct" was in the Neoclassical style (actually, more Renaissance Revival), making a powerful visual statement about the scale, permanence, and ideology of the proposed undertaking. For early engineering, the symbolism and myth making does not get much more intense.

Almost instantly, with a day of public spectacle and the creation of its first bridge, the B&O intimately associated itself with Classical Rome, the medieval concept of the city gate as portal to civilization, the traditions of Freemasonry, the founding of the Republic, and the American political ideology of "Life, Liberty, and the Pursuit of Happiness" (which to them meant economic opportunity). By adopting these myths as its own, the B&O acquired credibility, increased its influence, and sent powerful messages to its audiences and stakeholders. In important ways also, it was signaling that this was a true public work, undertaken to benefit the city and state as a whole.

Making Monuments. Benjamin H. Latrobe, who later became the Chief Engineer for the B&O, early in his career was the designer and construction engineer of the bridge carrying the Washington Branch across the Patapsco River Valley. At that time, the B&O still intended to build stone bridges at its major river crossings. This location was particularly scenic and worthy of a grand structure.

Latrobe's father was Benjamin Henry Latrobe, the English engineer/architect often credited with formally introducing the United States to classical revival architecture. Unsurprisingly, his 24 year old son chose an explicitly classical design for his first major structure. Yet while the British model for the Thomas Viaduct was on a tangent, he laid out his 612 foot long bridge on a four degree curve. Critics suggested that the viaduct could not possibly stand and called it "Latrobe's Folly."

The prime contractor was John McCartney, an Irish immigrant stone worker with a contentious disposition and a serious drinking problem, which made for an odd working relationship with the young Latrobe. McCartney was up to the challenge of building one of the largest stone bridges yet attempted in this country—notwithstanding the fact that larger and more elaborate stone structures were routinely built in ancient Rome 2000 years previous. McCartney understood the symbolism of this bridge.

As the viaduct neared completion in early 1835, it was memorialized in four distinct ways. First, the railroad declared the viaduct "officially" complete on July 4—the secular patriotic American holiday. Second, the directors of the company named it the Thomas Viaduct, in honor of Philip E. Thomas, president of the railroad from 1827 through 1836. Third, Thomas had Benjamin H. Latrobe commission artist Thomas Campbell to create a lithograph of the viaduct for public distribution throughout the United States. Finally, John McCartney erected his own monument at the east end of the bridge, dedicated to the B&O Railroad, its officers and Directors, and to himself.

There is a great deal of cultural production associated with this bridge. In fact, if there is any one structure which illuminates the interplay between engineering and culture, or the sophistication with which the B&O understood the value of an instant monument, it is the Thomas Viaduct. The son of "America's first civil engineer" was assigned to create a high visibility project–the grandest bridge in the country--intended to bring the benefits of the railroad to Washington, DC, the seat of the national government. Construction of the viaduct officially began July 4, 1832, on the patriotic holiday in the centennial year of George Washington's birth. The bridge itself was evocative of the great Roman aqueducts which brought life-sustaining water to the major classical cities. It also was a signifier of the B&O's imperial aspirations.

Naming it after Thomas was a gesture of respect and thanks from Directors pleased with his stewardship of the company through its difficult first eight years. The act continued the tradition established in 1829, when the company named its first major bridge the Carrollton Viaduct, after the B&O founder who also was the only living signer of the Declaration of Independence. That is all straightforward, and the naming of civil engineering works for deserving individuals is an old practice.

But there is even more going on here. By naming the viaduct, the railroad explicitly turned a bridge–a productive corporate asset–into a monument. The fact that it already was monumental in scale and classical in form made that even more effective. As a monument, it had additional work to do in culture and society. For example, it lent credibility to the new enterprise by seeming old, permanent, sturdy, and in the spirit of the founding of the Republic. The sometimes heated squabbles between Latrobe and the directors (over such things as cost and whether it would have a stone coping or iron railing) indicate that the principals knew well that they were creating a lasting statement. The symbolism was obvious, useful, and well understood.

Figure 13. John H.B. Latrobe's drawing may be read many ways, but it emphasizes the novel curvature of the viaduct. It derives from the emerging Hudson River School of American art and brims with meaning and allusion.

When Thomas asked Benjamin H. Latrobe to commission a lithograph for distribution, he was ensuring that the intended meanings for the bridge were conveyed to the widest possible audiences. Latrobe asked his brother, John H.B. Latrobe, to make the detailed drawing from which the lithographer cut the stone to make the prints. J.H.B. Latrobe at that point was one of the B&O's attorneys, and he would become the railroad's chief counsel in the future. But he also was a trained civil engineer (a graduate of West Point), a superb draughtsman, and deeply interested in history. In fact, J.H.B. Latrobe later helped found the Maryland Historical Society.

For the perspective, the Latrobe brothers chose a dramatic low angle from the southeast, which allowed John H.B. Latrobe to exaggerate sun and shade to clearly show the curvature of the bridge. Overall, the work conformed to the Romantic style then in favor in Europe. It was indeed a strikingly beautiful rendering, and Campbell took great pains to reproduce the original drawing in all its detail and subtlety. The lithograph included the principle dimensions of the viaduct, something which ordinary works of fine art usually omit.

As for McCartney, for the price of a little stone and some hard labor he was able to associate himself forever with the great men who built the rest of the railroad. Benjamin H. Latrobe designed the monument as a simple square stepped pedestal with an Egyptian obelisk (this was only a few decades after Napoleon dragged a few genuine Egyptian obelisks back to Paris). Carefully cut into the limestone were the names of the national leaders, the railroad's directors and staff, and of course, John McCartney. Like the bridge itself, his obelisk remains intact and only a little worse for wear. Its meaning is lost on those who see it from the daily commuter trains.

Figure 14. *Left*. Latrobe's multipurpose obelisk for McCartney is a monument to an engineering monument. *Right*. B&O president Philip E. Thomas clearly understood the vocabulary of symbolism available to him.

Inventing History.   Most people today with an interest in technology or railroading (and almost everyone born before the mid-twentieth century) are familiar in passing with the "Tom Thumb," the first locomotive in America. The myth has been thoroughly standardized and officially sanctioned, and it serves both as a parable of persistence and as a marker of American technological progress.

The standard version is that Peter Cooper came to Baltimore to build a locomotive capable of negotiating the B&O's sharp curves so that the railroad would not lose faith in steam power. He hammered a small engine together in a shed at Mt. Clare in the summer of 1830. It worked well on its test run, but on the way back from Ellicott's Mills, the driver of a horse-drawn passenger car on the adjacent track challenged Cooper to a race. The "Tom Thumb" pulled ahead until the blower belt slipped and the horse won this race–but as everyone knows, "the steam locomotive won in the end." Having proved his point, Cooper returned to New York to found the Cooper Union and become a revered philanthropist–an "Honest Man," as his biographer put it.

We shall leave aside the faint scraps of secondhand evidence that there was a race, and that it would have happened in a completely different fashion. Cooper did not call it the "Tom Thumb." It was built by the finest contractors and craftsmen Cooper's money could buy, and his motives had little to do with curvature. As I describe later, Peter Cooper wished to establish a locomotive building firm, and this was his demonstration project. It remained in sporadic service for at least six months.

An artist by the name of William H. Brown happened to be present at the First Stone ceremony in 1828, and he became one of the world's earliest railroad enthusiasts. After the Civil War, he set about writing a book entitled *The History of the First Locomotives in America.* It was a hagiographic account of the heroes who brought this amazing technology into being, and it was dedicated to none other than Peter Cooper.

In 1868, John H.B. Latrobe–by then the company's Chief Counsel of longstanding and increasingly active in the Maryland Historical Society and other cultural philanthropies–gave an address in Baltimore. His subject was the early history of the B&O from his perspective as a participant, and he described in great and stirring detail the story of the race between "this `Tom Thumb' of an engine" and a gallant gray horse. That apparently was the first formal account of Cooper's engine, and the published lecture caught the eye of William H. Brown, then researching his book.

Brown wrote to Latrobe, who suggested he write to Cooper, who told Brown that he basically couldn't remember a thing about the episode (it had happened nearly forty years earlier). If Latrobe said that is how it happened, then it was good enough for him. Ross Winans and a few other old timers added details and came up with a rough schematic of what the little machine looked like.

When the book was published in 1871, many people were growing wealthy (or at least making a good living) from the postwar railroad boom. Cooper was feted as the "Father of the Locomotive" and went about giving speeches and interviews based almost entirely on J.H.B. Latrobe's account, as embellished in Brown's book. Of course, because Cooper himself was telling the story, it certainly was authentic and had to be true. For a country proudly approaching its centennial and enthralled with the rush of technical progress, Peter Cooper and his B&O story were irresistible.

The City of Baltimore celebrated its 150th anniversary in 1880. After several false starts, a civic committee managed to arrange a large parade commemorating the city's founding. This was another of those complex events which may be understood in all sorts of symbolic ways. As with the original B&O founding parade, there was a great deal of cultural production being done by the railroad, both directly and indirectly.

The theme of the B&O's participation was "how the railroad saved Baltimore from commercial extinction." The company hauled Peter Cooper out of retirement in New York, and placed him at the position of honor at the head of the parade. In one section, the railroad had floats with tableaux illustrating important events from the company's earliest years. One of these featured a half-size wood and iron "replica" of Cooper's locomotive. The B&O was making what had been a legend "real" by crafting a physical object to represent it.

Farther back in the parade, the railroad's publicity agent–Joseph G. Pangborn, a classic late nineteenth century newspaperman/storyteller–had cooked up another float. This one featured a printing press cranking out "official" broadsheet histories of the B&O which were handed out to the crowd. That history featured an almost completely mythical version of Cooper's engine, patterned after Aesop's fable of the Hare and the Tortoise. The locomotive illustrated in Pangborn's broadsheet bore little resemblance to the "replica" further up in the parade, but no matter. Having its original inventor as the guest of honor gave legitimacy to Pangborn's mythic version of the story, too.

For the B&O's massive exhibit treating the entire history of land transportation at the World's Columbian Exposition in Chicago thirteen years later, Pangborn (by then the B&O's official historian) created a full size "replica." It conformed neither to any of the reliable sketches, nor to the 1880 model representing the Latrobe/Winans version, but to his 1880 newspaper wood cut. It was represented as a true copy of the "Tom Thumb," the locomotive that changed America. The B&O's exhibit was on display continuously for eleven years at two world's fairs and the Field Columbian Museum in Chicago, and seen by at least twenty million people. The myth/legend/story of the "Tom Thumb"–now ever further burnished by a man who never let the facts stand in the way of a good story–became part of the American cultural canon.

In 1926, as the B&O prepared for its centennial the next year, Pangborn's wood "replica" served as the basis for an operating "replica" of the Tom Thumb. The craftsmen at Mt. Clare Shops in Baltimore faithfully copied Pangborn's dimensions and form and somehow made it work as a locomotive. It was a star of the huge "Fair of the Iron Horse" centenary pageant and exhibition near Baltimore in 1927, and has been in demonstration service and on exhibit ever since. Millions of people have seen the "real" Tom Thumb, no longer even reliably understood as a "replica."

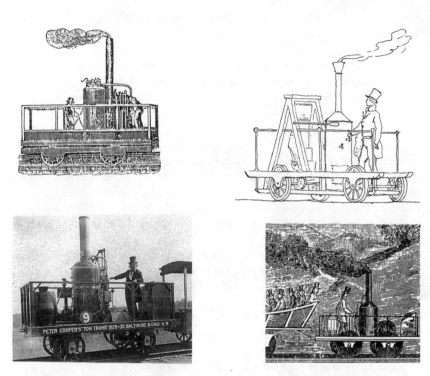

Figure 15. The many incarnations of Tom Thumb, clockwise from top left: An 1831 sketch from a contractor's ad; the 1880 Winans/Latrobe model; Pangborn's 1880 broadsheet version; the 1926 "operating replica" based on the wood 1893 "replica." Take your pick–at this point, all are real.

So what? Isn't this another egregious manipulation of history in the service of corporate public relations? Hasn't the "real" story of Cooper's engine ceased to be relevant? Doesn't the "Tom Thumb" story work better as an allegory for American inventive genius and the progress in technology? Who cares if the true story of the B&O's experience with Peter Cooper and early locomotives is stranger than the fiction?

Perhaps that is the whole point of this paper in a nutshell. The history and significance of the B&O Railroad has always been presented (knowingly or not) as a curious, but plausible, combination of fact and fiction. The railroad understood the power of myth and didn't hesitate to buff its accomplishments up a bit. American culture worked over that history pretty freely to suit its needs of the moment, and men and women of character and good faith found themselves promoting approved history which bore only passing resemblance to verifiable fact.

Part IV: The B&O Railroad and Antebellum Civil Engineering

Locating the B&O in the context of engineering in general would be a project unto itself. All we can hope to consider here are the outlines of antebellum industrialization and some of its salient issues, and then locate the B&O accordingly. Other railroads played important roles, and one must be careful not to ascribe sole agency to the B&O. But the railroad context is critical: Never before had a single human endeavor grown so large, complex, and changed people's "mental geography" so quickly. This one company was at the center of its adoption in the United States.

It is a commonplace to suggest that engineering projects do not happen for no reason. Works of engineering on the scale of railroads are responses to a combination of initial conditions, perceived need, and the conviction that an expensive engineering solution is appropriate. In particular, great works of engineering are called into being by extraordinary circumstances. The B&O was a disarmingly simple case of the right technology, in the right place, at the right time.

One way to fix the B&O's antebellum importance is to locate the company with respect to the series of nineteenth century transitions customarily described as the American Industrial Revolution. Participation in change almost always increases the significance of a historical "actor," whether that actor is a person, corporation, country, or technology. Sometimes with intent, sometimes by accident, and often unwittingly, the B&O found itself at or near the center of the American Industrial Revolution. Its participation in those transitions forms the basis for much of its historical legitimacy.

Creation of the American Railroad Industry. The most significant transition in which the B&O Railroad took a leadership role is the creation of the railroad industry itself. In 1827, there were a few short mineral tramways in operation and mere proposals for portage railroads. Steam locomotion was discussed, but not yet thought practical in this country. There were no railway suppliers, bodies of practice, mechanisms for financing, or laws particular to railroading. The railroad industry did not yet exist.

Three decades later, railroading was the largest single industry in the country. At the cusp of the Civil War, over 30,000 miles of track (of admittedly differing gauges and degrees of completeness) defined a railroad network east of the Mississippi. *Ashcroft's Railway Directory* listed over 400 operating companies in the US and Canada. Most were short and modest in scope, but a dozen railroads exceeded 350 miles in length. Capitalization stood at $1.15 billion, or roughly $24.4 billion in constant dollars. The value of the railway supply industry (the backbone of the emerging American iron and steel, machine tools, wood products, and metal fabrication industries) added a couple of hundred million to that total. The B&O ranked only fourth in terms of mileage, but it remained one of the four so-called "Trunk Lines" connecting the Midwest with the seaboard. And by 1860, only sectional politics delayed the start of a transcontinental railroad project.

By the time the B&O reached the Ohio River, the railroad had emerged as the most powerful tool available for opening the wilderness, creating markets, enhancing commerce, annihilating time and distance, and allowing Americans generally to impose their will on what was by then a truly continental nation. In the words of management analyst Peter Drucker, "The railroad was the truly revolutionary element of the Industrial Revolution. . . . For the first time in history human beings had true mobility"

No matter how that transition is measured, the results were great in degree and far-reaching in effect. The founding of the ASCE is one of the indirect consequences. Others were the creation of basic industrial capacity, the emergence of new systems of corporate management and control, a vast and complex body of case and statute law, and a cascade of new practices and technologies. The rapidity with which the United States embraced both the railroad and industrialization reveals railroading's inherent usefulness, and its role as a agent of change and modernization.

The First Industrial Revolution in America. After attaining independence, the United States faced what it understood as a stark choice: to pattern its future development along the lines of industrializing Great Britain, or to reject a future of poet William Blake's "dark Satanic mills" and remain a nation of "yeoman farmers" and craftspeople. Alexander Hamilton argued for a strong domestic industrial base. Thomas Jefferson advocated an agrarian nation with Great Britain as our "workshop."

Both views prevailed, in the sense that the United States was then, and remains, an agricultural powerhouse. But slowly after the War of 1812, and in earnest by the late 1820s, we began a period of development known as the First Industrial Revolution (to distinguish it from later clusters of technology adoption). Historians commonly date the beginning of Great Britain's IR to between 1700 and 1730. The standard model has it diffusing steadily throughout Europe and the Americas, and unevenly throughout the rest of the world. The facile term "First Industrial Revolution" masks a complex of great and subtle forces at work. The description following is a gross simplification.

Basically, the First IR comprised a series of technological advances such as the substitution of iron for wood and stone as a structural material; the substitution of mineral fuels for vegetable fuels; the replacement of animal power by mechanical power; and the general adoption of mechanized processes in favor of manual operations. Steam power tended to supplant water power, and we discern the early stages of mass production, standardization, and in general the adoption of rational and scientific decision making. The trend was away from craft production and toward industrial production (in factories), and there were corresponding social dislocations.

As the first major American railroad, the B&O played a large and direct role in America's Industrial Revolution. Most of its importance derives simply from the fact of its existence. It happened to be one of the earliest railroads, it was for a time the largest, and it was located in an area ripe for industrial development. The project itself was an immensely complex technological endeavor.

That is an example of historical contingency. The B&O's founders did not articulate a vision of Hamiltonian industrial democracy based on controlled industrialization. They did not, as a policy matter, plan a role as a pioneer in the American Industrial Revolution or directly engage any kind of formal economic development ethos. That would come later, once the power of the railroad to influence economic behavior became apparent.

The company's aims were much more focused: they had to build, and then operate, a railroad. They understood that to have direct and indirect consequences, and they were not shy about using whatever information, influence, or power they had available to achieve that goal. Shaping the Industrial Revolution was a collateral effect. In part, this is what it looked like.

The Evolution of Engineering Administration. No previous public works or engineering projects provided a suitable model for the building of a railroad on this scale, or for its operation once completed. Alfred D. Chandler, Jr. argued that the railroad industry was "America's first big business." Throughout this period, the B&O consistently was one of the longest, largest, most complex, and most analyzed of all of the early railroad projects. Partly that was because the B&O invented, tested, or refined so many of the business aspects of railroading and engineering. Two examples will illustrate the point.

The B&O's managers were familiar with the accounting practices of the day, and earlier engineering works (such as the Erie Canal) refined the collection and maintenance of survey and construction data. But it became apparent that the B&O would need to gather data continuously, process it in a timely manner, maintain various types of information in a variety of formats for extended periods of time, and understand how to use it constructively. This included administrative and financial data, records of assets and financial instruments, commercial information, operating information, planning materials, engineering and asset management data, and of course, the mass of raw and processed information associated with construction.

For the most part, these were new requirements and had to be devised and refined as the railroad went along. Information management is one of those topics whose history is ignored generally, but without which no large scale engineering project can succeed. Engineering initiatives such as the B&O took place within a larger framework of decision making, command and control functions, and a body of information comprising the growing institutional memory of the project. Unlike most engineering or entrepreneurial projects of that time, information on the B&O was not under the control of one person or maintained informally. It was regularized, routinized, made permanent, and understood as something which transcended any individual. The railroad clearly borrowed that ethos from the military and from law and government, but this was an early, and important, application to an engineering project.

Another early episode illustrates the evolving nature and sophistication of the project's engineering administration. I may have inadvertently conveyed the impression that the B&O's management was either uninformed about emerging technology or naive. That was not the case at all. In general, the company went about acquiring both knowledge and technology in effective and systematic ways. Setting aside the story of the "Tom Thumb, the first locomotive in America," the reality of Cooper's involvement (so far as I have been able to document it) is a useful, and perhaps even surprising, example. Think of this as the reality hiding behind the "Tom Thumb" myth.

Cooper was a successful New York businessman and gifted inventor who had purchased 3000 acres on the east side of Baltimore harbor for what we would understand as modern intermodal marine terminal. By early 1830, he had begun mining and smelting iron on his property. Cooper also was intrigued by the possibility of using a peculiar ratchet and pawl device of his own invention for steam locomotion. As an intuitive mechanician (the nineteenth century term corresponding loosely to mechanical engineer), he believed that half the energy available was lost in the motion of a common crank. His device, odd though it was, seemed to avoid that loss of motion.

Cooper was at that point a wealthy man. He came to Baltimore in the summer of 1830, engaged the most talented craftsmen in the city, and assembled a prototype locomotive in the machine shop of Mayger & Washington. His "crank substitute" did not work as hoped, but with a common crank the locomotive operated reasonably well. It began demonstration runs in August, 1830, and was tested sporadically over the next six months to a year. Cooper built at least one additional locomotive for which no record exists. One direct outcome was the series of locomotive trials held by the B&O in the fall of 1831 and based on the British Liverpool & Manchester trials of 1829.

Earlier in 1831, the railroad and Cooper entered into a contract calling for him to build six locomotives at $4000 each in less than two years. He was to construct the first one according to performance standards drawn up by Jonathan Knight, including speed, hauling capacity, reliability, fuel, and weight (all to conform to the limitations of the tracks then being laid). This locomotive was to be a test bed. Upon successfully remaining in continuous service for ninety days (during which both the B&O and Cooper would experiment and make tests), the resulting modifications and improvements would be incorporated into the design of the remaining five locomotives.

Cooper's intent was to establish, vertically integrate, and take a major position in the locomotive business. He had bog iron and hardwood to make charcoal on his property, with which he hoped to smelt, refine, and roll his own iron. The resulting locomotives essentially would have been mass produced with interchangeable parts. Cooper initially proposed retaining ownership of the engines, instead selling "moving power" to the B&O on the basis of passenger train miles operated and ton miles of freight hauled. The precedents were the contracts various livery operators had to supply horses to the B&O for its initial operations. The B&O declined his innovative proposal, preferring to own its locomotives and control its own costs–and savings.

Figure 16. Cooper's ideas resulted in the "Grasshopper" design, which proved to the B&O (and other railroads) that American locomotives were effective on American track. Some lasted over 50 years in switching service.

Peter Cooper was unsuccessful in his locomotive building venture, and the company released him from the contract and purchased his patent for a multi-tube boiler. But in effect, Cooper and the railroad had worked out an amazingly prescient program of equipment design, prototype testing, and integrated production. He proposed an operating contract whereby he retained responsibility for the initial capital costs and maintenance, while the B&O paid only for the motive power it actually used–at a premium, of course. That is the innovation claimed by General Motors and General Electric in the 1980s with their "power by the hour" locomotive leasing programs.

The point here is that while the B&O was busy "inventing" the railroad, it was not doing so in an unsophisticated or random manner. It used the best methodologies available and approached problems in a comprehensive, rational fashion. Throughout the antebellum period, the B&O made similar efforts in bridge design, operating practices, track and roadbed design, rolling stock standards, and other areas. In effect, the B&O had a research and development component built in to its engineering administration. That is one of the justifications for the company's reputation as an innovative work, and for the respect and deference afforded it by other railroads.

Engineering Transitions. The B&O was a framework within which many important engineering transitions played out. Again, this was not by design, and the founders could not have imagined how complex and changeable the field would so quickly become. But nearly all of the issues associated with the creation of the professional discipline of civil engineering found some expression on the antebellum B&O, and to a degree and depth greater than other railroads.

One of the first transitions was from military to civil engineering. The B&O was conceived at about the same time as the earliest civil engineering schools and courses offered in the United States. Until then, the only available formal training in engineering was at West Point, and most of the men who undertook civil or public works projects were either trained in Europe or self taught. Only rudimentary formal instruction in advanced mathematics was available domestically.

The 1824 Survey Act made Army engineers available to civilian projects if the Secretary of War judged them to be in the national interest. Ten topographical engineers helped survey the early B&O Railroad (some assisted with the C&O Canal, also). One of the Army engineers, Col. Stephen H. Long, served on the B&O's Board of Engineers. Most people subscribed to the notion that in peacetime, the Army (as the country's only reservoir of formal engineering ability) should make its skills available to the country at large. This was an expression of the ideology of national expansion.

In this very real way, the B&O represents the separation of civil engineering from its military origins. The B&O commenced design with regular Army officers as its engineering corps and concluded construction with an entirely civilian, newly professionalizing engineering staff. And the company helped a number of Army-trained engineers make the transition from government service to private practice. Many of those Topographical Engineers either remained with the B&O as civilians, or went on to other railroad projects as civil engineers.

Long himself used the B&O as a training ground and laboratory. In 1829 he published his *Rail Road Manual*, one of the first, and most influential, American civil engineering texts. Based largely on his B&O experience, Long devised or refined three bedrock principles of railway engineering: the two-step process of reconnaissance/survey for locating a railway; the theory of Equated Distance, which made it possible to reliably account for variables of curvature and grade when calculating the cost and length of proposed railroad lines; and the basic method for laying out railway curves.

It is useful to recall that in the mid-1820s, there were probably fewer than fifty men in the country who remotely could be called civil engineers, and not that many more Army engineers. Americans in general were not familiar with the idea of civil engineering, much less prepared to embrace it and assign rewards and respect to the profession (such as authority, compensation, and expert status). Building simple bridges and clearing roads were one thing; actually calculating stresses was quite another.

Figure 17. Running surveys in the 1830s was a combination of arduous measurement and computation, rough camping, and the exhilaration of exploration. This romanticized view of the B&O survey party in the Cheat River Valley is complete with woodsman in coonskin cap and fringes, view to the west, and the idea of progress coming to the frontier.

Antebellum railroad projects, and especially the B&O, shaped public perceptions of what engineering was, how it was practiced, and what the individual and collective benefits might be. Railroads offered a different set of lessons than canals or turnpikes because of their distinct challenges and much greater complexity. And because they were so new and noteworthy, they became a template for the broader discussion and adoption of technology.

The B&O also reinforced the international character of civil engineering. One of the first acts of the new company was to send three of its engineers–Jonathan Knight, William Gibbs McNeill and George Washington Whistler–to Great Britain for five months. They studied every aspect of British practice and spent time with eminent engineers (including John and Robert Stephenson). Ross Winans, one of the B&O's self-taught mechanical engineers, followed them in 1829 and stayed for the famous "Rainhill Trials" on the new Liverpool & Manchester railway in October of that year.

Engineers from Germany, Austria, and France made similar trips to the United States, touring the country and learning as much as possible about its railroads. The B&O figured prominently in all of these, and one of the resulting books–Carl Ghega's *The Baltimore and Ohio Railroad Crossing of the Alleghany Mountain Range*--was devoted solely to the B&O. References to the company were frequent in British and continental engineering journals.

Benjamin H. Latrobe, the B&O's Chief Engineer, spoke French and German, and the entire engineering corps was decidedly multilingual. The U.S. Military Academy did, after all, require fluency in French of its graduates and based its engineering curriculum on the French technical course. The B&O's influence extended even to Russia. Its ambassador had watched the progress of the B&O with interest, and the railroad maintained contact with the Czarist government. In 1840, the Winans family (always closely affiliated with the B&O) won a concession to build and operate a large part of the new Moscow to St. Petersburg Railway.

One of the most pivotal transitions made by the field in this era was the shift from what historian Monte Calvert termed the "shop culture" to a "school culture." Like all technologies evolving in the nineteenth century, engineering grew out of a tradition of practical and empirical knowledge. Gradually it adopted techniques of analysis and design based on mathematics, but as late as the 1840s, the B&O's Master of Road Wendel Bollman was testing designs for his all-iron suspension truss bridge by loading models to failure, then scaling up those dimensions which seemed most reliable.

All three of these transitions are discernable in the railroad careers of John Niernsee and Albert Fink. Although both Knight and Latrobe had instruction in mathematics, neither had an engineering education. Their B&O was a grand empirical experiment, increasingly informed by theoretically and rationally derived design methods. Niernsee and Fink are the vectors for that transition, bringing their rigorous European civil engineering training to bear on the B&O's challenges.

Two incidents involving Niernsee and Fink will illustrate the point. Carl Ghega published his book on the B&O in Vienna in 1844. While Ghega was in the US to do the research, it is difficult to imagine that he did not rely heavily on John R. Niernsee–the railroad's Office Engineer and himself an Austrian emigre. It would be safe to assume that Ghega formed a deeper and more contextualized understanding of the B&O by virtue of Niernsee's training, language, knowledge, and contacts.

Ghega's trip was sponsored by the Austrian government, which regarded the B&O as a model project for railroads in that mountainous region. His book appeared as Fink would have been starting the engineering curriculum at the Polytechnikum in Darmstadt (now the Darmstadt Technical University). The university likely would have had Ghega's book, and a student of Fink's caliber should have been familiar with it. That opens the possibility that Albert Fink knew more about the engineering issues of the B&O Railroad as a student in Europe than most of the practicing civil engineers in the United States–and that he sought employment with Latrobe because of it.

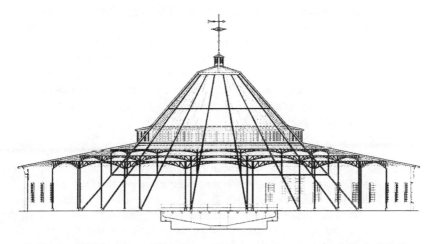

Figure 18. Architect Tim Yates delineated the preserved Fink B&O roundhouse at Martinsburg, WV. This is the latest National Historic Civil Engineering Landmark and the only one of six to survive.

As construction engineer for the extension to Wheeling, Fink had to create at least two, and potentially as many as four, intermediate shop complexes for rolling stock and locomotives. Fink designed an elegant, efficient, fire resistant iron-framed covered roundhouse which could be cast in sections at the company's shops in Baltimore, then shipped out for field erection. The design was novel, and much like a Plains Indian teepee, with inclined columns translating the central load of roof and cupola and creating a clear central space for the indoor turntable and radial tracks.

Equally noteworthy were Albert Fink's Tray Run and Buckeye Run Viaducts in the Cheat River Valley. To span ravines cut by sharply descending streams, he first built low stone arches to cross the runs and provide a solid base. On top of that foundation, he erected tall, slightly inclined cast iron columns to carry the track structure, with ample walkways for sightseeing. These were innovative and spectacular structures.

Fink adapted older British and German precedents of prefabricated cast iron framing to the American imperative for cheap, effective, easily-erected bridges and shop spaces. Fink's suspension trusses, viaducts, and structural work reveal an impressive command of tensile systems, compressive systems, composite structures, and cost issues for a young man at the beginning of his career. Literally, he was creating "machines in the garden," bringing the Industrial Revolution to near-wilderness areas like the new railroad towns of Piedmont and Grafton, Virginia. And he was putting into practice the plans and methods for mountain railroading his predecessor Niernsee described to Carl Ghega almost a decade previous.

One final point is worth mentioning. In the 1820s, when Baltimore and other Atlantic port cities were assessing the need and means to create effective transportation links with the west, the overarching question for the founders of the B&O was stark: Can this project be built? The issue was whether the concept of a railroad was viable, and if such a complex civil engineering work could be created in the United States.

Thirty years later, much had changed. In the December 8, 1854 issue of the *American Railroad Journal*, editor Henry V. Poor held forth in a lengthy article entitled "Railway Morals and Railway Policy."

"We have in this country a much better standard of engineering than in England. The greatest aim of American engineers is to secure the greatest returns on the investments. The construction of many of our more important lines is due to the skill of the engineer, among which may be named the Atlantic and St. Lawrence and the Baltimore and Ohio, the routes of both of which presented great difficulties, which have been surmounted with consummate skill. With *ordinary* engineering neither road could have been built." "The proper test of engineering is the *result*. . . .The first object to be gained is to make the road useful and *profitable*."

There was a subtext in Henry Poor's lengthy and insightful article. All of his comments and analysis rested on the offhand assumption that railroads could, and would, be built anywhere they were a paying proposition. No longer was there any question as to *whether* a project might be done. The argument centered on whether it would be a *good investment* and going concern. By 1854, two years after the ASCE's founding, civil engineering no longer was the problem. It was part of the solution.

The Range of Physical Transitions. No other single railroad–and perhaps, no American company–so vividly illustrates the physical transitions of the American Industrial Revolution in general, or the antebellum railroad in particular. In the period before the Civil War, the B&O shifted from a physical form used in Great Britain almost a century previous to one still in use in the twenty first century. Roman engineers could have built the initial B&O. Within reason, present day CSX equipment could have run on B.H. Latrobe's right of way and tracks of 1857.

As originally planned, the railroad was to be almost entirely of stone–masonry buildings, stone bridges, stone tracks. The railroad was to be built for permanence and durability. Partly the directors wished to reassure its stakeholders that this was a traditional–and thus feasible–construction project. The stone track reached as far as Sykesville, 31 miles out into the Patapsco River valley, before the B&O adopted a track of wood stringers topped with bar iron. At about the same time, the railroad gave up on stone viaducts and began building wood truss bridges. Not long after came the realization that opening the line had to come first. Permanence would come in time.

By the time the B&O reached Wheeling, its track resembled the "American standard" of hardwood crossties set in rock ballast with some version of a "T" rail spiked directly to the ties. Wood bridges were superceded by composite, modular, prefabricated spans of cast and wrought iron. The B&O began operations with horses hauling freight and passenger cars. The railroad committed to steam locomotion in 1834, and by the War had the largest locomotive fleet in the country. Perhaps most importantly, after the first twenty five years the railroad understood itself as a complex linear technological system, instead of as an elaboration of a traditional highway.

Up and down the railroad, the changes were of that magnitude. The first rolling stock literally was crafted by hand, with only a few crude machine tools. By the 1850s, the B&O had dozens of shops with equipment of varying ages and capabilities. But in the aggregate, the company had one of the largest single industrial plants in the country with one of the most comprehensive complements of machine tools. The railroad had its own foundry, rolling mill, and wood mills. It manufactured its own bridges and built virtually all of its own rolling stock.

One of the most painful lessons resulted from having only an early version of Stephen Long's theory of Equated Distance. At the beginning, with the cost of grading by hand so high and the proposed speeds so low (either by horsepower or slow locomotive power), Jonathan Knight and his engineers decided to optimize the location for mild gradients and minimum earthwork. The result was a reasonably level alignment with sharp curves, greater length, and low maximum speed.

Successful steam locomotives changed the B&O's thinking, and the railroad revised its location criteria in the mid-1830s, optimizing for broader curves at the price of steeper grades, more bridges, and heavier construction. That was epitomized by the line from Cumberland to Wheeling. The work of revising the antebellum B&O to a more modern configuration continues to this day, although at a vastly reduced level. The hard-won conclusion was that a railroad is never completed–it merely pauses between construction phases.

Figure 19. Of the two dozen or so civil engineers who helped create the antebellum B&O, three men–Jonathan Knight, Benjamin H. Latrobe, and Albert Fink–were primarily responsible for its form and completion.

Conclusion

The Baltimore and Ohio Railroad itself ceased to exist in 1987, when CSX Corporation surrendered the B&O's charter to the State of Maryland and absorbed the company's assets and operating rights into its railroad subsidiary. With that bit of legal housekeeping, the end came for one of America's oldest firms and the oldest continuously operating railroad company in the world.

It has not been an independent railroad for almost forty years. By the late 1950s, the B&O had traffic but little capital, while the Chesapeake and Ohio Railway had capital and wanted a broader revenue base. The railroads were complementary, and from the early 1960s forward the B&O operated under the control of the C&O or as part of the Chessie System Inc. Further mergers created CSX Transportation. Once again the B&O found itself on the leading edge of industry-wide change. The C&O/B&O merger initiated the modern railroad merger movement, which is in its end stages today.

Yet at the beginning of the twenty first century, the company remains in the cultural consciousness. "B&O" is represented disproportionately in electric train sets, among collectors of antiques, and throughout the railroad enthusiast field. In Baltimore, one of the country's largest and most respected railroad museums is dedicated to its history. One of the four railroads on the Monopoly board is the B&O. Among historians, the B&O enjoys significant stature as an "important" project. Whether they do economic, cultural, labor, political, technological, or social history, scholars recognize the pivotal role played by this one company.

I have alluded to the most immediate and powerful effect of railroading: its ability to alter radically a society's notions of space, time, distance, and intensity. In fact this was one of the most-remarked effects of the first railroads, and the cause for both awe and discomfort. German historian Wolfgang Schivelbusch explores the range of those shifts in attitude and assumption in his seminal work *The Railway Journey: Trains and Travel in the 19th Century*.

The B&O was the first railroad to begin regular service (in January, 1830), the first to successfully penetrate the mountains, one of the first to experiment with and operate steam locomotives, the first railroad to serve Washington, and an early participant in all aspects of what came to be called the Transportation Revolution. In and of themselves, these claims of priority or innovation mean very little. What is important is the cumulative effect of placing this one company in association with all of the fundamental shifts in attitude described by Schivelbusch.

The B&O thus played a critical role in a significant transformation not yet fully explored by historians. The antebellum railroad imposed "patterns" on the United States–physical, mental, perceptual, economic, and even emotional. As what I characterize a "patterning agency," the nineteenth century railroad ranks with government and religion as one of the three most powerful forces of antebellum change and development.

This is, I realize, a broad claim. I do not suggest that "the railroad" had that kind of effect as an explicit goal, or even that there was a conscious logic at work. In fact I argue largely the opposite: there were implicit and explicit needs for some agency to facilitate the physical ordering of the country, and to establish basic "ground rules" and templates for a rapidly evolving proto-industrial society. The country needed physical mobility, as well as social and economic mobility. The railroad happened to fit the bill.

We take those sorts of rules and behaviors for granted today. Americans before the Civil War had to learn, for example, to account for time and be governed by a clock, rather then the sun or weather. They had to become comfortable with the idea that large corporations might take their property, alter their family economy, or require them to conform to an industrial logic. They had to adapt to the virtual shrinking of physical distance, the increase in mobility, the extended reach of markets, and the reality of instantaneous communication over great distances. The telegraph–first constructed along the B&O between Baltimore and Washington in 1844–was so disconcerting that many Americans had trouble simply accepting its possibility.

More to the point, the B&O changed people's economic status and quality of life. As the line opened, it lowered the cost of transportation and created new markets. One of its most striking effects was to simultaneously halve the cost of passenger travel, increase its comfort and reliability, and boost its speed by an order of magnitude. When the railroad opened the Washington Branch between Baltimore and Washington in 1835, a ride of two hours replaced a full day's hard traveling by stage. That kind of effect had great power to evoke awe, admiration, hope, and all of the hostile emotions.

What this all adds up to (and there is vastly more to it) is a wholesale reconfiguration of American society based on railroad transportation. Cheap transport made intense urbanization possible, spurred industrial development, aided immigration and internal resettlement, helped create one common national market, opened the country to further regional specialization, and unlocked the value and development potential of millions of acres of farm and natural resource land.

Because it was such a powerful and useful tool, Americans willingly changed their habits and attitudes to accommodate the needs of the railroad. In turn, the railroad solved acute problems of geography and resource allocation. French engineer Michel Chevalier summed up his take on that phenomenon in a letter written to a Paris newspaper while he was on his United States inspection tour in 1834:

> "The American type . . . is devoured with a passion for movement, he cannot stay in one place. . . . The Americans have railroads in the water, in the bowels of the earth, and in the air. . . . The benefits of the invention are so palpable to their practical good sense that they endeavor to make an application of it everywhere and to everything, rightly or wrongly."

One of the most powerful statements an engineering project can make–in reality, its most lasting and significant contribution--is somehow to have changed the way people think. That is what distinguishes a competent project from a visionary work. It is what transforms engineering into culture and conditions a society to understand technology as good, bad, or indifferent.

The B&O offered some of the first evidence Americans would see of engineering's immense power to alter space, change duration, and completely rework what had been the natural order of things. Canals were simply long, thin lakes, and even improved roads were merely elaborations of long-existing paths. On them, travel and transport still took place with the speed, intensity, and form of Ancient Rome. Railroads were different, and their potential was both frightening and enticing.

* * *

The position we find ourselves in when trying to assess, or even define, the significance of this railroad recalls the ancient Buddhist parable of the blind men and the elephant. For the last 150 years, individuals, society, and the company itself approached the history of the early B&O with differing attitudes and expectations, and took from it whatever lessons and conclusions they needed at that time. The unfortunate result was that the simple question "what was the significance of the antebellum B&O Railroad" came to have a numbingly complex, and somewhat elusive, answer.

It may be that the real significance of the B&O Railroad lies somewhat outside an elaborately argued conclusion about myth making and engineering statistics. Truly great works of engineering ultimately exist in a state of being not of the engineer's making and far beyond the manager's control. They become part of the messy world of culture, perceptions, memory, and myth.

The antebellum B&O understood that. Distilled from the rhetoric and stripped to its essential two ideas, this is what the Directors of the B&O Railroad had to say after a quarter century of hard work, upon attaining its goal of the Ohio River:

"At length the great object is accomplished.
Man has triumphed over the mountains. . ."

Selected References

The formal literature on the subject is substantial and diverse. The following is a small sample of the work treating antebellum engineering and the B&O Railroad.

Baer, C. T. (1981). *Canals and Railroads of the Mid-Atlantic States, 1800-1860.* Eleutherian Mill-Hagley Foundation Wilmington, DE.

Brown, W. H. (1871). *The History of the First Locomotives in America.* D. Appleton, New York.

Calhoun, D. H. (1960). *The American Civil Engineer: Origins and Conflict.* Massachusetts Institute of Technology Press, Cambridge.

Calvert, M. A. (1967). *The Mechanical Engineer in America, 1830-1910: Professional Cultures in Conflict.* Johns Hopkins University Press, Baltimore.

Caplinger, M. (1997). *Bridges Over Time: A Technological Context for the Baltimore and Ohio Railroad Main Stem at Harpers Ferry, West Virginia.* Monograph Series, IV. Institute for the History of Technology and Industrial Archeology, Morgantown, WV.

Condit, C. W. (1960). *American Building Art: The Nineteenth Century.* Oxford University Press, New York.

Dilts, J. D. (1993). *The Great Road: The Building of the Baltimore & Ohio, the Nation's First Railroad, 1828-1855.* Stanford University Press, Stanford.

Ferguson, E. S. ed. (1965). *Early Engineering Reminiscences (1815-40) of George Escol Sellers.* U.S.N.M. Bulletin 238. Smithsonian Institution, Washington.

Ferguson, E. S. (1992). *Engineering and the Mind's Eye.* MIT Press, Cambridge.

Fogel, R.W. (1964). *Railroads and American Economic Growth: Essays in Econometric History.* Johns Hopkins University, Baltimore.

Ghega, C. (1844). *Die Baltimore-Ohio-Eisenbahn Uber das Alleghany-Gebirg, mit Besonderer Berucksichtigung der Steigungs- und Krummungsverhaltnisse* [The Baltimore and Ohio Railroad Crossing of the Alleghany Mountains, With Particular Consideration of the Grades and Curvatures]. Kaulfuss Witwe, Prandel, Vienna.

Grayson, L. P. (1993). *The Making of an Engineer: An Illustrated History of Engineering Education in the United States and Canada.* John Wiley, New York.

Hall, R. A. (1961). "Engineering and the Scientific Revolution," *Technology and Culture* (2), 333-340.

Harwood, H. H. (1979). *Impossible Challenge: The Baltimore and Ohio Railroad in Maryland.* Barnard, Roberts, Baltimore.

Hill, F. G. (1957). *Roads, Rails and Waterways: The Army Engineers and Early Transportation.* University of Oklahoma Press, Norman.

Hindle, B. and Lubar, S. (1986). *Engines of Change: The American Industrial Evolution, 1790-1860.* Smithsonian Institution Press, Washington.

Hollis, J. R. and Roberts, C. S. (1992). *East End: Harpers Ferry to Cumberland, 1842-1992.* Barnard, Roberts, Baltimore.

Hungerford, E. (1928). *The Story of the Baltimore and Ohio.* G. P. Putnam's Sons, New York.

Layton, E. T. Jr., ed. (1973). *Technology and Social Change in America.* Harper & Row, New York.

Long, S. H. (1829). *Rail Road Manual.* W. Wooddy, Baltimore.

Marx, L. (1964) *Machine in the Garden.* Oxford University Press, New York.

Meier, H. (1957). "Technology and Democracy, 1800-1860." *Mississippi Valley Historical Review* (43), 618-640.

Meinig, D. W. (1993). *The Shaping of America: A Geographical Perspective on 500 Years of American History. V. 2: Continental America, 1800-1867.* Yale, New Haven.

Merritt, R. S. (1969) *Engineering in American Society, 1850-1875.* University Press of Kentucky, Lexington.

Peters, T. F. (1996) *Building the Nineteenth Century.* MIT Press, Cambridge.

Reynolds, T. S. (1991). "The Engineer in 19th-Century America," in *The Engineer in America: A Historical Anthology from "Technology and Culture,"* ed. T. S. Reynolds.

University of Chicago Press, Chicago.

Ringwalt, J. L. (1888). *Development of Transportation Systems in the United States.* The Author, Philadelphia.

Rubin, J. (1961). "Canal or Railroad? Imitation and Innovation in the Response to the Erie Canal in Philadelphia, Baltimore, and Boston. *Transactions of the American Philosophical Society.* (57) 1, 3-106.

Sanford, C. L. (1958). "The Intellectual Origins and New Worldliness of American Industry." *Journal of Economic History* 18), 1-16.

Schivelbusch, W. (1977). *The Railway Journey: Trains and Travel in the 19th Century.* Urizen Books, New York.

[Smith, W. P.] (1853). *History and Description of the Baltimore and Ohio Railroad.* Murphy & Co., Baltimore.

Smith, W. P. (1858). *The Book of the Great Railway Celebrations of 1857.* D. Appleton, New York.

Stapleton, D. (1981). "Benjamin Henry Latrobe and the Transfer of Technology," in *Technology in America: A History of Individuals and Ideas,* ed. Carroll W. Pursell, Jr. MIT Press, Cambridge.

Stover, J. S. (1987). *History of the Baltimore and Ohio Railroad.* Purdue University Press, Lafayette.

Stover, J. S. (1961). *American Railroads.* University of Chicago Press, Chicago.

Stuart, C. B. (1871). *Lives and Works of Civil and military Engineers of America.* Van Nostrand, New York.

Taylor, G. R. (1951). *The Transportation Revolution, 1815-1860.* Rinehart, New York.

Vance, J. E. Jr. (1995). *The North American Railroad: Its Origin, Evolution, and Geography.* Johns Hopkins University, Baltimore.

Vogel, R. M. (1964) "The Engineering Contributions of Wendel Bollman." *Contributions from the Museum of History and Technology: Paper 36.* Smithsonian Institution, Washington.

The History of Building Material and Methods: Cement and Steel

HYDRAULIC CEMENT: The Magic Powder

Emory L. Kemp, F.A.S.C.E.[1]

Introduction

Natural cement was indeed a magic powder during most of the 19th century when it was used extensively for a myriad of public works and buildings but it had an especially significant synergistic relationship with the construction of canals. The network of canals provided a relatively inexpensive means of shipping cement to market; on the other hand hydraulic cements were essential for the construction of locks, dams, and other canal structures. The synergistic relationship is explored in terms of the Erie, Lehigh, Delaware and Hudson, Chesapeake and Ohio canals and the navigation on the Ohio River.

The secret of the "magic powder" rested upon its ability to "set" under water and equally important, the hydrated material was waterproof. Thus, in contrast with lime mortar, hydraulic cement could be used for foundation work and especially for hydraulic structures such as bridge piers and abutments and a wide range of canal structures.

Although there were significant pioneering efforts in America by Hyatt, 1816-1901, and in France by Joseph Monier, 1823-1906, Lambot, patent 1855, and Francois Coigent, patent 1861, to overcome the inherent weakness of concrete in tension by the addition of metal reinforcement, the era of reinforced concrete was to wait until the introduction of Portland cement at the end of the 19th and early part of the 20th centuries. (Straub 1964, 208) As a result, natural cement found its greatest application in hydraulic mortar in brickwork and stone masonry. Thus, the discussion will rest on the use of hydraulic cements as a bonding agent in mortar and not in reinforced concrete structural components. The material aspects of hydraulic cements, including Portland cement, are linked to the history of 19th century transportation and industry.

---

[1] Professor Emeritus of History and of Civil Engineering, Director of the Institute for the History of Technology and Industrial Archaeology, West Virginia University

An Ancient Lineage

Before considering the 19th century history of hydraulic cement it is well to identify the origins of this material in the context of other materials used in the ancient world. This is not an attempt at a comprehensive presentation of the subject, but rather to establish the lineage of natural, and later, Portland cement.

The most appropriate place to start is with mud. From prehistoric times, dating back thousands of years, mud was used in the form of sun dried bricks as well as mortar. Because the materials were not fired in a kiln few artifacts of this type of construction survive. In mud brick construction straw was mixed with the clay to control shrinkage. There are extant examples however of rammed earth construction in both Europe and America. (Singer 1954, Vol.1, 459-473)

The clay technology existed in parallel with the use of lime based mortars. Before, however, dealing with the use of lime it is important to note the use of several developments in the 4th millennium B.C.E. Rammed earth walls of a massive character were built and enhanced with the facing of cones into the walls, producing a closely spaced pattern which may well also have strengthened the mud wall. (Singer 1954, Vol. 1, 402) Later, in the 3rd millennium, we have the first archeological evidence of the true, as contrasted with the corbelled, arch in Mesopotamia. (Singer 1954, Vol. 1, 472) The origin of the arch is not a discovery utilized in grand architecture, but rather, arches were employed in drains. Stone was almost nonexistent in the great plain surrounding the twin rivers of Mesopotamia. Timber was scarce so the builders of that era depended on clay. With asphalt exuding to the surface in the area this material was used as both mortar and as stucco to provide a waterproof coating on a rammed earth or mud brick structure.

Mud and rammed earth technology dominated domestic building in Egypt, a land long associated with monumental architecture in stone. With unfired mud brick it is not surprising that little evidence remains of this vernacular architecture. Unlike Mesopotamia, Egypt was blessed with an abundant supply of stone for building together with extensive deposits of gypsum (derived from the Greek word for chalk.) It appears likely that the Egyptians would lay foundations for stone masonry without mortar. Gypsum mixed with water was apparently used to lubricate the positioning of large stone blocks after they had been dressed, a good example are the numerous pyramids. Gypsum ($CaSO_4+2H_2O$) was easily produced by heating the material and driving off SO_2. This is the equivalent of quick lime. Gypsum is still widely used for plaster and plaster board in modern construction on a worldwide basis.

Lime forms, an essential component in all hydraulic cements except those based upon bauxite, were unknown in ancient construction techniques in both Mesopotamia and Egypt. For the origins of the applications of lime we must turn to the Hellenistic, and later, Roman civilizations.

So far we have seen various materials used in the ancient world in the building arts. It is necessary now to turn to the origin of lime as the antecedent of the discovery and widespread use of hydraulic lime based on mortar used first in the Hellenistic world and transferred subsequently in the hands of Roman engineers later in the classical period. Manufactured lime from ancient times was little used as a principal ingredient in mortar or concrete nor did it serve as the primary component of hydraulic limes until the classical period of Greece and Rome.

In forming a direct linkage with natural cement it is therefore important that one consider, albeit briefly, the origins of lime materials based largely on archaeological evidence.

Limestone, calcium carbonate, is converted to quicklime by calcining at about 900°C. which removes calcium dioxide, CO_2, leaving calcium oxide, CaO, commonly known as quicklime. When mixed with water the lime slakes producing a chain of calcium hydroxide with a large expansion in volume and considerable amount of heat. These chains of calcium hydroxide very gradually recombine with atmospheric carbon dioxide under exposed conditions. Mortar made from lime has certain advantages recognized by ancient builders, in particular, the ease of laying up brickwork and stone masonry with lime mortar, that is, lime and sand. The resulting mortar beds are elastic under moderate loads and not so hard as to induce cracking in the brickwork or masonry. The exposed face of mortar deteriorates with time in the atmosphere and must be re-pointed. The decided disadvantage of this material is that the mortar is not waterproof and thus unsuitable for use in hydraulic structures such as foundations and indeed in any application in a moist environment. Nevertheless lime mortar was used continuously until modern times. One well remembers the necessity of tuck pointing old brick walls in the 20th century.

Although clay and asphalting mortars were use as early as the 3rd millennium B.C.E., lime was used in Babylon during the reign of Nebuchadnezzar, 605-562, in building construction. Archaeologist H. Frankfort Jacobsen and C. Preosser discovered a lime kiln at Khafage in the Middle East dated as early as 2450 B.C.E. (Davey 1961, 127) As early as 700 B.C.E. pozzolanic lime mortar was used by the Phoenicians. Such a hydraulic mortar may well have unknowingly resulted in the mixture of lime and a naturally occurring sand with an argillaceous component, usually clay. In any case it is the earliest example of the forerunner of hydraulic cement.

Greek and Roman Contributions

The triumph of Roman civil engineering is well known and well recorded particularly the achievements during the Roman Empire. What is much less known is the Hellenistic Greek contributions in the field, two of which are related to the development of hydraulic mortar. Volcanic rock of the island of Thera (a.k.a. Santorini) formed the basic component of hydraulic mortar used by the Greeks, particularly in their aqueducts.

The Romans are justly famous for the aqueducts that, as it turns out, are based on Greek precedent. One only has to understand that the information given by Vitruvius is based largely on Greek technology. (Rowland 2002) It was in the construction of aqueducts that both Greek and Roman engineers developed the highest skills in the use of hydraulic mortars for a wide variety of hydraulic engineering structures.

The ca. 6th century B.C. E. and the ca. 180 B.C.E. aqueducts are vivid reminders of the skill of Greek engineers. The Samos aqueduct consisted of both an aqueduct bridge and tunnel, while the later aqueduct at Pergamon features inverted syphons. (Singer 1954, Vol.2, 667-668) The Roman historian Herodotus names the engineer as Ecpalinus of Megara. As a result, he is the earliest known hydraulic engineer. As in modern tunneling technology, the Samos aqueduct system employed over one thousand men and was begun at both ends. Use of hydraulic materials and the mastery of the hydraulics of the system depended on surveying using various leveling devices. The Samos Aqueduct, however, resulted in a misalignment at the middle of the tunnel of sixteen and four tenths feet, or five meters, necessitating a sharp bend in the alignment which really did not affect the hydraulic grade line. Herodotus declares the Samos aqueduct as one of the three great Greek public works. The Pergamon aqueduct featured the Greek invention of the inverted siphon. By using this device a direct alignment could be maintained but resulted in pressures of twenty atmospheres which required pressure seals in the joints. The solution was to use a quick lime and oil mixture which would expand around the pipes, sealing each joint against the pressures that result when the grade line dips below the hydraulic grade line. Vitruvius, a 1st century architect and engineer, well known for his ten books on architecture, reported on the technology of building and particularly the use of hydraulic cements in brickwork and stone masonry. Writing in the 1st century B.C.E., his books reflected the technology of the Greeks as well as current Roman practice. (Rowland, 2002)

In terms of cementitious materials and their use as a bonding agent, Vitruvius deals first with sand in terms of a suitable material which should be sound, clear and well graded. The addition of ordinary sand to lime produces a mortar which found widespread use as an inert material that reduces the cost of the resulting mortar and, equally important, reduces shrinkage compared to the use of the neat lime, or so-called grout. Superior performing mortar results in using well-graded sand, where the interstices formed with a coarse sand are filled with finer materials giving a maximum density. Such a gradation requires less lime or cement and results in a stronger material with minimum shrinkage. Vitruvius states "there is also a type of powder that brings about marvelous things naturally. It occurs in the regions of Baiae (near the modern city of Naples) and in the countryside that belongs to the towns of Mount Vesuvius. Mixed with lime and rubble it lends strength to all other sorts of construction but, in addition, when molds employing this powder are set into the sea they solidify under water." (Singer 1954, Vol.2, 407) (Davey 1961, 121)

It is important to note in this statement the addition of potsherds, the technique of mixing pulverized brick or potsherds with lime, is still practiced in India and elsewhere. Since quick lime can be produced at a comparatively low temperature vis a vis natural or Portland cement, this is an economical method to produce a substance superior to pure lime in terms of mortar, which exhibits hydraulic properties.

In the 1st century B.C.E., concrete played a leading role in Roman architecture and engineering. The Pantheon, in Rome, is a superb example. Not only vaults and domes were cast in concrete, but also walls and other more mundane structures. In the case of vaults, one technique was to cast walls behind masonry which served as the formwork. Another technique was to use formwork and produce a concrete with layers of stone which were then filled in with lime mortar, often hydraulic lime mortar, layer after layer.

From the time of Vitruvius in the 1st century until nearly four centuries later there was a great surge in building to accompany the vast empire. In the spirit of Vitruvius and Frontinus, a certain Farentinius wrote these lines in the fourth century C.E. "at the mixing let one part of lime be added to two parts of sand but if to the river sand you add a third part of powdered earthenware it will give the work an unbelievable solidity." This produces a mildly hydraulic cement which is still widely used in underdeveloped countries. (Plommer 1973, 55)

Before leaving the Roman engineering world it is instructive to present information on the means of securing watertight channels for Roman aqueducts. For this purpose, the work of Malinowski sheds new light on an ancient technology. (Malinowski 1979) The method used for aqueduct liners is the application of a smooth watertight interior surface which would prevent leakage and improve the hydraulic flow by reducing surface roughness. In the moist atmosphere of an aqueduct both shrinkage and temperature changes were largely eliminated. The real test for the Roman engineers was to achieve a watertight surface in the channel of long elevated aqueducts built without expansion joints. In these cases the builders relied on old and proven empirical systems consisting of multilayer plaster systems. Using the Roman elevated aqueduct at Caesarea in Israel, Malinowski reports on a six-layer plaster consisting of the following mortar layers beginning with the internal to the external. A grey material with charcoal particles based on lime followed by a layer of white grout consisting of crushed marble, a hydraulic lime consisting of finely crushed brick with lime followed by another grey layer, a layer with white marble, followed by another red layer of crushed brick. The mortar layers become thinner and thinner as they approach the surface. With this system the grey mortar forms a slightly absorbing base resulting in good adherence of the fresh white and following red mortar. This adherence is of special importance for the construction of vertical surfaces along the wall of the aqueduct channel. The grey mortar as applied had tightness and strength while the white layer shows unusual hardness and great tightness and the red one is also tight and very strong. The surface of each of the layers was roughened at the end of each application. Both the composition and the polishing resulted in the extraordinary properties of Roman aqueducts. (Malinowski 1979, 69)

Middle Ages to the Industrial Revolution

The fall of the Roman Empire as depicted dramatically by Gibbons, 1737-1794, brought with it the collapse of Roman building technology, especially the use of hydraulic cements discussed above. As Kenneth Clark comments in his popular television series called *Civilisation,* just how close European civilization was to being snuffed out. Nevertheless, the architecture of the former Roman Empire was transformed into the Romanesque style on a modest scale. This movement later led to the towering cathedrals of the high Gothic period. The transformation came with the introduction of rib vaulting as seen at Durham cathedral which was begun in 1096, with vaulting supported by large circular Romanesque columns faced with masonry and filled with rubble stone-type concrete of limited strength. (Fletcher 1954, 354-356) (Batsford 1954, 29-32) These massive columns were inferior to Roman building and did not advance the use of hydraulic cements. Without benefit then of strong hydraulic cements and concrete medieval and renaissance architects and builders relied on finely jointed ashlar masonry needing only a limited use of mortar, such as the rib vaulting at Durham cathedral.

Industrial Revolution

It is not the purpose of this narrative to debate the beginning of the industrial revolution but only to note that the surge of new industries coupled with new transportation systems put new demands on architects and engineers, beginning in the 18th century. Suitable materials were needed for the construction of foundations and other hydraulic structures, such as locks and dams, subjected to water.

Most buildings and other structures above ground were set in fat lime and sand and left to air dry. This precluded the use of lime in foundations and hydraulic structures and equally important in less expensive building using rubble masonry with wide joints. (Straub 1964, 206)

Although records are lacking, there was a limited use of hydraulic limes using Dutch trass which continued the tradition of pozzolanic ash of the Roman world. One notable example which used hydraulic lime or natural cement was the Pont Royal in Paris. Completed in 1685 to a design by Z.H. Mansard, with an elaborate specification including details of the masonry to be laid up in a pozzolanic mortar, it is one of the earliest sources of information on hydraulic cement. Although details are lacking, it may well have been the first use of such mortar in France. (Singer 1954, Vol.3, 425) In addition, there are probably further examples in Europe dating before the industrial revolution.

The need for hydraulic cement as the binder in mortar and concrete greatly increased from the middle of the 18th century. The beginning of the modern period is appropriately associated with John Smeaton, 1724-1792, builder of the Eddystone lighthouse. As a leading engineer of his day, Smeaton combined a practical builder's outlook with a curiosity for experimentation such as his notable work on waterwheels. He established a society for engineers and was amongst the first to use the term civil engineer. Smeaton's opportunity came from the unexpected fire which destroyed the Eddystone lighthouse in 1755. Located several miles out to sea and some fourteen miles south of Plymouth on a rock ledge exposed to severe action of wind and wave, it served as a beacon for navigation and it was essential that the lighthouse be rebuilt and the task fell onto Smeaton. With his experimental bent, Smeaton undertook tests on the properties of various lime mortars available to him. In the end he decided upon the well known pozzolanic ash from Italy. In the course of these tests he determined that the hydraulic properties of the various limes were dependent upon the amount of clay (argillaceous) matter. Unfortunately, his research was not widely known until 1790 when his comprehensive report on building the Eddystone light was published. (Skempton 2002, 621-622) Shortly after Smeaton's account, Bruce Parker was granted a British patent for a hydraulic cement which he named "Roman cement". The principal component consisted of nodules found in London clay on the English south coast. These were essentially nodules of limestone with a clay component.

As a leading engineer with the French Corps des Ingenieurs des Ponts et Chaussées, Louis Joseph Vicat, 1786-1861, combined practical engineering design with his extensive research. Amongst his inquiries was his effort on wire cable suspension bridges first introduced in France by Marc Seguin of Tournon. It was Vicat who invented the spinning of cables for wire suspension bridges. He studied the full range of lime-based materials from fat limes to moderately and strongly hydraulic limes determining the optimum calcining temperature for each one. Coining the term *hydraulic cement*, he classified natural and artificial hydraulic cements. In the latter case artificial cement was obtained by the conscious mixing of calcareous and argillaceous materials. With his work he laid the groundwork for scientific understanding of cements. (Straub 1964, 176, 206-207) (Mahan 1880, 21, 41) (Bauer 1949, 3) (Williams 1966, 406, 463-464)

The issuing of British patent #5022 in 1824 to Joseph Aspdin has often been cited as the beginning of the Portland cement industry. Aspdin called his patented product "Portland Cement". His patent however did not mention an essential element in the production of Portland cement in that the blended ingredients had to be calcined at a high heat approaching a cindering temperature. His claim was challenged by L. C. Johnson, 1811-1911, who stated later that he discovered the so-called over-burnt clinker, which when pulverized, produced a superior cement. In any case Aspdin's son did not open a manufacturing operation until 1843. (Barfoot 1974) (Skempton 2002, 22-23)

In the American case, as we shall see later, Portland cement did not become available as a domestic product until the latter years of the century beginning in the late 1880s. Thus the history of hydraulic cement for the first three quarters of the nineteenth century relied on the use of natural cement. Its significance in large Victorian public works and buildings must be considered in the history of building. The use of hydraulic cement in one form or another gained momentum in both Britain and Europe during the 19th century. One of the greatest early uses of Portland cement occurred in building the great embankment along the Thames River in London. This significant public works project consisted of a large sewer hidden in the embankment. Under the direction of Sir Joseph Bazalgette, 1819-1891, the project took from 1858 until 1875 to complete and used 70,000 tons of cement. (Williams 1960, 426-427) (Singer 1954, Vol. 4, 448)

The beginning of the employment of natural cement in America was on a much more modest scale closely associated with the developing network of canals. By 1811, and following the appearance of Parker's Roman cement in 1796, James Frost, c. 1774-1851, a builder from Norwich in East Anglia, England, began investigating natural cement works on the Essex coast possibly as early as 1807, but only in earnest after the expiration of the Parker patent in 1810. He experimented with clay in lime together with other ingredients, and visited Vicat in France in the 1820s; later, in 1833, he immigrated to the US. He combined clay and limestone in a slurry which was the predecessor of the wet mill process used in a number of American cement mills to produce Portland cement. Frost's wet mill process was not a success nor did he reach the high temperatures needed to produce a truly Portland cement. (Skempton 2002, 239)

In early American canal building attempts were made to use quick lime which was found to be quite unsatisfactory for use under water or in damp conditions. The well-known Huguenot émigré Benjamin Henry Latrobe, 1764-1820, apparently used common lime with imported trass, a volcanic earth, which imparted a certain degree of hydraulic characteristics in his Chesapeake and Delaware canal structures in the early 1800s. (Skempton 2002, 394-395) Loammi Baldwin, 1745-1807, was chief engineer and director of the Middlesex canal, under construction from 1796-1797. It is not clear if his mortar had any hydraulic properties of if he imported hydraulic cement. (Fitzsimons 1972, 6-7) Building the Erie Canal not only ushered in the canal mania in America but also the beginning of the natural cement era which persisted until World War II, with increasing competition from Portland cement in the last decades of the 19th century and throughout the first half of the 20th century. Had it not been for the demand for natural cement in building a network of canals in the early part of the century, builders might well have continued to use lime mortar and found it unnecessary to import natural cement from Britain and Europe. The subsequent establishing of natural cement industry in America in all likelihood delayed the introduction of Portland cement.

In 1817 Canvass White, 1790-1834, was employed by Benjamin Wright, 1770-1842, (Fitzsimons 1972, 132-134) as an engineer on the Erie canal, and sailed to England to examine canal structures including the use of hydraulic cement. (McKee 1961) (McKee 1975) (Fitzsimons 1972, 126-128) Although Wright recommended the use of imported natural cement the canal commissioners had insufficient funds so that during 1818 and first part of 1819 stones for locks and other hydraulic structures were laid up with common lime mortar. The result was early evidence of failure. Fortuitously cement rock was discovered by White in the course of excavating the central portion of the Erie Canal and in so doing rock was found to be quite suitable for the production of natural cement. The official record of the discovery appeared in the report of the canal commissioners. The canal engineers responded with alacrity to the manufacture of truly hydraulic cement obtained from so called cement rock rather than importing such materials. During the construction of the canal 500,000 bushels of natural cement were used.

The Magic Powder

Born at Black River, Ohio in 1825, Quincy Adams Gillmore, while serving as an Army officer made the most influencial contribution to the understanding of applications of natural cement through his monumental series of publications. (Fitzsimons 1972, 50) Published by the U.S. Engineer Department in 1863, Gillmore's 331 page study entitled *Practical Treatise on Limes, Hydraulic Cements and Mortars* became a milestone in the study of cementitious materials resulting in eleven editions between 1863-1896. These various editions dominated the field until the end of the century. (Gillmore 1863)

In his well-known treatise on civil engineering, D.H. Mahan devotes a large portion of his popular text to the physical characteristics of cementitious materials. (Mahan 1880)   Shortly after the eleventh and last edition of Gillmore's work, Homer A. Reid and Uriah Cummings published books on cement and concrete. (Reid 1908) (Cummings 1898)   As a raw material, cement rock drew the attention of geologists in several states. A noteworthy contribution was made by Edwin C. Eckel to the 54th Annual Report of the Regents of the University of the State of New York in 1902. (Eckel 1902) His interests ranged from the nature and location of limestone in New York to details of testing cement samples.   He later published a study on the resources of western Virginia. (Eckel 1909) As the market for natural cement waned the attention of researchers was re-directed to Portland cement. Thus, the above references represent the flourishing period of publications on natural cements.

As we have seen the use of calcium compounds for a wide variety of uses such as mortar, stucco and concrete dates from ancient times.   Before Smeaton identified the role of clay or the research of Vicat on the composition of cements all of the applications were on an empirical basis without an understanding of the chemistry involved.

While not a discourse on the chemistry of cement, an understanding of the reactions that take place can help in unraveling the historical development of cement and clarify the various terms employed.   Grout is simply a mixture of cement or lime with water which sets to form a solid mass with considerable compressive strength but lacks in tensile capacity. With the addition sand, mortar is obtained and has been used widely from ancient times to provide a bed between the brickwork or stone masonry. A mixture of cement or lime sand, coarse aggregate such as gravel forms concrete following the addition of water which will hydrate the cement or slake the lime, uniting the compounds into a solid mass.

Before discussing manufacturing methods, an attempt should be made to classify various calcium compounds according to their engineering properties. Categories can be divided into those hydraulic and non-hydraulic cements. Both lime and gypsum plaster are non-hydraulic and thus will not set under water nor are they waterproof. Gypsum is and has been widely used for interior plaster work applied to wood or metal lath or supplied in large sheets, typically in America the most common is four feet by eight feet. Gypsum is a sulfurous compound of calcium designated $CaSO_4 + 2H_2O$. If one heats gypsum and drives off a portion of the water the result is Plaster of Paris ($CaSO_4 + \frac{1}{2} H_2O$), also known as Keene's cement or cement plaster. Adding water causes it to slake with an expansion of volume and the liberation of heat.

Common lime is obtained by heating, i.e. calcining, pure or nearly pure limestone. Such heating produces quicklime, CaO, obtained from $CaCO3$ by driving off carbon dioxide at a temperature below $1,000°C$. Like Plaster of Paris the addition of water causes quicklime, i.e. caustic lime, to slake with the liberation of heat and a gain in volume. The reaction produces a chain of calcium hydroxide $Ca OH_2$. This reaction will occur only in the air and not under water. With lime exposed to the elements some of the mortar might reabsorb carbon dioxide from the atmosphere and return calcium carbonate. This, however, is not very common. More common is rain or moisture or even rising damp in foundations which will leech out the lime leaving behind just the sand since the lime is soluble in water. Thus, the need to repoint brickwork and stone masonry periodically.

The secret of understanding various properties of cementitious materials such as hydraulic lime or natural cement is the percentage of argillaceous material usually in the form of clay, shale or slate. A useful table was published by D.H. Mahan in 1880, which relates the properties to the percentage of clay, table 1. (Mahan 1880, 19) One must understand however, that in the case of Portland cement it also is necessary to calcine the material at incipient fusion which is considerably higher than the temperatures used to produce hydraulic limes or cements.

Early producers of lime products depended upon naturally occurring limestone with less than 5% impurities to produce fat limes. These were preferred by both builders and farmers. Those limestones with an excess of 5% impurities were classified as meager or lean lime in which case they do not slake as readily as fat lime and do not produce as much heat with the addition of water. To use a Victorian term, fat limes produced an unctuous material, and with the addition of fine sand a very workable mortar. To a lesser extent this is also true of meager limes unless argillaceous materials are present which may impart hydraulic properties.

Many decry the lack of hydraulic properties of lime/sand mortars but they do represent a superior bonding agent especially for brickwork. Lime/sand mortar is more elastic than mortars composed of Portland cement and sand, and thus is favored by bricklayers for its ease of application. From a structural point of view this product yields elastically with any movement in say a wall and will prevent cracking of the brickwork.

If a calcium lime with argillaceous components, not exceeding about 10%, is heated to a little less than 1,000° C a slightly hydraulic lime is the result. At that temperature carbon dioxide is driven off as in the production of lime but not high enough to sinter the limestone. Besides producing a lime component only, the calcium oxide will unite with a silica present to form a calcium silicate. At this temperature most of the resulting chemical ingredients are of the dicalcium silicate variety with little tricalcium silicate being formed and probably not any significant amount of calcium aluminate. It is these latter compounds which provide the rapid set of modern cement.

With the clay component exceeding 10% a limited amount of silicates will form, thus producing a material with reduced hydraulic properties compared to either natural cement or later Portland cement. Pozzolanic cement differs from hydraulic lime in its properties and production methods, dating from ancient times and still employed in places like India. Pozzolanic cement is obtained by producing lime at a comparatively low temperature and mixing the hydrated lime with pozzolanic earth associated with volcanic areas or mixed with soft burned brick or tile giving the argillaceous component necessary to produce hydraulic properties.

Pozzolanic cement differs not only from fat lime but from natural cement which is made from naturally occurring limestone located by a trial and error method in the field. Small samples of what appears to be a suitable limestone is heated at a forge, ground and tested in water. Cement rock has been discovered in many of the canal constructions across the eastern part of the United States and has produced a material which is considerably cheaper than importing cements from Europe.

Because natural cement rock varied in composition so did the resulting cement in its physical and chemical properties. There is little evidence that mill owners attempted to enhance these components by the addition of clay before firing. Typically the argillaceous portion of the rock varied from 13 to 35% of which silica ranged from 10-22%. In many cement rocks used to produce natural cement magnesium was present. At the low temperatures encountered in producing natural cement MgO acts interchangeably with quicklime. It was claimed by many to improve the strength and durability of natural cement. The presence of magnesium is a benefit in the case of natural cement, but as we will encounter later magnesium is deleterious in the case of Portland cement production. Natural cement is calcined at somewhat higher temperatures say 1100°-1200°C than quicklime but just high enough to drive off carbon dioxide from the limestone.

The cement is ground into a fine powder and stored in bulk, in barrels and bags. Needless to say the product varies not only from quarry to quarry but also within a given site. At the Round Top mill along the C&O canal the rock from different parts of the mine was blended to achieve a more uniform and higher quality cement.

The approach to producing Portland cement was to use a predetermined blend of suitable elements and to fire them at temperatures high enough to ensure that the rock reached a heat of incipient fusion. As a result some cement rock has the properties necessary to produce Portland cement but most mixtures are the result of a blend of carefully controlled ingredients. In Portland cement production the ingredients are ground into a fine powder before firing. The resulting clinker, which is like old fashioned cinders, is ground to a very fine powder with gypsum added in the correct proportions to control the rate of set. With a firing temperature in the range of 1600°-1700°C tricalcium silicates and aluminates are produced which influence the setting time. In fact, modern high early cement has an increased percentage of these tricalcium silicates and aluminates. With the adjustment of ingredients it is also possible to provide a sulfur-resistant cement. High magnesium limestone or the presence of excessive amounts of lime could yield a cement which has self-destructive tendencies by delaying the hydration of either of these elements. It appears likely that attempts to produce Portland cement at Rosendale were unsuccessful because the cement rock had a high magnesium content.

In the beginning of the industry in the 19th century, first intermittent and then continual vertical kilns were employed, figures 1,2,3,4,6,7,8. These early kilns were unable to produce the necessary heat to produce Portland cement. To supplement the figures on kilns, Ries provides a useful description:

> The old kilns are made of stone, with the
> interior either round, oval or rectangular in

> *cross-section, and lined with fire brick. They*
> *are open at the top, and taper at the bottom to*
> *an opening, through which the burned stone is*
> *discharged. When the material is not being*
> *drawn, this hole is sometimes kept covered by*
> *grate bars...the height of all is 34 feet. The*
> *more modern kilns are cylindric in shape, made*
> *of boiler iron. They are from 40-45 feet high*
> *and lined with fire brick.* (Reis 1901, 687)

But they were highly effective in the production of natural cement. This led to the introduction of the Schoefer vertical kiln at Copley in an attempt to make Portland cement, figure 12. These were introduced in the 1890s and phased out early in the 20th century being replaced by rotary kilns invented in England and remain an industry standard.

Grist mills were easily converted to grind cement with both continuous and intermittent kilns closely resembling either lime or charcoal iron furnaces. The usual shape of the interior is a truncated cone, figure 8. Like the charcoal furnaces they were often built against the bank to enable the kilns to be charged more easily. In the case of the intermittent kiln the fuel and limestone were charged in alternate layers from the top which rested on a stone arch. Andrews provides a useful description:

> *When filled and well rounded over at the top*
> *they were ready for firing. About eight cords of*
> *wood were needed for a kiln and this had*
> *already been placed about the opening of the*
> *flues within easy reaching distance. As it took*
> *two days and one night to burn a kiln, the firing*
> *was started in the morning, continued through*
> *the day and night, and slowed down about dark*
> *of the second day. After firing it took three or*
> *four days to cool the mass sufficiently for*
> *handling before it could be taken out and*
> *wheeled into the upper story of the mill a short*
> *distance away.* (Andrews 1924, 15-16)

This type of kiln required a continuous supply of fuel at the bottom and a great deal of heat was lost each time the kiln was discharged. Although short-lived in the industry, the final development of the vertical kiln first used by Saylor in 1892 was the Schoefer kiln and the only type of furnace achieving the high temperatures required to produce Portland cement clinker. Separating the fuel from the limestone charge the Schoefer kiln largely eliminated any pollution of the clinker.

Frederick Rasome of England invented the rotary kiln which represents the final development of the cement kiln in 1885, figure 14. Saylor installed rotary kilns to replace the Schoefer kilns at Copley in 1904.

For natural cement production once the cement rock was either quarried or mined the rubble stone was reduced to a small size in typical "cracker", figure 9. Once the kiln was charged Ries describes the operation:

> *In burning natural cement rock, the fire is first started in the bottom of the kiln, and on this are spread alternating layers of coal and rock. The coal is of pea or chestnut size commonly. As the burned stone is drawn from the bottom, fresh stone and fuel are added at the top. The kilns are commonly built on a hillside, or where the ground is flat, five six or more in a row, and in either case tracks are laid at the top to facilitate the delivery of the stone and fuel. The yield of these kilns is large, being 50-120 barrels of cement per ton of coal.* (Reis 1901, 368)

The cross-section of a typical natural cement mill is shown in figure 5.

Grinding cement was done with traditional buhr stones mounted horizontally as in a grist mill or sometimes edge rollers moving vertically in a circle around a fixed base, figure 10. These traditional means of grinding could not cope with the hard clinker associated with Portland cement in fact the wear on the stones in producing natural cement required constant dressing. One of the solutions lay in the use of a ball mill where steel or iron balls rotated in a cylinder, figure 13.

Reinforced, and later, pre-stressed concrete became a universal building material in the 20th century based on developments to overcome concrete's inherent weakness in tension. It was not until the beginning of the 20th century that structural concrete made a significant influence on the building arts. As a result Portland cement, because of its superior strength and uniformity, came to dominate the field, table 3. In fact, nowhere in the literature do we find natural cement used in reinforced concrete construction.

While natural cement dominated the field of public works throughout the 19th century the mainstay of the industry revolved around the construction and operation of canals and river navigations. On large navigation projects such as the canalization of the Great Kanawha natural cement was specified not only for the first moveable dams which began in the last quarter of the 19th century but for their replacements in the 1930s, in connection with one of the earliest and largest of the New Deal public works. The original moveable dams were replaced by four high lift roller gated dams in which natural cement was preferred by the U.S. Army Corps of Engineers over Portland cement for massive plain concrete locks and dams.

Nevertheless, natural cement production, beginning at the end of the 19th century, steadily declined until it was used for the production of the brick mortar mixes as substantially the only market left, table 3. Even earlier the canal system so long associated with natural cement suffered an earlier and more rapid decline as towpath canals were largely abandoned as commercial enterprises in the 20th century and certainly by the beginning of the Great Depression. In an effort to resuscitate the trade the New York barge canal, built at the time of the First World War, was not a commercial success because the nation's transportation system was increasingly committed to railways, highways, and river navigations such as the Ohio River.

As a result of the abandonment of the canal the historical significance of natural cement has been largely forgotten. An attempt to bring the significance of the industry to the attention of historians and the public at large means much work needs to be undertaken in areas long a favorite of canal enthusiasts such as the Lehigh Valley, Ulster County New York, Balcony Falls Virginia on the James River or the great cement industry at Louisville Kentucky. It is a rich field of endeavor which should appeal to engineers as well as local historians.

The following list is intended to demonstrate the extent of the industry and the relationship with canals.

A Selected Gazetteer of Natural Cement Mills

Together with his brother, Hugh, Canvass White established a cement works based upon his patent. This launched White on a career as a leading canal engineer and also a pioneering manufacturer of natural cement. This industry expanded greatly during the first half of the 19th century. By 1899 there were 76 natural cement mills producing 9,868,179 barrels (300 lbs.) By the end of the century several of the earlier mills had gone out of production so the total number active during the 19th century is considerably in excess of the 76, table 2.

Completed in 1825 the Erie Canal one of the great public works in antebellum America not only was it an engineering triumph but also a financial one as well. The canal corporation paid handsome dividends to investors from the beginning. Thus, it is not surprising it initiated the canal mania similar to an earlier experience in Britain. Despite Gallatin's report of 1808, the era was characterized not by a co-operative system of developing a national network, but one that was highly competitive. (Gallatin, 1808)

The Lehigh Canal and Navigation Company improved the Lehigh River with a series of bear-trap dams forming navigation pools behind each dam. Collapsing the moveable bear trap dam caused a freshet which flushed the coal arks downstream. This invention attributed to Hazard and White provided only downstream navigation thus the arks were broken apart and sold for lumber in Philadelphia after unloading their coal.

With the triumph of the Erie Canal, completed in 1825, the owners of the Lehigh Canal and Navigation Company decided to form a slackwater navigation and, where necessary, a parallel dug canal to bring coal vessels as far as eastern Pennsylvania where the Lehigh Canal joined the Delaware Canal. It is not surprising that Canvass White was selected to undertake this work; his design featured conventional pound locks with two sets of gates constructed with stone masonry set in natural cement. White found sufficient cement rock in the valley of such quality and quantity that it was to become a major industry later in the century. At first the canal company itself became the primary user of the locally produced natural cement. This early development led to the first production of Portland cement in America. The grinding of a calcined cement rock at J.K. Siegfried's grist mill in 1828 in Northampton Borough launched the natural cement industry in the Lehigh Valley. At Copley, across the Lehigh River form Siegfried's mill, David O. Saylor found a superior cement rock which served as the basis of his Anchor Brand cement. Nevertheless, British Portland cement because of its superior strength, durability, and uniformity captured an increasing share of the market, thus Saylor began his investigation of Portland cement with a view of producing his own brand. Although his original investigation was not successful he persisted till a suitable product was obtained. Using British techniques he ground the cement rock formed them into briquettes and fired them at a higher temperature. It should be noted that the ingredients were not an artificial blend of calcareous and argillaceous materials but rather the native rock found in the area. Fortunately deleterious ingredients such as iron components and magnesium were absent in much the same way in which the ore used by Bessemer and Mushet was free of deleterious materials. Having perfected his manufacturing method sufficiently by 1870 he applied for and was granted a patent in 1871. John Eckert, a chemist, worked with Saylor in the production of Portland cement. Their work received a gold medal at the Philadelphia Centennial Exposition in 1876. In the production of natural cement Saylor used vertical kilns first the British bottle kilns, figure 11, and then the Danish Schoefer kiln, a modification of an original German design. With greatly increased demand however, more efficient kilns were needed. The answer came from the invention by Frederick Ransom in England in 1885 of a rotary kiln which is still the industry standard, figure 14. At a

Keystone Cement plant in the Lehigh Valley Jose F. DeNavarro experimented with this new device and in 1895 the company was producing Portland cement with the use of a rotary kiln. The clear superiority of the rotary kiln resulted in the Copley Mill getting their first rotary kiln in 1899 and by 1904 abandoning the vertical Schoefer kilns. Thus by the beginning of the 20th century all the elements of production were in place for the rapid expansion of the industry. So extensive was the development of the industry on a world wide basis that concrete became the universal building material with the Lehigh Valley being the birthplace of the American Portland cement industry. (Metz 1996) (Hellerich 1977) (Roberts 1914)

Due to the technical and financial success and pioneering work of Saylor and DeNavarro new cement works were established in the Lehigh Valley. The location was favorable to east coast markets and blessed with suitable raw materials. Such mills as the Nazareth cement company, the Dexter Portland cement company, the Dixie cement company, and the Penn Atlas cement company dotted the Lehigh Valley.

Turning to natural cement in America and its link to canal construction and operation, the Delaware and Hudson canal is a good example. Even before the opening of the entire length of the Erie canal in 1825 the Delaware and Hudson Canal was begun in 1823. Unlike the Erie which provided a much needed link between the Hudson River and the east coast on the one hand, and the Great Lakes on the other, a number of canals of relatively short length were proposed and selected ones built to haul bulky products to urban centers. Perhaps the most prominent in this category was the Delaware and Hudson organized by Maurice and William Wertz of Philadelphia to provide a water route from their coal holdings above the Delaware River along the Lackawaxen River to the Hudson and thence to New York City. (Shaw 1990, 84-86)

John B. Jervis, an experienced canal engineer from upstate New York, was appointed chief engineer of this new enterprise. Benjamin Wright whose sobriquet is the father of American civil engineers, surveyed the route and undoubtedly recommended Jervis who had earlier served with him on the Erie Canal. During his residency on the D&H, Jervis designed inclines, invented the leading truck for locomotives to enable them to operate on rough track, and imported the Stevenson locomotive from Britain the Sturbridge lion. Jervis later gained fame as the engineer in charge of constructing the mighty Croton Aqueduct from upstate New York to the center of New York City.

Although the tonnage carried in the initial operation of the D&H was rather small, businessmen financed a large re-building for 130-ton coal boats to replace the original 20 ton designs. John Roebling is associated with this canal when during its enlargement he built suspension bridge aqueducts in three locations the most notable and only surviving one is at Lackawaxen over the Delaware River. In both the original construction and the enlargement of the 1850s natural cement played a crucial role in the course of construction of the D&H. Cement rock was encountered at High Falls and recognized by the assistant canal engineer James S. McEntee. As a result of this discovery natural cement could be produced along the line of the canal saving both time and money.

John Littlejohn signed a contract for supplying local natural cement. After erecting his first kiln he began supplying cement to the canal as early as 1827. Other mills were also located elsewhere along the canal and not just in Rosendale Township nevertheless with the completion of the canal, Littlejohn closed his mill. A detailed account of the Rosendale cement in Ulster County near High Falls has been published by Ann Gilchrist. (Gilchrist 1976)

After a brief hiatus cement production was reestablished by Judge Ehrendorf, with his partner Watson E. Lawrence, when they founded in 1827 the Rosendale Manufacturing Company, later re-named the Rosendale Cement Company, using British-type bottle or pot kilns which were of the intermediate type and of limited production, figure 11. By intermediate one understands that after charging these kilns and heating the ingredients from the bottom of the kiln when the cement rock was sufficiently calcined the kiln was shut down and the contents removed and then the procedure started over again. These early kilns were later replaced with vertical continuous draw kilns, figure 8. Ehrendorf later founded a partnership with John P. Austin who, as a former federal employee, managed to secure government grants, notably the Croton aqueduct of which John P. Jervis was chief engineer and also the Brooklyn Navy Yard. Under the superintendence of Ezekiel Maynard the entire site was upgraded at great expense to the company. With the Depression of 1837 Lawrence was forced into bankruptcy and had to sell the Rosendale Cement Company to his creditors.

Putting this failure behind him Lawrence founded the Lawrence's Cement Works a third of a mile from the Rosendale Cement location. In a complex web of financial dealings the company was incorporated in 1858 as the Lawrenceville Manufacturing Company only to be dissolved in 1861. The former superintendent William N. Beech incorporated the Lawrenceville Cement Company in 1862. In 1881 the company apparently became the property of R. and C.I. Lefever. It should be noted that a competitor the Rosendale Lime and Cement Company founded in 1850 was also located in Lawrenceville. Earlier, from 1848 to 1853 Andrew J. Snider operated a quarry in the area and was the founder of the A.J. Snider and Son Cement Company with a loading dock on the D&H Canal.

At Bruceville, near High Falls was the cement works of Nathaniel Bruce which was later transferred to James H. and Jacob D. Vandenmark. Also in the same area one could find the cement mill of Delfield and Baxter. One of the most noteworthy events occurred in 1899 in Rosendale when the quarries of the New York and Rosendale cement company collapsed. While all of the workers escaped injury, an estimated 25,000 dollars in damage was sustained. Earlier the company had supplied cement for the Brooklyn Bridge, which spread the fame of cement from the Rosendale area as a superior product.

The James cement company as a near neighbor supplied natural cement to various locations via the D&H Canal. Also in the township one would have found not only the Rosendale Cement works but also the Norton Cement mill named for the owner F.O. Norton of New York City. In the same area the Abbey Lime and Cement Company located their mills. In the Rosendale Township the Lawrence Cement Company had their mill at Eddyville with quarries at Hickory Branch. The local cement rock was also exploited by the Warren Lime and Cement Company of Troy New York and later called the Capital Lime and Cement Company.

Hugh White, brother of Canvass White, operated a mill at Whiteport. The operation was later transferred to Louis W. Memsfield, Hugh's brother-in-law, and in 1847 the property was sold to a young entrepreneur who operated under the name of the Newark and Rosendale Lime and Cement Company. The transport of cement was a crucial economic factor in determining the price per barrel of cement delivered. To reduce costs the Newark and Rosendale Lime and Cement Company together with the Lawrence Cement Company built a wood plank road from their quarries to their cement mills. Like many early plank roads this was not durable and was replaced in 1869 by a horse drawn tramway.

Additional mills were located in Ulster County by Conley and Schaeffer Cement Works just south of the Hudson River Cement Works. Located at Lefever Falls two more mills added to the plethora of such operations in such a limited area, namely, the Martin and Clearwater Cement Company and the New York Cement Company. It is noteworthy that the latter company had the only testing laboratory in the entire region.

As often happens, as industries mature, consolidations occur. The cement industry like iron and steel underwent consolidation. In 1907 under the organizing activity of Samuel B. Carendale, who established the Carendale Rosendale Cement Company this consolidation occurred at a time when Portland cement production was rapidly overtaking the traditional natural cement market. The only local companies not included in this merger were the New York Cement Company and the Andrew J. Snider Company.

The Internal Improvements Movement predated the foundation of the New Republic in Virginia. For example colonial leaders, notably George Washington, promoted the idea of connecting tidewater Virginia with the Ohio River by improving the James River and connecting it by canal with the Great Kanawha River a major tributary of the Ohio River. Farther north efforts were to be made to improve navigation with a skirting canal around the falls of the Potomac and other navigational improvements. It was not, however, until after the success of the Erie Canal that work started in earnest on both of these projects.

These two ventures were intended to provide links to the heartland of America and not simply to provide a short haul means of supplying coal to urban markets. The Erie Canal sparked an intensely competitive movement with each East coast port engaged in constructing a canal/river navigation to the west. The only cities outflanked were Boston and Baltimore which later opted to build railways. Besides the Erie Canal, Pennsylvania interests undertook the construction of the Pennsylvania Mainline Canal and Railway. Whereas, Washington D.C. in like manner backed the construction of the Chesapeake and Ohio canal and Richmond built the James and Kanawha Canal. Each of these engineering adventures made copious use of natural cement.

In the case of the C&O canal a series of eleven independent cement mills dotted the course of the canal from Washington to Cumberland Maryland, figure 15. The *Cement Mills Along the Potomac River* provides details of all of these mills, with special reference to the Shepherdstown mill. (Hahn 1994) The first mills were along the lower reaches of the canal while others appeared as the construction progressed upstream. Thus, the cement industry in the Potomac valley owes its origin to the canal both in providing a suitable market for constructing locks dams and other hydraulic structures but also providing a cheap and efficient means of transporting the cement to markets, notably in Washington and Baltimore.

If one were traveling upstream in the canal packet along the C&O canal the first mill to be observed would have been the Seneca Mill located on or near Seneca Creek. Little is known of this mill, even the location is uncertain despite field investigation to locate either the mill or the quarry. It did however supply cement in the early days of canal construction as witnessed by the reports of the canal commissioners.

One of the earliest mills that remain is the Carrollton Cement Mill in Frederick County Maryland. It furnished the C&O Canal Company more than 40,000 bushels of natural cement. The mill was operated by Bracket and

Gray at Carrollton Manor, apparently owned by Charles Carroll a signer of the Declaration of Independence and a strong supporter of the Baltimore and Ohio Railroad. Except for a large battery of six cement kilns very little is known about this company, more survives in the field than in the literature.

Moving upstream a packet boat passenger would next encounter the Tuscarora Cement Mill also in Frederick County Maryland. From records it has been established that the mill began operations in 1830 with a commitment to supplying 2,500 bushels in May followed by up to 3,000 bushels per week at 11.75 cents per bushel. Shortly after it opened under Mosher and Eggleston a pair of canal contractors Tompkins and Bardick managed the operations for the canal company to ensure a steady supply of cement for canal construction. Nevertheless the practical upshot was that the cement was of inferior quality possibly because of a lack of argillaceous components in the cement rock. Unfortunately no chemical analysis of the raw material survives, nor representative samples of mortar made from this cement mill. Thus, this product was used sparingly with preference given to the Shepherdstown Mill located just upstream.

In a letter dated January 14, 1828, Doctor Henry Boteler wrote to Charles F. Mercer president of the C&O canal informing him of the presence of what proved to be a deposit of cement rock in great quantity and of sufficient quality.

Unlike the earlier mills downstream the Shepherdstown mill is well documented with archeological and archival information in the antebellum period. From 1828 to 1837 this mill was the primary producer for the canal and for a number of well known buildings in Washington D.C. This mill was begun as a grist mill where at the beginning both flour and cement were produced on the same buhr stones. It is the only mill prominently featured in the Civil War. Destroyed by federal forces to prevent it falling into the hands of the Confederates, it was the scene of what was called the battle of the cement mill as confederate forces were withdrawing from the bloodshed encountered at Antietam and crossed the Potomac river just ahead of the pursuing Union army. Needless to say the war disrupted commercial activities at the mill which did not reopen until 1867.

When production resumed flour was no longer produced on the buhr stones. Various attempts were made to inject new life into the operation including the development of water power to run the mill, but because of low water in the Potomac a steam engine was put into service and the canal had to be used to supply coal from Cumberland or purchased from the nearby railway.

Although cement rock limestone was the same on the Maryland side as on the Virginia shore at Shepherdstown, no Maryland mill was begun because the owner of Ferry Hill plantation John Blackford resisted development on his land. Thus, it was not until 1888 that William H. Blackford, who then owned land, built a cement mill called the Antietam Cement Company and later changed the name to Potomac Cement Company. It consisted of three large kilns capable of producing 500 barrels of cement per day for each kiln. In the first season the mill produced 5000 barrels of cement. A flood the following year, 1889, ruined 1,500 barrels of cement, destroyed the Cooper's shop and shifted the warehouse built alongside the Shenandoah Valley railroad off its foundations. The cement mill did not last many years but no record of the closing of the mill has been located. Hooks mill supplied cement for the canal from 1835 to 1837 and again in 1840. From canal company records we know that the cement from the mill found its way into aqueduct number seven and lock fifty-two for a mere 7 cents per bushel. There is no later record of the mill; it apparently ceased operation before the canal reached Cumberland in 1852. Its precise location is unknown despite a field survey.

Canal engineers discovered a copious amount of cement rock adjacent to the canal at Round Top just upstream in Hancock Maryland in 1837 and by the next year George Schaeffer had made an arrangement with the canal company to rent the land alongside the canal and obtain waterpower rights to run his mill machinery. The mill supplied cement for canal construction above mile 134 on the canal, mile 1 being in Georgetown Maryland.

> The Round Top Cement Works are located
> about three miles above Hancock, Md., on the
> Chesapeake and Ohio Canal. The mill, which
> stands on the tow-path between the Potomac
> River and canal, comprises two run of four feet
> French burrs, driven by a forty-horse water-
> power, derived from the discharge of the water
> of the canal into the river. The kilns resemble
> those at Shepherdstown and Cumberland coal
> is used for burning.
>
> The cement layers at this place crop out on the
> left bank of the Potomac, and have been cut off
> for the excavating of the canal. They are
> exceedingly crooked and tortuous, bending up
> and down, and doubling upon each other in a
> very complex manner. Their aggregate
> thickness is about 48 to 50 feet, comprising
> eleven distinct layers, each possessing marked
> and peculiar properties. (Gillmore 1863, 55)

Fire was the nemesis of the Round Top Cement Company. The first occurred in 1846 which Schaeffer rebuilt shortly thereafter. During the Civil War he sold the plant to Robert Bridges and Charles Henderson. The cement plant burned again in 1897 taking with it the bridge across the canal which collapsed into the canal prism. The new owners rebuilt the mill, this time a three story building, 120 feet long by 30 feet wide. The mill burned again in 1903 causing considerable damage to the machinery and destroying the buildings. Not surprisingly the company struggled on but was insolvent four years later. Little remains of the mill but the kilns are a notable feature on the side of the canal. The compelling reason for the closure of the company was the intrusion of the Western Maryland Railroad which passed above the site and curtailed quarrying cement rock. Round Top Cement became well known in the region supplying cement for the construction of the United States capital, the Washington Monument, the War, State, and Navy buildings, in addition to the Washington Reservoir, the boundary sewer, and the bridge over the Potomac. From an engineering point of view the most noteworthy structure was the monumental Cabin John Bridge, part of the Washington aqueduct. Numerous structures in Baltimore also used Round Top cement as did the Baltimore and Ohio Railroad stretching from Baltimore nearly 400 miles to Wheeling Virginia (later West Virginia).

Journeying upstream along the canal one would see Leopard's mill, also operated by George Schaeffer to supply cement to canal contractors during the period 1835 to 1841, apparently going out of business shortly after 1841. Like all cement mills farther downstream the Cumberland Hydraulic Cement and Manufacturing Company in 1836 built a mill supplying natural cement to the canal. Even though the company had invested in an expansion of cement making from 500 to 1,000 barrels a day in 1898 it was a short lived operation and was out of business early in the 20th century. During its period of operation the mill provided the highest quality cement of any on the C& O Canal. A latecomer to natural cement production despite the decline in natural cement in favor of Portland cement was the mill at Pinto about 5 miles southwest of Cumberland. The plant complete with an extensive battery of kilns, which still exist, was established by the Cumberland and Potomac Cement company in 1891 remaining in production to 1898 and possibly later. Little is known about the company or the plant itself. At about the same time like its neighbor, Cedar Cliff's mill was established in 1891. The following description published by the West Virginia Geological survey states:

> *The mill is a frame building, two and one-half*
> *stores high, containing a 120 H.P. engine, four*
> *boilers, a crusher and four runs of buhrs. It is*

*located on the hillside above the railroad track
and not far from the Potomac River. There are
six stone kilns twenty-eight feet high and nine
feet in diameter, in which the rock and fuel
were placed in alternate layers. The capacity
was 37-50 barrels of cement per kiln daily.
The cement cost about 25 cents a barrel to
manufacture, and it reached 115 to 140 pounds
tensile strength in twenty-four hours when
tested neat (without sand). The rock was taken
from three tunnels above the mill level. These
tunnels have been worked under the hill for a
distance of 200 feet, running a little east of
south. The rock dips at a high angle and the
layers vary in composition. Some of the layers
can be used alone, making a quick-setting
cement, but the best results were attained by a
mixture of three parts of rock in the second
tunnel with one from the first tunnel. Some of
the layers have no cement properties and have
to be thrown aside or left in the mine. The
variation in these layers, which come close
together, required careful attention and skilled
workmen.* (Grimsley 1906, 501-503)

The kilns associated with the Cedar Cliffs Company display some of the finest cut stone masonry of any in the Potomac valley, figure 1,2,&3.

From a study of these eleven cement mills the synergistic relationship between canal construction and the rise of a number of mills whose essential purpose was to supply natural cement to the building of the C&O canal is clear. With the failure of the entire Potomac valley to achieve the level of industrialization enjoyed in the Lehigh Valley and elsewhere only the Shepherdstown and Round Top mills supplied significant amounts to Washington, Baltimore and elsewhere. The rest of the mills struggled to market their product after the canal was completed to Cumberland in 1852. These mills quietly slipped into oblivion leaving behind only ruins and, in some cases, nothing at all to remind people of the once flourishing industry.

As part of the Internal Improvement Movement the C&O canal was intended to join the Ohio River at Pittsburgh but terminated at Cumberland and instead of being a major transportation artery to the west became a coal hauling canal along with the transport of other bulky items. Unlike the canals in New York and Pennsylvania when the C&O canal reached Cumberland after a long struggle, in 1852 it proceeded no farther since the B&O railroad reached Wheeling on the Ohio River also in 1852 and thus completed the much longed for connection between Chesapeake Bay and the Ohio River. The great 1927 flood served as the death knell of the C&O canal. Today it enjoys a considerable reputation as a national park.

In quite the same manner champions of the James River and Kanawha Canal sought to connect tidewater Virginia with the Ohio system incorporating a canal connecting the James and the Kanawha Rivers. (Kemp 2000) It was a tantalizing prospect as only 33 miles separated the headwaters of the James from the westward flowing Kanawha river system. Incorporated to carry out this grand scheme the James River and Kanawha Canal received a charter in 1785. Starting with a flourish the private company found the project beyond its resources and the state purchased the company but later returned it again to private ownership. Through both public and private ownership the canal finally reached Buchannon 196 miles above Richmond in 1856. From Buchannon a turnpike road made the connection with the Great Kanawha river system. Later in the 19th century a great vision entitled the Central Water Line failed to gain federal support. The final blow occurred when the C&O railway bought the canal and in a symbolic and practical move laid tracks on the tow-path of the canal beginning in 1881. As an essential material needed in canal construction, natural cement was first purchased from mills in the Rosendale area of New York in September 1827. With encouragement from Benjamin Wright, John H. Cocke Jr. found cement rock at Balcony Falls. Earlier in 1826 the first cement mill in Virginia was located 7 miles above the James River on its tributary the Maury River. After the reorganization of the company in 1836 as a private venture, Judge John Kinsey, who had been successful in locating cement rock on the Morris canal, working on behalf of the canal company, located four sites. His investigation included calcining rock samples so as to identify suitable geologic formations. This investigation marks the beginning of cement production at Balcony Falls.

> *The James River Cement Works are located at Balcony Falls, Rockbridge county, Virginia, on the James river, and the James River and Kanawha Canal. The mill stands on the tow-path, and contains two crackers and four run of French burr-stones of medium size, driven by water-power derived from a dam across James River, erected by the Canal Company. The power is deemed sufficient to turn six run of stone. Six kilns are located at the mills. The*

> *quarries, of which there are two opened in the same stratum, are on the margin of the river, about one mile above the mill, from which point the stone is transported to the kilns in boats, on the slack water of the dam. This deposit is generally known in Virginia as the "Blue Ridge Quarry"* (Gillmore 1863, 57)

Lured away from the Cumberland Hydraulic Cement Manufacturing Company in 1848 Charles H. Locher, was employed to operate the James River and Kanawha canal cement interest. By 1850 he had rebuilt the mill and the records also indicate that Locher owned six kilns just above the J.R.& K. cement mill, which he sold in 1850 and that may explain why he rebuilt the earlier mill. It was certainly this mill in which he installed steam power, since the JR&K cement mill located on the canal at the Blue Ridge dam used water power rather than steam. From archival research it appears that cement making at Balcony Falls lasted until 1895, long after the C&O railway laid its tracks on the tow-path.

Many hazards including shoals and snags inhibited navigation on the Ohio River from the earliest days of steam navigation but the greatest obstacle was at the falls of the Ohio near Louisville, Kentucky. To circumvent the falls, a skirting canal was constructed from 1825 to 1830. While the Louisville and Portland canal was under construction a suitable cement rock was discovered and became the source of a large cement industry in the area. A description in 1858 stated:

> *Near Louisville, Kentucky, at the foot of the falls of the Ohio River, there is a deposit of cement stone, which for many years has been extensively used throughout the West, and particularly along the Mississippi River. The deposit is six feet thick; the stone is burnt in the ordinary draw-kilns, anthracite coal being used for fuel. The mill contains one pair of four and a half feet French burrs, driven by water-power.* (Gillmore 1863, 59)

Legend has it that Canvass White discovered the cement rock reef which formed the falls. The discovery date was 1826 and White, as a consultant, did not appear on the scene until 1828. Although not confirmed it probably was David S. Bates (1777-1839) the canal's chief engineer from 1825-1827 who made the discovery.

Beginning 1830 John Hulme founded the Louisville Cement Industry when he purchased and converted the Tarascon Mill at Shippingport Island and converted it from a grist mill to one which ground calcined cement. Hulme later, in 1833, became the canal superintendent and served in this position until 1852. The next owner, Theodore Scowden serving as chief engineer for the construction of the Scowden Lock bought the Tarascon Company mill in 1865 and operated it until 1872 when the Speed interest, a well-known family in the Louisville area, acquired it and renamed it the Louisville Cement Company. Expanding the business, James Speed purchased quarries in Indiana and built additional mills around the town of Speed in Indiana. It appears that until 1867 the Speed interests had a virtual monopoly on cement production when the Union Cement Company built a mill at Shippingport Island not far from the Tarascon site. From all accounts the area produced superior cement which was shipped up and down the Ohio River and as far as the lower Mississippi.

> *Hulme's Hydraulic Cement is superior to any made in the U.S. and by testing is found superior to the famous imported Roman Cement. It becomes solid as stone when exposed to water and is in great demand through the West and South.* (Louisville Journal, Feb. 10, 1847)

The Louisville Journal article of May 11, 1868 noted that the steamboat *Meinotte* had a load of 1,500 barrels of cement which it took on board at New Albany while the steamboat *America* shipped 600 barrels to Cincinnati. In a headline of August 30, 1892 the newspaper reported that a disastrous fire had destroyed the Tarascon Mill. Nevertheless the Louisville Cement Company continued in the business of producing natural cement until well in after World War II. Their principle product was a mortar mix for brickwork which was advertised as being superior to Portland cement which was generally too strong for brickwork or even stone masonry. It was too hard and unyielding. The problem is overcome today by mortar mixes using Portland cement and up to 40% quicklime.

This gazetteer gives one an idea of the extent of the industry. There were mills farther west and in locations throughout the middle Atlantic and northeastern states. Until more information is forthcoming one is unable to give a comprehensive view of the history of natural cement in the nation and its transformation into the Portland cement industries in the late 19th century.

LIME	CLAY	RESULTING PRODUCTS	DISTINCTIVE CHARACTERS OF THE PRODUCTS
100	0	Very fat lime.	Incapable of hardening in water.
90	10	Lime a little hydraulic.	Slakes like pure lime, when properly calcined, and hardens under water.
80	20	Lime Quite hydraulic.	Slakes like pure lime, when properly calcined, and hardens under water.
70	30	Lime quite hydraulic.	Slakes like pure lime, when properly calcined, and hardens under water.
60	40	Plastic, or hydraulic cement.	Does not slake under any circumstances, and hardens under water with rapidity.
50	50	Plastic, or hydraulic cement.	Does not slake under any circumstances, and hardens under water with rapidity.
40	60	Plastic, or hydraulic cement.	Does not slake under any circumstances, and hardens under water with rapidity.
30	70	Calcareous pozzolano (brick).	Does not slake or harden under water, unless mixed with a fat or a hydraulic lime.
20	80	Calcareous pozzolano (brick).	Does not slake or harden under water, unless mixed with a fat or a hydraulic lime.
10	90	Calcareous pozzolano (brick).	Does not slake or harden under water, unless mixed with a fat or a hydraulic lime.
0	100	Pozzolano or pure clay brick.	Same as the preceding.

Table 1. Classification of various cements according to Petot.
(Source: Mahan 1880.)

	NO. OF WORKS	PRODUCT BARRELS (300 Pounds)	VALUE
Florida (1898)	1	7,500	7,500
Georgia	1	13,000	9,750
Illinois	3	537,094	187,983
Indiana & Kentucky	19	2,922,453	1,022,858
Kansas	2	150,000	60,000
Maryland	4	362,000	144,800
Minnesota	2	113,986	56,793
New York	29	4,689,167	2,813,500
Ohio	3	34,557	17,279
Pennsylvania	5	511,404	255,702
Tennessee	1	10,000	8,000
Texas	1	12,000	20,400
Virginia	3	63,500	38,100
West Virginia	1	52,727	21,090
Wisconsin	1	396,291	151,992
TOTAL	76	9,868,179	$4,814,771

Table 2. Natural cement production in the United States in 1899.
(Source: Reid 1908.)

YEARS	NATURAL CEMENT (Barrels)	PORTLAND CEMENT (Barrels)
1818-1829	25,000	---
1830-1839	100,000	---
1840-1849	425,000	---
1850-1859	1,100,000	---
1860-1869	1,642,000	---
1870-1879	2,200,000	8,000
1880-1889	4,346,000	147,000
1890-1899	8,070,000	1,728,000
1900-1909	5,050,000	33,383,000
1910-1919	757,000	83,995,000
1920-1929	1,520,000	143,224,000
1930	1,792,000	161,197,000
1931	1,227,000	125,100,000
1933	432,000	63,473,000
1935	1,006,000	76,742,000
1936	1,819,000	112,650,000
1940	2,535,000	130,217,000

Table 3. Comparison of the production of natural and Portland cement in the United States, 1818-1940.
(Source: Draffin 1976.)

Figure 1. Battery of natural cement kilns at Cedar Cliff.
(Source: Hahn collection, IHTIA.)

Figure 2. Top view of one of the Cedar Cliff natural cement kilns.
(Source: Hahn collection, IHTIA.)

Figure 3. Draw pit at Cedar Cliff kilns.
(Source: Hahn collection, IHTIA.)

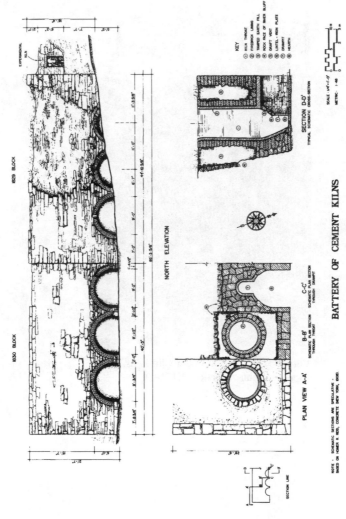

Figure 4. Battery of continuous cement kilns at Shepherdstown, WV. (Source: Kemp 1996.)

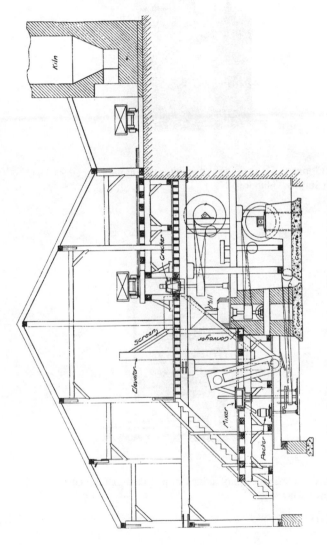

Figure 5. Cross section of a natural cement plant. (Source: Bleininger 1904.)

Figure 6. Cross section of an intermittent natural cement kiln.
(Source: Gillmore 1872.)

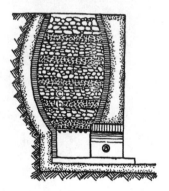

Figure 7. Cross section of a draw or perpetual natural cement kiln.
(Source: Davey.)

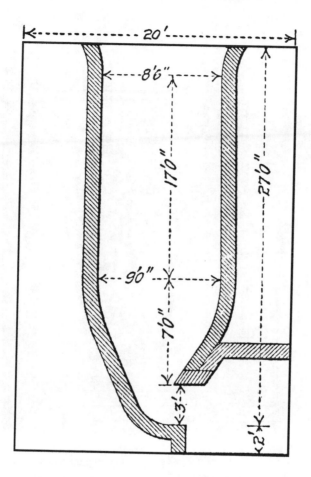

Figure 8. Cross section of a perpetual kiln for producing natural cement. (Source: Reid 1908.)

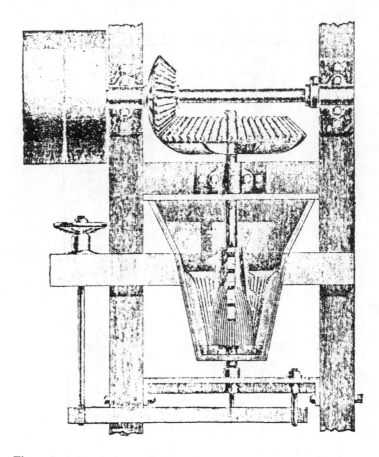

Figure 9. A pot cracker used reduce cement rock to a manageable size before calcining. (Source: Reid 1908.)

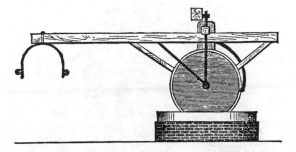

Figure 10. Horse powered vertical stone grinder.
(Source: Gillmore 1872.)

Figure 11. A bottle kiln at Aspdin's cement works in England. It is the oldest extant Portland cement kiln. (Source Stanley 1979.)

Figure 12. Schoefer Portland cement kilns at Coplay, Pennsylvania. (Source: Hahn collection, IHTIA.)

Figure 13. A Gates ball mill used to grind cement clinker. (Source: Bleininger 1904.)

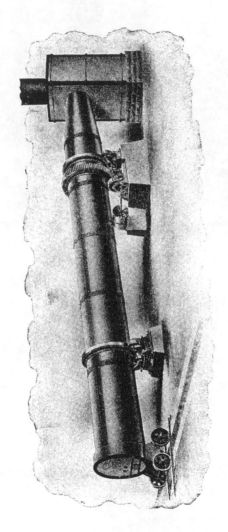

Figure 14. An Allis-Chalmers rotary kiln for Portland cement production. (Source: Bleininger 1904.)

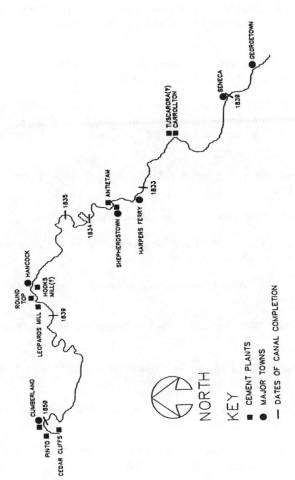

Figure 15. Natural cement mills along the Chesapeake and Ohio Canal. (Source: Hahn & Kemp 1994.)

References

Andrews, Frank D. (1924) *History of the Discovery of Water-Limestone and Early Manufacture of Cement at Southington, Conn.*, Vineland, Vineland, N.J.

Barfoot, R.J. (1974) "Joseph, James and William – The Aspdin Jigsaw." *Concrete*, Vol. 8, No. 8, August 1974, Concrete Society, London.

Bastoni, Gerald (1982) "Episodes from the Life of Canvass White Pioneer Civil Engineer." *Proceeding of the Canal History and Technology Symposium*, Vol. 1, January 30, Center for Canal History and Technology, Easton, PA.

Batsford, Harry and Fry, Charles (1954) *The Cathedrals of England.* Batsford, London.

Bauer, Edward E. (1949) *Plain Concrete.* McGraw-Hill, New York.

Bleininger, Albert Victor (1904) *The Manufacture of Hydraulic Cements.* Legislature of Ohio, Columbus, OH.

Cloos, Ernst (1951) *Mineral Resources of Washington County.* State of Maryland, Baltimore.

Cummings, Uriah (1898) *American Cements.* Rogers and Manson, Boston, USA.

Davey, Norman (1961) *A History of Building Materials.* Phoenix House, London.

Derry, T.K. and Williams, Trevor I. (1960) *A Short History of Technology.* Oxford University Press, London.

Eckel, Edwin (1902) *54th Annual Report of the Regents.* New York State Museum. Vol. 3, University of State of New York, Albany.

Eckel, Edwin (1909) *The Cement Resources of Virginia West of the Blue Ridge.* University of Virginia, Charlottesville.

Fletcher, Banister (1954) *A History of Architecture on the Comparative Method.* 16th edition, Batsford, London.

Fitzsimons, Neal (1972) *A Biographical Dictionary of American Civil Engineers.* American Society of Civil Engineers, New York.

Gallatin, Albert (1808) *Public Roads and Canals.* Reprint Kelley, New York, 1968.

Gilchrist, Ann (1976) *Footsteps Across Cement.* Published privately, Rosendale, New York.

Gillmore, Q. A. (1863) *Practical Treatise on Limes, Hydraulic Cements and Mortars.* Van Nostrand, New York.

Gillmore, Q. A. (1871) *Practical Treatise on Coignet-Beton.* Van Nostrand, New York.

Grimsley, George P. (1906) *West Virginia Geological Survey: Clays, Limestones and Cements.* WV Geological Survey, Morgantown, W.V.

Hahn, Thomas F., and Kemp, Emory L. (1994) *Cement Mills Along the Potomac River.* Institute for the History of Technology and Industrial Archaeology, Morgantown, W.V.

Hellerich, Mahlon H. (1977) "Saylor Park Cement Industry Museum." Unpublished note, Copley, PA.

Kemp, Emory L. (1982) *History of Concrete.* Bibliography, No. 14, American Concrete Institute, Detroit.

Kemp, Emory L. (1996) *Industrial Archaeology Techniques.* Krieger, Malabar, FL.

Kemp, Emory L. (2000) *The Great Kanawha Navigation.* University of Pittsburgh Press, Pittsburgh, PA.

Mahan, D.H. (1880) *A Treatise on Civil Engineering.* Wiley, New York.

Malinowski, Roman (1979) "Concretes and Mortars in Ancient Aqueducts." *American Concrete Institute,* Detroit.

McKee, Harley J. (1961) "Canvass White and Natural Cement, 1818-1834." *Journal of the Society of Architectural Historians.* Vol. 20, No. 4.

McKee, Harley J. (1975) "Historical Development of Hydraulic Cement," presented at the American Concrete Institute Symposium, April 10, Boston MA and unpublished.

Metz, Lance E. (1996) *A Brief History of the Cement Industry of the Lehigh Valley and Nazareth.*, Nazareth Keepsakes, Nazareth, PA.

Plommer, Hugh (1973) *Vitruvius and Later Roman Building Manuals.* Cambridge University Press, Cambridge, England.

Reid, Homer A. (1908) *Concrete and Reinforced Concrete Construction,* Clark, New York.

Ries, Kenneth (1901) "Lime and Cement Industries of New York." *Bulletin of the New York State Museum,* Albany, New York.

Roberts, Charles R., Stoudt, John B., Krick, Thomas H., Kietrich, William J., (1914) *History of Lehigh County, Pennsylvania and a Geneological and Biographical Record of its Families.* Lehigh Valley Publication Company, Allentown, PA.

Rowland, Ingrid D. and Howe, Thomas N., eds. (2002) *Vitruvius Ten Books on Architecture,* Cambridge University Press, Cambridge, England.

Shaw, Ronald E. (1990) *Canals for a Nation.* Kentucky, Lexington, KY.

Singer, Charles, Holmyard, E.G., and Hall, A.R. (1954) *A History of Technology.* Vol. 1, Oxford University Press, London.

Skempton, A.W. (2002) *Biographical Dictionary of Civil Engineers.* Thomas Telford Press, London.

Stanley, Christoper C. (1979) *The History of Concrete.* C & CA, Wexham Springs, England.

Straub, Hans (1964) *A History of Civil Engineering.* MIT Press, Boston, MA.

Wertime, Theodorea and Wertime, Steven F., eds. (1982) *The Evolution of the First Fire-Using Industries.* Smithsonian Institution, Washington, D.C.

Development of Reinforced Concrete Arch Bridges in the U.S.: 1894-1904

Dario A.Gasparini[1]

Abstract

In 1894 Fritz Von Emperger presented a paper on concrete-iron bridges at the International Engineering Congress. Von Emperger showed examples of European reinforced concrete arch bridges and extolled the qualities of German cement. He described tests that showed the superior performance of the Melan reinforcing system, which was patented in the U.S. in 1893. Late in 1894, Von Emperger built the first Melan arch in the U.S. at Rock Rapids, Iowa. Although a few reinforced concrete arch bridges had been built in the U.S. prior to 1894, Von Emperger's efforts in promoting the Melan system as a "permanent", cost-effective alternative to products of bridge companies initiated a sustained, rapid development of technologies related to reinforced concrete arch bridges. Reinforced concrete systems were enthusiastically adopted and entrepreneurial designers such as Edwin Thacher and Daniel Luten developed and patented their own proprietary systems. Engineering literature of the period contains revealing opinions on the nature of composite action, advantages and disadvantages of various arch forms, "proper" shapes for arch axes and suitable design methods. The evolution of reinforcing systems is followed and pioneering examples of reinforced concrete arch bridges are given.

Introduction

The introduction of reinforced concrete for arch bridges and other parts of the infrastructure at the end of the 19th century required the concurrent development of a daunting number of technologies:

   a) Cement technology
   b) Concrete technology
   c) Reinforcement technology
   d) Analysis and design methods for reinforced concrete composite materials and elements
   e) Analysis and design methods for reinforced concrete arches
   f) Construction technologies

Each of these technologies had numerous unresolved technical issues. Appropriate component materials and processes for manufacturing Portland cement were still developing. Understanding of hydration was incomplete. An experimental or

[1] Professor, Department of Civil Engineering, Case Western Reserve University, Cleveland, Ohio 44106; dag6@po.cwru.edu; Phone 216-368-2699

scientific basis for control and design of concrete mixes did not exist. There was a paucity of data on the properties of hardened concrete, especially creep, shrinkage and freeze-thaw resistance. There were many, often proprietary, reinforcing systems, which used smooth or deformed bars, expanded metal, rolled rail or standard sections, or fabricated, truss-like assemblies. The behavior of steel-concrete composite sections was unresolved. Issues of bond, anchorage of steel, shear and flexural strength of beams were in debate, as well as appropriate modeling and design of arch forms for a monolithic material like concrete. The issues included the proper variation of cross-sectional area and moment of inertia with position, proper shapes for the centroidal axis, proper rise to span ratios, effects of temperature changes, creep and shrinkage, and the relative advantages of different arch forms (fixed-fixed, hinged, tied, etc.). Construction technologies such as forming, placement of steel, placement and compaction of concrete and curing of concrete were being developed.

The breadth of technologies involved in the realization of reinforced concrete arches and the effective race that existed between European and American engineers to develop and exploit them make comprehensive historical assessments on reinforced concrete difficult. In addition to identifying European and American contributions to each technology, identifying contributions of individuals and firms, correctly identifying significant design "firsts", there are other, broader issues that are important to the history of civil engineering. One is the relative influence of property rights of inventors versus standardization on the exploitation of a technology. Another is the way reinforced concrete bridges were designed and constructed versus the way iron and steel bridges were realized through bridge companies. A third issue is field versus shop control of quality. Finally the issue of the relationships between systems used for buildings and those used for bridges and other objects is intriguing.

This paper does not attempt an overview of the growth of all the technologies nor a study of the main issues that affect their development. Rather, it focuses specifically on the development of reinforcing systems used in the U.S. for reinforced concrete arches and discusses some of the early bridge designs. It discusses primarily the systems of Jean Monier and Joseph Melan and the contributions of Friedrich (Fritz) Von Emperger, Edwin Thacher and William Mueser. It focuses on the period from 1894 to 1904, when reinforcing systems were simplified and pioneering arch bridge designs were done.

Reinforced Concrete Bridges in the U.S. prior to 1894

One of the earliest U.S. patents for a reinforcing system is that granted to Ernest Ransome in 1884 for his twisted or torqued bars. In 1889, Ransome built what is recognized as the first reinforced concrete arch in the U.S., the Alvord Lake bridge in Golden Gate Park in San Francisco. Snyder and Mikesell (1994) describe the span but do not give details of the reinforcement used. Ransome subsequently focused his activities on building systems and his bridge appears to be a singular achievement that did not lead to further developments.

A second reinforced concrete arch is described in the September 7, 1893 issue of Engineering News. The bridge, shown in Fig. 1, was built over Pennypack Creek in Philadelphia. It was designed by Carl A. Frick, Superintendent of Bridges. The bridge had two 25'spans with rises of 6'6". The arch was reinforced with " 1-1/2inch mesh wire nets...placed about 2ft apart, horizontally and vertically, throughout the concrete. The diameter of the wire in this netting is ¼ inch" (Engineering News, 1893). Fritz Von Emperger mentioned the bridge in the November 16, 1893 issue of Engineering News, commenting that "the beginning has been made".

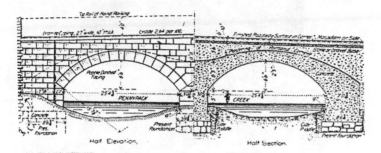

Figure 1. Reinforced concrete bridge over Pennypack Creek, Philadelphia
(Engineering News 1893a)

The Reinforcing System of Jean Monier and his November 22, 1892 U.S. Patent (486,535)

It is well known that Jean Monier, a "Paris gardener", began fabricating horticultural objects in the1860's using concrete reinforced with nets of wires. Monier patented his system in 1867 but it became widely known and used when G.A. Wayss, in about 1878-9, became the owner, or "sole representative" of the Monier system in Germany and neighboring countries. The system is essentially the same as current practices. It consists of one or more grids of reinforcement. The grid dimensions varied from 2" to 10". The larger, or "carrying" wires, were typically 0.39", while the smaller, orthogonal, "distribution" wires were typically 0.28". The rods were wired together at their intersections, much like today.

A Monier arch, built at the train station in Matzleindorf, Austria, is mentioned in the February 1, 1890 issue of Engineering News. The bridge had a span of 32.8ft, a rise of 3.3ft and an arch thickness that varied from 6" at the crown to about 8" at the abutments. Results of tests on the bridge are reported in the August 2, 1890 issue of Engineering News. An equivalent uniformly distributed load of 1741 lbs/ft² was applied before the abutments "gave way".

Other Monier arch bridges are described in the May 23, 1891 Engineering News. One consists of three arches, each with a span of about 29'-6", spanning over railroad tracks at Moedling, near Vienna, Austria. It was constructed in 1899-90 and tested in

September, 1890. The same article describes a bolder design at Wildegg, Switzerland. The bridge, shown in Fig. 2, was built in the autumn of 1890 and tested on November 14, 1890. The span was 122ft and the rise was only 11.5ft. Two grids of reinforcement were used, one near the intrados and another near the extrados. The arch thickness varied from 7.9" at the crown to 25.6" at the abutments.

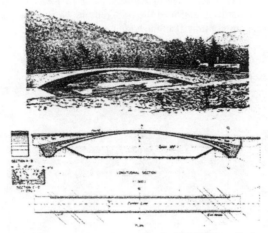

Figure 2. Melan reinforced concrete arch, Wildegg, Switzerland
(Engineering News, 1891)

Engineering News of February 16, 1893 describes three Monier arch bridges, one of 50ft span, one of 65.6ft span and another of 114.8ft span. The article notes that the "principal objections brought forward were: (1) That the wire would eventually corrode by coming into contact with the moist concrete; (2) that the concrete would not adhere to the smooth surface of the iron, and thus the two materials would not act together, and, (3) that the iron wire would expand and contract under changes of the temperature at a different rate than the concrete, and thus cause cracks in the latter material". The article notes some favorable results from corrosion experiments and states that, according to Bonniceau, a French author, the coefficient of thermal expansion for concrete is "0.0000143 for a change of temperature of 1° C. and that of iron is 0.0000145, or practically the same".

In his influential 1894 paper in the ASCE, Fritz Von Emperger describes three additional Monier arches, including one of 132ft span, built in 1890 "for show purposes only" for a fair at Bremen. In 1894, Von Emperger was representing the competing Melan system and thus notes that the Monier reinforcement is "not stiff itself". He cites a failure and "distrust" of the Monier system and notes that "this difficulty has been overcome by the other systems now in use, which use rolled shapes instead of wire netting". The American discussions of the article clearly reveal that the principal experience in the U.S. to 1894 was with building floors. Many of the respondents expressed surprise that the coefficients of thermal expansion of steel and concrete were the same.

The September and December 1899 issues of Cement and Engineering News contain a series of articles by E.Lec Heidenreich on the Monier system. Heidenreich admits that he is a "representative of the patentee for Monier construction in the United States". Revealingly, he focuses on the uses of the Monier system for tanks, storage silos, building floors and culverts but does not emphasize bridges.

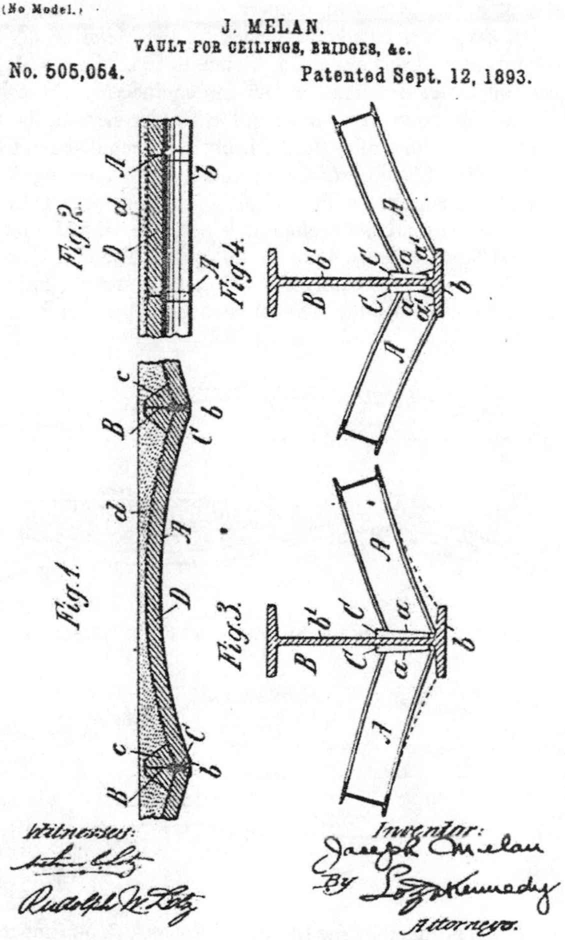

Figure 3. Joseph Melan U.S. Patent 505, 054

The Reinforcing System of Joseph Melan and his September 12, 1893 U.S. Patent (505,054)

Fig. 3 shows the drawings submitted with the Joseph Melan U.S. patent. They show that the system consists of centrally placed stiff beams, denoted by the letter A, in parallel with "rammed concrete" between them. That is, the system does not rely on steel-concrete bond and composite behavior. The steel is centrally placed, not where tensile stresses occur in the concrete. Rather than a concrete reinforcing system, it is more appropriately steel in parallel with concrete. The Melan system was represented in the United States by Fritz Von Emperger. In his 1894 article in the ASCE Transactions, Von Emperger clearly advocates for the Melan system. He pointedly states that: "The inventor of this system, J. Melan, is, in contrast to J. Monier, a well-known engineer and writer. He ranks in German engineering circles as an expert in arch construction…".He notes that, unlike in the Monier system, the steel beams are stiff and that: "They are a kind of centering itself". He then discusses tests performed for the Austrian Society of Engineers and Architects for evaluating different systems. The test specimens were arches with a 13.5ft span. The systems compared were a brick arch, a plain concrete arch, a Monier arch with one grid of reinforcement and a Melan arch with 3-1/8inch I beams 40inches on center. The half span loading used is shown in Fig. 4 . The tests predictably showed that the stength of the melan system was about four times that of the Monier system, which only had one grid of reinforcement.

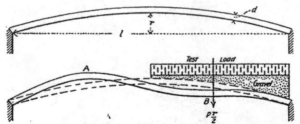

Figure 4. Loading used for Austrian tests (Von Emperger, 1894)

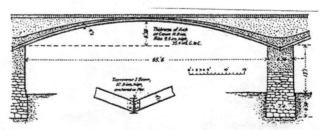

Figure 5. Melan bridge over the Moldau at Rostock (Von Emperger, 1894)

Von Emperger cites three completed Melan arches having 23.5ft, 39.5ft and 65.6ft spans. Fig. 5 shows the longer arch, which appears not to have any cover for the steel beams, although Von Emperger, in his closure of the discussions states that "..in the Melan arch all iron is covered... I have to state this definitely, because my designs do not show the coating which is supposed to be applied according to European practice, and are, therefore, somewhat misleading." Von Emperger also discusses his design approach. Letting the total load, Q, be expressed as the sum of that carried by the concrete, Q_c , and the steel, Q_s , Emperger states:

$$Q_c/Q_s = (E_cI_c)/E_sI_s) \tag{1}$$

Emperger's equation is not strictly correct because it is only the ratio of the moments that is dependent on the moments of inertia, and not the ratio of the "loads". Emperger also gives formulas for total steel and concrete stresses that depend only on the flexural stiffness ratio given in Eq.(1). Those formulas are also not strictly correct because in an arch the total axial force is shared in proportion to E_cA_c/E_sA_s and the total moment is shared in proportion to E_cI_c/E_sI_s. Von Emperger also gives values for the horizontal thrust for the case of a uniformly distributed live load over half span and over the full span. His formulas are consistent with a statically determinate, three-hinged model of the Melan arch.

In his discussion of Von Emperger's paper, W.R. Hutton gave a very perceptive comment on the Melan system: " The Melan system appears to be an entirely different thing. In it no attempt is made to supply directly with metal the deficiency in tensile strength of concrete. It is a combination in one structure of materials entirely distinct in their characteristics, in which combination the moment of inertia of the sections is inversely proportioned to the modulus of elasticity of each material."

The Melan and Monier systems presented American designers with two distinct choices. The Monier system, with its loose bars, relied on adhesion and composite behavior. The Melan system simply used steel in parallel with concrete, with no reliance on bond and composite behavior.

Other European Reinforcing Systems

The November 16, 1893 issue and the April 12, 1894 issue of Engineering News describe the reinforcing system of R. Wuensch, of Budapest, Hungary, patented in 1884. Fig. 6 shows a 16.6ft model of the system used for a Hungarian government test. Von Emperger also describes the Wuensch system in his 1894 paper. He lists six highway bridges, with spans ranging from 13.5ft to 55.9ft that were build using the system. Fig. 6 indicates that Wuensch embedded a rigid frame of reinforcement within the concrete. Therefore the system probably behaves as a rigid frame with a haunched girder.

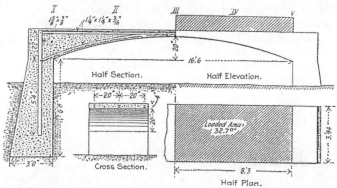

Figure 6. Test model of R. Wuensch reinforcing system (Von Emperger, 1894)

Of course the most widely used reinforcing system in Europe was that of Francois Hennebique, patented in 1892 (Cusack 1984). The Hennebique system, however, was used principally for buildings and reinforced concrete girder bridges, and had less influence on reinforced concrete arch bridges.

Reinforced Concrete Arch Designs in the United States from 1894 to 1904

Perhaps because of Von Emperger's influence, U.S. arch bridge designs of this period were dominated by Melan-type reinforcing systems. The Monier system of tied individual bars was largely relegated to other applications, although, over time, there was a gradual acceptance of composite behavior and systems similar to that of Monier evolved in the United States. In this period, four engineering firms were pre-eminent in the design of reinforced concrete arches in the U.S.:

a) F. Von Emperger, William Mueser and the Melan Arch Construction Company of New York
b) Keepers and Thacher
c) William Mueser, Edwin Thacher and the Concrete-Steel Engineering Company
d) Daniel B. Luten and the National Bridge Company, founded in 1902

The first three constitute a continuity of ideas and designs whereas the work of Daniel B.Luten represents an independent direction.

The first two Melan arches in the U.S. were built by Von Emperger in 1894-5. The first was a 32ft span arch built in Rock Rapids, Iowa. The second was the 70ft span, Eden Park Bridge in Cincinnati. The Rock Rapids Bridge is extant but has been moved to a park and no longer serves as a bridge. Its history and structural behavior have been documented by the Historic American Engineering Record (HAER). These studies are contained in two reports, IA-63 and IA-89, which are in the permanent HAER Collection in the Library of Congress. The Eden Park Bridge is also extant, in very good condition, and remains in service. The Rock Rapids arch is reinforced with either I or rail sections having 3" flanges, spaced 3ft on center. The Eden Park Bridge is reinforced with 9" I beams also spaced 3' on center.

Von Emperger describes another Melan arch, also built in 1895, in Engineering News (VOL. XXXIV, No. 19). It is a footbridge in Stockbridge, MA with a span of 100ft. Fig. 7 shows the very elegant design, which has a 9" arch thickness at the crown. The bridge is reinforced with 7"deep, 15lbs/ft, I-beam sections, 28" on center. William Mueser tested the bridge by applying an equivalent distributed load of 75lbs/ft^2.

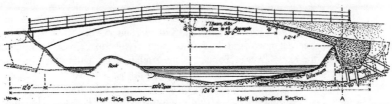

Figure 7. Melan arch for footbridge at Stockbridge, MA (Von Emperger, 1895)

In 1897, Von Emperger received a patent (No. 583,464) for a reinforcing system that used two reinforcement ribs, either trussed or simply spaced, one near the intrados and the other near the extrados. Von Emperger reasoned that: "The use of solid ribs for reinforcing the masonry arches has not given satisfaction in small spans where the iron should be at a smaller distance than four inches, which is the smallest I-iron in the market, while in larger spans the I-iron could not be well bent. The use of metal in the core of a vault or arch is useless, as only those portions of the ribs are called into action which are located near the intrados and extrados of the vault or arch, and as a correctly-constructed arch should increase in cross-section toward the haunches, the reinforcing-ribs should follow the shape of the arch in the same manner."

Edwin Thacher embraced the Melan system and quickly executed prominent, innovative designs. In the autumn of 1895 and spring of 1896, Thacher designed what very likely was the first reinforced concrete arch railroad bridge, the crossing of the Michigan Central Railroad over Southern Boulevard, in Detroit, MI. The bridge, shown in Fig. 8, had a skew span of 48ft and was reinforced with 41 "ribs" spaced 30" on center. The "ribs are built up of four angles 4in by 4in by ½ in., with a solid web 12ft long at the center, and laced with 5in. by ½ in bars between the center web plate and the end. They are 15in. deep at the crown and 24in.deep at the springing" (Railroad Gazette, 1899).

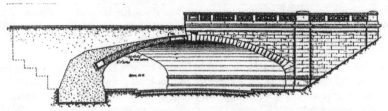

Figure 8. Melan arch for railroad bridge over Southern Blvd., Detroit, MI
(Railroad Gazette, 1899)

In 1896, Thacher patented a design for an *open spandrel* reinforced concrete arch
(U.S. Patent No. 570,239). From March 27, 1896 to January 12, 1898 the firm of
Keepers and Thacher designed and built the great five-span reinforced concrete
Melan bridge over the Kaw River at Topeka, Kansas again using latticed steel trusses,
centered on the arch axis, as reinforcement (Engineering Record 1898). On January
10, 1899 Thacher received U.S. patent N0. 617,615 for a reinforcing system which
recognized that fabrication of trussed reinforcing was not necessary. He simply
separated the intrados bar from the extrados bar: "The bars act as the flanges of
beams to resist bending moments, whereas the shearing stresses, which are small, are
taken by the concrete alone". He devised bars that were "readily bent" "readily and
cheaply spliced" and "can be stored or shipped in straight form". On September
1899, he published a very long and detailed article in Engineering News on
"Concrete-Steel Bridge Construction". Thacher discussed practically all the issues on
arch construction, including the relative benefits of open and closed-spandrel forms,
hinged arch forms, shapes of arch axes, proportioning of reinforcement, cement and
concrete technology and construction. Thacher included the Keepers and Thacher
Specifications for Reinforced Concrete Construction. In 1902 he received another
U.S. patent for improved reinforcing bars (U.S. Patent No. 714,971). At about this
time (Simmons, 1993), he also entered a partnership with William Mueser and
founded the Concrete-Steel Engineering Company. The Melan-Von Emperger-
Mueser-Thacher lineage is beautifully illustrated by Fig.9, an advertisement appended
to Homer Reid's 1907 concrete textbook. The flyer also illustrates a Melan arch built
on the Vanderbilt estate in Hyde Park in 1898 and the modern reinforcing bar,
patented by William Mueser.

Independently from the lineage culminating in the Concrete-Steel Engineering Co.,
Daniel B. Luten devised innovative, competing reinforced concrete arch forms and
reinforcing systems. Luten founded his National Bridge Company in 1902 in
Indianapolis. His reinforcing systems utilized separate round bars, bent to follow the
regions of potential tensile stresses, and thus rely on composite action. Luten devised
an innovative tied arch system, using a reinforced concrete tie *in the bed* of a stream.
Luten's accomplishments are beautifully described in James Cooper's book "Artistry
and Ingenuity in Artificial Stone"(Cooper, 1997).

Figure 9. Advertisement for the Concrete-Steel Engineering Co. (Reid, 1907)

The VonEmperger and Thacher Papers for the 1904 International Engineering Congress

Von Emperger's and Thacher's contributions to the International Engineering Congress of 1904, published in the ASCE Transactions, provide good perspectives on developments in reinforced concrete arch construction from 1894 to 1904. Von Emperger credits Thacher for his 1897 patent proposing " flat steel bars, independent of each other, near the intrados and extrados of the arch. These bars are subjected to tension and compression only. That part of the steel, which in the Melan system is represented by the web of the arched ribs, is saved, but the concrete must take all the shearing stresses without aid from the steel." Von Emperger then provides a history of the development of reinforced three-hinged arches in Europe from 1898. He states that "one of the most important advantages gained in the use of hinges is the greater independence of the structure from the execution of the work." Von Emperger also discusses the use of reinforced concrete girders for smaller spans.

Thacher, in his article, notes that since 1894, "the Concrete Steel Engineering Company, of New York City, and their predecessors have built, or are now building, under the Melan, Thacher, and Von Emperger patents, abut three hundred spans of concrete-steel bridges, distributed over nearly all parts of the United States". He notes that the longest reinforced concrete arch span in construction is 132ft but envisions open spandrel spans of 500ft or more.

Observations

Von Emperger's strong advocacy and active design practice after 1894 made the Melan system dominant for reinforced concrete arches in the U.S. The Melan system did not rely on composite action between concrete and steel but rather used steel in parallel with concrete. Edwin Thacher and other designers slowly realized that composite action between steel and concrete was achievable and that reinforcing systems could then be simplified. The new reinforcing systems that evolved essentially approached that of Jean Monier, who used individual rods, wired together to form grids, in the 1860's. Nonetheless, the reinforced concrete arch became common in the U.S., using primarily Melan-type reinforcing systems, in the period from 1894 to 1904. The following decade saw experimentation with hinged arch forms, the development of simpler systems for short span bridges and the use of open spandrel arches for larger spans.

References

Cooper, J.L. (1997) *Artistry and ingenuity in artificial stone*, Copyright by James L. Cooper, Greencastle, Indiana, ISBN 0-936631-13-9.

Cusack, D. (1984-5). "Francois Hennebique: The specialist organisation and the success of ferro-concrete: 1892-1909." Transactions of the Newcomen Society, Vol. 56, p. 71-85.

Engineering News (1890) Aug. 2, p. 91.

Engineering News (1891) "The Monier method of constructing arches" May 23, p. 499-500.

Engineering News (1893) "Some recent examples of Monier arch construction" Feb. 16, p. 148-149.

Engineering News (1893a) "Concrete arch highway bridge Philadelphia, PA" Sept. 7, p. 189-190.

Engineering News (1893b) "Concrete iron roadway bridge at Neuhausel, Hungary" Nov. 16, p. 391-392.

Engineering News (1894) "The Wuensch and Melan systems of fireproof floor construction" April 12, p. 305.

Engineering Record (1898) "Construction of the Topeka Melan bridge" Vol. XXXVII, No. 20, April 16, p. 426-428.

HAER No. IA-63 (1995). "Melan arch bridge." HAER Collection, Library of Congress, Washington, D.C.

HAER No. IA-89 (1996). "Reinforced concrete arch bridges." HAER Collection, Library of Congress, Washington, D.C.

Heidenreich, E.L. (1899a). "Monier system of constructing elevators." *Cement and Engineering News*, Sept., p. 36-38.

Heidenreich, E.L. (1899b). "Monier constructions." *Cement and Engineering News*, Dec., p. 85-88.

Melan, J. (1893). "Vault for ceilings, bridges, etc." U.S. Patent 505, 054, Sept. 12.

Monier, J. (1892). U.S. Patent 486, 535, Nov. 22.

Railroad Gazette (1899) "A Melan concrete-steel railroad bridge" Vol. 31, March 3, p. 151.

Reid, Homer (1907) *Concrete and reinforced concrete construction*, Myron C. Clark Publishing Co., New York.

Simmons, D.A. (1993). "Pioneer concrete arch bridge remains in Dayton." *Ohio County Engineer*, Fall issue, p. 10-11.

Snyder, J. and Mikesell (1994). "The consulting engineer and early concrete bridges in California." *Concrete International*, May, p. 38-44.

Thacher, E. (1896). "Bridge." U. S. Patent No. 570, 239, Oct. 27.

Thacher, E. (1899). "Concrete arch." U. S. Patent No. 617, 615, Jan. 10.

Thacher, E. (1899a). "Concrete-steel bridge construction." *Engineering News*, Sept. 21, p. 179-185.

Thacher, E. (1902). "Material of construction." U.S. Patent No. 714,971, Dec. 2.

Thacher, E. (1904). "Concrete and concrete-steel in the United States." Transactions of the American Society of Civil Engineers, International Engineering Congress, 1904, p. 425-458.

Von Emperger, F. (1894). "Tie development and recent improvement of concrete-iron highway bridges." Trans. American Society of Civil Engineers, Vol. XXXI, p. 437-488.

Von Emperger, F. (1895). "Melan concrete arch of 100-ft. span, Stockbridge, Mass." *Engineering News*, Vol. XXXIV, No. 19, p. 306.

Von Emperger, F. (1897). ""Vaulting for ceilings, bridges, etc." U.S. Patent No. 583, 464, June 1.

Von Emperger, F. (1904). "Concrete and concrete-steel." Transactions of the American Society of Civil Engineers, International Engineering Congress, 1904, p. 523-544.

THE ROLE OF JOHN FRITZ IN THE DEVELOPMENT OF THE
THREE-HIGH RAIL MILL, 1855-1863
by
Lance E. Metz
and
Donald Sayenga

In his 1974 work, *From Know-How to Nowhere: The Development of American Technology*,
Professor Elting E. Morrison of the Massachusetts Institute of Technology cited the 1857 development
of the "three-high rail mill" by John Fritz of the Cambria Iron Company in Johnstown, Pennsylvania,
as a seminal event in the development of American Technology and as an illustration of the effectiveness
of the trial and error methods of early American engineers.[1]

Other American technological historians have made similar statements concerning the significance
of the three-high rail mill. Invariably these accounts have been predicated on the examination of a limited
number of available sources, chief of which is the *Autobiography of John Fritz*. Historians have long
considered this 1911 work to be the most thorough and accurate account of the creation of the three-high
rail mill, and since it is also widely available, his autobiography has been utilized by modern scholars.

However, the recent availability of the Fritz family papers has made it possible for the first time to
reevaluate the creation of the three-high rail mill and its place in America's technological history.[2] In
the process of this reevaluation the authors have uncovered previously unknown primary source materials
which not only add additional detail to the traditional accounts of the creation of the three-high rail mill
but also have induced them to question the nature of Fritz's achievement and the overall importance of
the three-high rail mill as a technological milestone. Equally as important, the Fritz family papers shed
new light on the true role that William Kelly played in the birth of the pneumatic steel industry.

To begin this reevaluation of the development of the rail mill it is necessary to examine the early
career of its creator. John Fritz (1822-1913) was born in rural Chester County, Pennsylvania. His father
George, a farmer and a millwright, possessed a considerable local reputation as a skilled mechanic.
Although John attended school long enough to become literate and acquire a rudimentary knowledge of
various forms of mathematics, his true education was derived from visits with his father to the various
mills where the elder Fritz practiced his trade. In this manner he acquired a pragmatic knowledge of how
machinery functioned and an overwhelming desire to improve it.[3]

In 1838, at age 16, John Fritz became an apprentice of a blacksmith shop and repair facility in the
Chester County town of Parkesburg. Since blacksmiths of this era also served as primitive machinists,
he was able to participate in the fabrication or repair of an amazing range of mechanical devices. Within
a few years this type of work enabled him to acquire a considerable knowledge of such skills as
boilermaking, forging, gunsmithing, and iron rolling. John Fritz's training was also enhanced by his
close social association with the mechanics at the nearby locomotive repair shops of the Columbia and
Philadelphia Railroad. By allowing him to have the run of their shop and to participate in their repair
duties, the railroad mechanics enabled him to gain a thorough understanding of practical mechanical
engineering.

Seeking new challenges, John Fritz took a position in 1844 as a workman at the rolling mill operated
by the firm of Moore and Hooven in Norristown, Pennsylvania. There he was able to display his
burgeoning talent for mechanical innovation which would become the hallmark of his later career.
Assigned to help complete the erection of the rolling mill's machinery, he was able to examine and
analyze the weaknesses that were inherent in its design. When the rolling mill began operation producing
merchant bar stock, he was able to perceive its flaws and in the process of routine maintenance he was
able to design and construct many improvements. To better utilize his talents his employers made him

the mill's chief mechanic. Seeking to learn more about other aspects of iron product production, chief mechanic Fritz soon mastered the mill's foundry and puddling operations, and within a few months he was able to improve the performance of the furnaces which were essential to these operations. He also exhibited a great skill in winning the respect of his fellow workers. By 1845 he had risen to become in short succession the superintendent of the night and the day shifts.[4]

In 1849 John Fritz accepted a job offer from the firm of Reeves, Abbott and Company to supervise the erection and operation of an ironworks at Safe Harbor in Lancaster County, Pennsylvania. This change of employers was to prove particularly noteworthy because for the first time he would become involved with the management of a primary iron production rail rolling mill facility and also because he came into contact with David Reeves. The principal partner of Reeves, Abbott and Company, David Reeves was an individual who would later play a catalytic role in Fritz's creation of the three-high rail mill.

Although the anthracite-fueled blast furnace at Safe Harbor was placed in blast before he arrived, Fritz was soon able to greatly improve its productivity. He supervised the erection of the rail rolling mill; by the end of 1849 it had produced 5,567 tons of rails, most of which were sold to the newly formed Pennsylvania Railroad.[5] Unfortunately, Fritz's tenure at Safe Harbor proved to be all too brief and trying. Plunging into the task of keeping the Safe Harbor works operating at its peak efficiency, Fritz had paid scant attention to his health. Safe Harbor possessed a reputation as a notoriously unhealthy area during this period, so it is not surprising that he was soon stricken with a persistent fever and chills. Unable to work, he left Safe Harbor and returned to Chester County in search of a cure.[6]

After a period of convalescence, Fritz accepted a commission from Reeves, Abbott and Company to ascertain the potential of the recently discovered iron ore deposits of Michigan's upper peninsula. Surveying the area near Marquette, he was quite impressed with the enormous tracts of ore lands that he found. Unfortunately, he was unable to convince his employers that they should invest in opening mines in this region, which would eventually become the most productive in America.[7]

Upon returning to Safe Harbor, Fritz attempted to resume his duties at the ironworks. Unfortunately, his illness returned; within a few months it forced him to depart permanently from the area. After a further convalescence, he once again entered the employment of Moore and Hooven at Norristown. Still troubled by fever and chills, he consulted a noted Philadelphia doctor who finally cured his illness.[8]

David Reeves remained impressed with John Fritz's abilities; with his health restored, the young ironmaster was able to accept another assignment from his patron. He was given the task of rebuilding the anthracite-fueled Kunzie Furnace, which was located along the Schuylkill Navigation near Philadelphia.[9] Within a few months John Fritz had made many mechanical improvements to the furnace and its associated hot blast system, transforming it into one of the most productive ironmaking establishments in the region. Within a year, however, he once again sought new challenges and returned to his former position at Norristown. In 1851 he won the heart of Miss Ellen Maxwell of Norristown, and she and John Fritz were married during that same year. A kind and gentle woman, Ellen proved to be a tolerant and understanding wife. Their only child Gertrude died at the age of seven in 1860.[10]

With a wife to support, Fritz sought a position that would offer him a greater chance for financial advancement. Forming a partnership with his younger brother George and two of his brothers-in-law, B.F. Stroud and Isaac E. Chandler, he constructed a machine shop and foundry at Catasauqua, Pennsylvania.[11] Catasauqua was at the center of the Lehigh Valley's iron industry, which had recently become the most productive in America.[12] Realizing that many of the ironworks of the Lehigh Valley needed machinery and other products, the new enterprise seemed to be assured of a steady source of customers. Unfortunately, the foundry and machine shop were undercapitalized, and by 1854 Fritz and his partners were in financial difficulty. It was sold to a partnership headed by David Thomas, the pioneer of America's anthracite iron industry.[13]

John Fritz was rescued from his failure at Catasauqua by an offer from David Reeves to become general superintendent at the Cambria ironworks in Johnstown, Pennsylvania. Although Reeves had primarily focused his activities on the Phoenix Iron Company at Phoenixville, Pennsylvania, he also maintained extensive investments in other ironworks as a hedge against the failure of any one of his enterprises.[14] At the time of Fritz's hiring, Cambria was a very troubled enterprise and a major worry to Reeves.

The Cambria Iron Company was the outgrowth of the entrepreneurial activity of Johnstown merchant George S. King.[15] As the western terminus of the Allegheny Portage Railroad of the Pennsylvania Main Line Canal, Johnstown had become a prosperous regional transportation center. It was also situated near deposits of bituminous coal and iron ore, in a region containing hardwood forests that were harvested to prepare the charcoal needed to fuel blast furnaces. In 1842 King and his partner, Dr. Peter Shoenberger, placed into blast the Cambria Furnace which produced pig iron for sale at Pittsburgh.[16]

Due to favorable conditions caused by the high tariff on imported iron imposed in 1842, Schoenberger and King soon expanded their operations by building additional furnaces near Johnstown. By 1846 King and Schoenberger controlled four furnaces. Unfortunately, the downward revisions of import duties which occurred in 1846 allowed a flood of cheap, high-quality British iron products to enter the American market; many furnaces in America were closed because of this competition.[17]

Faced with this prolonged general depression in the American iron industry, the partners began to look for a new product that would restore their firms' fortunes. By 1852 the completion of the Pennsylvania Railroad between Harrisburg and Pittsburgh provided them with an opportunity. Like other American lines, the Pennsylvania Railroad was a major consumer of iron railroad rails; and, since it passed through Johnstown, King hoped it would become a major customer of a rail mill at this location. In February of 1852 he began his search to find the capital needed to build this facility.[18]

Traveling to the eastern financial centers, King had little success in New York, but he succeeded in attracting investors at Boston. After a period of intense financial and political maneuvering, he obtained a state charter for his new enterprise on June 29, 1852.[19] Failure by the Boston investors to fulfill their obligations forced him to again solicit the resources of the New York financial community. Stockbroker and banker Simeon Draper provided the nucleus of the needed funds and served as the catalyst for attracting additional investors. By August 29, 1852, the Cambria Iron Company had been fully organized and capitalized at $1,000,000.[20]

Although the Cambria Iron Company owned extensive properties totaling over 14,000 acres of timber, iron ore, and coal lands scattered over Cambria and its neighboring counties of Indiana, Somerset, and Westmoreland, the new concern focused its primary attention on a site in Johnstown adjacent to the Pennsylvania Railroad. By March of 1853 work had commenced on four coke-fueled blast furnaces, and initial work had begun on the rolling mill and other necessary production facilities.[21] Soon afterward, the Ohio & Toledo Railroad made a proposition to the young firm for the purchase of rails. Since the Ohio & Toledo Railroad was short of cash, its management offered to exchange $200,000 of their company's bonds for rails. The Cambria Iron Company accepted this offer even though the rail mill was still under construction. The initial price for the rails was fixed at $85 per ton, but after further negotiation and the sale by the railroad of the bonds, the final price was reduced to $55 per ton, even though the prevailing market price for rails was $80 per ton.[22] It soon became apparent that the management of the iron company underestimated the problems they faced in making their mill operational when they accepted this agreement.

The Cambria Iron Company was soon faced with rising construction costs that were met only by the sale of bonds. With the additional money raised by the sale of its bonds, the company was able to continue the construction of its rail mill and other production facilities.[23] Before the rail mill could be completed, the Cambria Iron Company suffered a financial reverse due to the combined effects of a general decline in the market price of pig iron and rails and the bankruptcy of Simeon Draper, the firm's chief backer. Since Draper had pledged the assets of the Cambria Iron Company as security for fulfilling the rail contract with the Ohio & Toledo Railroad and there was no money on hand to actually produce their rails, the company's plant was faced with confiscation by its creditors.[24] John Fritz had recently arrived, and he summarized the bleak outlook of the Cambria Iron Company at the time in the following passage from his autobiography: "The workmen were restless and threatened to quit work ... at all times the works was on the verge of bankruptcy."[25] Fritz was able to prevent the works from being seized by convincing the U.S. Marshal who had been sent by the Ohio & Toledo Railroad to allow him to roll rails for other customers who were paying the prevailing market price. By keeping the mill in production and the workers paid, Fritz was able to roll the required rails for the Ohio & Toledo.

The beleaguered management of the Cambria Iron Company began a desperate search for new capital. A group of Philadelphia creditors organized a committee to find ways to save their investments through the reorganization of the company. Headed by Daniel J. Morrell of the mercantile firm Martin, Morrell and Company, this committee recommended that additional capital be invested in order to keep the Cambria Iron Company in operation. A wealthy Philadelphia merchant, Matthew Newkirk, purchased a large percentage of the company's stock. He became President of the Cambria Iron Company, while at the same time George S. King relinquished any role in the company's management.

Matthew Newkirk was not successful in his efforts to revitalize the company. The almost desperate nature of his activities is clearly evident in the following excerpts from a series of recently discovered letters which chronicle the tribulations of the Cambria Iron Company during this period.

On August 12, 1854, Matthew Newkirk wrote to John Fritz:

We have purchased the Blowing Engine of [I.P.] Morris & Co. and it has been shipped to you immediately and when it arrives I want you to have it set up immediately. They will send you the drawings and have all ready. If Mr. David Reeves says anything to you about remaining you had better say to him he had better make his arrangements to leave for I am very positive in my decisions. I want you to turn your main attention to getting all the furnaces to making pig metal as soon as you can. Write me soon.[26]

The continuing pressure on Fritz to increase production was also evident in the following letter of August 18, 1854, which Newkirk's associate John Anderson wrote to Fritz.

I have examined the contract for furnishing iron for the C.P.& I. RR. Co. and find that weight of the iron to be supplied was to be about 56 lbs. to the yard. They were to furnish a model. Now Mr. Newkirk supposed that the rolls had been made under the direction of Mr. Rizner who knew all about it and supposed he had acted in accordance with the contract -- Will it be necessary to have a new set of Rolls to make the 56 lb. iron or can the rolls now in use to set any closer together so as to make the rails weigh 56 lbs., let us hear from you on this subject.

What does the present rails weigh to the lineal yard? Is there any model furnished by the C.P.& I. RR. Co. and if so is the rolls turned to make iron of the same pattern. If no model has been furnished let us quickly obtain one.[27]

On September 2, 1854, Newkirk once again wrote to Fritz dealing with production matters. Again, his concern for quick production is evident:

We have made a contract with the state for 3,500 tons of rails. You must now put the steam on the mill in good earnest or we shall be behind with our debts. I will send you a copy of the contract that you may know what to do. I would write you more at length, but I am so unwell I cannot write more at this time. I want you to consult Mr. Anderson fully on all your movement. You will find him safe and sound in his judgment and very candid. has Mr. [David] Reeves left? Write me as often as you can for I like to hear from you.[28]

The urgency of Newkirk's demand continued unabated as is reflected in the following letter that he wrote to Fritz on September 9, 1854. It also related some of the problems that Fritz faced in making the rail mill function properly:

I received your favor of the 5th instance and noted its contents. I am pleased to hear you say you are going to turn your attention particularly to the blast furnaces. Finish one as soon as you can and start her then follow with the others as fast as you can. We want you to finish the mill, work as hard as you can and we will furnish you with the necessary quantity of iron and of the kind you want if it can be found. For if we do not begin to press the matter of rails now we shall certainly fall behind with our engagements. We have sent you the contract with the State General Pomeroy who we think is a clever man. Get from him a model of the rail he wants you to roll as you must have one as your guide, that we may have no mistake after the rails are made and when you get the model if you can improve in the pattern, perhaps he will allow you to do so. Mr. Kelton called and said he had some good red short iron from Cornwall Ore. I told him to send you a sample of 50 tons to try which he agreed to do immediately. I will order a straightening machine immediately from Reeves, Buck and Co. of Phoenixville and a punch from Neal Matthews of Bush Hill. I am pleased to see that you are making arrangements to protect the rolling mill against fire as well as you can.

Mr. Anderson writes me you have trouble with the large flywheel. Tufts was here from Boston who made the wheel, he says there is no difficulty about the size of it if it is geared right as it is not intended to make more than 1/3 or 1/2 as many revolutions as the usual kind and makes the mill run much more steady for heavy work. If it will not answer write as soon as you can what you would recommend. We must not let the mill lie idle.[29]

Seeking to gain the maximum possible production from the Cambria plant, Matthew Newkirk wrote the following letter to John Fritz on September 14, 1854, in which he unequivocally grants Fritz exclusive authority over its machinery:

Mr. Anderson has arrived safe at home and made a general report of what is doing at Johnstown. I am satisfied you are doing all that any one can do placed in similar circumstances with yourself. I want you now to go forward and manage everything in the Rolling Mill, Foundry, Machine Shop, Pattern Shop, and Blacksmith Shop and allow no one to interfere with you. Mr. John Anderson will return on Monday and while he is there I want you to consult and advise with him about any important change or any difficulty that may arise from accidents or unexpected circumstances. Allow no one else to interfere with your arrangements or orders and if they do not obey promptly discharge them forthwith. Urge on the Blast furnaces as fast as you can by any advice or facility that you can give.[30]

Despite the best efforts of Fritz to produce a marketable product, the financial condition of the Cambria Iron Company continued to deteriorate during the autumn of 1854. Nowhere is this decline more clearly evident than in the excerpt from the following letter written by John Anderson to Fritz on November 25:

It is the wish of the Directors of the Company that every expense that can be dispensed with should be reduced — Money matters are very tight indeed — I wish you would if you and Mr. Perry can agree upon it suspend operation on the coke yard and if as Mr. Reeves thinks we can cast our pig metal in coke dust sand that had better be done.[31]

Additional evidence of the financial crisis that faced the Cambria Iron Company is contained in the following letter of December 1, 1854, from the firm of Morris and Town to John Anderson:

Your esteemed order of November 23 was duly received and we were about to send the iron when we were advised that your concern had suspended payments, therefore retain the articles until we know what arrangements you intend making for the payments of your claims. Awaiting your reply.[32]

The impending bankruptcy of the Cambria Iron Company spurred Matthew Newkirk to urge John Fritz to greater effort in a letter of December 15, 1854:
I do hope that you may have no accident to interrupt your works, drive the state order day and night and make the rails as fast as you can and manage your men in the best way that you can to keep things moving.[33]

The year 1855 brought little financial relief to the beleaguered Cambria Iron Company. Its workers became increasingly unsure of their pay and as a consequence became reluctant to work. This decline in employee morale is clearly a matter of great concern to Newkirk in the following passage from a January 24, 1855, letter to Fritz:
Say to all the mechanical and laboring men about the works and furnaces in the town and country to go on and work with full confidence that they are all going to be paid for their labor and under any and all circumstances that no man or set of them shall get any advantage over them. By the 20th of February we expect to receive enough cash to settle up our payroll.[34]

Despite his best efforts, arranging to pay the men by February 20 proved to be a false hope. This failure confronted Fritz with a very unpalatable task as is shown by a letter sent to him on February 13, 1855, by Newkirk:
We are trying to arrange for our payroll about the 20th. If we cannot raise funds enough to pay all up I wish you would give as a reason it is because we had to take old iron from the state in the place of Cash, which is the fact. You had better perhaps meet quietly with some of your leading men to let them know this, that they may be prepared for it. You may apprise them as soon as we can sell the old iron or pig metal or rails they will be paid and if we can make a loan we shall do so.[35]

The payroll was not met by February 20 and in a desperate effort Matthew Newkirk traveled to Harrisburg to attempt to secure advance payment for rails that had been ordered by the State of Pennsylvania. On March 5, 1855, he wrote the following pleading lines to Fritz in a final effort to keep the disgruntled laborers at work:
I feel very confident I have made an arrangement at Harrisburg to get an appropriation for $50,000 to pay for our rails to the state if we can have the iron made and ready to deliver, drive the mill as hard as you can and try and keep your puddlers and railmen at work, you may assure them in a short time that they will get their pay if they make the iron ... You must do all that you can to quiet the men and keep them at work.[36]

Although he had labored mightily to keep the Cambria Iron Company solvent, Newkirk's efforts ended in failure. By May 21, 1855, he had relinquished control of the enterprise to a new company which possessed greater credit resources. This new partnership was the firm of Wood, Morrell and Company; it was composed of Charles S. Wood, Richard D. Wood, David Reeves, Daniel J. Morrell, Wyatt W. Miller, William H. Oliver, and Thomas Conorroe. These partners leased the facilities of the Cambria Iron Company for a five-year period which would expire on June 30, 1860.[37]
It was decided that Daniel J. Morrell would leave Philadelphia to assume the overall management of the Johnstown operation where he, along with John Fritz, would play a central role in the development and exploitation of the three-high rail mill.

In his *Autobiography* John Fritz described Daniel J. Morrell in the following terms: "He was a very clever gentlemen, but knew nothing about the iron business."[38] Yet, this very lack of technical background may have been Morrell's greatest asset in his role as the general manager of the Cambria Iron Company. By concentrating on the business operation of this concern he laid the groundwork for its eventual success. Unlike Newkirk who was constantly distracted by the pressing financial need to concentrate on existing operations, Morrell planned to build for the future. He also possessed the insight and business acumen to exploit the technological breakthrough which John Fritz was about to develop.

Early evidence of the management skills of Daniel J. Morrell is supplied by the inventory that Fritz compiled for the new managing partners soon after they had entered into the lease of the property. This document was used by Fritz to summarize not only the current state of the Cambria Iron Company's physical plant but also the improvements that Fritz continued to make to it:

Agreeable to your request I would state that the improvements and alterations made at the Cambria Iron Works since you became the lessees have been recommended or approved by me and were in my opinion absolutely necessary for the successful and profitable operation of the works. The earlier expenditures have more than paid for themselves already and those in progress and contemplated it is believed will more than pay for themselves within one year of their completion. It is not claimed, however, that no blunders have been committed for in the management of work of the magnitude of these, it would require more than human foresight to avoid sometimes erring in judgment, but as evidence of the utility of what has been done in the way of improvements it is only necessary to compare the present productiveness of the works with what they were at the time that you took possession of them. The utmost capacity of the mill has been but 200 tons of rails per week and with the machinery and furnaces then in use, it was impossible to calculate with any degree of certainty upon making that average yield. Our present production is more than double that amount and when all of the improvements now in progress are completed with the iron working well, it will be so easy to make from 500 to 600 tons per week as it was then to produce one-third as much and by the great saving in fuel and labor by using double instead of single puddling furnaces and making steam from the waste heat the cast of production per ton is much reduced.

The mill contained when you leased it 20 single puddling furnaces, 2 top and bottom and about 6 rail furnaces, 2 pair of puddle rolls and 1 pair of top and bottom rolls, we now have 18 double puddling furnaces and expect by the end of the summer to have 12 more making 30 in all or about equal to 60 simple and 12 heating furnaces. With one more in contemplation, we have now 5 pair of puddle rolls, 1 pair top and bottom rolls and 1 pair of billet rolls and have added a new mill with 3 pair of rolls capable of rolling all sizes of merchant iron and have substituted 2 of the longest sized Burden squeezers for the Winston squeezers then in use. Have also put in 2 small engines to drive the saws and forms and have just completed a set of force and supply pumps of a capacity to supply the whole mill with water and we are now erecting a new water tank capable of holding about 11,000 gallons of water. This improvement was much needed, it will be a great service and prevent the whole mill from having to stop when a pump valve gives out as here-to-fore. The new puddling and heating furnaces are worked with Blast and each furnace is supplied with a 26 foot boiler by which means ample steam is produced to drive all the machinery in the mill without any expense for fuel or firing.

We are now building an entire new rail mill, engine and all to take the place of the one in use, the necessity for doing so is apparent to any practiced observer. The great size and weight of the flywheel of the present mill renders it extremely hazardous to run the engine at the speed necessary to produce good work and the unwieldy size of the rolls makes the labor of rolling so severe that man then cannot endure it any length of time. It is only by giving additional wages that men can be kept.[39]

It is clear from the above letter that the rail mill at Cambria remained the most troublesome of all of the company's manufacturing operations. In his *Autobiography* John Fritz bluntly stated that "To continue to run the mill as it was, I could see nothing ahead but a most disastrous failure."[40] To overcome this problem Fritz proposed to the management of Wood, Morrell and Company "to build a new train of rolls, three-high and twenty inches in diameter."[41] It was this decision that has been cited by historians as the beginning of a technological revolution. Yet, the questions must now be asked, to what extent was the three-high rail mill an innovation and from where did John Fritz derive his inspiration for its construction?

The need for high-quality, moderately priced rails had become apparent almost since the earliest days of American railroads. In 1826-1827 the Lehigh Coal and Navigation Company had experimented with cast iron plate rails based on an earlier English model at their anthracite coal mine near Summit Hill, Pennsylvania. However, America's pioneer railroad soon turned to English mills for their rail needs. As early as 1830 the extent of British rail imports was already quite high. A single order for 40,000 tons of rails from the Dowlais Iron Works in Wales was not uncommon for this period.[42] Since the road beds of American railroads were often poorly constructed over difficult terrain, wear on the rails was often exacerbated. This tendency was heightened by the fact that the often cash-starved American railroads quickly increased their traffic loads in order to gain the maximum possible amount of revenue to pay off their crushing financial debts. As a result, even the best British rails wore out quickly, often with disastrous results.[43] Throughout the 1840s and 1850s the demand for British rails for use as replacements and in new construction was prodigious with one scholar estimating that over fifty million dollars was spent on this product between 1840 and 1855.[44]

Americans were slow to produce rolled wrought iron rails. Over 4,185 miles of railroad line had been completed in this country before the first domestic manufacturers entered the market.[45] It was not until a large increase was placed on the tariff on British rails in 1842 that American rail mills began operation. It is probable that the first rails produced in America were manufactured by the Mount Savage Iron Works in Allegheny County, Maryland, in 1844 while the first modern T-type rails were rolled at the Montour Iron Works in Danville, Pennsylvania, later in that same year.[46]

By the end of the 1840s many rail mills were operating, the most notable of which were the Lackawanna Coal and Iron Company at what is now Scranton, Pennsylvania, and the Trenton Iron Company of Trenton, New Jersey. A decrease in British rail prices during the early 1850s caused a temporary setback in American rail production. By 1854 the industry began to revive. The May 6, 1854, issue of the *American Railroad Journal* listed the following annual production figures for American rail mills:

Montour	13,000 tons
Rough and Ready (Danville, Pa.)	4,000 tons
Lackawanna	16,000 tons
Phoenix Iron Works (Phoenixville, Pa.)	20,000 tons
Safe Harbor	15,000 tons
Great Western (Brady's Bend, Pa.)	12,000 tons
New Works (Pittsburgh, Pa.)	5,000 tons
Pottsville Iron Works (Pottsville, Pa.)	3,000 tons
Cambria	5,000 tons
Massachusetts Iron Works (Boston, Mass.)	15,000 tons
Mt. Savage	15,000 tons
Richmond Mill (Richmond, Pa.)	5,000 tons
Trenton Iron Works	15,000 tons
Washington Rolling Mill (Wheeling, Virginia,	

now West Virginia)	5,000 tons
New Mill (Portsmouth, Ohio)	5,000 tons
Total	100,000 tons[47]

Despite the rising American production of rails, the majority of rails used in the United States continued to be imported from British manufacturers. American mills were able to supply less than one quarter of the American demand. The *American Railroad Journal* attributed this predominance to the long-term credit that the British manufacturers could offer and the superior quality of their product.[48] However, the continued rapid expansion of the American railroad system could absorb increased American production if the problem of poor rail quality could be overcome. Poor quality was directly related to the American rail manufacturing process.

Almost all early American rail mills were two-high mills which operated in the following manner. A stack of wrought iron bars called a rail pile, heated and hammered together, was passed between two horizontally mounted rolls. Grooves on the rolls progressively reduced and shaped the pile, but once the pile had initially passed through the rolls, it was necessary to pull it back over the top roll aided by the roll rotation to feed it through the next set of grooves. This laborious, time-consuming action was necessary because the rolls turned continuously in only one direction; it was not possible to reverse the rolling operation without making drastic adjustments on the steam engines that drove the mills.[49]

Because the pile had to be dragged over the top roll, it often cooled before it could be completely transformed into a rail. This cooling often caused the pile to split or shatter. In the process it often damaged the rolls and their driving mechanisms. This occurrence was so common that almost all rail mills, including Cambria, were constructed with specially weakened parts that would break first in the hope this would prevent greater damage if a pile broke.[50]

Until the problems inherent in the rolling of rails could be solved, American production costs were high: large work areas were needed to drag the rail piles back over the top roll; the damage was often done to the rolling machinery by split piles; rerolling damaged rails was an added expense. The end result was very often a rail of inferior quality.

To solve the problems in manufacturing rails, the British and European producers had developed mills in which the direction of the rolls could be reversed to allow the piles to be fed back between the two rolls before it had a chance to cool. However, Americans lacked the ability to produce the powerful reversing steam engine needed to power such mills. As a result, American innovators were forced to try other approaches. In 1854 a Mr. William Harris of Pottsville, Pennsylvania, proposed a novel way to solve the problem of completing the rolling process before the pile could cool. Harris described a rail mill that "instead of having a single pair of horizontal rollers with nine separate grooves, it had nine vertical rollers each with a single groove. The rolls are arranged in a continuous line with close boxes or ducts between them so that the pile is fed in one end and comes out a finished bar of railroad iron at the other."[51] Although Harris's innovative mills could be considered the precursor of the modern continuous rolling mill, the complexity of driving it with the primitive engines and power transmissions of the 1850s must have been its Achilles heel because no more is heard of it.

As was noted earlier, John Fritz took a different approach by developing a three-high mill. In his conception a three-high rail mill consisted of one additional roll placed on the top of the conventional two-high mill. This made it possible to eliminate the laborious and time-consuming pull-over process; the pile could immediately be returned by being passed through the grooves of the top and middle rolls after it had been initially shaped by the grooves of the bottom and middle rolls. This process could be repeated until the rail was completely shaped. Since the process was much speedier than the conventional pull-over technique, the pile was able to be rolled at a higher heat, resulting in a superior product. It was also a less expensive one since it employed fewer men.[52]

In his *Autobiography*, John Fritz makes no mention of the source of his inspiration for the three-high mill. Yet it is possible, based on recently discovered materials, to make some conclusions about this question. It is known that the use of small three-high mills to produce various iron products in Europe was well known to many American ironmasters. The proprietor of the Trenton Iron Company, Abram S. Hewitt, wrote: "In France it is to be noted that Anzin three-high trains have been in use for rolling girders since June 1844. There is also a three-high plate mill at Le Creuset and the principle of the three-high mills appears to be perfectly well understood in Europe."[53]

The most likely source of John Fritz's knowledge of European three-high mills was *The Iron Manufacture of Great Britain*, by William Truran. This 1854 treatise states: "To economise time, in the rolling of the smallest bars, which otherwise would become comparatively cold before finishing, the guide and small merchant bar trains are furnished with three rolls to each pair of standards, the three being geared together by pinions; the bottom roll and lower side of the middle roll reduce the iron in the forward motion, while the top roll and upper side of the middle roll work in a contrary direction, and the bar is returned between these for reduction in the backward motion. The three-roll movement has been applied to rail and other large mills, but its most prominent advantages are limited to trains of small rolls."[54] Plate XVII of this volume also depicts in detail a three-high mill at Dowlais in Wales. nI essence Truran has described and illustrated the basic mill that Fritz would eventually construct at Cambria. Two items of this description are particularly significant: the mention of guides, which would become a key improvement of John Fritz's claims for an American patent on his development, and the fact that Truran plainly states that three-high mills were used for the production of rails. Another element supporting the authors' belief that *The Iron Manufacture* provided the inspiration for John Fritz's development of a three-high rail at Cambria is found in the existence of a copy of Truran's work in the collections of Hugh Moore Historical Park & Museums, Inc., at Easton, Pa. This particular copy bears the handwritten inscription of Reeves, Buck and Company on the inside of its front cover. Since it has been established that David Reeves was both John Fritz's long-time patron and a major partner in Wood, Morrell and Company, it is very likely that John Fritz was able to study this very book during one of his visits to the Philadelphia area in 1856.[55] Further evidence supporting this supposition is provided by the following letter of July 12, 1856, in which Charles Wood invites John Fritz to visit Phoenixville, the site of the Reeves, Buck and Company office and ironworks:

In Mr. Townsend's letter of the 11th inst. he gives a very doleful account of the working of the roll mill. I would be glad to go up and consult or have a talk with you about it, but I cannot leave the city at present. The first question I should ask is the iron at fault, which I suppose you will answer in the affirmative. The next is can we find a remedy. This question I know is more difficult, but if you give your mind to it I think that you can rectify the evil. Do you not overload the furnaces, what I mean is, are you using too large a quantity of ore for the fuel [coke] so that some of the iron fails to get sufficiently heated to draw off the impurities or comes out in too crude a state. These and many other things may have suggested themselves to your thinking mind, but as this and all the manufacturing department is under your particular charge with full authority to carry them out with good economy, we have to appeal to you when things go amiss. Perhaps you may get some new ideas by seeing and consulting with some old iron men. If you will come to the city soon after Mr. Morrell returns, I will go with you to Phoenixville to see Mr. David Reeves and Mr. Griffiths and have a talk with them and if any other person occurs to you who can give us any light on the subject we will pay them a visit ...[56]

Despite his convictions that a three-high rail mill could be successfully designed and constructed at Cambria, Fritz had great difficulty in convincing the managers of Wood, Morrell that such a large investment in new machinery had to be made.

Only Fritz's adamant refusal to rebuild the Cambria rail mill in a conventional two-high manner and his threat to resign his position induced Wood, Morrell and Company to accede to his plans.[57] Although he had wanted to build the new three-high mill with 20-inch diameter rolls, he did accept with misgivings their recommendation to make the rolls 18 inches in diameter. For several months he was able to devote his energies to construction of the new mills until a legal protest from the minority partners in the firm of Wood, Morrell and Company forced him to stop. Fearful of being held legally responsible for Fritz's actions, the majority partners ordered him to cease work on the new mill. Since Wood, Morrell and Company only leased the Cambria Iron Works, the minority partners felt that the company should concentrate on short-term profits. Long-term improvements, they believed, would benefit only the majority partners of Wood, Morrell and Company who were stockholders in Cambria Iron Company. The nature and intensity of these protests and threats can be readily discerned in the following letter of April 24, 1857, written by minority investor Wyatt Miller to the managing partners of Wood, Morrell and Company:

> It has come to my knowledge indirectly that you are either now or have in contemplation to make improvements or expenditures in altering or enlarging Cambria Iron Works. I have also ascertained from statements furnished me from your office that similar outlays have previously been made in the same way and that the amount thus expended considerably exceeds the limit to which the partnership articles permitted. According to my opinion, the powers assigned to you as managing partners of the firm of Wood, Morrell and Company are specifically defined and in exceeding them without full concurrence and assent of the partners you take upon yourselves a responsibility for which you may be held accountable...[58]

On May 9, 1857, the managing partners of Wood, Morrell and Company wrote to Fritz asking him to justify his expensive new improvements which included the three-high rail mill:

> We desire you to furnish us in writing your views in regard to the various improvements, repairs, and alterations which it has been decided expedient to make and which are in contemplation by us as lessees of the Cambria Iron Works, stating whether in your opinion such expenditures were and are justified by the prospective advantages to be derived from them and whether they will pay for themselves during our lease at the works...[59]

In his *Autobiography*, John Fritz described how he finally won the consent of the managing partners of Wood, Morrell and Company to finish the installation of the three-high rail mill:

> The next Sunday morning Mr. Townsend came to the mill where he found me in the midst of the regular Sunday repairs. After I was pretty well through with them he took me aside and showed me the protest. My hands being greasy, I asked him to read it to me, which he did. After all these years have passed, there is no person other than myself who can fully appreciate the trying position the managers were placed in. On the one hand, I was urging them to build a mill, on an untried plan, as a strong minority called it, this minority also legally notifying the managers that they would hold them personally responsible for the result. On the other hand, I was absolutely refusing to build the mill they wanted, and besides all this, they ridiculed the idea of adopting a new and untried method that was against all practice in this and the old country, from which at that time we obtained our most experienced iron workers. Moreover, the prominent ironmakers in all parts of the country had said to Mr. Morrell that the whole thing was a wild experiment and was sure to end in a failure, and that young, determined, crack-brained Fritz would ruin him. The heaters and rollers all opposed the three-high mill and appointed a committee to see the managers and say to them that the three-high mill would never work, and that they, themselves, would suffer by reason of its adoption, but that if the managers would put in a two-high geared train, which they said was the proper thing to do, the mill would go all right.

As I now look back to that eventful Sunday morning, many long years ago, sitting on a pile of discarded rails, with evidences of failure on every side, Mr. Townsend and myself quietly and seriously talking over the history of the past, the difficulties of the present, and the uncertainties of the future, I cannot but feel, in view of what since has come to pass, that was not only a critical epoch in the history of the Cambria Company, but that as well the future well-being of my life was in the balance. For, as Mr. Townsend was about to leave, after a full discussion of the Cambria Iron Company's condition at that time, he turned to me and said: "Fritz, go ahead and build the mill as you want it." I asked, "Do you say that officially?" to which he replied: "I will make it official," and he did so; and here I wish to say that to no other person so deservedly belongs the credit, not only of the introduction of the three-high-roll train but also of the wonderful prosperity that came to the Cambria Company, as it does to Mr. Edward Y. Townsend, then its Vice-President.

Notwithstanding I now had the consent of the Company to go on with my plan for the new mill, many of my warmest friends, some of whom were practical ironmen, came to me and urged me not try such an experiment. They said I had taken a wrong position in refusing to build the kind of mill the company wanted. "By so doing," they said, "you have assumed the entire responsibility and in all probability the mill that you are going to build will prove a failure, and being a young man your reputation will be ruined for life." To this I replied that possibly they were right, but that I had given the subject the most careful consideration and was willing to take my chances on the result.[60]

As construction continued on the three-high rail mill, it became known to the managing partners of Wood, Morrell and Company that Fritz planned to dispense with the practice of using breakable parts that would easily shatter under the strain of a warped or split pile and thus protect the mill from serious damage. Fritz had made this decision because he wished to avoid the multitude of delays caused by the replacement of these "break-up pieces." He also believed that the system of guides he had designed would prevent the rail piles from binding and that the rapidity with which his new mill would complete the rolling process would prevent the cooling and consequent brittleness of the piles. Despite the entreaties of managing partners at Wood, Morrell and Company, and other individuals associated with the iron trade, John Fritz could not be swayed. Aided by his younger brother George, who had joined him at Cambria, he rapidly completed work on the new mill. On July 3, 1857, America's first three-high rail mill was placed in operation despite the continuing reluctance of Cambria's workers. The scene is vividly described by John Fritz in the following passage from his *Autobiography*:

The train was now practically completed, with all breaking devices abandoned. The old mill was stopped on the evening of the 3rd of July, 1857, and after the 4th I commenced to tear the old mill out, and get ready to put the new one in, and also to put the new engine in place at the same time. Everything in the rail department was remodeled and the floor line raised two feet. On the 29th of the month everything was completed and the mill was ready to start. I need not tell you that it was an extremely anxious time for me, nor need I add that no engraved cards of invitation were sent out, that not being the custom in the early days of iron making; had it been, it would not have been observed on that occasion.

As the heaters to a man were opposed to the new kind of mill, we did not want them about at the start. We secured one, however, out of the lot, who was the most reasonable one amongst them, to heat the piles for us. We had kept the furnace smoking for several days as a blind. At last, everything being ready, we charged six piles. At about ten o'clock in the morning the first pile was drawn, and it went through the rolls without the least hitch of any kind, making a perfect rail. You can judge what my feelings were as I looked upon that perfect and first rail ever made on a three-high mill, and you may know in part how grateful I felt toward the few faithful and anxious men who were about me and who stood by me during all my trials and difficulties, among them were Alexander Hamilton, the Superintendent of the mill, Thomas Lapsley, who had charge of the rail department, William Canam, and my brother, George.

We next proceeded to roll the other five piles. When two more perfect rails were rolled we were obliged to stop the engine, as the men were all so intently watching the rolls that the engine had been neglected, and, being new, the eccentric had heated and bent the eccentric rod so that the engine could no longer be worked. As it would have taken some time to straighten the rod and rest the valves, the remaining piles were drawn out of the furnace onto the mill floor. About this time the heaters, hearing the exhaust of the engine, came into the mill in a body, and from the opposite end to where the rails were. Seeing the unrolled piles lying on the mill floor, they took it for granted that the new train was a failure, and their remarks about it were far from being in the least complimentary. Mr. Hamilton, coming along about that time and hearing what they were saying about the mill, turned around, and in language more forcible than polite told the heaters, who were Welsh, that if they would go down to the other end of the mill they would see three handsomer rails than had ever been made in Wales, where the greater part of the rails used in this country at that time came from, as well as the heaters who were so bitterly opposed to the three-high mill."[61]

Another description of John Fritz's triumph is in a recently discovered letter written in 1902 by a former Cambria worker, Joseph Graham, who was present at the first day of the three-high rail mill's operation:

In looking back to the latter part of the year 1857 and for your effort there to make such improvements in manufacturing rails as would and [did] give Wood, Morrell and Company in the manufacturing of rails which you did by the three-high rolls—you certainly remember the third and last effort I think in June 1857 when J.H. Lapsley the roller made success. I heated the first heat ever successfully rolled on a three-high rolls. I remember distinctly of Lapsley telling me in the morning that he had ordered my furnace light up and he wanted me to make the heat, he was just 8½ days from the time he commenced putting the mill in order until I was ready to draw my heat which I did successfully. The field was then open for your ingenuity which you and your brother George carried through successfully. About one month after starting the mill burnt down and it seemed as if the Company was afraid to start but you did start and before the Company finally agreed (I can't give the date) and told me you wanted me to take my own time so I could be ready to draw a heat. I think that you said your engine should run up to 12 [revolutions] per minute (it did not run very fast) and a gentleman from Philadelphia stood on No. 1 with a watch in hand timing the rolling of the heads. After they examined the rails he came back to me and handed by a $10 bill, he gave Lapsley $20. Well Mr. Fritz I have been thinking of you many times and I feel as if I am not able to go to Bethlehem now physically or financially, if you come out our way will be please call and see me. My eyesight is almost gone and I can hardly read the lines I write on. Please answer.[62]

Soon after it had successfully begun operation the three-high rail mill was damaged by a fire which completely destroyed the building housing the mill. Through the herculean efforts of John Fritz and his workers, the mill was placed back in operation within twenty-eight days of the fire.[63] Further information about the extent of this catastrophe is provided by the following letter sent to Daniel J. Morrell by Charles Wood on August 3, 1857:

The insurance companies will require an estimate of the loss and as no one is so competent as Mr. Fritz he had better make an estimate of the loss on the building and on the machinery separately and in his estimate of the loss to the machinery let him separate that belonging to our firm from that belonging to the Cambria Iron Company. He cannot come at the loss accurately but let him give it according to his best knowledge and belief. I expect the insurance companies will send an agent or agents who will make examination and report. Please let Mr. Fritz make this as early as he can. The Cambria Company has insurance on the mill of $8,000 and on the machinery of $8,000. Our insurance on the mill and machinery is $35,000, half on each. They ought to settle promptly, but I fear with some of them we shall have trouble.[64]

By January 1, 1858, the three-high rail mill which had been operating in the open air, was enclosed in a fire-resistant brick building almost 1,000 feet in length and 100 feet in width.[65] With its completion, the technological part of the three-high rail mill creation was complete. Now the task facing John Fritz and Wood, Morrell and Company was to exploit this achievement.

Financial and commercial repercussions of the 1857 Panic were devastating but short-lived. As soon as the demand for iron rails was restored, Morrell, Townsend, and the rest of the group leasing the mill at Cambria conceived a plan to sell the technology developed by the Fritz brothers and, at the same time, stabilize the rail market. To achieve this end, they proposed the new rail mill should be patented. Once granted, a patent becomes public knowledge, but fees can be charged for use of the ideas. This was an unusual step, allowing an employee to register a personal patent for work done on company time with company funds, but it was part of a larger strategy intended to keep the Fritz brothers at work on the improvements.[66]

To obtain patent rights, Fritz took a model, with thirty dollars in gold, to Washington D.C. in early June, 1858. He hired patent agent A.B. Stoughton to process an application describing the new mill. It was filed with the government office on June 22. Fritz outlined a "new manner" of "rolling rails, beams, bars, etc., of iron." After extremely cursory examination the Patent Office rejected his application the following day.[67]

The patent examiner declared nothing novel was presented in the description of the three-high Fritz mill. He cited similar rolls, pictured in Truran's book; a previous application made by William Borrow in 1853 which had been rejected; a new patent of the same sort that had been issued three weeks earlier to Stephens & Jenkins, and the previous use of guides "to retain the rolled article in shape" in an 1846 patent. The examiner was not at all impressed by the Fritz claims.

The inability to gain a patent at first stymied hopes of the Cambria group to form a patent pool. Meanwhile, word was being spread about formation of the pool, so that Cambria lessors were contacted regularly by representatives of other rolling mills hoping to gain rights. Many of those interested were smaller companies on the fringe of the railroad network, but it was hoped the big mills, like those in Danville or Lackawanna, or even Cooper & Hewitt's Trenton Iron Company, would become intrigued.

With the cooperation of the Fritz brothers, inventors Stephens & Jenkins were contacted and a statement was obtained from them disclaiming rights to specific parts of the Fritz mill. The draft was rewritten, placing the words "three high rolls" in quotation marks. After a thorough patent search, funded by the Cambria sponsors, Stoughton sent the papers back to the Patent Office in September, accompanied by a passionate defense of the Fritz mill's novelty. "Now certainly there must be invention in what he has done, somewhere," Stoughton insisted. "Mr. Fritz does not claim to have invented "three high rolls" ... neither Truran or Borrow show what Fritz claims ... the fact of guides having been used to direct the bar into, and out of the rolls, does not meet the claim. Mr. Fritz does not claim a guide, but a special kind of guide, viz: a yielding one." The agent anguished over further delays, begging "... reconsideration of the application, knowing from the interest which iron manufacturers express in his invention and its wonderful production, that there is not only novelty in it, but novelty of a very high and very valuable character."

The careful rewording and thorough searching reversed all opposition. Fritz was awarded patent protection on October 5 for two claims: an arrangement of three-high rolls specifically to roll down fins, thus saving handling time, and the yielding guides that allowed faster rolling. He said these improvements would permit rolling rails "one third longer" than normal, but agent Stoughton was more ebullient, asserting "twice as long." After awhile, when the claim was used to sell licenses, the advantage was advertised as "fourfold increase."

Enthusiasm for the improvement was not general, however. Fritz said later: "We had no money, and at that time the ironmen were looked upon as paupers. The banks would not loan them any money ... the ironmen got but little, and that little only for a short time, the bankers fearing they would fail, as in the early days of rail making they were likely to do."[68] One of the iron men who failed in 1857, but lived on to play a larger but more controversial role was William Kelly of Eddyville, Kentucky.

Originally from Pittsburgh, Kelly was born in 1811. After making a little money in a drygoods business with his brother, Kelly met and married a girl in Eddyville. Her wealthy father induced Kelly to go into the ironmaking business at that town. In 1846 Kelly tried to make iron in a charcoal blast furnace without any great success. He then tried to save fuel costs by developing various arrangements for blowing air directly into the liquid metal. This practice was called "burning iron" in the industry. Nothing that Kelly attempted in his blast furnace achieved any spectacular results, but he did keep records of his tests starting in 1847, and he did make and sell something he called "Air Boiled" iron in the early 1850s.

In the days before metallurgical analyses, allowing air into contact with liquid iron inside a furnace was scrupulously avoided by ironmakers of the era because of a prejudice about resultant poor quality. A treatise by the great English ironmaker Richard Johnson warned about allowing the presence of air during the final stages of the puddling process. Describing iron heated until liquid as mercury, Johnson said:

> When it has reached this point it experiences a violent agitation, technically called 'the boil' which is produced no doubt by the oxidation of the carbon and the escape of the carbonic oxide then generated.... This part of the operation requires great skill in the puddler; for nearly the whole of the carbon has been oxidized, or as it is technically termed 'burnt'; and thus not only does great loss ensue in the quantity of malleable iron produced, but also the iron containing a certain quantity of oxide of iron is brittle, and of bad quality.[69]

Kelly's tinkering as a novice was insignificant. He might have passed from history without notice, except for an incredible series of blunders in England which brought attention to that which otherwise was worthless. Kelly was not alone in deliberately burning iron. At least two others had tried the same. Joseph Martien had done much the same experiment at the Renton works in Newark and also at the Ebbw Vale company in Wales. By trial and error, the same result was attained by an English inventor named Henry Bessemer who was working on an armaments project.

The denouement is almost comical. None of the three had a process that actually worked, but in the end all three filed for U.S. patent protection for a process generally condemned by all who tried it.

Bessemer, a successful inventor without any background in the iron business, proceeded methodically as was his habit. His various efforts did not satisfy him although he produced many different varieties of iron. Ultimately, he sought the advice of George Rennie, a noted engineer. It was Rennie, who, after examining Bessemer's results, jumped to the conclusion the inventor had accidentally devised a way to make high-quality malleable iron directly from hot metal, by-passing the puddling process, an enormous cost-saving. Convinced a great discovery had been made, Rennie badgered Bessemer into announcing this achievement to the ironmaking community.

In 1856, Bessemer wrote and delivered a technical paper "On The Manufacture Of Malleable Iron Without Fuel." He also routinely applied for English patent protection, as was his normal custom. All evidence indicates he was astonished by the enthusiastic reaction to his paper. Dozens of ironmakers applied for licenses to use his fuel-saving method, even though the inventor himself was by no means certain just exactly what it was he had discovered. Nonetheless he granted rights to others, collecting handsome fees, and used the money to continue his experiments. He also continued to file for additional patents.

Bessemer obtained U.S. patent protection, (No. 16,082; November 11, 1856) describing a special cylinder for "conversion" of "molten crude iron or remelted pig iron or finery iron into steel, or malleable iron, without the use of fuel for reheating or continuing to heat the crude molten metal." Although he did not claim to have been first to inject air into molten iron, the publication of his patents aroused both Kelly and Martien, who immediately filed counterclaims. Kelly won, but his patent, (No. 17,628; July 23, 1857) covered only the use of air blasts for "decarbonizing" molten iron "in the hearth of a blast furnace."[70]

In Wales, Ebbw Vale refused to buy a Bessemer license. Instead they employed a metallurgist, Robert Mushet, to explain why Martien's experiment did not work. This was the decisive action resulting in the advent of a process for cheap steel. To remedy the defective quality of iron produced by the air blowing action, Mushet, without hesitation arranged to add a corrective substance "known in Prussia as Spiegeleisen" to burnt iron in a converter vessel.[71]

Mushet knew exactly what he was about, in contrast to the groping of Kelly, Martien, and Bessemer: "For the purpose of remedying these defects in cast-iron purified by the action of a blast of air, and to convert it into malleable iron or steel, I add and combine with such purified iron, while in a melted state, a triple compound or material consisting of or containing iron, manganese, and carbon... The proportion of the triple compound to be added to the purified cast-iron may vary as the circumstances may require...."[72]

Manganese, an iron-like metal which is a normal constituent of some iron ores, is usually quite brittle in the pure state at room temperatures. When combined with molten iron, however, manganese exhibits a preferential ability to unite with oxygen and sulfur. Also, a certain amount of manganese will alloy with molten iron, improving its mechanical properties as a solid-solution strengthener. The beneficial effects of manganese in iron had been established many years earlier by the crucible steel industry.

In modern metallurgical terms, what Mushet accomplished in his experiment was to activate the transposition of undesirable ferrous oxide and sulfide into corresponding manganous compounds which were only slightly soluble and thus more easily floated off as slag. At the same time carbon and manganese from the spiegeleisen (or ferromanganese), when mixed with the purified molten iron remaining in the converter, created a new metallic alloy with superior mechanical properties.[73]

Mushet filed for an English patent on his manganese alloying technique in September, 1856 and applied for a U.S. patent on the 25th day of March, 1857. He also met with Bessemer, an historic confrontation. If Bessemer knew that his own patent was puffery, he apparently gave no indication to Mushet. Meanwhile, for some reason the owners at Ebbw Vale became convinced all the experiments were fakery, even though a single rail made by this method was demonstrating exceptional wearing qualities on the Midland R.R. They ceased paying fees on Mushet's application. The British application lapsed. Incredibly, Mushet's great discovery became public property in England, about the same time he was granted patent protection (No. 17,389; May 26, 1857) in the United States.

Abram Hewitt of Trenton Iron reacted to Bessemer's paper by asking about a license. He then tested the process without success. Hewitt next encouraged Martien to proceed with construction of a converter at Newark, but Martien's efforts were no more successful than the techniques used by Bessemer and Kelly. By August 1857, Hewitt was becoming anxious, asking Martien: "How do you succeed in making good iron? Give us some facts to judge by."[74] Then the Panic of 1857 swept the nation causing Hewitt to lose interest.

In 1858, Goran Goransson, a Swedish ironmaker who had taken a Bessemer license, succeeded in making good metal with a converter using local ore in Sweden. It was ore with unusually high manganese content, so that in effect he had produced Mushet's cheap steel directly from the ore, proving it was possible to do so without using Mushet's additive technique. Following this, Bessemer delivered another technical paper in May, 1859, implying he had known about this aspect of the technology all along. This strengthened his posture with the original licensees, all of whom were free to use Mushet's technique.

After his business failed in Kentucky, Kelly returned to Pittsburgh and transferred his patent to his own father to avoid having it pass over to his creditors. Next, he contacted the owners of other furnaces seeking a place to continue his trials. These additional experiments came to the attention of Morrell at Cambria, who reasoned it would be a good thing to bring Kelly together with the Fritz brothers. Kelly began experimenting in cooperation with Cambria, but accomplished very little, as can be seen in the following passage from a letter Kelly wrote to John Fritz from Pittsburgh, on November 5, 1859:

> I had an interview this afternoon with Mr. Morrell and was informing him of the results of my experiments at Oliphant's furnace (located near Uniontown, Pa.) with my new process, when he suggested I would communicate to you in detail the results ... My object was to take the metal from the furnace and put it into an air boiling furnace and after refining it perfectly inject a stream of carbonic acid into the iron so as to bring the iron 'to nature' and then run it into ingots ready for the rolls.[75]

It is now obvious, from the content of this letter, that recarburizing by use of carbonic acid (carbon dioxide) could never solve the problem. In essence, as of late 1859 Kelly, like Bessemer, was merely groping. Equally as important, this letter also proves that not all of his experiments were conducted at Johnstown. Nevertheless, Kelly was an honest and dedicated man, and Morrell continued to back his studies, hoping to obtain a converter that would actually work.

John Fritz, however, was presented simultaneously with another opportunity which he could not refuse. He was induced to return to the Lehigh Valley in July of 1860 and start up a new ironworks in Bethlehem, Pennsylvania, for the specific purpose of providing better rails to the Lehigh Valley RR.[76] At the time Fritz relocated to Bethlehem Iron, he was already convinced the converter notion was a sham, and it is certain nothing done by Kelly prior to 1860 could have convinced anyone otherwise. He afterward said he had "in some way learned that phosphorus was not permissible in the manufacture of good wrought iron" and he saw no way for the converter to reduce phosphorus.[77]

George Fritz remained at Cambria and continued to work out refinements in the new rail mill. He and his brother remained in constant contact. Together they had applied for an additional patent on the mill (No. 25565; September 27, 1859). The two patents were placed in the care of a Philadelphia lawyer named John Kennedy who structured a holding entity, The Rolling Mill Improvement Co., to shepherd them.

In the United States domestic wrought iron rail business was booming on the eve of the Civil War. Production had grown eightfold, from 24,000 tons per year in 1849 to over 200,000 tons per year in 1860. Even so, considerable amounts of rails were still being imported from England. Cambria was producing good malleable iron rails which lasted ten years or more in severe service. When the Civil War began, demand for iron rails at first fell off. The price dropped very quickly, from about 65 dollars per long ton in the early 1850s, down to 42 dollars or less in the early part of the war. Then inflation brought the prices back up to over 100 dollars, and the boom resumed.

Meanwhile, efforts to form the Fritz mill patent pool encountered roadblocks. For example, William Hancock and John Foley, who operated the famous Rough and Ready mill in Danville, flatly rejected all approaches, insisting: "We are entirely in ignorance of what you claim or wish us to do. As far as we know three high rolls have been in use for fifty years."[78] However, with Morrell negotiating an agreement, Hancock & Foley soon joined forces with Cambria. By 1863, Morrell was ready to complete an agreement among all those mills in the pool which were chartered under the laws of Pennsylvania. These included Cambria, Bethlehem Iron, Lackawanna Iron, Palo Alto Iron, the Allentown Rolling Mill, and Hancock & Foley at Danville.[79]

About this time Morrell came into contact with the only other man in the United States who continued to have faith in the notion of making good malleable iron or steel by burning iron in a converter. He was Eber B. Ward, son of a poor lighthouse keeper, who had made a great fortune as a shipowner on the Great Lakes. In 1844-45 when government surveyors discovered evidence of large iron ore deposits near the north coast of Upper Michigan, Ward established a rolling mill at Wyandotte, where the Detroit River enters the lakes. He obtained some of the very first iron smelted from the new ores in February, 1848.

Almost alone, Eber Ward visualized ironmaking and ironrolling on the shores of the Great Lakes, at the new lake ports such as Chicago, away from the existing canal network. He devoted his entire fortune to making this vision a reality. Ward aligned himself with an experienced ironmaker, Zoheth Durfee, whom he sent to England to make a thorough study of the latest technology. On January 7, 1863, Zoheth's cousin William Durfee made inquiry about the Fritz mill patents:

The proprietors of the "Chicago Rolling Mill" in whose behalf I write are about extending their works and contemplate putting in rolls sufficiently large to roll girders from fifteen to twenty four inches deep should occasion require. The principal business of the mill however will be the rolling of rails. What will be your price for the use of your patent machinery for the above purposes including a detailed drawing of the rolls housings guide etc. and what size rolls would you recommend? An answer to this directed to the undersigned care of Captain E.B. Ward, Detroit, Michigan, would greatly oblige.[80]

While in England, Zoheth Durfee appears to have been the person who unraveled the mysteries of why the Bessemer process did not work at first, and the curious situation involving Mushet's crucial patent. Meanwhile, at Cambria, Morrell hired a young chemist, Robert Hunt, who set up a laboratory in 1860 for the study of iron chemistry, achieving a more complete understanding of the phenomena than had been previously available. Shortly after Hunt arrived, Kelly's fruitless experiments came to an end.[81]

As an outcome of the Fritz mill contact, Ward became aware Morrell had been sponsoring Kelly's continued experiments. They then apparently conspired together, reasoning that if there was any future for the steel converter in the rail business, Bessemer would be powerless to act against them because his original patent had no value in the face of Kelly's prior claim. The combination of successful steelmaking techniques with the Fritz mill patent pool was visualized accurately as the foundation of a great industry.

During the war, Ward and Morrell, together with other investors, created something called Kelly Pneumatic Process Company, more or less a secret enterprise. They obtained the rights to use Kelly's patent, in exchange for a rather minimal royalty. Kelly himself had very little to contribute beyond his name. There is no evidence he played an active role. He later went into the axe manufacturing business.[82]

William Durfee built a converter shop at Wyandotte which secretly duplicated Bessemer's process at Sheffield. Wyandotte was chosen over Cambria because it was on the outer periphery of the American ironmaking community, where secrecy could be maintained, yet closest to the Lake Superior ore deposits. Hunt relocated there to help. In September 1864, cheap steel was manufactured at Wyandotte by a composite process, never fully revealed, which must have copied a goodly number of Bessemer's techniques, plus the Mushet recarburizing. Simultaneously, Zoheth Durfee went back to England and offered Robert Mushet a partnership in the new company in exchange for the exclusive right to use his patent in the United States. Mushet assented on October 24, 1864, a date which marks the true birth of the American steel industry and the culmination of a process that began with the development of John Fritz's three-high rail mill.[83]

NOTES

1. Elting E. Morrison, *From Know-How to Nowhere: The Development of American Technology* (New York, N.Y.: Basic Books, 1974), 68-82.

2. Between 1986 and 1988 the remaining personal papers of John Fritz and his brothers George and William have been donated by the late Mrs. Ellen Fritz Sahlin Hartshorne and her daughter Mrs. Penelope Hartshorne Batcheler to the Hugh Moore Historical Park & Museums, Inc. The Fritz collection is currently housed at the Museum Support Center of the Hugh Moore Historical Park & Museums, Inc., where they have been conserved, catalogued, and made available for scholarly research.

3. The best accounts of John Fritz's early life are contained in *The Autobiography of John Fritz* (New York, N.Y.: John Wiley and Sons, 1912), *John Fritz Pioneer in Iron and Steel* (West Chester, Pa.: Chester County Historical Society, 1954), and Lance E. Metz, *John Fritz (1822-1913) His Role in the Development of the American Iron and Steel Industry and His Legacy to the Bethlehem Community* (Easton, Pa.: Hugh Moore Historical Park and Museums, Inc., 1987).

4. *Autobiography of John Fritz*, op.cit, 49-56.

5. For a description of the operation of the Safe Harbor complex, see Ernest T. Schuleen, "Two Rivers and a Village: The Story of Safe Harbor," *Journal of the Lancaster County Historical Society*, Vol. 85, No. 3, 1981, 89-91, John W.W. Loose, "Anthracite Iron Blast Furnaces in Lancaster County, 1840-1900," *Journal of the Lancaster County Historical Society*, Vol. 86, No. 3, 1982, 101-104, and John W.W. Loose "The Anthracite Iron Industry of Lancaster County: Rolling Mills, 1850-1890," *Journal of the Lancaster County Historical Society*, Vol. 86, No. 4, 1982, 133-134.

6. W. Raymond Rossiter and Henry S. Drinker, "Biographical Notice of John Fritz," *Transactions of the American Institute of Mining Engineers*, 1913, 3.

7. Metz, op.cit., 6.

8. *Autobiography*, op. cit., 72-75.

9. See J.P. Lesley, *The Iron Manufacturers Guide* (New York, N.Y., John Wiley and Sons, 1859) 13, for information on Kunzie Furnace, (later Spring Mill Furnace).

10. Metz, op.cit., 7-8.

11. *Centennial History of Catasauqua, Pennsylvania, 1853-1953* (Catasauqua, Pa.: Catasauqua Publicity and Historical Committee, 1953), 61.

12. Craig L. Bartholomew and Lance E. Metz (Ann Bartholomew, ed.), *The Anthracite Iron Industry of the Lehigh Valley* (Easton, Pa.: Center for Canal History and Technology, 1988), 41-42.

13. *Centennial History of Catasauqua*, op.cit., 61-62.

14. Peter Temin, *Iron and Steel in Nineteenth Century America: An Economic Inquiry* (Cambridge, Massachusetts: M.I.T. Press, 1964), 104.

15. Sharon Brown, "The Cambria Iron Company of Johnstown, Pennsylvania," *Canal History and Technology Proceedings*, Vol. VII, 1988, 21.

16. Henry Wilson Storey, *History of Cambria County Pennsylvania*, Vol. I (New York, N.Y.: Lewis Publishing Company, 1907), 402.

17. Brown, op. cit., 22.

18. James M. Swank, "Early Iron Enterprises in Cambria, Somerset, Westmoreland, and Indiana Counties," *Johnstown Centennial, 1800-1900*, 67.

19. Storey, op.cit., 410.

20. George D. Thackery, "Brief History of the Cambria Plant of the Bethlehem Johnstown, Pennsylvania," 1925 (unpublished manuscript in the Bethlehem Steel Corporation Collection, Hugh Moore Historical Park & Museums., Inc., Easton, Pa.), 17.

21. Nathan Shapee, "A History of Johnstown and the Great Flood of 1889: A Study of Disaster and Rehabilitation," (unpublished Ph.D. dissertation, University of Pittsburgh, 1940), 63.
22. Brown, op.cit., 24-25.
23. S.A. Cox, "Engineers Report to the President of the Cambria Iron Company," December 15, 1853, contained in Bethlehem Steel Corporation Collection, Hugh Moore Historical Park & Museums, Inc., Easton, Pa.
24. *Autobiography of John Fritz*, op.cit., 101-102.
25. Ibid., 102-103.
26. Letter from Matthew Newkirk of Philadelphia to John Fritz at Johnstown, August 12, 1854, contained in Fritz Collection of the Hugh Moore Historical Park & Museums, Inc., Easton, Pa.
27. Letter from John Anderson of Philadelphia to John Fritz at Johnstown, August 18, 1854, contained in the Fritz Collection of the Hugh Moore Historical Park & Museums, Inc., Easton, Pa.
28. Letter from Matthew Newkirk of Philadelphia to John Fritz at Johnstown, September 2, 1854, contained in the Fritz Collection of the Hugh Moore Historical Park & Museums, Inc., Easton, Pa.
29. Letter from Matthew Newkirk of Philadelphia to John Fritz at Johnstown, September 9, 1854, contained in the Fritz Collection of the Hugh Moore Historical Park & Museums, Inc., Easton, Pa.
30. Letter from Matthew Newkirk of Philadelphia to John Fritz at Johnstown, September 9, 1854, contained in the Fritz Collection of the Hugh Moore Historical Park & Museums, Inc., Easton, Pa.
31. Letter from John Anderson of Philadelphia to John Fritz at Johnstown, November 25, 1854, contained in the Fritz Collection of the Hugh Moore Historical Park & Museums, Inc., Easton, Pa.
32. Letter from Morris and Town to John Anderson both of Philadelphia, December 1, 1854, contained in the Fritz Collection of the Hugh Moore Historical Park & Museums, Inc., Easton, Pa.
33. Letter from Matthew Newkirk of Philadelphia to John Fritz at Johnstown, December 15, 1854, contained in the Fritz Collection of the Hugh Moore Historical Park & Museums, Inc., Easton, Pa.
34. Letter from Matthew Newkirk of Philadelphia to John Fritz at Johnstown, January 24, 1855, contained in the Fritz Collection of the Hugh Moore Historical Park & Museums, Inc., Easton, Pa.
35. Letter from Matthew Newkirk of Philadelphia to John Fritz at Johnstown, February 13, 1855, contained in the Fritz Collection of the Hugh Moore Historical Park & Museums, Inc., Easton, Pa.
36. Letter from Matthew Newkirk of Philadelphia to John Fritz at Johnstown, March 5, 1855, contained in the Fritz Collection of the Hugh Moore Historical Park & Museums, Inc., Easton, Pa.
37. *The Manufacturies and Manufacturers of Pennsylvania of the Nineteenth Century* (Philadelphia, Pa.: Galaxy Publishing Co., 1875) 221. See also Storey, Vol. I, op.cit., 413-414.
38. *Autobiography of John Fritz*, op.cit., 107.
39. Letter from John Fritz at Johnstown to firm of Wood, Morrell and Company of Philadelphia (circa 1856), contained in the Fritz Collection of Hugh Moore Historical Park & Museums, Inc., Easton, Pa.
40. *Autobiography of John Fritz*, op.cit., 108.
41. Ibid.
42. Kenneth Warren, *The American Steel Industry, 1850-1970: A Geographic Interpretation* (Oxford, England: Clarenden Press, 1973), 50.
43. See *American Railroad Journal* of December 26, 1856, for a summary of the duration of railroad rails — "On the first introduction of railroads it was confidently assumed that rails would last for infinite periods, but experience has soon demonstrated that railway bars were subject to delamination and disintegration from heavy loads. Their duration in some instances has not exceeded two or three years.
44. Ibid., see also Albright Zimmerman, "Iron for American Railroads, 1830 to 1850," *Canal History and Technology Proceedings*, Vol. V, 1986, 63-65.
45. Warren, op.cit., 50.

46. William T. Hogan, *Economic History of the Iron and Steel Industry in the United States*, Vol. I (Boston, Massachusetts: D.C. Heath and Company, 1971, 38-40.

47. The annual production figures quoted for the various rail mills is derived from an article, "Manufacture of Railroad Iron in the United States" in the May 6, 1854, issue of *The American Railroad Journal*.

48. *American Railroad Journal*, August 2, 1856.

49. Hogan, Vol. I, op.cit., 38.

50. *Autobiography of John Fritz*, op.cit., 112.

51. "New Process of Rolling Railroad Iron," *American Railroad Journal*, August 19, 1854.

52. *Autobiography of John Fritz*, op.cit., 112-113.

53. Abram S. Hewitt, "The Production of Iron and Steel in Its Economic and Social Relations," *Selected Writings of Abram S. Hewitt*, Allen Nevins, ed. (New York, N.Y.: Columbia University Press), 45.

54. William Truran, *The Iron Manufacture of Great Britain* (London, England, 1856), 154.

55. *Autobiography of John Fritz*, op.cit., 103. Although John Fritz states that he had already developed the idea of the three-high rail mill before he journeyed to Philadelphia, one must remember that his autobiography was dictated during his eighty-eighth year and that an examination of the original manuscript of the *Autobiography* in the Archives of Lehigh University in Bethlehem, Pa. reveal it to have been largely edited and partially composed by N. Emory, a Lehigh University administrator.

56. Letter from Charles Wood of Philadelphia to John Fritz at Johnstown, July 12, 1856, contained in the Fritz Collection of the Hugh Moore Historical Park & Museums, Inc., Easton, Pa.

57. *Autobiography of John Fritz*, op.cit., 109.

58. Letter from Wyatt W. Miller of the Safe Harbor Iron Works to the Managing Partners of Wood, Morrell and Company, Philadelphia, April 29, 1857, contained in the Fritz Collection of the Hugh Moore Historical Park & Museums, Inc., Easton, Pa.

59. Letter from Wood, Morrell and Company at Philadelphia to John Fritz at Johnstown, May 4, 1857, contained in the Fritz Collection of the Hugh Moore Historical Park & Museums, Inc., Easton, Pa.

60. *Autobiography of John Fritz*, op.cit., 110-111.

61. Ibid., 114-115.

62. Letter from Patrick Graham at Johnstown to John Fritz at Bethlehem, January 29, 1902, contained in the Fritz Collection of the Hugh Moore Historical Park & Museums, Inc., Easton, Pa.

63. *Autobiography of John Fritz*, op.cit., 119.

64. Letter from Charles Wood at Philadelphia to Daniel J. Morrell at Johnstown, August 13, 1837, contained in the Fritz Collection of the Hugh Moore Historical Park & Museums, Inc., Easton, Pa.

65. *Autobiography of John Fritz*, op.cit., 120.

66. It is clear that by 1857-1858 John Fritz was owed considerable salary from both the Cambria Iron Company and Wood, Morrell and Company. In a document contained in the Fritz Collection of the Hugh Moore Historical Park & Museums, Inc., Easton, Pa., John Fritz totaled the money due him for the period between May 21, 1855, to January 1, 1858, and reached a sum of $3,786.10. It is quite possible that the assignment of the patent to him was intended to be means of settling this debt.

67. All quotations are from the patent Case File, U.S. Patent No. 21,666 "Rolling Railroad Iron," National Archives, Suitland, Maryland.

68. *The Autobiography of John Fritz*, (New York, 1912), 132.

69. This detailed description of iron chemistry and reactions, by F.C. Calvert and Richard Johnson was reprinted in the *Journal of the Franklin Institute* November 1857, 339-345.

70. Bessemer obtained two U.S. patents in one week, a crucial development which has been overlooked by most American scholars. His other patent for "Improvement In Smelting Iron Ore" (No. 16083,

November 18, 1856) did not confine him to "any particular form of furnace" but claimed to be a new process to make iron "without the employment of ordinary carbonaceous fuel." This is what Kelly opposed.

71. A very complete description of Mushet's work including the text of all his patents is given by Fred Osborn, *The Story of the Mushets* (London, 1952). The idea of adding manganese to crucible steel was well established at the time. See the discussion by C. Sanderson, *Scientific American* Vol XI No. 5 October 13, 1855, 40.

72. Improvement in the Manufacture of Iron and Steel, U.S. Patent 17,389, May 26, 1857.

73. Brick, Pense, & Gordon, *Structure And Properties Of Engineering Materials - 4th Ed* (New York, 1978) 265

74. Letter from Abram Hewitt at Trenton to Joseph Martien at Newark August 11, 1857, in the Cooper & Hewitt manuscript collection, New York Historical Society, NYC. Some of the earlier efforts at the Renton works in Newark are described in *Scientific American* Vol IX No. 22 February 11, 1854, 1.

75. Letter from William Kelly at Pittsburgh to Fritz at Johnstown Nov. 5, 1859, contained in the Fritz Collection of the Hugh Moore Historical Park & Museums, Inc., Easton, Pa.

76. Agreement between John Fritz and Bethlehem Iron Company, July 10, 1860, contained in the Fritz Collection of the Hugh Moore Historical Park & Museums, Inc., Easton, Pa. This agreement specified that John Fritz was to receive an annual salary of $5,000 per year to serve as general manager and superintendent of the Works of the Bethlehem Iron Company. He would also be granted 100 shares of stock over a four-year period in return for the rights to his patents and his expertise in the design and construction of furnaces and rolling mills. This agreement should be interpreted in light of Bethlehem's later participation in the patent pool because the two agreements are linked together. John Fritz remained at the Bethlehem Iron Company until his retirement in 1892-1893. Along with his brother George and Alexander Holley, John Fritz played a prominent role in the development of the Bessemer steel industry in America, and through his introduction of heavy steel forging technology, he did much to create the basis for the modern U.S. military/industrial complex. He remained the elder statesman of the American Iron and Steel Industry until his death at Bethlehem in 1913.

77. *Autobiography of John Fritz*, op.cit., 146.

78. Letter from Hancock & Foley at Danville to John Fritz at Johnstown December 15, 1862, contained in Fritz Collection of Hugh Moore Historical Park and Museums, Inc., Easton, Pa.

79. Draft of the 1863 Fritz mill patent agreement, contained in Fritz Collection of Hugh Moore Historical Park and Museums, Inc., Easton, Pa. By 1866, the Fritz mill had been universally adopted by all rolling mill companies, all of which ultimately obtained proper licenses. R. W. Hunt "The Evolution Of American Rolling Mills," *Transactions*, Vol. XIII, ASME (1892), 51.

80. William F. Durfee at Chicago to John Fritz at Johnstown, Pa., January 7, 1863 contained in Fritz Collection of the Hugh Moore Historical Park and Museums, Inc., Easton, Pa.

81. Leaving behind a great mystery, a small Bessemer-type converter, which Kelly is said to have used in tests at Cambria, has survived. There is no documentation to support this claim. The converter first attracted public notice after the devastation of the Cambria plant by the disastrous Johnstown flood of 1889. Recently the converter was scientifically analyzed for the first time. Professor Robert B. Gordon of Yale University states on Pages 776-777 of his research note "The Kelly Converter," which appeared in *Technology and Culture*, Volume 33, No. 4, 1992, that "for a further test of how this converter had been used, one of the firebricks of the lining was removed and samples of the brick were identified by examining thin sections of the samples with a petrographic microscope; their compositions were determined with an electron microscope. The slag consists of silica-rich glass containing abundant magnetite crystals and no trace of metal.

Droplets of metal are ordinarily found in firebrick that has been in contact with liquid steel; none were found in the converter. These observations indicate that the converter was used in a steel-making attempt and that the charge of pig iron was overblown and that useful metal was made." The converter is currently a part of "A Material World" exhibit at the Smithsonian Institution's National Museum of American History. It remains the property of the Bethlehem Steel Corporation.

82. When the Kelly Process Company was formed, Kelly didn't even control his own patent. His father had died and the patent had passed into the hands of other members of his family. See Chapter IV of John Boucher's *William Kelly*, Greensburg, Pa., 1924.

83. There was but one stumbling block, a very large one. By chance an enterprising young American, Alexander Holley, obtained a license from Bessemer. He built a small converter for ironmaker John Griswold at Troy, NY. Morrell was forced to reach an accommodation with Griswold. Details are given by John Bergenthal, "The Troy Steel Company and Its Predecessors, 1987, (an unpublished manuscript that is an enlargement of a senior thesis at Rennselaer Polytechnic Institute, Troy, New York) and by Jeanne McHugh, *Alexander Holley and the Makers of Steel*, Baltimore, Maryland, John Hopkins University Press, 1980. See also *Introduction to a History of Ironmaking and Coal Mining in Pennsylvania*, James W. Swank (Philadelphia, 1878), 84.

Historic Water Supply Systems

The Philadelphia Water Department:

203 Years of Serving a Community's Needs With Engineering Solutions and Environmental Protection

Ed Grusheski

In June of 1975 the American Society of Civil Engineers honored the Philadelphia Water System as a "National Historic Civil Engineering Landmark...a tribute to its great early engineering years" (PWD 1975). I take particular pleasure, therefore, in being able to share the history of the Philadelphia Water Department with the Society during its Sesquicentennial Celebration. I will focus on the early years of water supply in Philadelphia, but will bring you up to the present discussing both the engineering achievements and environmental issues which have shaped the Department's development.

It would be a delight for me to recount a story that reflects rational political decision-making, and successful planning by engineers. The history of the development of Philadelphia's water system is not one of those stories. Like the development of water systems in most other 18th and 19th century urban centers, political decisions were made only after the community was in crisis. Solutions were engineered with some success, only after much trial and error. And, needless to say, those engineered solutions created unanticipated problems of their own. But somewhere in all of this, rational political decisions were made, and successful solutions – even elegant ones – engineered.

If Philadelphia had grown as the "Greene Country Towne" William Penn planned – a perfectly rational grid extending from the Delaware River to the Schuylkill River, with ample open space around commercial, industrial and residential structures – the city's drinking water supplied from wells would have lasted longer (Weigley, p. 226). However, many of those immigrating to Philadelphia came from crowded European urban centers, and took a certain amount of comfort in living in close quarters. By the end of the 1730's, 30,000 people lived east of 7th Street between South and Race Streets. The city's groundwater was being severely compromised by the privies which met the needs of the 30,000 inhabitants. In addition, industries – such as tanneries – which filled great pits in the ground with caustic materials were polluting the water supply. Philadelphians came from European cities where one did not drink the water. There were substitutes. There was no need to invest the large sums of money to provide the city with safe drinking water. Moreover, municipal pumps were available to fight fires (Weigley, p. 79), so even the ever-present threat of conflagrations destroying whole sections of 18th century cities seemed not to move Philadelphia's city fathers to plan for an urban water delivery system.

Only after a series of devastating yellow fever epidemics in the 1790's were Philadelphia's leaders compelled to finance and design the nation's first water delivery system meant to supply an entire city (Stapleton p.29). In one year alone, one-sixth of the population of Philadelphia succumbed to this terrible epidemic. Yellow fever, as we now know, is carried by mosquitoes. At that time, however, many felt that it was waterborne. Others believed that it was a result of the filth which filled the streets of the city. And so, a source of clean fresh water, both for drinking and to periodically wash the streets, was needed.

Benjamin Henry Latrobe's Centre Square Waterworks, Philadelphia

Benjamin Henry Latrobe, Engineer, who was in Philadelphia in 1798 to build the Bank of Pennsylvania, was asked by city officials to propose a solution. In December of that year he presented them with the pamphlet *View of the Practicability and Means of Supplying the City of Philadelphia with Wholesome Water.* In drawing from his knowledge of London's steam-powered waterworks of 1783 (Stapleton p.27), Latrobe proposed two pumping stations, one on Chestnut Street at the Schuykill River, the other at Centre Square. Each station housed a Boulton and Watt type low-pressure steam engine, which operated an 18" diameter vertical pump. Water was directed from the Schuylkill River into a settling basin. From there it was pumped up into a 2000 ft. long brick conduit through which it flowed by gravity to a basin beneath the Centre Square pumping Station. At Centre Square the water was lifted by the second pump to one of two elevated wooden reservoirs in the dome of this elegant Neoclassical structure. From this height water flowed by gravity through a distribution system of wooden pipes to the homes and businesses of subscribers, and to hydrants (Latrobe, 1799). In January of 1799, the Watering Committee was formed to act on the Latrobe plan. The Centre Square Waterworks went into operation in 1801.

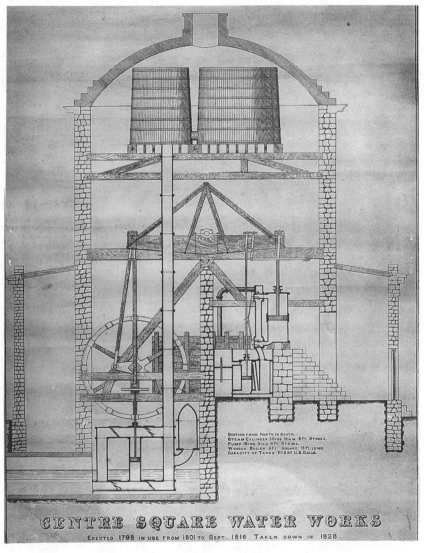

Philadelphia's Centre Square Waterworks, Interior

Starting with Latrobe, the engineers who developed Philadelphia's water system have always been keenly aware of the environmental issues which affect the success of a water delivery system. Already, at the end of the 18th century, the Delaware River was being visibly polluted by the increasing development of Philadelphia along its banks. The City's drinking water would be supplied, therefore, from the "uncommonly pure waters" – as Latrobe characterized them in his pamphlet – of the yet-to-be developed Schuylkill River.

The Centre Square Water Works was only a temporary solution. The boilers, made of wood and lined with metal, broke down and were inefficient. If one engine went down, the entire system went down. If the pumps were not operational, there was only a twenty-five minute supply of water in the domed reservoir. The subscribers were not happy.

As early as 1811, the Watering Committee charged Frederick Graff who had apprenticed with Latrobe and was now the chief engineer of the Centre Square Waterworks to find a solution. Graff and his associate John Davis selected a site just beyond the city limits in the Spring Garden District, the highest point close to the city called Faire Mount. At the base of the "mount" a pumping station would be erected with construction of a reservoir at the top, 56 feet above the highest ground in the city (Gibson pp.11-12). To avoid the interruptions of service now legendary with the Centre Square Works, the new pumping station would house two steam engines, two pumps and two boilers. Using one engine at a time, the second would always be in reserve in case of a mechanical breakdown, or boiler explosion. In addition, the reservoirs atop Fairmount would hold many days' supply of water for the city.

The Fairmount Water Works went into operation in 1815. It housed a low pressure Boulton and Watt-type steam engine, as well as a new high pressure steam engine designed by Oliver Evans. The still "uncommonly pure" water of the Schuylkill was pumped up to the great height of the reservoirs – actually a series of reservoirs, which also acted as sedimentation basins. From there, it flowed by gravity into the old Centre Square distribution system. By 1817, with an expanded 32 mile distribution system of spruce and yellow pine pipes, the Fairmount Water Works supplied 3,500 customers and three hundred public pumps (Koeppel, p. 107).

As impressive as this may sound, there were still many problems. In 1818 and again in 1821, the boiler for the high pressure steam engine exploded, killing three workers. Moreover, the annual cost for fuel for the two steam engines was in the neighborhood of $30,000.00, lagging far behind revenues. By 1819, the decision was made to abandon steam power in favor of water power. Between 1819 and 1822, what was then the longest dam in the world was built across the Schuylkill River at Fairmount. The River was essentially redirected into a forebay – or mill race – behind a new pumping station. The water was diverted onto breastwheels 15 feet wide by 16 to 18 feet in diameter. The turning wheels operated pumps, which raised water through iron force mains up the hill to the reservoirs (Gibson, pp.18-19). In 1822, the first year of operation, three of the planned eight wheels were in use. Each wheel worked a pump forcing over a million gallons of water a day to the reservoirs. The city went from paying more than four hundred dollars to pump a million gallons of water with steam power, to less than four dollars to pump a million gallons with water power (Watering Committee, 1822).

The "Golden Age" of Fairmount began. By 1837, six waterwheels were in operation; twenty thousand households and businesses generating over one hundred thousand dollars in annual revenue, received over three million gallons of water a day (Koeppel, p. 109). By 1841 when Charles Dickens came to Philadelphia, all eight wheels were operational. So, impressed with Fairmount was Dickens – and there was very little about America that impressed the English novelist – that he declared in his **American Notes** of 1842:

"Philadelphia is most bountifully provided with fresh water, which is showered and jerked about, and turned on, and poured off everywhere. The Water Works, which are on a height near the city, are no less ornamental than useful, being tastefully laid out as a public garden, and kept in the best and neatest order. The river is dammed at this point, and forced by its own power into certain high tanks or reservoirs, whence the whole city, to the top stories of the houses is supplied at a very trifling expense" (Dickens, p. 47).

Frederick Graff's Fairmount Water Works, Philadelphia

The Fairmount Water Works became one of the most visited American sites, and one of the most reproduced American images. Europeans were enamored with its success – a marriage of nature and industry for the good of Philadelphia's citizenry. Of course, Philadelphia had carefully tried to avoid the mistakes which European cities had made with regard to their water systems. Indeed, in the case of Fairmount, rational political decisions had resulted in a successful and elegantly engineered solution.

During this period, also, a vast improvement was made to the distribution system. Philadelphia's grid layout presented a particular problem for the wooden pipe distribution system – turning right angles with wooden pipes caused dramatic drops in pressure. In addition, the small size of the borehole in a wooden pipe limited the quantity of water, which could be delivered. As early as 1819, therefore, Philadelphia began to install only cast-iron pipes. Frederick Graff designed pipes, which gently curved at corners, and had considerably larger diameters than the wooden pipes they replaced (Gibson, p. 15). By the 1840's the city was served chiefly by a cast iron pipe distribution system. During his visit, Dickens was genuinely impressed with the high water pressure on the 4th floor of his hotel, an experience he had in no other city.

One of the more interesting aspects of Philadelphia water system was the creation of Fairmount Park, a very early attempt at watershed protection. From the beginning the engineers of Philadelphia's water system understood that residential, commercial and industrial development along the banks of the Schuylkill River, its chief source of drinking water, threatened the quality of that supply. They, therefore, supported acquisition of land above the Fairmount Water Works to be set aside for a public park (Gibson, p. 31). The North Garden in the 1830's, Lemon Hill in the 1840's, Sedgley in the 1850's and in 1867 huge tracts of land on the east and west banks of the Schuylkill, upstream of Fairmount were purchased as park land – 4,000 acres in all (Weigley, p. 376). This attempt at watershed management worked for a time to control the quality of the water in the Schuylkill River.

During the Civil War, however, communities and industries upstream of Philadelphia on the Schuylkill River grew exponentially. Not the least of which was the coal industry, which fueled the war efforts of the North. By the 1870's and 80's the quality of Philadelphia's drinking water was being compromised by the effluent of the communities and industries upstream. Water Department engineers began to champion filtration, which was being used successfully in European cities. But the high cost of building the facilities to supply such a large population with filtered water presented an immediate barrier to the politicians. The filtration debate continued until the later 1890's when, finally, the quality of Philadelphia's drinking water threatened public health. At the end of the nineteenth century Philadelphia suffered the worst typhoid epidemics of any American city (Weigley, p. 496). By this time the Schuylkill River was running black with coal culm, and scum rose to the water's surface, as it does today in a waste water treatment facility.

As with the yellow fever epidemics a century before, it took a major public health crisis to build the political will to take action. Between 1900 and 1911 Philadelphia built, what was then, the world's largest filtered water system. It included five major slow sand filter plants, four on the Schuylkill River, one on the Delaware. Three were later converted to rapid sand filters. Four heavy duty pumping stations were built to move water through the system efficiently. The size of the filters – the Torresdale plant on the Delaware covered 58 acres – meant that the five plants were at the edge of the city where cheaper land was available. A huge project of laying the distribution mains to bring the filtered water into the built sections of the city was undertaken at the same time. The cost of the new system was nearly $28 million dollars. Before filtration 95% of the city's water was taken from the Schuylkill. After filtration 70% of the city's water was taken from the Delaware. It was felt that relying on the larger, faster moving Delaware River would improve the water quality situation (Bureau of Water, 1909). As each of the plants went on line, the typhoid death rates dropped steadily. In 1912 with the addition of chlorine to the filtered water, the annual typhoid and cholera epidemics were brought to a halt.

Construction of the Distribution System to Bring Filtered Water to Philadelphia

Now that Philadelphia could deliver water to its citizens that would not make them sick, the city turned its back on the rivers. We could filter the polluted water. Making it safe to drink. No longer was there the need to try to protect the rivers.

Indeed, in the first half of the 20th century pollution in both the Delaware and Schuylkill Rivers increased. Both rivers became open sewers (Lewis, 1924). The problem with this turn of events, was that although one would not die from drinking Philadelphia's tap water, the source was so polluted that the taste and odor of the water made it a less than desirable beverage. No longer "water of uncommon purity" as Latrobe first characterized it; Philadelphia's water was now called "Schuylkill Punch".

In 1951, Philadelphia's sewer system became the responsibility of the Water Department. In many respects this was a good political move. The Water Department had an incentive to get a handle on the city's sewage. Through most of the twentieth century, Philadelphia's Department of Public Works, in which the responsibility for sewage resided, struggled to fund the sewage collection and treatment plan which had been state-mandated in 1914. That plan was a remarkable engineering undertaking. It included building enormous interceptor sewers along the banks of the Delaware and Schuylkill Rivers which would direct the city's wastewater to one of three planned treatment plants. World War I, the Depression, World War II, as well as "sexier" City projects like a subway system that vied with the sewer projects for bond issues - all conspired to slow progress in improving the quality of Philadelphia's drinking water sources. Under the Water Department, and with City Council approval to raise sewer rents (Weigley, p. 627), the 1914 plan was finally realized in 1957 with the completion of the City's three primary wastewater treatment plants.

The Clean Water Act of 1972 set the stage for further improvements. By 1984, Philadelphia was on-line with secondary wastewater treatment. And by the 1990's most communities upstream of the city has gone to secondary treatment, as well. This has resulted in a remarkable turn around in the quality of Philadelphia's source water. Dissolved oxygen has returned to the levels of 1880. More than eighty varieties of fish have returned to the rivers at Philadelphia. Rowers can now overturn without the fear of having to receive painful tetanus shots. And, of course, the aesthetic quality of Philadelphia's drinking has improved with improvement of its source water. Today the city's drinking water is no longer the "Schuylkill Punch" of years past, but rather is more in line with Latrobe's original characterization of "water of uncommon purity".

As much credit as we can give to the success resulting from the highly engineered solution of secondary wastewater treatment, more needs to be done to improve and protect the quality of our drinking water. Today, 80 to 90% of the pollution in the Delaware and Schuylkill Rivers is the result of non-point sources, stormwater runoff.

Today, the Philadelphia Water Department, through neighborhood watershed improvement projects undertaken by its Office of Watersheds, and through its award-winning public education programs at the Fairmount Water Works Interpretive Center, is attempting to improve water quality by encouraging rational political decisions that will lead to successful – perhaps even elegant – solutions. But that is another story for another time.

Philadelphia From the Schuylkill River,
the Fairmount Water Works in the Middle-ground

Bibliography

Annual Reports (AR) of the *Watering Committee* (1801 – 1850), L. R. Bailey, Philadelphia

Annual Reports (AR) of the *Water Department* (1852 – 1886), Dunlap Printing Co., Philadelphia

Annual Reports (AR) of the *Bureau of Water* (1887 - 1919), Dunlap Printing Co., Philadelphia

Bureau of Water (1898). *Documents Relating to the Pollution of the Schulkill River*, Dunlap, Philadelphia

Department of Public Works, Bureau of Water (1909). *Description of the Filtration Works and Pumping Stations*, Philadelphia

Dickens, Charles (1842). **American Notes**, Chapman and Hall, London

Gibson, Jane Mork (1988). "The Fairmount Waterworks", *Bulletin*, Volume 84, Numbers 360, 361, Philadelphia Museum of Art, Philadelphia

Koeppel, Gerard T. (2000). **Water For Gotham: A History**, Princeton University Press, Princeton, New Jersey

Latrobe, B. Henry (1799). *View of the Practicability and Means of Supplying the City of Philadelphia With Wholesome Water*, Zachariah Poulson, Jr., Philadelphia

Lewis, John Frederick (1924). **The Redemption of the Lower Schuylkill,** The City Parks Associated, Philadelphia

Philadelphia Water Department (1975). *ASCE Landmark Designation Program*, Philadelphia

Stapleton, Darwin H. (1999). "Introducing...Clean Water", *Invention and Technology*, Volume 14, Number 3, American Heritage, New York

Weigley, Russell F., Ed. (1982). **Philadelphia: A 300-Year History**, W.W. Norton, New York

John B. Jervis and the Development of New York City's Water Supply System

Robert A. Olmsted[1], P.E., F.ASCE

Abstract

New York City became the nation's largest city at the end of the 18th century. Its population exploded after the completion of the Erie Canal in 1825 opened its protected port to midwestern markets. But without a dependable, adequate water supply system, the city could not grow. That water supply system was due to the engineering genius of John Bloomfield Jervis (1795-1885). Jervis, the leading American civil engineer of the antebellum period, built or managed three canals, seven railroads and two major urban water supply systems. In addition to the original Croton Water Supply System, Jervis engineered other projects important to the city, including the Delaware and Hudson Canal and the Hudson River Railroad. His career spanned more than half a century. The Croton Water Supply System (1842) was the most outstanding municipal water supply system in the United States at the time and was the prototype for later projects throughout the world.

Introduction

New York City is the nation's largest city. In the 2000 census, for the first time the city's population exceeded 8 million souls, while the metropolitan region is home to some 20 million people, A vast infrastructure, designed and built by civil engineers, is needed to support the economy of the city: 374 km (231 mi) of subway routes, 14 underwater subway tunnels, four rail tunnels, four vehicular tunnels, over 2,000 bridges, nearly 10,400 km (6,400 mi) of streets, 11,300 ha (28,000 acres) of parks, 14 waste water treatment plants, more than 10,200 km (6,300 mi) of sewers, and most importantly, the world's best water supply system reaching over 200 km (125 mi) to the headwaters of the Delaware River in the Catskill Mountains. Not surprisingly, many of these projects are ASCE National Historic Civil Engineering Landmarks (NHCEL).

Early Water Supply

The nation's metropolis had humble beginnings. Following Henry Hudson's exploratory excursion up his namesake river in 1609, the Dutch West India

---

[1] Transportation Consultant, Chair, ASCE Metropolitan Section History and Heritage Committee, 33-04 91st Street, Jackson Heights, NY 11372

Company established a small trading post, which they called Nieuw Amsterdam, at the tip of Manhattan Island in 1625. Taken over by the British in 1664 and renamed New York, this company town grew into the largest metropolis in. and the United States. By 1800, New York had surpassed Philadelphia as the nation's largest city.

Early Manhattanites got their water from rainwater collected in cisterns, privately owned shallow wells and later from street pumps. Well water was important to the Dutch as they needed fresh water to brew Nieuw Amsterdam's favorite beverages, beer and Dutch hot chocolate. A few public wells were dug in 1660s and 1670s, but the first systematic effort began in 1677. Lower Manhattan's greatest natural feature of the time was the 18-m (60-ft) deep, 19-ha (48-acre) spring-fed pond known as the Fresh Water Pond or "Collect". As the city grew in size, its citizens drew water from the pond and nearby wells. It was on Collect Pond that in 1796, John Fitch successfully demonstrated his steamboat a decade before Robert Fulton. But New Yorkers soon began filling in the pond, drained by a canal along today's Canal Street, by dumping their garbage, debris and animal carcasses into the lake, and then wondered why they got sick. It was filled in by 1811. Early sketches show the Collect Pond as a rustic place. One observer wrote, "...there was no more beautiful spot on the lower island." Too bad the City Beautiful and Parks movements came a century too late to save this idyllic spot for posterity. Instead, the pond's legacy gives today's civil engineers headaches in unstable building foundations and leaking subway tunnels.

The English apparently were more concerned with tea than beer, and "tea water" became a euphemism for good water. Of course, the fact that water had to be boiled to make tea may have helped prevent disease. Some tea water was carted in from the "suburbs" in casks, but several entrepreneurs ("tea-water men") dug their own wells and marketed tea water. The most famous was the Tea Water Pump near Chatham Square, which was the city's main source of water until the late 1700s. But as the city grew and water became more polluted, the need for a better system was apparent. In 1774, Christopher Colles (1739-1816), an engineer, proposed building a water system "for furnishing the City of New York with a constant supply of fresh water" consisting of a large well north of town, a reservoir able to store 4,500 m^3 (1,200,000 gal), later increased to over 7,500 m^3 (2 million gal,), steam pumps capable of raising 750 m^3 (200,000 gal) per day, and a distribution system of hollow wooden pipes. Work began, but the Revolutionary War intervened before the plan was completed. Incorporated in 1799, the Manhattan Company built the city's first real, but limited, water supply system in the early 1800s. It dug a well near Reade and Centre Streets, built an impressive 2,100-m^3 (550,000-gal) reservoir on the north side of Chambers Street and a distribution system consisting of 40 km (25 mi) of hollow wooden mains. The Manhattan Company's franchise was a subterfuge. Promoted by Aaron Burr, a political opponent of Alexander Hamilton, the company really wanted to establish a bank and had legislation passed to allow them to build a more politically acceptable water supply system and incidentally engage in other activities. The company became the Chase Manhattan Bank. Remains of the company's well were unearthed while excavating for an office building in 1926.

Fires, disease and an expanding population made the need for a new water supply system urgent. Yellow fever epidemics in 1798 and 1822 and cholera epidemics in 1832 and 1834 killed thousands. No one was sure what caused cholera. It was not until 1849 that a London doctor concluded that polluted water was the culprit. The Great Fire of 1835 destroyed 674 buildings and leveled a good chunk of the city. But there had been major fires in 1776 and 1828, and smaller fires almost yearly. Water for fighting fires, or just cleaning streets of filth, was badly needed.

Finally, the city began investigating proposals in earnest. City officials began looking at streams north of Manhattan in Westchester County, including the Bronx River, Saw Mill River and Croton River as sources. The New York City Common Council appointed a commission 1833 to plan the Croton system. Major David Bates Douglass, a West Point professor and War of 1812 veteran, initiated the design and began surveys for the route for the aqueduct. The voters of the City of New York approved the Croton proposal in April 1835 and Douglass was appointed chief engineer, a position he held until replaced by John B. Jervis in 1836. The replacement was controversial. While there may have been political overtones, it was said that Douglass was too professorial and lacked practical experience, while Jervis had earned a "can do" reputation.

John B. Jervis

John B. Jervis, said to be America's greatest civil engineer of the antebellum period, whose career spanned over half a century, left his imprint on the American landscape: a legacy of canals, railroads, bridges and perhaps most importantly, the Croton Water Supply System. He was involved in an early unsuccessful attempt to found a national civil engineering society in 1839, 13 years before ASCE was founded in New York City in 1852.

John Bloomfield Jervis, the son of a carpenter, was born in Huntington, Long Island, N.Y. on December 14, 1795. When he was three, his family moved to Fort Stanwix in upstate New York, where he worked in his father's sawmill and farm. Fort Stanwix was a frontier town, a military outpost in the French and Indian War (1754-1763) and the site of a Revolutionary War battle in 1777. An event of historical engineering significance took place on July 4, 1817 near Fort Stanwix, by then known as Rome: the beginning of construction of the Erie Canal. Lacking trained civil engineers, the 590-km (365-mi) Erie Canal was the first engineering "school" in America where many of the young nation's early engineers got their training through practical experience gained on the canal's construction. Young John Jervis was one of them. His engineering "education" and career began in 1817 when the canal's chief engineer, Benjamin Wright, a fellow Fort Stanwix resident, hired young Jervis as an axman for a canal survey party. He worked his way up the ladder to rodman in 1818 (where he was paid $12 a month), and later became resident engineer in charge of several sections of the canal. ASCE calls Benjamin Wright (1770-1842), "the Father of American Civil Engineering".

Delaware and Hudson Canal

When the Erie Canal was completed in 1825, Benjamin Wright hired Jervis as Principal Assistant on another canal that was important to New York's economy, the Delaware and Hudson Canal. The 170-km (105-mi) D&H Canal, which opened in 1829, was built to convey coal from anthracite mines near Honesdale, Pennsylvania, to New York City. When Wright moved on to other work in 1827, Jervis became the Delaware and Hudson's Chief Engineer. The D&H was Jervis' first exposure to the new technology of railroads, a technology that soon made most canals obsolete. Jervis built the 26-km (16-mi) "gravity" railroad, a unique system of inclined planes, to carry coal from the mines near Carbondale to the beginning of the canal at Honesdale. Four steam locomotives were purchased from England; the Stourbridge Lion was the first locomotive to run on a track in the United States. But it proved to be too heavy for the track (or the track too light for the locomotive) and the locomotive became a museum piece. One of Jervis' notable achievements was the development of a steerable truck for locomotives to enable them to negotiate sharp curves found on American railroads.

The D&H Canal was abandoned in 1898. Its outstanding remaining feature is the Delaware Aqueduct, a wire suspension bridge built by another great engineer, John Roebling, in 1848 to carry the canal across the Delaware River. An ASCE NHCEL since 1972, it is Roebling's oldest surviving suspension bridge. It was restored by the National Park Service in 1983.

After the D&H, Jervis continued his career as chief engineer of the first railroad in New York State, the Mohawk and Hudson RR between Albany and Schenectady; the Schenectady and Saratoga RR; the upstate New York Chenango Canal between Binghamton and Utica; and was consulted on he first enlargement of the Erie Canal. On the Chenango Canal project, Jervis devised improved rain and stream gauges to determine the relationship between rainfall and runoff more precisely, a hallmark in American hydrology. In 1836, he embarked on a new career as chief engineer of the Croton Water Supply System.

The Croton Aqueduct and Water Supply System

The Croton Water Supply System was a remarkable achievement for its day. The main features of the system were a dam across the Croton River, a tributary of the Hudson, about seven miles upstream from its mouth near Ossining; a 66-km (41-mi) gravity aqueduct, essentially an enclosed canal, to convey the water to the city; the impressive High Bridge; and three reservoirs, the Croton Reservoir, a receiving reservoir in what is now Central Park, and a distributing reservoir at 42nd Street.

Work on the dam, the first large masonry and earth-fill dam in the nation, began in 1837. The dam was nearly washed out in a flood in 1841. To protect against future floods, Jervis devised a now common innovative feature. He designed the masonry spillway with a reverse or ogee curve, and added a stilling basin to dissipate the water's energy. The 15-m (50-ft)-high dam was completed in January 1843. It created Croton Reservoir, an eight km (five-mile)-long, 160 ha (400-acre) lake with a storage capacity of about 225,000 m³ (600 million gal).

Construction of the aqueduct began in May 1837. The work was divided into four divisions with 96 subdivisions. The horseshoe-shaped stone and brick aqueduct is about 2.6 m (8 ½ ft) high and about 2.3 m (7 ½ ft) wide. The profile drops at about 20 cm per km (13 inches per mile). The alignment follows the contour of the land where possible, but several tunnels, embankments and bridges were needed to penetrate ridges or to cross over valleys. For example, the 23-m (76-ft)-high stone arch that carries the aqueduct across Sing Sing Kill in the village of Ossining spans 27 m (88 ft). It is known locally as the "Double Arch" because the aqueduct arch spans a second arch that carries a village street (Broadway) across the same stream. A visitor's center, which houses an exhibit on the aqueduct, is nearby. In order to prevent air pressure from building up in the enclosed aqueduct, thirty-three 4.3-m (14-ft) high stone ventilators were built at about 1.6 km (one mi) intervals. Six waste weirs provide overflow protection. Meanwhile, the two Manhattan reservoirs were placed under construction. The 12-m (38-ft)-high earth embankments of the 14-ha (35-acre) Receiving (or Yorkville) Reservoir had a storage capacity of 68,000 m³ (180 million gallons). Discontinued in 1890, the reservoir has been replaced by the Great Lawn in Central Park. The Murray Hill Distributing Reservoir at 42nd Street and Fifth Avenue was the end of the line, 68 km (42 mi) from the Croton Dam. This masonry structure reservoir, designed in an Egyptian architectural style, held 90,000 m³ (24 million gallons) of pure water. Since 1911, the site has been occupied by the New York Public Library.

Perhaps the most impressive feature of the Croton system is the signature High Bridge across the Harlem River. Douglass and others had proposed a high viaduct to cross the Harlem, but Jervis initially argued for a low-level bridge carrying an inverted siphon on the basis of cost. Nevertheless, Jervis, studied both options: a 442 m (1,450-foot)-long high bridge just below the hydraulic grade 42 m (138 ft) above high water, or a 24-m (80-ft)-long arch 15 m (50 ft) above the river. At half the cost, the water commissioners preferred the low bridge, but there was concern about the impact on navigation. (The courts had just declared the Harlem River navigable in 1839.) Jervis was caught in the middle. In 1839, the state legislature dictated either a high bridge or a tunnel. Jervis prepared estimates for building a tunnel using cofferdams, and a high bridge. Although he believed that the tunnel would be cheaper, he was concerned over the many uncertainties and contingencies in building such a tunnel, and recommended the high bridge. Since either proposal would take longer to build than the aqueduct itself, temporary pipes were placed across the river. Construction on High Bridge commenced in 1839. The bridge consisted of 15 Roman-style semi-circular masonry arches, eight of which spanned

24 m (80 ft) and the remainder (over land) 15 m (50 ft). Finally, in 1848, six years after the Croton Aqueduct was placed in service, water flowed through High Bridge's two 91-cm (3-ft) diameter iron pipes, completing the Aqueduct project.

High Bridge was a favorite destination for New Yorkers for many years. The pedestrian walkway on top of the bridge was especially popular. A 2.3-m (90-inch), pipe was added in 1860. In 1927, five of the stone arches were replaced with a 98-m (322-ft) steel arch to improve navigation. The 442-m (1,450-ft)-long High Bridge, still standing but unused, is New York's oldest extant bridge. Instead, two deep-level water tunnels now distribute water throughout the city, with a third under construction.

Water was first introduced into the Croton Aqueduct on June 27 1842. The city celebrated with fireworks, high-flying balloons and a huge parade on October 14, 1842. A 15-m (50-ft)-foot high geyser of crystal pure Croton water gushed from the fountain at City Hall. Historian Edward Spann said, "Without the Croton System there could not have been a doubling of the narrow island's [Manhattan] population from 1845to 1855. Nor could the city's economy have developed as rapidly…"

A new Croton Aqueduct Department was formed in 1849 to plan improvements to the Croton System and, in addition, build sewers and pave streets in the city. It was in the office of Alfred W. Craven (1810-1879), Chief Engineer of the Croton Aqueduct Department, that the *American Society of Civil Engineers* was founded on November 5, 1852 in New York City.

Hudson River Railroad and Later Career

Building on the reputation he gained on the Croton project, Jervis was retained as a consultant on the Boston Water supply system (1845-1848). Between 1845 and 1850, he was chief engineer of the 227-km (140-mi) Hudson River Railroad, which was built on the east bank of the Hudson River from New York City to East Greenbush, opposite Albany. Jervis was well suited for that job because of his familiarity with the terrain since the railroad ran parallel to the Croton Aqueduct between Yonkers and Croton. The railroad became an important link in the New York Central System's "water-level route" from New York to Chicago, route of the famed *Twentieth Century Limited*, and is now part of the CSX system. After 150 years of service, Amtrak's Empire Service trains and MTA Metro-North Railroad Hudson Line commuter trains still use the railroad's tracks. For the next fifteen years, Jervis was engaged on several midwestern railroad projects.

New York's Water Supply System Today

As impressive as it was, the Old Croton system became inadequate within a few years because of increasing population and per capita water consumption due to the growing popularity of conveniences that we now take for granted like flush toilets and private baths, and increased commercial use of water. The city began to expand the Croton system in the 1870s and 1880s. The New Croton Dam (Cornell Dam), a masonry gravity dam that was the highest dam in the world at the time, was completed in 1906 (Alphonse Fteley, engineer, ASCE President 1898). The Old Croton Dam is now submerged in the enlarged reservoir. A New Croton Aqueduct, three times the size of the old aqueduct and partly tunneled through hard rock, was built between 1885 and 1893 and placed in service in 1890. Today the Croton system has 12 reservoirs and three controlled lakes having a capacity of 86.6 billion gallons. The Croton system supplies the city with about 10 percent of its water needs. Largely out of service since 1955, the Old Croton Aqueduct is a now a popular hiking and biking trailway, *The Old Croton Aqueduct State Historic Park.* The Croton Aqueduct is both a NHCEL and a National Historic Landmark.

Two other water supply systems were built in the 20th century: the Catskill and Delaware systems. The Catskill system taps streams in the Catskill Mountains, 160 km (100 mi) northwest of the city. Its two reservoirs, Ashokan and Schoharie, have a capacity of 53,000,000 m^3 (140 billion gallons). The 149 km (92-mi), Catskill Aqueduct, largely a cut-and-cover conduit, conveys water to reservoirs north of the city. The Catskill system supplies about 40 percent of the city's needs. The Delaware system taps headwaters of the Delaware River about 200 km (125 mi) northwest of the city. Its four reservoirs, Roundout, Pepacton, Cannonsville and Neversink, have a capacity of 121,000,000 m^3 (323 billion gal), and supply about 50 percent of the city's drinking water. The 138-km (85-mi)-long Delaware Aqueduct, a deep rock tunnel, is one of the longest tunnels in the world. In order to protect the purity of the supply, and to avoid the high cost of building a filtration plant, the city has developed a long-range watershed protection program. The keystone is the 1997 Watershed Memorandum of Agreement between the city and the upstate farming and residential watershed communities. The water supply system, which serves 9 million people (8 million in the city, 1 million in suburbs along the aqueducts) performs well, but is strained at times. For example, during the drought emergency in early 2002, the reservoir levels dropped to less than 50 percent of capacity. Although spring rains brought the level to 89 percent of capacity, storage was still 8 percent less than normal before dropping again (at the time of writing). Conservation, rather than new sources, which are almost impossible to implement, is the key for the future.

Postscript

Jervis returned to Rome in 1864 where he continued an active life for two decades. He became a Member of the ASCE in 1867 and in 1868 he became one of the society's first Honorary Members. He organized the Merchants Iron Mill (now Rome Iron Mill) and wrote several books and papers, among them: *The Question of Labour and Capital* (1877), which extolled the virtues of hard work and honest living, and, at the age of 83, a lecture on *Industrial Economy.* He died in Rome on January 12, 1885 at the age of 89 and bequeathed his home to the city for a public library, which opened in 1895 as the Jervis Public Library. His papers are kept at the library in 79 boxes arranged chronologically. In addition, the collection includes 367 drawings, 38 of which are downloadable on the library's website. It is said to be one of the most complete set of early railroad and canal reports, engineering drawings and maps in existence. With permission, they are available to qualified researchers.

Jervis' legacy lives on. The city of Port Jervis, N.Y. on the Delaware and Hudson Canal, is named for him. The Old Croton Aqueduct, now a state historic park, is a popular public hiking and biking trail. High Bridge, although altered and unused, New York City's oldest extant bridge, still stands, while Rome's Jervis Public Library preserves his heritage. He demonstrated that even octogenarians can lead an active life – he was awarded a LL.D degree from Hamilton College at 83. He truly left his mark on the city and state of New York.

References

American Society of Civil Engineers (1972) *A Biographical Dictionary of American Civil Engineers,* New York, N.Y.

American Society of Civil Engineers, Metropolitan Section (1997) *A Guide to Civil Engineering Projects in and Around New York City,* New York, N.Y.

Blake, Nelson M. (1956) *Water For The Cities: A History of the Urban Water Supply Problem in the United States,* Syracuse University Press, Syracuse, N.Y.

Fitzsimons, Neal, Editor (1971) *The Reminiscences of John B. Jervis: Engineer of the Old Croton,* Syracuse University Press, Syracuse, N.Y.

Galusha, Diana (1999) *Liquid Assets: A History of New York City's Water System,* Purple Mountain Press, Fleischmanns, N.Y.

Hall, Edward H. (1917) *Water For New York City,* Hope Farms Press, Saugerties, N.Y., reprint 1993.

Jervis Public Library, Rome, N.Y. Website

Koeppel, Gerard T. (2000) *Water For Gotham: A History,* Princeton University Press, Princeton, N.J.

Larkin, F. Daniel (1990) *John B. Jervis, An American Engineering Pioneer*, Iowa State University Press, Ames, Iowa

New York City Department of Environmental Conservation (1992) *The Old Croton Aqueduct: 150th Anniversary Edition,* New York

Tompkins, Christopher R. (2000) *Images of America: The Croton Dams and Aqueduct,* Arcadia Publishing, Charleston, S.C. (Many pictures)

Worthy of the Nation: The Washington Aqueduct and its Chief Engineer, Montgomery C. Meigs

Thomas P. Jacobus[1]
Patricia A. Gamby

Abstract

Washington Aqueduct is the name given by Montgomery C. Meigs in November 1853 to the water supply for the nation's capital. Meigs was the principal individual responsible for the development of this significant engineering achievement. This paper will describe how Meigs and his team of engineers designed and built the original elements of the water system that would serve Washington, DC from 1860 to 1900 with little modification. It will also examine the farsightedness of the design and how it has been successfully and economically adapted to meet the ever-increasing need to provide abundant water to the national capital area. As we approach the 150th anniversary of the founding of the Washington Aqueduct in 2003, with many of the original structures still in use, the current challenge for engineers assigned to the Washington Aqueduct is to continue to accommodate the future of water treatment into these historical elements.

Introduction

In 1853, Congress gave the mission of building an adequate water supply for the nation's capital to the US Army Corps of Engineers. This was logical as not only did the Corps possess much of the nation's engineering expertise, but also the Army was part of the administrative organization that oversaw the provision of services to the federal city. There was a pressing need for the development of a proper water supply as local wells and stream diversions were insufficient to provide fire protection, and the sanitation practices of the day contributed to a wide range of diseases.

A remarkable engineer Montgomery C. Meigs and a team that he assembled laid out the original plans which consisted of a low dam on the Potomac River at Great Falls which would divert water into a 10 mile long closed aqueduct and deliver it by gravity to a receiving reservoir created at a location known as Dalecarlia on the District of Columbia boundary with Maryland. In that receiving reservoir, some sediment naturally settled out. Then, the water continued in another conduit two miles further to the distributing reservoir at Georgetown where additional

---

[1] Chief, Washington Aqueduct, 5900 MacArthur Boulevard, Washington, DC 20016-2514; 202-764-2753; thomas.p.jacobus@usace.army.mil

sedimentation occurred. Relying on gravity, the water then traveled through a series of pipes to the Capitol and the surrounding area.

All of that original system is still in use with some necessary modifications to accommodate the emergence of chemical and mechanical treatment processes and the expansion necessary to meet the modern demand for treated water.

Throughout the 150 years of its existence, the Washington Aqueduct organization has remained part of the Corps of Engineers – which makes it unique in our nation that a federal agency functions as a public water utility for a metropolitan area. What has changed is that since the early 1900's the Washington Aqueduct has operated on an independent financial basis with all funds for operations, maintenance and capital improvement coming from the sale of water. Even though the current employees are part of the federal civil service and the real estate and structures and all property are held in the name of the United States government, there is no public money (i.e., federal appropriations) involved.

Over the years, Washington Aqueduct has emerged as a wholesaler with minimal distribution system responsibilities. Its customers are the District of Columbia, Arlington County, Virginia, and the City of Falls Church, Virginia. It serves approximately 180 square miles and 1,000,000 people in the District of Columbia and Northern Virginia. The U.S. Environmental Protection Agency, Region 3, regulates it since the District of Columbia does not have primacy for water under the Safe Drinking Water Act.

It is governed by a Wholesale Customer Board, which approves operating and capital budgets and, by extension, the wholesale rate for themselves. However, as a federal entity, Washington Aqueduct is subject to all federal regulations. As a result, it must comply with Section 106 of the National Historic Preservation Act and consult with the Maryland Historic Trust and the Maryland State Historic Preservation Office, the District of Columbia State Historic Preservation Office, and the Advisory Commission on Historic Preservation. Additionally, other federal statutes require it to coordinate construction projects (new construction or renovations) with the National Capital Planning Commission and the Commission of Fine Arts.

The Need for Water

Wells, springs and cisterns were the sources of water for the residents in the nation's capital in the 1850's. A large spring known as Smith Spring was owned by the federal government, and it had supplied water to some federal buildings including the capital since 1837. The other major springs were directed to the area of Massachusetts and Pennsylvania Avenues and served federal buildings through wooden and cast iron pipes. Some of these had been in service since the early 1800's.

In 1852, the population of Washington was about 58,000, and the springs and wells were insufficient to meet the need. The frequent contamination of the product delivered from these springs and wells coupled with the unattractiveness of the often muddy Potomac River and the lack of a means to bring that Potomac River water to the residents formed the basis of the need to improve the water infrastructure. There were schemes proposed to develop a public water supply system based on local creeks or the creation of a basin on the mall with water from the Chesapeake and Ohio Canal. But these options were not developed.

In September 1850, Congress had appropriated $500 for the War Department to conduct an examination and survey of the best methods of supplying the city with pure water. That report was completed in January 1851, but a follow-on appropriation of $1,000 was insufficient to develop the options fully.

The seminal event for the establishment of the Washington Aqueduct occurred on Christmas Eve 1851. A spark from a stove ignited a fire that engulfed the room in the Capitol Building that housed the Library of Congress. As there was insufficient water available to extinguish the flames, many irreplaceable items were lost.

In 1852, Congress appropriated $5,000 for the necessary surveys and the task was assigned to Brigadier General Joseph G. Totten, the Chief Engineer of the Army. In November 1852, Totten recommended the appointment of Lieutenant Montgomery C. Meigs to take on the job of conducting the survey.

The Man who Built the Washington Aqueduct: Montgomery C. Meigs

Montgomery Meigs was born in May 1816 in Georgia but was raised in Philadelphia. His father was a physician. His grandfather was a teacher and college president who ultimately became the commissioner of the United States General Land Office. As a child, Meigs showed many of the characteristics he would later display in his management of the design and construction of the Washington Aqueduct. He held firmly to his beliefs, was persevering, and highly principled.

Meigs left the University of Pennsylvania in 1832 to attend the United States Military Academy at West Point. West Point was the nation's engineering school and the Corps of Engineers was engaged with a major responsibility for building important elements of the national infrastructure. He graduated from West Point in 1836 number five in a class of 49. An early assignment for Meigs was a survey of the Mississippi River under the leadership of Robert E. Lee, West Point class of 1829. Meigs' first assignment in Washington was in 1849 as an assistant to the Chief of the Army Corps of Engineers General Totten. It was he three years later who would recommend Meigs to the Secretary of War to undertake the studies Congress had authorized in 1852 for the development of a water supply system for the nation's capital.

Family was important to Meigs. He had three sons and two daughters. Throughout the difficulties of military service he continually focused on the welfare of his family. One of his sons John Rodgers Meigs attended West Point and served in the army with distinction before being killed in the Civil War.

While this discussion relates how Meigs' talents were focused on the development of the Washington water supply, it is remarkable to note that at essentially the same time he oversaw work on the extension of the US Capitol Building and the completion of its dome and the extension of the general post office building. Upon completion of the work on the Aqueduct, Meigs designed and oversaw construction of the Pension Building in Washington, which is now known as the National Building Museum. While still serving as Quartermaster General of the Army in 1876, he contributed a design for the new national museum in Washington. Today, this structure is known as the Smithsonian Institution's Arts and Industries Building on the mall. Meigs also participated in the project to complete the Washington Monument after the Civil War.

Meigs' engineering and architectural structures incorporated many novel features. His idea to use the water transmission mains as part of the support structure for one of the major bridges in the Aqueduct system is an example of that thinking. He took meticulous care with all aspects of his design and financial accounting. To improve his efficiency, he even developed his own system of shorthand.

The Plan

Having briefly served in Washington in 1849, Meigs returned to Washington in November 1852 and was assigned by General Totten to take charge of the work on the water supply planning. It took him only nine days to survey the situation and submit his first report to Totten. A scant three months later, he submitted the final report, which Totten quickly recommended to the War Department for construction.

In his report, Meigs evaluated the water supplies of New York, Boston and Philadelphia. He also was influenced by the Roman aqueducts. The options he explored were in Rock Creek (a stream entering the city from the north) and on the Potomac River at both Little Falls and Great Falls. He logically evaluated the engineering advantages and disadvantages of each as well as producing detailed cost estimates. His report recommended the Great Falls option calculating it would supply 36 million gallons per day using a seven-foot diameter conduit. Interestingly, Meigs parenthetically noted that if the conduit's diameter were increased to nine feet that the capacity would double.

In his decision to recommend construction of the system according to Meigs' plan, Totten indicated that the nine-foot diameter should be used. As it turned out the city's population grew more rapidly than had been projected in 1852; and had the seven-foot diameter been adopted, the demand would have exceeded the capacity by 1890. Instead, the nine-foot diameter served the city until 1927.

The detailed plan moved forward quickly. By June 1853, a plan and profile drawing of the entire system was approved and signed by General Totten, Secretary of War Jefferson Davis, and President Franklin Pierce. This drawing is in essence the "charter" of the Washington Aqueduct. An exact copy made in 1885 hangs in the main conference room of the administration building of the Washington Aqueduct today. It is a reminder and an inspiration to the current staff of the engineering excellence that preceded them.

The plan called for a succession of elements that needed to be constructed. It began with a low dam on the Potomac River at Great Falls, Maryland, to divert the river flow into the conduit. From the riverbank, the conduit passed under the Chesapeake and Ohio Canal, which was built a few years earlier, and entered a gatehouse building with 20 sluice gates. Those controls maintained proper water levels in the 10-mile long conduit that took the water to the receiving reservoir at Dalecarlia. That conduit was a combination of tunneling and cut and cover construction with a slope of nine and a half inches per mile. The Dalecarlia receiving reservoir was at the Maryland – DC boundary about five miles west of the Capitol. Two more miles of conduit took the water to a site on the palisades of the Potomac above Georgetown. The reservoir constructed there was known as the distributing reservoir. Distribution pipes carried the water as far as the Navy Yard on the Anacostia River. The total length of the system was 18 and one half miles. Figure 1 below shows the relationship of those elements in the context of the modern service area.

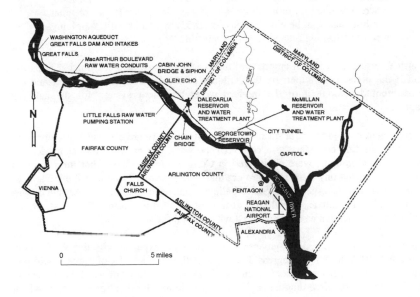

Figure 1. Current Washington Aqueduct Service Area with Location of Infrastructure

The Work Begins

Being politically adept and knowing that this undertaking would require a substantial amount of financial support over a number of years, Meigs decided on a celebration to commemorate the start of work. Assembled at Great Falls on November 8, 1853, were President Pierce, Secretary of War Davis, Senator Steven A. Douglas, Mayor and Council of Washington, plus guests for a total attendance exceeding 200. The location of their dinner that evening still exists as the Great Falls Tavern at lock 20 of the Chesapeake and Ohio Canal. It is operated now as a museum by the National Park Service.

Recently, on May 11, 1996, a letter to the editor of *The Washington Post* written by Mr. Harry C. Ways was printed. It read:

> "When President Clinton spoke on the environment from the roof of the Washington Aqueduct intake structure at Great Falls, Maryland, he was not the first U.S. President to speak at that site. President Franklin Pierce spoke at almost exactly the same spot at the groundbreaking ceremony for the new water supply system for the national capital on November 8, 1853. Other speakers were Secretary of War Jefferson Davis, Senator Stephen A. Douglas and Captain Montgomery C. Meigs, the builder of the facility, who took the occasion to name it the Washington Aqueduct. Inside the building where President Clinton spoke, water was flowing into the original conduit that was begun that day in 1853, which is still supplying water to Washington and Northern Virginia."

Harry Ways served as chief engineer of the Washington Aqueduct from 1972 to 1991 and is the preeminent expert on the building of the Washington Aqueduct and the subsequent development of the modern water treatment system for the nation's capital.

Work began with great enthusiasm but proceeded slowly largely because there were delays in funding necessary to accomplish the work. In an argument as modern as today, Meigs' budgets were challenged in Congress on the basis that the requested federal appropriations were for a project that would primarily benefit the citizens of Washington and Georgetown. Some believed that even though the Aqueduct was badly needed, the citizens should pay at least one-half if not more of the cost. In the end, Meigs' structures were constructed with federal funds.

Until July 1880, the federal government funded all expenses for the Washington Aqueduct. But in 1880, Congress directed that the District of Columbia would pay one-half of all expenses for construction and operation of the Aqueduct thereafter. In 1916, Congress required that the city be responsible for all operating expenses, which were to be paid from water revenues of the District. Until 1927, one-half of construction costs were shared between the District of Columbia and federal funds.

After 1927, all costs, both operating and construction, were paid from revenues derived from the sale of the water produced by the Aqueduct.

Work on the Aqueduct required not only men and materials, but also "horsepower," often in the form of mules. Frequent advertisements appeared in newspapers in the east requesting bids for materials, laborers, and animals. Keeping the experienced foremen was a challenge, and Meigs was personally involved in getting permission to raise pay as necessary.

By 1857, the conduit was about 20 percent complete and work had just been started on the major bridge over Cabin John Creek. In all, there were 11 tunnels, 26 culverts and four stone bridges necessary to maintain the grade of the conduit from Great Falls to Dalecarlia. In figure 2 below, Meigs is pictured at the Washington Aqueduct wharf on the Potomac River in Georgetown where major items were off loaded for transit to the project site.

Figure 2. Montgomery Meigs posing with Washington Aqueduct valves

1857 and 1858 were very productive years for work on the project due to the flow of significant appropriations, but in 1859 work was suspended for lack of funds.

As a result of a political dispute, in September 1860 Meigs was temporarily relieved of his duties as chief engineer of the Washington Aqueduct and was assigned to Fort Jefferson, Florida. In January 1861, Meigs returned to reassume his role as chief engineer and the project moved to completion.

Completion

Completion of the Cabin John Bridge (the longest and most challenging of the bridges) was necessary to bring water from the Potomac to the Capitol. Even though the rest of the infrastructure was in place, it took until 1863 to finish the bridge; and until that was done no Potomac River water could enter the system.   However, the Dalecarlia receiving reservoir was adjacent to Little Falls Branch, so diverting that stream and filling the reservoir was a temporary measure to allow the system to begin operating.

But even without the Cabin John Bridge, with eight and a half miles of 12-inch pipeline laid to the Capitol and on the Navy Yard, and with the use of a temporary main laid across Rock Creek, Meigs delivered water to a fountain at the base of Capitol Hill on January 3, 1859. Great acclaim arose from the assembled crowd as water jetted 100 feet into the air.

The Cabin John Bridge

The singular engineering achievement, which is celebrated to this day, is the Cabin John Bridge. Meigs intentionally set about to make this truly a remarkable structure. Graceful in appearance and functional in design, it carries the nine-foot diameter conduit across a clear span of 220 feet. Originally known as the Union Arch Bridge it is popularly known today as the Cabin John Bridge. In the four year period between his first report describing the bridge in 1853 and the time work started in 1857, Meigs modified the design. The original concept was to be six arches. However, Meigs believed that any increased cost of constructing a single arch to span the gap might be made up from savings elsewhere on the Washington Aqueduct project, and in building the single arch, he would in effect be creating a monument to his own prowess as an engineer. Upon its completion in 1863, it was the longest masonry arch span in the world and remained so for more than 40 years. The cut stone arch of Quincy granite was imported by ship from Massachusetts. The sandstone parapet walls were quarried from the local Seneca quarry, which was the source of the stone used on the Smithsonian castle on the national mall in Washington.   To get the stone to the bridge site, Meigs dammed Cabin John Creek and construed a lock system to tie into the Chesapeake and Ohio Canal so that the stones could be floated up from the docks in Georgetown. The construction period for this magnificent structure coincided with hostilities of the Civil War; and for that reason, construction was delayed as the construction trestle and centering arch made of wood was removed for a time to keep Confederate forces from burning it. Figure 3 shows the completed bridge with a Civil War wagon train passing over it.

Figure 3. Cabin John Bridge with Civil War Wagon Train, c. 1864

Alfred Rives and Charles Talcott ably assisted Meigs in his engineering, but Meigs took individual credit for the design. Not only is the bridge part of Washington Aqueduct's National Historic Landmark designation, it has also been designated a National Historical Civil Engineering Landmark by the American Society of Civil Engineers. The American Water Works Association also designated the bridge a historic water landmark, significant in the history of water supply.

Today, it not only delivers nearly 100 million gallons per day of raw water to the treatment plant, but also it has become a major link in the suburban highway transportation system carrying commuters to the District.

An inspection of the facilities designed and constructed by Meigs reveals his penchant for marking them with celebratory phrases as well as his own name. At the Georgetown distributing reservoir a large underground vault provided access to the discharge pipes and control valves. The pipes were reached by descending a circular staircase consisting of 39 steps. Cast in the tread of each step was "Washington Aqueduct" and each riser was cast in the shape of "M.C. MEIGS" as shown in figure 4. On the Cabin John Bridge carved into a stone in the arch is an engraving that reads: "M.C. Meigs, CHIEF ENGINEER, WASHINGTON AQUEDUCT, AD 1859 FECIT." A stone plaque in the east abutment reads: "UNION ARCH, Chief Engineer, Capt. Montgomery C. Meigs, U.S. Corps of Engineers, Esto Perpetua." A carved stone plaque in the west abutment is engraved to read: "WASHINGTON AQUEDUCT, "Begun AD 1853: President of the U.S. Franklin Pierce; Secretary of War Jefferson Davis; Finished AD 186-; President of the U.S. Abraham Lincoln; Secretary of War Simon Cameron." In 1862, Jefferson Davis' name was removed from the stone. In 1908, President Theodore Roosevelt ordered it restored.

Figure 4. Risers of the circular stairway at the Georgetown vault

The Rock Creek Bridge

The original concept to get the pipelines across Rock Creek in order to preserve canal boat navigation was to construct them to pass under the creek. However, in 1855, Meigs worked on a design to use the pipelines themselves as arches to support the bridge. What came of this was a bridge with two 48-inch cast iron pipes with a 200-foot clear span and a 20-foot center rise as shown in figure 5. Upon its completion in 1860, it replaced the temporary pipes Meigs had laid in 1859 to inaugurate the system. Though not readily visible today, the original pipes reside in recesses between new concrete arches that support the weight of the modern traffic, but those pipes continue to deliver water to the city.

Figure 5. The Rock Creek Bridge with 48-inch water mains as support arches

Arlington National Cemetery

Montgomery Meigs died in 1892 and is buried in Arlington National Cemetery. His memorial stone is marked: "Soldier, engineer, architect, scientist, patriot." While serving as Quartermaster General of the Army, it was Meigs who acquired this property once owned by Robert E. Lee, as a cemetery for Civil War dead. In front of Meigs' memorial in Arlington is a half-sized bronze likeness of his son, John, killed in the Civil War on October 3, 1864. John's remains were transferred to Arlington in 1892 upon the death of his father.

Farsightedness of the Design: Adaptable for the Future

As important as the basic engineering achievement of successfully bringing the water of the Potomac River to the city of Washington was in 1864, Meigs should be credited with creating a system that was both robust and adaptable. Whether he could envision future treatment facilities or the present-day expansion of the service area into Northern Virginia is certainly a matter of conjecture, but the natural evolution of the water system has fit hand-in-glove with Meigs' original concept.

In the late 1890's, plans were made to construct the first treatment works. A site on North Capitol Street near the Old Soldiers' Home and Howard University was chosen. It was the site of Smith Spring that was still supplying water in addition to that coming from the aqueduct system. Washington Aqueduct engineers determined that if a four and one half mile new tunnel were constructed from the Georgetown distributing reservoir through Rock Creek Park, the Potomac's waters could be delivered to a new reservoir to be known as the McMillan Reservoir, which could be used as the source of settled water for a slow sand filtration treatment plant. The topography allowed gravity to take the water through the new tunnel to the new McMillan Reservoir. Meigs' concept of gravity flow was retained.

The McMillan treatment plant was in service by 1905, and it marked a major step forward in providing safe and reliable water for the city. In addition to the pioneering work with slow sand filtration, the McMillan plant grounds received a masterful touch in a landscape architecture design by Fredrick Law Olmstead, Jr. The photo at figure 6 shows the grounds as a place where residents could stroll and enjoy the interesting species of plants and the symmetry of the design. Unfortunately the grounds were closed to the public at the beginning of World War II. In 1985 the slow sand plant was replaced with a new facility incorporating rapid sand filtration, and an impeller was added to pull more water through the tunnel from Georgetown. For nearly 100 years this treatment location has benefited from Meigs' original work.

Figure 6. The McMillan Filtration Plant c. 1930

In the mid-1920's the growth of the city was projected to exceed the capacity of the water supply infrastructure. Since this growth was primarily on the western side of the city, and just across the Potomac River from Arlington, Virginia, engineers set out to find a location for a new treatment plant. Land adjacent to the Dalecarlia receiving reservoir was obtained. However, the capacity of Meigs' original conduit from Great Falls was insufficient to feed the new plant.

The solution was simple: a second conduit was laid parallel to the first one. Its construction was far less elegant and it was designed in a way that did not incorporate the artistry in the bridges and conduits that accompanied the first one. At Cabin John Creek, a functional but plain concrete siphon was the means across the gap. Nevertheless, the water from that second conduit still goes to the original receiving reservoir. The treatment works at Dalecarlia were designed as rapid sand filtration, and the facility was constructed of brick and incorporated many interesting aesthetic details.

The Dalecarlia treatment plant site has been expanded twice since its construction; and even though the capacity of Meigs' original conduit was modified in the 1930's to achieve 100 million gallons per day, the combined 200 million gallons per day of the two conduits was insufficient by 1950. A new pumping station on the Potomac River at Little Falls (one of the original sites that Meigs had investigated) delivers additional water into the Dalecarlia receiving reservoir to be treated at one of the two plants.

As it has turned out, Meigs' Great Falls option for the conduit intake was exactly the right solution for the Washington Aqueduct's growth. Had the intake been placed at Little Falls, the easy expansion of the system based on gravity would not have been possible. The Rock Creek option would have been exhausted long ago.

The major growth of the metropolitan Washington area has put increased demands on the Potomac River. To accommodate that, upstream storage to augment the natural flow of the river was constructed in the 1980's. Estimates are that an adequate supply will be available until at least 2030, and the current Washington Aqueduct infrastructure exists to treat considerably more water than the projected demand.

National Landmark Status

To celebrate Washington Aqueduct's important engineering and architectural history and heritage, the Meigs elements have been designated a National Historic Landmark, and many of the newer additions fall under the category of contributing resources.

Washington Aqueduct initially submitted its National Historic Landmark nomination to the National Park Service in 1973. While a very good start, additional refinement was necessary. Also, it needed to comply with a provision of the National Historic Preservation Act and complete a Cultural Resource Management Plan, which was done in 1998. It serves as an everyday reference and makes current actions much easier to document and submit.

One issue that Washington Aqueduct has had to deal with is what to do with structures that are covered under the National Historic Preservation Act but no longer serve in their original capacity. A tension can exist between senior managers who might wish to remove the structures to reduce maintenance costs and historical compliance program staff that get very strong guidance from the historic preservation agencies to retain them. The approach has been to take a long-term view. We have stabilized most to prevent further deterioration, incorporated some into current operations, and removed only a few.

Challenges for the Future: Accommodating the Past

Washington Aqueduct has adopted a strategy of "full and enthusiastic compliance" with all aspects of the National Historic Preservation Act. We have taken what some may see as just another unfunded federal mandate and attempted to use it to our advantage. In our particular case of being in the nation's capital, our compliance with the National Historic Preservation Act which necessitates our coordination with the State Historic Preservation Officers usually also leads to conformance with the National Capital Planning Act which requires that projects are reviewed by the National Capital Planning Commission and the Commission of Fine Arts. We have found that, without exception, this consultation adds value to our projects. Alternative ideas and solutions emerge that are better than what we could have achieved only within our organization.

So how does that square with the principal business objectives of the Washington Aqueduct water production operations, which are to deliver a safe product, do so reliably and do it in a cost-effective manner? What does the business get out of this compliance process and the time, effort and money spent in dealing with it?

We have found that public confidence in the water supply is largely gained from an external vantage point. When you have facilities that the public sees frequently in their travels around their community, or when the news media comes to do a report on some EPA program or a drought or for any reason, the architecture and physical condition and appearance of the treatment facilities either send a message of a well-run treatment operation or one where water quality is in doubt. Clearly, we choose to send the former message.

We also know that the success or failure of the water production mission relies on how well the employees do their jobs. Employees produce superior results if they are operating in quality surroundings and are motivated to achieve greatness. Some of that motivation comes from a sense of tradition and history – that what they are doing today furthers the work of their predecessors and that they too are leaving a legacy for the future. If you are fortunate to have some very interesting and renown American architectural treasures originally laid out by Montgomery C. Meigs, a Frederick Law Olmsted, Jr. landscape design, and a 150 year history of providing water to the nation's capital, that certainly can be the basis for pride in current operations. Compliance with the National Historic Preservation Act is a furtherance of that tradition and a motivating factor for the organization.

More and more citizens are taking an interest in the water treatment process and the inherent quality of the product delivered to them. Our focus on the historic aspects of the facilities and our opportunity to celebrate them as part of the ongoing water production mission provide us new opportunities to tell the water story.

As an enabling part of our strategy for full and enthusiastic compliance, we have made sure to have key, responsible employees well-trained and experienced in dealing with the various parties. Along the way, we have gotten very good at the consultation process. Our individual work and association with experienced contractors as well as with the state historic preservation offices have taught us preservation techniques. Most importantly, we have integrated our compliance with the National Historic Preservation Act into the core of our engineering and maintenance staff. Compliance is not an afterthought or an appendage. It is considered from the outset part of every action we take. This includes routine maintenance, for some well-meaning but thoughtless act can destroy or seriously damage some piece of historic fabric of the facilities.

Conclusions

Washington Aqueduct managers and employees have been good stewards of the cultural resources entrusted to their care. We believe that we have achieved a level of cultural stewardship that goes well beyond "lip service" that might be paid to one of many federal mandates. It has become ingrained in the fabric of the organization, much like our basic commitment to consistently providing safe drinking water. We have actively looked for ways to keep the historic thread pulled through current daily operations. Finding new uses for the old and taking the opportunity to blend the old with the new have paid off. In doing so, we have improved the aesthetics of our treatment plants and ancillary facilities. We are less cluttered and disjointed that we might otherwise have been. Anecdotal evidence gathered from informal conversations with employees shows that while it may not be their job to directly assist in the historic preservation projects, they have noticed and have taken pride in the outcome. We believe that the pride is reflected in their daily jobs on behalf of the modern operations.

In most cases, the physical strength and toughness of the older architecture and equipment is superior to modern construction. Once the investment has been made again in the substantial, historically accurate materials, we have found that our maintenance costs are slightly less and life span has increased.

We have earned the respect and trust of the State Historic Preservation Offices and the other historic advisory agencies. As a result, they require less exhaustive reviews that are expensive and time consuming for us because they know we are out not only to satisfy the requirements but also to go beyond them.

We have also garnered important positive publicity and public interest. Feature articles in the print media about the contributions of the Cabin John Bridge and the original aqueduct structure to the development of the modern metropolis and the functioning of the national government are an indication that what we do matters to others.

In July 2001 we closed the Cabin John Bridge to traffic while historically accurate upgrades and renovations of the deck were accomplished. In November 2001, the reopening ceremony of the Cabin John Bridge drew local and state and national political figures to help celebrate the perpetuation of this piece of American history. It's not a bad thing for our public relations or our reputation when the Maryland congressional representative comments, "This is the way all federal projects should be executed."

Politically as involved as Montgomery Meigs was to achieve his engineering vision in the Washington Aqueduct, he would have been proud.

Preservation Case Studies

LOST INCA TRAIL: PREHISTORIC CIVIL ENGINEERING

Kenneth R. Wright, P.E., Fellow[1]
Ruth M. Wright, J.D.[2]
Alfredo Valencia Zegarra, Ph.D.[3]

INTRODUCTION

When the young Yale history professor Hiram Bingham came upon Machu Picchu in 1911, he was impressed with the quality of engineering that had gone into the site some four centuries before by early Americans (Figure 1). Bingham went back to Machu Picchu in 1912 as director of a Yale/National Geographic Society expedition to clear and document the ruins (Bingham 1930). Bingham's description of his explorations filled the entire April 1913 issue of *National Geographic Magazine* and brought Machu Picchu to the world's attention; Bingham opined that, "The Inca were good engineers." He could also have stated that the Inca civil engineers were good road builders because by the time Conquistador Pizarro arrived, the Inca were operating and maintaining some 14,000 miles of road system that stretched from Ecuador to Chile and from the Pacific Ocean to high Andean settlements after crossing mountain passes with 18,000-foot elevations. Bingham described the main Inca Trail leading into Machu Picchu. It was a masterpiece of prehistoric Inca road building.

One example of a civil engineering road-building feat that escaped Bingham's sharp eye, however, was right there at Machu Picchu, but it lay undetected in the forest cover of the lower east flank for another 87 years before being uncovered in 1999 (Wright, Valencia and Crowley 2000). The Machu Picchu Inca Trail extension proved to be a fine example of route selection, slope stabilization, environmental design and construction; all accomplished without a written language, iron, steel or the use of the wheel (Moseley 1992). Machu Picchu was described in ASCE Press's *Machu Picchu: A Civil Engineering Marvel* (Wright and Valencia 2000); however, the book dealt with the new Inca Trail only in passing, because data on the trail was still being processed at the time of its writing.

Machu Picchu was a royal estate of the Inca Emperor Pachacuti, some 80 kilometers (km) from the capital of Cusco at an elevation of 2,458 meters (m) and 1,400 km south of the Equator. It lies near the headwaters of the Amazon River at longitude 72°32' west and latitude 13°9' south on a ridge between two jutting peaks—Machu Picchu Mountain and Huayna Picchu Mountain. Machu Picchu dates from AD 1450

---

[1]CFO and Chief Engineer, Wright Water Engineers, Inc., 2490 W. 26th Ave., Ste. 100A, Denver, CO 80211; phone (303) 480-1700; krw@wrightwater.com
[2]Member, Board of Directors, Northern Colorado Water Conservancy District, Loveland, CO; Member, Colorado State Legislature (1981-94); Member, American Bar Association; ruthw@webaccess.net
[3]Professor, Department of Anthropology, Universidad de San Antonio, Cusco, Peru; rumi@unsaac.edu.pe

to 1540 and was finally abandoned in AD 1572 (Rowe 1990). Steep topographic relief is a result of tectonic activity, faults, and downcutting erosion from the Urubamba River that nearly encircles it (Wright, Witt and Valencia 1997a). Fortunately, the geologic activity provided the abundant granite bedrock upon which Machu Picchu is built and that the Inca used for building stones.

Discovery of the Inca Trail on the east flank of Machu Picchu in 1999 answered a long standing question about the road that led from the far away Inca capital of Cusco and seemingly ended at Machu Picchu. Exiting from Machu Picchu there were only three known minor trails leading beyond to the Urubamba River, some 525 m below that would provide access to the Amazon lowlands. Could the main Inca Trail have just terminated at Machu Picchu? Now we know that the answer is no! It continued through Machu Picchu and from there it descended 525 m with a slope distance of 1.8 km to the left bank of the Urubamba River at an average slope of 35 percent. The name of the river changes from Urubamba to Vilcanote in the Machu Picchu area. For purposes of simplicity, we will continue to call it the Urubamba River.

Initial Exploration

A paleohydrological study of Machu Picchu was commenced in 1994 (Wright 1996). By the third field session in 1995, with the water oriented studies well ahead of schedule, our machete crews were directed to the east flank where team archeologist, Dr. Alfredo Valencia Zegarra, had reported potential fountains and mapped five terrace systems back in 1969 following a forest fire. One fountain was found in 1995, another in 1996, and three more fountains were identified early in 1997 on the lower east flank. Based on these findings, it was concluded that a long buried trail must have existed that would have connected these sites to Machu Picchu.

In November 1997, an attempt was made to trace a trail from the long granite stairway in Conjunto 13 (a conjunto is an enclosure or group of buildings) of Machu Picchu proper, but to no avail (Figure 2). The likely trail route was cleared for only 45 m to where an Inca security station was found. Archeological documentation of the structure was performed, but our allotted field time came to an end. Finding the elusive trail would need to wait.

Planning for the 1998 field season included securing an excavation permit for the five already discovered fountains of the lower east flank. The Instituto Nacional de Cultura (INC) tightly controls the actual excavation of archaeological sites so that all protocol is carefully followed and all artifacts found are recorded. The subsequent field excavations uncovered the five fountains representing four different layouts and designs. Fountains 3 and 4 were ceremonial fountains of classic Machu Picchu Inca design. Once excavated and their channels cleared, the fountains burst into operation with water jets discharging from water channels that have a Venturi-like channel spout into the newly, cleared out stone basins lying below. Once the water cleared, the fountains were utilized as a field water supply for the excavation crews. The water was pure and thirst quenching (Figure 3).

Fountain 1 also flowed when it was excavated and its orifice cleared out; however, Fountain 1 was judged to be a utilitarian water source for field agricultural workers with little or no ceremonial purpose. Nearby was a finely constructed granite gate, an entrance through the outer perimeter wall for a secondary trail leading up from the river below. Fountain 2 was found in and amongst a series of terraces overlooking the Urubamba River far below beyond a rock cliff. It, too, was an ancient fieldworker water supply.

Fountain 5 was discovered at a site that had been severely damaged by a huge rockfall from the steep cliffs of Huayna Picchu. It was evident that the site had been an important one; it had an undisturbed walled cave with a carefully crafted water tunnel below it that penetrated the mountainside to intercept groundwater. The water source flow ranged between 0 and 40 liters per minute (L/min). Beneath and downhill of the fountain area was an accumulation of Inca potsherds that allowed us to date the site to the classic Inca period of about AD 1400.

Connecting Fountain 5 with Fountains 3 and 4 we found a wide granite path some 122 m long that contained fine stairways, all supported both uphill and downhill by elaborate terrace systems, which provided much needed stability on the steep 2:1 sloped mountainside (Figure 4). This short section of trail coupled with the three fountains on its uphill side raised the question of where did the trail start and where did it end? It was obvious from the limited excavation data then in hand that the 122 m of trail represented an extraordinary example of prehistoric civil engineering, but was it a part of a larger and more extensive Inca trail? To answer these questions, the author applied for and obtained another archeological exploration permit from the INC for the September 1999 field season (Wright, Valencia and Crowley 2000).

Inca Trail 1999 Exploration

Archaeological exploration from the main Machu Picchu site to the Urubamba River through a dense forest on average slopes of 2:1 would require a crew of local Quechua Indian macheteros. After inquiring about the availability of laborers in Aguas Calientes, the town at the foot of the Machu Picchu mountainside, it was recommended that we attend an important funeral in the town square and follow the procession to the rocky and steep graveyard at the edge of town. Following the internment, we sat on gravestones and interviewed willing workers before darkness fell. Eight strong Indians were signed up for two weeks of work. They would join our team of two registered Peruvian archaeologists and the four of us from Denver's Wright Water Engineers, Inc.

Two exploration teams were assembled, one that would start from the top at the 1997 Inca security station and the second that would work uphill from the river. In 1998, a brief reconnaissance of the mountainside discovered the remains of a left riverbank stone wall above which many Inca terraces were noted in dense forest and undergrowth. It would be here that the second crew would begin after crossing the fast flowing river (Figure 5).

The trail exploration technique adopted was effective. It consisted of finding buried granite by probing a machete into the forest floor. We would work up and down the slope until the unmistakable "clunk" was heard of steel on granite. Then the rock would be uncovered and cleared. Usually, it was just a random rock, but when the rock was specially shaped, further clearing would be performed, likely finding other shaped stones and then the trail itself. Once the trail segment was cleared and the direction noted, the macheteros would merely follow the trail with forest clearing, excavation and clearing off of the actual granite trail or stairways. When the trail was lost, the procedure of working the hillside with machetes would be repeated. Observing the engineering features of the trail along with the fine workmanship provided pleasure and satisfaction to the Quechua Indians each step of the way because they knew they were participating in an important discovery that would contribute to the knowledge of their cultural heritage.

TRAIL EXPLORATION

The Machu Picchu east flank trail was identified and mapped to the landing on the left bank of the Urubamba River at a point immediately upstream of a series of cascades situated at the base of the Huayna Picchu cliff. It proved to be a masterpiece of ancient civil engineering performed with a high standard of care to insure functionality and longevity under harsh and challenging conditions. The river landing is 1.2 km upstream of Mandor Pampa.

From the river landing, the trail follows a torturous zigzag route uphill over a distance of 1.80 km and a rise of 525 m to the qolqas (storehouses) of Machu Picchu in Conjunto 11. The trail is illustrated on the U.S. Air Force 1963 aerial photograph in Figure 6.

The east flank trail was judged to be a continuation of the Cusco–Machu Picchu Inca Trail. It led to Mandor Pampa and points downstream, as well as providing direct access to the Urubamba River's right bank trail from Ollantaytambo. Of all the other known trails out of Machu Picchu, this is the only one of such high quality, directness, width, modest slope and infrastructure that would qualify it as the extension of the Inca trail out of Machu Picchu. During Inca times, the trail also served as the main access to the numerous east flank tombs that were identified by Hiram Bingham in his 1912 exploratory work (1913, 1930) (Figure 7).

The ceremonial fountains, the extensive high quality terrace systems, the numerous wide granite staircases, the formalized rest stops with special views, the ability to provide line-of-sight security control from various control points and the abundance of tombs along its route all testify to the east flank trail being an important Inca trail.

Surveys

To document the trail from the Urubamba River to Conjunto 11 in Machu Picchu, our two Denver civil engineers performed a tape and compass survey of the newly cleared route using a Brunton compass both for azimuth and slope. The vertical elevation gain and horizontal distance of each segment was computed and recorded. Due to the fact that the compass independently determined the azimuth of each segment, there are no accumulated angle errors in the resulting trail alignment. Furthermore, compass bearings to known independent points provided a procedure for checking locations at various intervals along the trail. The vertical distance was confirmed by checking the original topographic survey prepared by Civil Engineer Robert Stephenson in 1912 under Bingham's direction (Figure 7).

Results of the field instrument surveying of the trail are summarized in Table 1 and are presented on the profile in Figure 8.

TABLE 1
Inca Trail Measurements

Slope length	1.80 km
Horizontal length	1.45 km
Point-to-point map distance	0.95 km
Vertical rise	525 m
Average slope	35 percent (3:1)

The trail width ranges from 1.0 to 3.3 m with an average of about 1.8 m. Larger widths are often found on special staircases (Figure 9). The trail details are shown on Figure 10.

Special features such as terrace walls, stairways, ramp walls and rest stops were documented in the field notes as part of the route survey for the subsequent mapping effort. A hydrographic survey was also made at the Urubamba River landing. These features represented prehistoric civil engineering achievements that were exposed to the light of day for the first time in four and a half centuries.

Trail Views

Several viewing platforms are associated with the trail; however, one deserves special attention. It is located immediately above Fountains 3 and 4 (Figure 11). The direction of the view is N20°E. The view is perfectly framed by Putucusi Peak on the right, Huayna Picchu on the left and Mount Yanantin in the center, with the river below (Figure 12).

Trail Connections in Machu Picchu

A steep granite staircase emerges from the forest just to the east of Building 8 and 9 of Conjunto 13, as shown in Figure 2. The staircase angles to its left to a rock ramp just below a path. Continuing on this path leads to Intimachay in Machu Picchu. To the north of this staircase is a series of steep terraces. Along the rock ramp, the path curves northwest to a very steep granite staircase with terraces on both sides. This staircase leads to the base of Conjunto 11, the storehouses.

Conjunto 11 consists of six qolqas, each two stories high. There are two entrances for each of the qolqas: one at the lower level and one at the upper level. There are horizontal paths between the qolqas for easy access. From the base of Conjunto 11, a long stairway between Conjuntos 10 and 11 provide access to all of the qolqa entrance pathways, completing the connection between the qolqas and the east flank trail. Agricultural products grown on the east flank terraces and on the floodplain of the Urubamba River could easily be transported to the qolqas without passing through other parts of Machu Picchu.

Fountains

Fountains 3, 4 and 5 were excavated under an archeological permit in September 1998. A description of the 1998 excavations of the three fountains is presented below in detail, and drawings are presented in Figures 13 and 14.

Excavation of Fountains 3, 4 and 5

Fountains 3, 4 and 5, as well as a lateral drainage channel, the trail and the terrace, were covered with a thick layer of topsoil and dense vegetation along with a series of landslides coming from above. We cut the vegetation and cleared the surface of this area. Objectives were as follows:

a. To determine the engineering form of the structures, their relationships and other associated elements.

b. To study the stratigraphy and pre-Hispanic cultural artifact content.

c. To uncover buried structures.

d. To determine the engineering related to the use of the water through location of the spring and its channels and distribution of the fountains and their drainage routes.

Fountain 3

Fountain 3 is in the form of a cube surrounded by four walls, being accessible through a small opening located on the north side. A small channel on the western side of the fountain takes the water from above to the edge of a monolithic carved granite stone. This stone leans slightly inward towards the basin in such a manner that the water could fall freely from the spout without adhering to the surface of the stone—the free fall of the water in a thin stream allowing the collection of water in ceramic vessels.

In its south and east walls are small trapezoidal niches. The basin of the fountain is quadrangular, bordered by three large stones placed on the level, with no stone base.

The water was emptied through a corner of the basin towards the exterior drainage channel. The fountain was built entirely of carved granite stones, which were well fitted with a few wedges and joined by a fine clay mortar. It is in an ordinary state of preservation, although the side walls have lost their original inclination and are in danger of collapse, for which reason it is urgent to carry out restoration. This fountain is adjacent to the east flank trail.

Fountain 4

Fountain 4 is located immediately above Fountain 3. Both are hydraulically linked through the flow of the water coming from a common spring (Figure 3).

This fountain is slightly larger than Fountain 3. It is rectangular and bordered on its west side by a small terrace that is a viewing platform. The south wall has a small niche (about 28 centimeters [cm]), which is above a drainage outlet. On the north is a doorway, and the east wall has two niches in trapezoidal form. Spring water is collected and carried to the fountain by means of a narrow channel running along the south side of the upper terrace.

On the west side of the fountain is a large carved stone with a small channel carved in it through which the water falls to the quadrangular basin below. In the base of the fountain floor is a small channel that conducts the water to Fountain 3 on the level immediately below.

There is a narrow doorway in an average state of preservation next to the stairway serving the fountains. The north wall is in a poor state of conservation and is about 50 percent collapsed. The east wall has a niche and is also starting to collapse at the upper part.

Near the southwest corner is a drainage channel about 10 cm wide that served to control the volume of water used in the fountains. Its walls were constructed with some care with stones joined with clay mortar.

The workmanship in the construction of the fountain was average for Machu Picchu, but its stone components were worked with care and joined together with clay mortar.

The small viewing platform is above and west of the fountains. It was covered by a layer of surface humus of a dark brown color, branches and dry leaves, stones that had fallen from the adjacent walls and a large sized stone that had fallen from above. The excavation revealed the following:

> Near the northeast edge we found a green diorite stone instrument (10 by 8 cm), which possibly was a tool used by those who worked there. In the northeast corner we found one step from the stairway for access to the platform.

In the southeast corner at a depth of 0.15 cm we found a channel that goes along the whole east profile with an average width of 10 cm. Next to the southeast corner we found the spring that furnishes the water for the channel, which is connected to Fountains 3 and 4 and with a drain on the east side that crosses the retaining wall.

Fountain 5

The Fountain 5 area includes a water tunnel, terraces and a walled cave, as shown on Figure 14.

The presence of large rocks in the middle of the fountain area made the excavation difficult. These rocks came from a rockslide from the hillside above, likely from the vicinity of the pass where the trail to Huayna Picchu Mountain starts. The excavation revealed three layers as follows.

> Layer I is made up of large, medium and small granite stones that cover nearly all the surface of the area. Removal of this layer was very complex because of the enormous dimensions of the stones and for safety reasons, since loose stones could produce major rockslides below. Nevertheless, we were able to penetrate below some of the stones and observed a small terrace in the middle of the structure, which forms part of this collapsed structure along which the duct continues to the terrace immediately below.

> Layer II is made up of topsoil, rotted vegetal material and loose soil that contains some small stones.

> Layer III is made up of lightly compacted brown soil directly covering the surface of the excavated area. There, next to a large rock, the skull of a rodent was found, seemingly crushed by the rockslide. This skull was well preserved and was a greenish color.

In regard to the rodent skull, George Eaton reported that he found bones belonging to at least eight rodents in the burial caves and trash heaps he excavated in Machu Picchu (Eaton 1916). Later he added that of all this material, he selected a skull from the genus Agouti, found in an average state of preservation from a small trash heap generated from kitchen waste, on the eastern edge of the urban center of Machu Picchu in the upper part of the eastern slope terraces. Eaton states that ". . . it is different than all of the species described up to this time under the genus Agouti. Therefore, I have selected this skull (C.O. 3227) as the type of the new species, which I have the honor of naming *Agouti Thomasii* in honor of Oldfield Thomas of the British Museum (Natural History)." (Eaton 1916)

In the Fountain 5 excavation, we also found a skull from this new type of rodent described for the first time by Eaton. The skull we found was deteriorated at the back and the lower mandible was not found; nevertheless, the front part was in good condition. It belonged to an adult animal with a great deal of wear and tear on its eight teeth (four at each side of the upper mandible). We believe the study of this animal is of great interest to the natural sciences, because this unusual animal may be in the process of extinction due to the fires that have gravely affected the ecological equilibrium of Machu Picchu.

The water tunnel is large and was built beneath the terrace reaching to the face of a large rock where the source of the spring water is located. It has a transversal quadrangular cross section and was built on a base of large stones, with enormous lintels in its upper part.

The water flow is irregular. In September 1999, the spring was completely dry, while in 1998 in the same season, about 40 L/min of water were flowing. In 1998, water also poured out through a corner and upper part of the terrace located immediately below the water tunnel.

Immediately above the water tunnel and fountains is a small walled cave. Exploration allowed us to confirm that the rock serving as its base is the same as that found in the bottom of the water tunnel where the spring flows out. The tunnel is formed by a thick wall of granite stones with fitted joints

Fountain 6

Fountain 6 is situated downhill near the river at Station 0+75. It was not excavated. This fountain was flowing. It appeared to be a simple structure for use of a natural water source to provide a routine drinking water supply for workers and travelers. The stonework has deteriorated.

Hydraulic Innovation

Three notable features represent the hydraulic characteristics and care associated with the east flank Inca Trail:

a. The drainage channel and conduit south of Fountains 3 and 4 represent special engineering care taken to manage trail drainage near the midpoint of the Inca Trail.

b. The water tunnel of cut stones at Fountain 5 demonstrates special technology used in an effort to improve the reliability of the groundwater supply. This is exemplified by the presence of stone walls that form the tunnel with a flat cut slab roof. The height of the downstream portion of the water tunnel is estimated at 1.6 m. Unfortunately, a rockfall has severely damaged the tunnel and has destroyed the fountain leaving only a walled cave, tunnel remains, terraces and the damaged Inca Trail.

c. Most notable are the two approach channels for Fountains 3 and 4. Both approach channels were designed using the Venturi principle to improve flow characteristics of the water jet during low flow conditions. In Fountain 4, the width of the channel decreases from 6.0 to 4.0 cm over a length of 20 cm, serving as a form of nozzle that causes the water velocity to increase, which relates to a basic hydraulic formula of velocity = discharge ÷ area. Just upstream of the Fountain 4 approach is a channel with a unique bifurcation built inside the wall so that excess flow is diverted to a wasteway that discharges into an adjacent drainage channel. During times of excess flow, the bifurcation limits the amount of water into the Venturi channel for optimum operation and water jetting. The operating fountain is shown in Figure 15.

Water Quality

The water quality of the fountains along the east flank trail was determined by laboratory testing to be good, as shown in Table 2. The low dissolved solids that range from 36 to 56 milligrams per liter (mg/L) represent high quality water as would be expected close to the recharge area of the granite bedrock

TABLE 2
East Flank Water Quality Sample Results from September 26, 1998

Parameters Tested	Units	Fountain 1	Fountains 3 & 4	Fountain 6
pH	---	7.0±	7.0±	---
Flow	L/min	1-10 L/min	8 L/min	10 L/min
Inorganics				
Total Dissolved Solids	mg/L	36.0	56.0	24
Total Alkalinity	mg/L	9	16.2	NT[1]
Chloride	mg/L	<0.25	0.41	NT[1]
Sulfate	mg/L	<0.25	0.49	NT[1]
Dissolved Metals				
Sodium	mg/L	2.6	3.1	NT[1]
Potassium	mg/L	U[2]	0.98	NT[1]
Calcium	mg/L	1.4	3.0	NT[1]
Magnesium	mg/L	0.35	1.3	NT[1]

[1]NT = not tested.
[2]U = not detected at the reporting limit.

River Hydraulics at Landing

The documented lower terminus of the Inca Trail (Figure 16) is approximately 1.2 km upstream of Mandor Pampa and about 1.5 km downstream from the bridge at the abandoned Ruinas railroad station at a location where crossing on foot can be made during periods of low flow of the Urubamba River.

The river discharge on September 26, 1999 was estimated to be 72 cubic meters per second (m³/sec). Based on observed high water lines, the 1998-99 peak flow was estimated at 230 m³/sec (Figure 17). During the peak flow period of 1998-99, the original Inca wall was overtopped by the river water level by approximately 10 cm (Figure 16).

The primary natural features of the Inca river landing are as follows:

a. A large rock and earth debris deposit from the Huayna Picchu gully forms the left bank cove.

b. A boulder bar deposit at the base of the Inca Trail stairway provides a large flat landing during most of the year.

c. A steep downstream river slope, terminating in several cascades, creates supercritical flow velocities and a dampening of upstream water level variations, particularly during high flow periods.

d. The large rocks deposited at the base of the gully form a hydraulic control section immediately downstream of the landing. The control section provides benefits as follows:

- It limits the magnitude of river water level fluctuations.

- It creates a pooling phenomenon upstream.

e. A vertical wall cliff immediately upstream of the Inca landing helps to direct flow straight downstream by providing a deflection wall for the river currents. The vertical wall also provided for improved Inca Trail security control.

Trail Stability

The use of terraces for erosion protection and trail stability was evident in Conjuntos 4 and 6. The successful system of terracing at Conjunto 4 is illustrated in Figure 18, which is a view from the top of Huayna Picchu.

While not as concentrated as in Conjunto 4, terracing in Conjunto 6 is likely nearly continuous from the lowest gully to Cave 3. Here, the steep slope at the base of the Huayna Picchu cliff created a confined location for the trail with the cliff on the north and the steep drop off to the river on the south.

Trail Summary Description

The east flank trail is a magnificent work of engineering and workmanship performed with a high standard of care. It is a continuous whole; however, for presenting a trail summary description, it is divided into six parts as described below.

The reader should note that the description begins at Machu Picchu, although the survey starts at the river and proceeds uphill according to standard mapping practice. This is done to enable the reader to begin at a well-known point of reference. Therefore, the stationing for locations is from 14+50 at the qolqas to 0+00 at the river (Figure 8).

The First Section (Qolqas at 14+50 to the Inca House at 13+00)

Between 14+50 and 13+50, most of the trail is well known and long established as shown on Figure 2. The long and steep granite staircase is sturdy and well constructed. This first section goes to the Inca security station at Station 13+00 that is preserved only to a height of 0.5 m. This enclosure (Figure 19) has a somewhat irregular rectangular plan with a length of 5.0 m and a width of 2.7 m. The walls are 0.50 m wide. It has a rock for support of its back wall where there also is a cave. There is no evidence of a front wall, so likely it was a wayrona such as those seen along all the Inca trails between Huayllabamba and Machu Picchu. A wayrona is a rectangular building open on one of its long sides. There is no evidence of any niches or windows, probably due to its poor state of preservation. This small enclosure is adjacent to and above the trail. In 1998 we found a stone peg nearby as evidence of a gable roof, as well as a short stretch of stairway, which indicates that the original trail passed below this building.

The stone elements of the long granite staircase are granite blocks in an irregular parallelepiped form. Each step consists of three to five carefully placed stones— sometimes a large stone and another smaller one together. Steps vary in accordance with the slope of the terrain. There are wide ones from 35 to 40 cm with a similar height and with very narrow surfaces where, at times, one can only place a foot sideways. One also encounters short ramps or inclined planes.

The function of this section is to connect the trail with the urban center of Machu Picchu and for control of access by the Inca security station previously described.

The Second Section (13+00 to 10+88)

Between the Inca security station and the stairway is the beginning of a short secondary trail to a small conjunto of terraces, which we called 3A. This small conjunto of terraces consists of some 12 terraces connected by means of "sarutas," or long stones cantilevered out from the walls of each one of the terraces (sometimes called "flying stairways"). At the foot of the lower terrace is an old pathway that joins the exterior outer wall of the lower east flank Conjunto 1 with Conjuntos 3 and 4, which were explored in previous years.

Down the primary trail from the Inca security station is a spiral staircase (12+75 to 12+50) with four turns (Figure 20). Then, at 11+50, there is a long rock paved ramp with a buttress wall on the north side. A stairway we named the Grand Staircase has 10 steps, is located at 10+92 and is 4 m long, 1.40 m high and 1.78 m wide. The steps have an average height of 0.30 ms, and the stairway is made up entirely of large stone elements, nearly monolithic. The large blocks are well set, and the stones have been fitted using a clay mortar, which has almost disappeared through erosion.

The Third Section (10+88 to 8+60)

Section 3 of the trail goes from the Grand Staircase to Conjunto 4. After coming down the Grand Staircase, one arrives at a large ramp at 10+47. It is constructed with a retaining wall with flat well-seated rocks. This inclined plane is in a good state of preservation and represents good Inca engineering.

In the lower part of the ramp there is an old rockfall that has caused serious destruction of part of the trail. One also sees enormous rocks covered with vegetation of lianas and evidence of broken trees, which indicate that this rockfall occurred some time during the 1980s. A narrow gully follows the rockfall. At 10+11 is a branch trail to the west that represents a possible connection to the area of the Sacred Rock in Machu Picchu; however, the branch trail remains unexplored.

At 9+72 begins a unique "S" staircase to 9+55 that is laid out with engineered curves and individual steps carefully positioned to form smooth curves (Figure 21). From this point, the trail descends in a straight manner via a narrow stairway in a poor state of preservation until it arrives at the site of a group of enormous terraces that form the central part of Conjunto 4 at Station 8+61.

Several meters above Conjunto 4 (at 9+00) on the north side, there is an excavation 2.50 by 2.50 m with a depth of 1 m. It is difficult to identify its cultural affiliation. It could be an incomplete Inca work or an excavation by the Bingham party from 1912.

The Fourth Section (8+60 to 7+28)

This fourth section extends between Station 8+60 at Conjunto 4, passing by Fountains 3 and 4 to Conjunto 5. It is an extraordinary example of Inca engineering and the best-preserved section of the whole trail. This trail extends along the upper west side of the terraces of this conjunto following a pathway approximately from east to west. The trail runs between two extensive terraces, the trail having an average width of 2 m with steps some 2.80 m wide. It connects Fountains 3 and 4 to Fountain 5 found at the west end of Conjunto 4.

One of the stairways on this trail segment has nine steps—the upper one being 3.3 m wide with a height of .25 cm. It is made of carefully placed stones. For example, the lower step has a tread of 0.40 cm, a height of 0.26 cm and consists of five well-placed stones. This section of the trail can be described as magnificent with three fountains and a viewing platform with a fine view (Figure 12).

The Fifth Section (7+28 to 3+40)

This section extends between Fountain 5 at the north end of Conjunto 4 at Station 7+28 to Cave 3 of Conjunto 5 at Station 3+40. It connects Conjunto 4 with Conjunto 5. At Station 7+00, twin retaining walls some 19 m long begins, followed by 93 m of survey line where the trail evidence was lost. At Station 5+87, the trail is again evident with 10 staggered boulder steps, and at Station 5+76 twin retaining walls again appears above the trail. Seven meters further downhill commences a 38-m-long section of random boulder placement at Station 5+60.

Uphill of the survey line at Station 5+10 is a rockslide that is considered hazardous. Beyond the rockslide and after a 2-m-high wall, one comes to Cave 2 at Station 4+60. This cave has a trapezoidal floor plan, and three of its interior sides have been almost entirely covered in fine masonry. One also sees small niches and several loose carved stones on the interior. Some sections of the wall have been destroyed. From Cave 2, the survey route descends the steep hillside north of the terrace Conjunto 5 that Dr. Valencia documented in 1969.

At Station 4+27, the survey route and the trail take a sharp right angle bend to the north, where they lead to the cathedral-like Cave 3 at Station 3+40 (Figure 22). Cave 3 has a rectangular floor plan and is limited along the east by means of some terraces and along the south by a conjunto of medium-sized rocks; along the north and west sides one sees the granite massif fissure, which is likely a geologic fault. Along its interior wall, there are two large carved rocks placed in a manner of large seats. One of them has a stepped form on one of its sides; the other is rectangular. From here, one has a marvelous view of the narrow Urubamba canyon, the base of Putucusi Mountain and part of Mandor Pampa. A highlight of the Cave 3 area is a curved wall that is 2 m high with a near perfect radius of 21 m (Figure 23). The wall was not completely cleared and, therefore, the extent of the wall is not known in its entirety.

The Sixth Section (3+40 to 00+00)

Downhill from Cave 3 are five terrace walls commencing at Station 2+80. The walls are well constructed and in good condition. This is an impressive set of terraces that we named Conjunto 6. It is believed there are many terraces not yet cleared and surveyed that lie under the forest.

The trail lies at the base of the Huayna Picchu cliff for some 120 m until Station 1+50 where there are additional terraces and a 15-m-long earth ramp that leads to an elaborate area of construction south of a small Huayna Picchu gully.

The complex of terraces, Fountain 6 and an in-situ monolithic carved stairway lie between Stations 1+30 and 0+50 (Figure 24). The monolithic stairway was completely covered with vegetation and soil. After clearing work, we verified that it has 14 steps. This portion of the trail is one of the most fascinating because of the challenges faced by the Inca engineers and the carving of the trail into the granite bedrock.

Finally, at the base of the trail (Station 0+00) are the remains of a stairway and wall that have been subject to the ravages of time and the forces of the Urubamba River during periods of flood (Figure 16). It is likely that the trail continued downstream along the left bank of the Urubamba River several hundred yards to a suitable Inca bridge crossing site; however, there are no visible remains.

MANDOR PAMPA AND THE ARTEAGA DWELLING PLACE

On September 11, 1999, the banks of the Urubamba River were explored downstream of the trail landing as far as Mandor Pampa. Mandor Pampa represents a valley bottom agricultural area from which foodstuffs would have been carried on the east flank trail and, therefore, it was important to our engineering explorations. Present residents of Mandor Pampa, Lorenzo Alagon Santos and his wife, Angelica Jauregui de Alagon, were interviewed. They asked one of their farm workers to take us to "Arteaga's house." Hiram Bingham said that in 1911 he passed by a hut with a straw roof, whose owner was Melchor Arteaga. This hut, in its turn, was an inn for the occasional travelers who passed by the site. Bingham said that "At dawn on July 24, 1911 there was an icy rain...when we asked him where the ruins were, he pointed directly toward the top of the mountain...I left the tent at 10:00 in the morning." (Bingham 1930)

The Arteaga dwelling place at present is a ruin. Its floor plan is in the form of a rectangle with two doorways. It is 9.9 m long and 5.2 m wide with a preserved height of 1 m. The wall has an average width of 70 cm and has sloped walls. The dimensions of the building, fragments of pre-Hispanic ceramics, finely worked and carved stones and a stone mortar led to the conclusion that Arteaga's house was an Inca building (Figure 25). Mandor Pampa today is rundown, but it still has many avocado plants; our pollen testing adjacent to the east flank trail showed evidence of avocado pollen from Inca times.

CONCLUSIONS

The Inca trail on the east flank of Machu Picchu served a vital communication and economic function. The trail discovered by the 1998-99 archaeological exploration is the continuation of the Inca Trail that joined Cusco with Machu Picchu and which then continued to Mandor Pampa, joining with the Inca Trail on the right bank of the Urubamba River in such a way that one could continue on downriver as far as the deepest part of the Amazon jungle. Likewise, one could return to the upper part of the valley as far as Cusichaca, Ollantaytambo or Cusco by another route.

The east flank trial which joins Conjunto 11 (qolqas) in Machu Picchu with the left bank of the Urubamba River has several branches like the one leaving from Conjunto 4, which went toward Conjuntos 1, 2 and 3 and the gate entrance, which faced the trail on the east flank. One also sees other small branches that connected, by means of stairways, with several sectors of each of the conjuntos of terraces.

This east flank trail filled an important economic function by connecting the six large conjuntos of agricultural terraces and other smaller ones, assuring the cultivation and harvest of useful food plants such as corn, potatoes and likely beans. It also assured the provision of water by means of five springs along the main trail and its branch trails located in Conjuntos 1, 3, 4 and 6 and by providing easy direct access to the water in the river below.

The trail also fulfilled a ceremonial and sacred role since it joined the Machu Picchu urban sector with two burial groups. The first group was located northeast of the point where the east flank trail begins at the stairway to the qolqas, where Hiram Bingham found many tombs. Dr. Bingham found the second group to the southeast. The caves, which are found along the trail, may be the structures of some of these tombs.

Traversing the trail from Machu Picchu to the river provides striking examples of prehistoric civil engineering that ranged from extraordinary geotechnical slope stabilization to granite trail stairways, all having withstood the ravages of time, heavy precipitation and steep slopes. The water supply engineering provided strategically situated fountains for the workers and weary travelers to refresh themselves. Environmental engineering incorporated unsurpassed scenic views. Planning included the placement of check stations and lookouts where the Inca military forces could clearly view key portions of the trail to ensure security. Road building by the Inca civil engineers encompassed all phases of a complete transportation infrastructure to support a far-flung empire that was connected by roads for the purpose of moving goods, communication by means of runners and the movement of military personnel. Strategically situated food storehouses ensured adequate nutrition for them.

The newly discovered east flank Inca Trail of Machu Picchu provides a good example for study of Inca technology because it lay buried and untouched for nearly five centuries. A prehistoric ruin, untouched by humans since its abandonment, provides and unparalleled opportunity for engineers to examine ancient technology. Our research demonstrates that Professor Hiram Bingham was correct in 1913 when he stated that the Inca were good engineers.

REFERENCES

Bingham, H. (1913). "In the Wonderland of Peru." *National Geographic Magazine,* April 23: 387-574.

Bingham, H. (1930). *Machu Picchu, a Citadel of the Incas.* Report of the Explorations and Excavations Made in 1911, 1912 and 1915 Under the Auspices of Yale University and the National Geographic Society. New Haven, CT: Yale University Press.

Eaton, George F. (1916). *The Collection of Osteological Material from Machu Picchu.* Memoirs of the Connecticut Academy of Arts and Sciences, Volume V, May.

Moseley, M. E. (1992). *The Incas and Their Ancestors: The Archaeology of Peru.* London: Thames and Hudson.

Rowe, J. H. (1990). "Machu Picchu a la Luz de Documentos de Siglo XVI." *Historica*, 14(1) 139-154.

Valencia Zegarra, A. and A. Gibaja Oviedo (1992). *Machu Picchu, La investigación y Conservación del Monumento Arqueológico despuès de Hiram Bingham.* Cusco, Peru: Municipalidad del Qosqo.

Wright, K. R. (1996). "The Unseen Machu Picchu: A Study by Modern Engineers." *South American Explorer,* 46(Winter) 4-16.

Wright, K. R., G. D. Witt and A. Valencia Zegarra. 1997b. "Hydrogeology and Paleohydrology of Ancient Machu Picchu." *Ground Water*, 35(4) 660-666.

Wright, K. R. and A. Valencia Zegarra (2000). *Machu Picchu: A Civil Engineering Marvel.* Washington, DC: ASCE Press.

Wright, K. R., A. Valencia Zegarra and C. M. Crowley (2000). *Archaeological Exploration Of The Inca Trail, East Flank Of Machu Picchu and Palynology Of Terraces.* Denver, CO: Wright Paleohydrological Institute.

Figure 1. The civil engineering genius of the prehistoric Inca Empire resulted in the building of Machu Picchu in only 90 years. This *National Geographic Magazine* (May 2002) photograph demonstrates that the work of the Inca engineers has endured, under harsh environmental conditions, for nearly five centuries.

Figure 2. Until 1999, this long granite staircase led to nowhere except impenetrable thick forest. Now, it is known to have connected to the Inca Trail. The terraces in the foreground were found to contain pollen of maize, avocados and medicinal plants.

Figure 3. Fountain 4 was buried for nearly five centuries, but when it was excavated and the channels cleaned, it burst into operation attesting to the excellent work of the Inca hydraulic engineers. A common spring serves both hydraulic structures.

Figure 4. The newly discovered Inca Trail on a steep 2:1 mountain slope was built to last for the ages by careful stabilization work that included a good foundation and protective terraces both uphill and downhill.

Figure 5. The Urubamba River route provided access to the Inca Trail river landing, but its fast flowing currents and downstream cascades proved too hazardous for crossing after the first day of exploration.

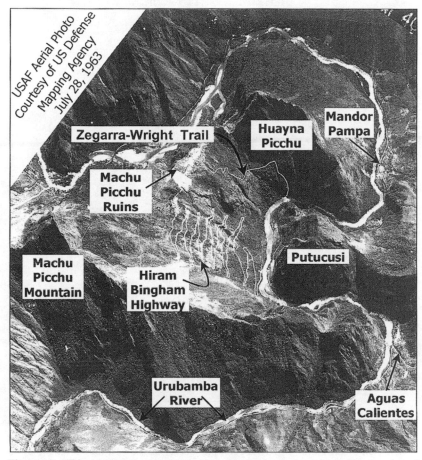

Figure 6. This 1963 U.S. Air Force photograph shows the Machu Picchu rugged setting, nearly surrounded by the Urubamba River. The newly discovered Inca Trail route is shown in the upper portion of the photograph.

THE PERUVIAN EXPEDITION OF 1912
UNDER THE AUSPICES OF
YALE UNIVERSITY & THE NATIONAL GEOGRAPHIC SOCIETY
HIRAM BINGHAM, DIRECTOR
MACHU PICCHU & VICINITY

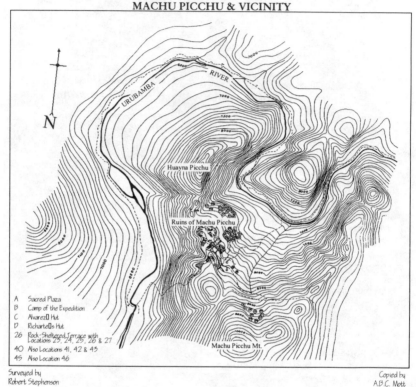

A Sacred Plaza
B Camp of the Expedition
C Alvarez⬚ Hut
D Richarte⬚s Hut
26 Rock-Sheltered Terrace with
    Locations 23, 24, 25, 26 & 27
40 Also Locations 41, 42 & 43
45 Also Location 46

Surveyed by
Robert Stephenson

Copied by
A.B.C. Mott

Figure 7. The 1912 survey map of the Machu Picchu area by Civil Engineer Robert Stephenson of the Hiram Bingham party still serves as the best map available of the environs of Machu Picchu.

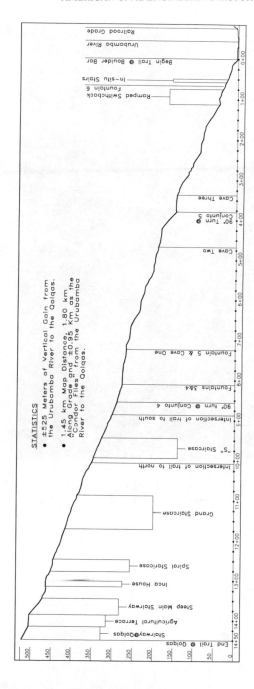

Figure 8. The profile of the east flank extension of the Inca Trail is shown rising 525 meters over a horizontal distance of 1,450 meters. Features a ong the trail are described along with their location. The survey was performed using a Brunton compass and tape.

Figure 9.   The prehistoric stairways of the Inca Trail provide testimony to the great capabilities of the Inca civil engineers to build for the ages. Here, Coauthor Ruth Wright is shown at Station 8+26 near Fountains 3 and 4.

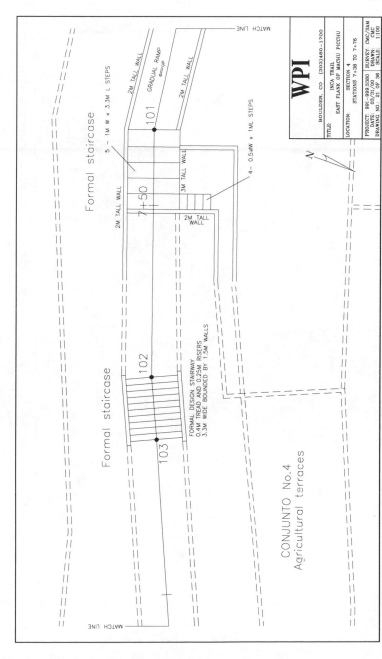

Figure 10. The Inca Trail details demonstrate the high standard of care used by the prehistoric civil engineers. The 3.3-meter-wide stairway on the right is complimented with adjoining ramps, well built stone walls and, to the left, a wide staircase with 11 stairs having uniform risers and treads.

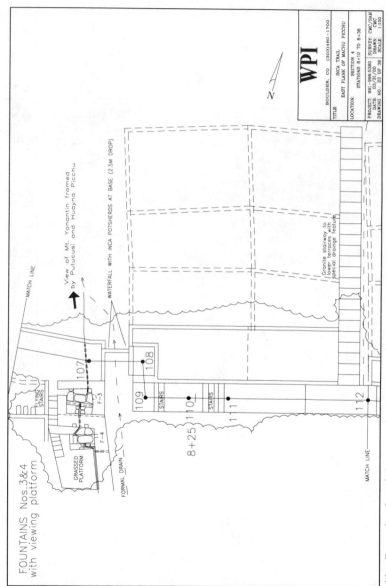

Figure 11. Fountains 3 and 4 were enhanced with a grassed viewing platform that provided a breathtaking view of three holy mountains and the river below. Inca pottery was found at the base of the formal drain.

Figure 12. With Mount Yanantin in the center, Putucusi on the right and Huayna Picchu on the left, the Inca civil engineers provided the traveler with a spectacular view from Fountains 3 and 4. Their attention to environmental design is legendary.

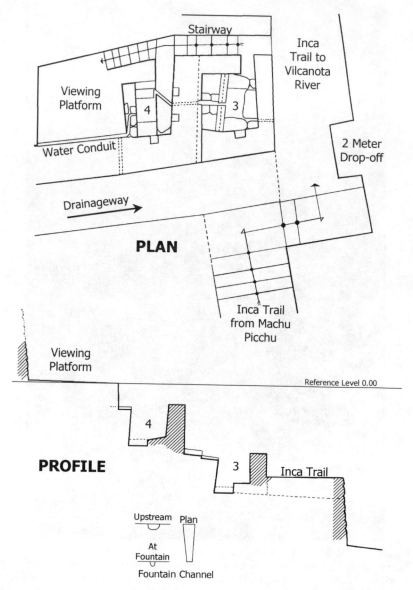

Figure 13. The archaeological excavation of Fountains 3 and 4 was documented by the authors for permanent record and use by Andean scholars. These two fountains were ceremonial in design. The fountains are in operation as of November 2002.

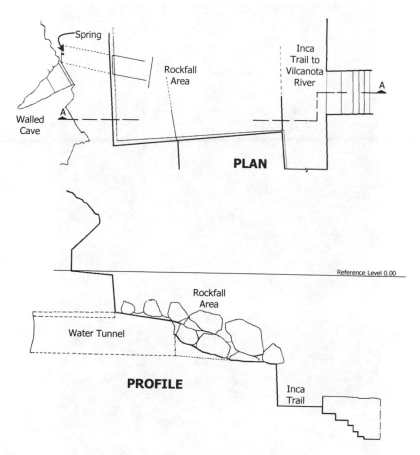

Figure 14. The unusual Fountain 5 area showed special Inca building care and innovative engineering. A water tunnel formed with shaped stone intercepted groundwater and a walled cave helped create stability above. Exploration of Fountain 5 was terminated when poisonous snakes let it be known that they already occupied the site.

Figure 15. The Inca civil engineers accounted for periodic low spring flow by carving the fountain's (Fountains 3 and 4) channel nozzles with a Venturi-like shape to increase the exit velocity. Here, Ruth Wright points to the special hydraulic shape.

Figure 16. The remarkable remains of the Inca wall at the river landing on the left bank of the Urubamba River has withstood some five centuries of flood waters. Here, civil engineer Chris Crowley shows the elevation of the 1998-99 high water level some 10 centimeters above the apparent top of wall.

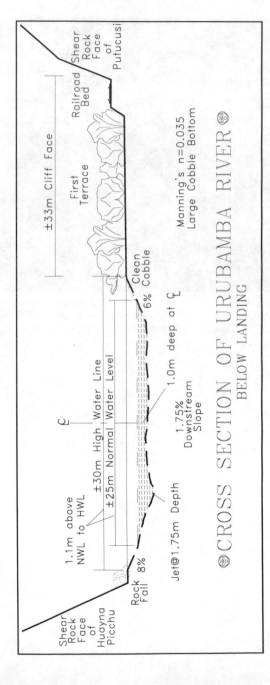

Figure 17. A cross section of the Urubamba River at the Inca Trail landing enabled the authors to estimate the 1998-99 peak flow at 230 centimeters. This tributary to the Amazon River is steep and fast flowing with high sediment transport capability.

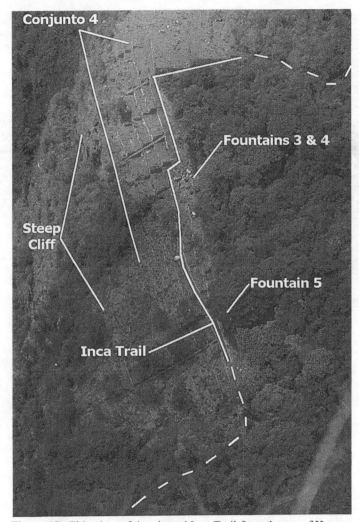

Figure 18. This view of the cleared Inca Trail from the top of Huayna Picchu Mountain shows Fountains 3, 4 and 5 along with the slope stabilization terraces lying just to the right of the steep cliff that falls off to the river below.

Figure 19. An Inca security station near the upper long granite staircase was found in 1997; however, the trail route was still unknown. The building is 5.0 x 2.7 meters in size and was likely a three walled enclosure with an open front.

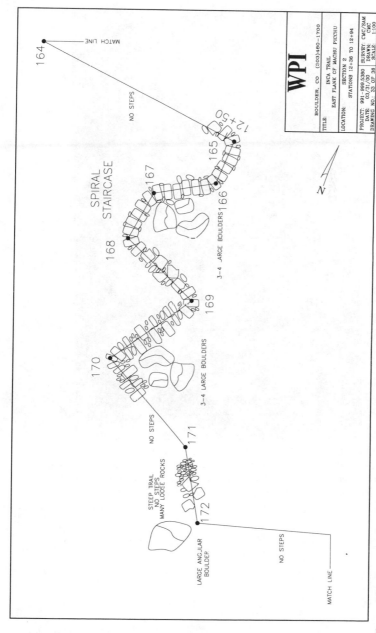

Figure 20. A masterpiece of stairway construction was discovered at Station 12+50. The spiral stairway has four turns with steps that are granite. Climbing this stairway for the first time in nearly five hundred years provided an insight into the quality work of the prehistoric civil engineers.

Figure 21. The "S" Staircase at Station 9+55 was laid out with engineered curves and individual granite steps that were carefully positioned to form smooth curves.   Peruvian Archaeologist Dr. Alfredo Valencia is at the far left.

Figure 22. A cathedral-like cave is situated at Station 3+40. From the granite seats, the Inca traveler had a fine view of the Urubamba valley and the mountains beyond; the same as that being enjoyed by the archaeological exploration crew shown above.

Figure 23. Near the cathedral-like cave (Figure 22) is a long curved 2-meter-high retaining wall that had a radius of 21 meters. The full extent of the wall was not determined; it was not fully explored.

Figure 24. A major treat for the civil engineering exploration team was the excavation of the in-situ monolithic carved stairway at Station 0+50. Fourteen steps were carved into the bedrock granite with a fine stone retaining wall above it.

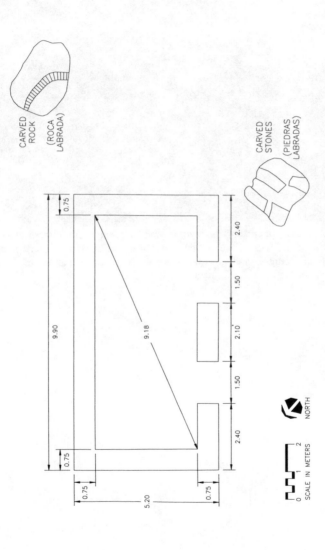

Figure 25. Along the Urubamba River downstream of the Inca Trail landing is Mandor Pampa from where Hiram Bingham first heard of Machu Picchu in 1912 from Melchor Arteaga, a local Quechua Indian. Arteaga's dwelling place was an Inca building ruin with a size of 5.2 x 9.9 meters. Had it not been for Arteaga, Bingham would likely not have found Machu Picchu. Mandor Pampa was an Inca agricultural area to support Machu Picchu with foodstuffs.

Benjamin Wright, Bad Stone, Poor Cement and One Hundred Miles to Go: Building the Monocacy Aqueduct of the Chesapeake and Ohio Canal

Robert J. Kapsch, Ph.D., M.ASCE[1]

Abstract

The Chesapeake and Ohio Canal was constructed between Washington, D.C. and Cumberland, Maryland from 1828 until 1850. This construction was an early marker of the beginning of American civil engineering. The Chief Engineer of the Chesapeake and Ohio Canal was Benjamin Wright, the Father of American Civil Engineering. The legislative charter for the new canal company stipulated that 100 miles of the canal had to be constructed in the first five years. There were two obstacles for the company to meet this mandate: the legal blockade put in its way by the Baltimore and Ohio Railroad and the construction of the Monocacy Aqueduct, one of the largest bridges built in the United States at that time. Wright couldn't do anything about the legal battles of the company but he could about the Monocacy Aqueduct. He designed the Monocacy Aqueduct in the autumn of 1828, only a few months after he started working for the company. Construction began the next year. Wright visited the construction site in July 1830. What he found was that the first three piers had been constructed of inferior stone. This inferior stone had not been detected by Wright's resident engineer, Herman Böye, as Böye was on his deathbed in Georgetown. Nor had it been detected by the Assistant Engineer, Charles Ellet. Ellet would go on to a distinguished engineering career but in 1830 was only nineteen years old and inexperienced. The bad stone should have been detected by Wright's supervisor of masonry, Robert Leckie, but wasn't. Leckie would resign a short time after. Wright ordered the piers torn down and rebuilt. Two of the five years to build the Monocacy Aqueduct had thus been lost. More problems hindered construction. The first two contractors, Hovey and Osborne, walked away from the project. Hydraulic cement was in desperate short supply and efforts to develop alternative sources proved inadequate. These and other problems were overcome by the Chesapeake and Ohio Canal and the Monocacy Aqueduct was opened in the required five years. The canal company thus met the requirements of their legislative charter. On seeing the aqueduct, Robert McFarland, the Superintendent of Masonry, wrote, "...in point of beauty and perfection of workmanship, (it) will ... compare with any work of the kind either in this Country or in Europe..."

---

[1] Senior Scholar in Historic Architecture and Historic Engineering, National Park Service, 1100 Ohio Drive, S.W., Washington, D.C. 20242; telephone 202-619-6370; Robert_Kapsch@nps.gov.

I.   The Beginning of the Chesapeake and Ohio Canal (1828)

The Chesapeake and Ohio Canal began with the demise of the Potomac Canal.[1] The Potomac Canal was a river navigation system with bypass canals around the major Potomac River obstacles (Great Falls with a drop of 76 feet; Little Falls with a drop of 37 feet). It was begun by George Washington in 1785 and relied principally on sluices dug or blasted out of the river bed to provide adequate river navigation. By 1823 the Potomac Canal was widely viewed as a failure. It was a financial failure in that it only paid one dividend to its investors; it was a legal failure in that it did not meet the legal requirements of the enabling legislation passed by the legislatures of Maryland and Virginia; and it was an engineering failure in that it only provided satisfactory navigation 33 to 45 days a year and because many of the "improvements" made by the Potomac Company to the river bed to improve navigation actually made the river more difficult and dangerous to the navigation that it was intended to help.

Instead of a river navigation system, it was decided that a stillwater canal would be needed to provide water transportation between tidewater, at Georgetown in the District of Columbia, and Cumberland, Maryland – 184 miles distant. The promoters of this stillwater canal envisioned the canal continuing from Cumberland over the mountains to the Ohio River. The difference between a stillwater canal and a river navigation system was that a stillwater canal didn't use the river for transportation. Instead, the canal would be constructed parallel but separate from the river. Low head dams of five or so feet across the Potomac River would divert water into the stillwater canal via feeder locks. Instead of dangerous sluices in the river bed, the stillwater canal would rely on lift locks, hydraulic devices that could lift 20 ton canal boats eight feet in elevation in three minutes, sometime less. Canal boats would be propelled by mules or horses on placid "levels" – areas of smooth water between locks. Stillwater canals were much better than river navigation systems. As Isaac Briggs, then the state engineer for the Board of Public Works of the State of Virginia wrote:

> Of all modes of internal navigation, this is the most perfect, is liable to the fewest interruptions and afford a transportation cheaper than any other mode of inland conveyance yet discovered.[2]

A stillwater canal required detailed planning and design. This required a detailed engineering study for the length of the proposed canal. The Chesapeake and Ohio Canal had not one but two such studies. The first of these studies was undertaken by the U.S. Army Topological Engineers and submitted in 1826 through the Board of Engineers and the Secretary of War.[3] This report recommended that the Eastern Section of the proposed Chesapeake and Ohio Canal (the section between Georgetown in the District of Columbia and Cumberland, Maryland) be routed up the Potomac valley on the Maryland side of the river as, "the obstacles are generally of less magnitude than on the Virginia side; the exposure is more favorable, and will cause in the Spring an earlier, in the Fall a later, navigation; no aqueduct will become necessary at Cumberland ..."[4]

The report by the U.S. Topological Engineers found that the proposed Chesapeake and Ohio Canal was feasible but expensive – over $8 million for the eastern section, from Georgetown to Cumberland.[5] The supporters of the proposed Chesapeake and Ohio Canal called for a second study, to be undertaken by civilians trained on the Erie Canal.

In 1828, civil engineers James Geddes and Nathan Roberts submitted their study of the proposed canal.[6] Like the earlier study by the U.S. Topological Engineers, they located the canal on the Maryland side of the river. Unlike the earlier study, they estimated that only $4 million would be required to construct the canal from Georgetown to Cumberland – one half of the earlier study.[7]

The proposed stillwater canal would have to cross a number of rivers and streams entering the Potomac. The largest such crossing would be the Monocacy River on the border of Frederick and Montgomery Counties, Maryland. Here the Monocacy was in excess of 500 feet in width. Here Geddes and Roberts estimated that the aqueduct needed to be only 133 feet long, and constructed of a wood trunk supported by stone abutments and piers for a total cost of $20,000.[8]

Geddes' and Roberts' report was enthusiastically received by the supporters of the Chesapeake and Ohio Canal. It was the information the promoters of the canal needed to enact the necessary enabling legislation. The Chesapeake and Ohio Canal Company was brought into existence.

II.   Beginning the Work (1828)

Congressman Charles Fenton Mercer was elected President of the new Chesapeake and Ohio Canal. One of his first acts was to invite Benjamin Wright to Washington to become the Chief Engineer of the new canal. Wright was paid $5,000 per year.[9]

> A gentleman of great reputation who had been pre-eminent in the valuable corps of practical Civil Engineers formed by the New York Canals, had been earnestly, though informally invited to Washington, … for the organization of the Company … which brought to Washington Judge Wright …"[10]

Wright, assisted by John Martineau, began working on the design of the canal on June 25, 1828. They worked on the design seven days a week, "suspend(ind) their labors for the 4th of July only."[11]

Ground was broken for the new canal on July 4, 1828, at Little Falls above Georgetown.[12] Wright and Martineau laid out the first seventeen miles of the canal in one half-mile sections. The first contracts were issued by the company in August 1828.[13] A second group of contracts were issued by the company, covering some additional twenty-five miles, were issued in October 1828.[14]

· ELEVATION ·

Figure 1 and 2. Monocacy Aqueduct, downstream face, Chesapeake and Ohio Canal. Constructed 1829-1833, Benjamin Wright, engineer. Largest of the eleven aqueducts on the Chesapeake and Ohio Canal. The aqueduct had an overall length of 516 feet and consisted of seven masonry arches of 54 feet span each. The legislative mandate of the Chesapeake and Ohio Canal Company required the first 100 miles of the canal be completed in the first five years. This mandate could not be fulfilled unless the Monocacy Aqueduct was complete.

Figure 1: Downstream elevation, Monocacy Aqueduct, Chesapeake and Ohio Canal National Historical Park, near Dickerson, Maryland. L. Ewald, Jr., delineator, not dated (ca. 1936). Historic American Buildings Survey, Prints and Photographs Division, Library of Congress, HABS MD-19, Sheet 1 of 2.
Figure 2: Downstream elevation, Monocacy Aqueduct, Chesapeake and Ohio Canal National Historical Park, Near Dickerson, Maryland. Jack E. Boucher, photographer, 1959. Historic American Buildings Survey, Prints and Photographs Division, Library of Congress, Prints and Photographs Division, HABS, MD,11-____,37.

Wright and the Chesapeake and Ohio Canal Company were in a hurry. The charter for the company required that the company complete 100 miles within five years of its organization. The company was confident that it could meet or exceed this requirement:

> The charter of the Company requires that one hundred miles of the canal shall be finished in five years from the organization of the Company, and the experience of several States of this Union demonstrates that, with adequate funds, the entire eastern section can be completed, as economy recommends that it should be, in the shorter period of three years from its commencement, the time which the Board had early assigned for that labor.[15]

After the first forty-two miles were placed under contract, Wright turned to the design of the Monocacy Aqueduct, the single largest and most expensive structure that needed to be built in the first 100 miles of the canal. The Monocacy Aqueduct was located at milepost 40 above Georgetown. If the Monocacy Aqueduct couldn't be completed within five years, then a large middle section of the 100 miles could not be watered and the company would not meet its legislative mandate.

III.   Designing the Monocacy Aqueduct (1828-1829)

In the autumn of 1828, Wright turned to the design of the Monocacy Aqueduct. There were several approaches Wright could have taken. One was to ferry canal boats across the Monocacy River. Many of the early American canals used locks to lock canals boats down to the river, ferry the canal boats across river, and then again used locks to lock the canal boats back up to the canal level on the other bank. This was less expensive than building an aqueduct. Another alternative, identified in the Geddes and Roberts survey, was to build the piers of masonry and the aqueduct trunk of wood. These wooden aqueducts were inexpensive and quick to build. But Wright determined that the Monocacy Aqueduct (and the Seneca Aqueduct, which he was also designing) were to be constructed of masonry. A masonry aqueduct was the most durable and long standing construction solution to cross a river but it was also the most expensive and took the longest time to build. By October 1828 Wright had completed ground plans and elevations of the proposed masonry aqueduct:

> "I have prepared ground Plans & Elevations of the Aqueducts at Seneca Creek & Monocacy River which are ready to lay before you."[16]

Wright's design for the Monocacy Aqueduct was very close to what was constructed.

> "The Plan for the Monocacy Aqueduct is drawn with a water way 19 feet wide at bottom, 20 feet at Top. The Towpath Parapet 8 feet wide and the other wall 6 feet wide. I have drawn the Plan to 7 Arches of 54 feet span each and 6 piers and two Abutments: the piers are 10 feet thick and a Pilaster at each end of the Pier projecting one or two feet & 7 feet wide."[17]

Wright described the work that was to be undertaken on the Monocacy Aqueduct:

By calculation there will be about 8,500 perches including wing walls, and the price of 6-3/4 dollars per perch I think no more than a fair price – if it is done on that solid substantial manner which I have described to Mr. Hovey. The whole of the arches are to be cut to patterns so that every stone is cut thro and thro the arch. The piers and abutments are to be cut and rusticated if required. The arch ring stone to be also rusticated two inches. The spandrel; and parapet walls and inside of the trunk to be cut and bottom flagged with cut stone if so directed or the bottom covered with well jointed plank if it shall be considered best. Coping to extend across the whole of parapet and project one foot if required.[18]

In these heady, early days of the Chesapeake and Ohio Canal Company, the Board of Directors even considered making the Monocacy Aqueduct double the width proposed by Wright which would have allowed boats to pass each other, but also would have greatly increased the cost and the time required to build it.[19] An aqueduct wide enough for a single boat, as designed by Wright, was decided upon.

By January 1829, Wright and Nathan Roberts had designed the wooden false centering that would be required to support the ring stones until the masonry arch was closed. They also prepared the quantity take offs of the lumber needed to build this centering.[20]

Wright also designed the administrative apparatus under which the Monocacy Aqueduct would be constructed. A contractor would be selected to build the aqueduct but would work under the supervision of a Resident Engineer, aided by an Assistant Resident Engineer. It would be the Resident Engineer and Assistant Resident Engineer that would oversee the contractor's work and certify the quantities of work put in place so that the contractor could be paid. Wright laid out the first fifty miles of the Chesapeake and Ohio Canal into five engineering residencies – the Monocacy Aqueduct fell under the Fifth Residency. For the Fifth Residency Herman Böye, a Danish engineer who had previously worked in Richmond on the first map of Virginia funded by the state, was selected as resident engineer.[21] Böye was a sensitive young man who very much missed his friends in Richmond and his musical pursuits. For much of the time that he was Resident Engineer of the Fifth Residency Böye was quite ill, perhaps with cholera. He would die on March 20, 1830 at age thirty-eight.

Charles Ellet Jr. was selected as the Assistant Engineer for the Fifth Residency. Ellet would become a distinguished civil engineer. He would later construct the Fairmount Park Suspension Bridge in Philadelphia and the Wheeling Suspension Bridge in Wheeling, West Virginia, as well as many other structures. But in 1829, at the outset of construction of the Monocacy Aqueduct, Ellet was only nineteen with little experience. He, like Böye, would also become ill on the banks of the Monocacy. Ellet, however, recovered.

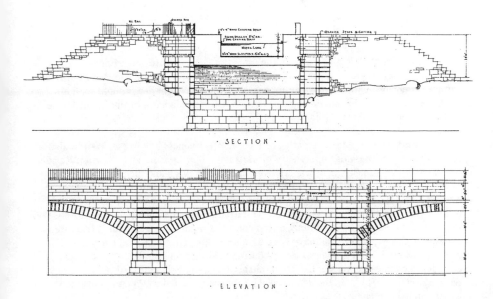

· SECTION ·

· ELEVATION ·

Figure 3. Section and elevation of a typical arch, Monocacy Aqueduct. After two years, Wright found that the stone used in the piers was inadequate and ordered them demolished and rebuilt. The Chesapeake and Ohio Canal Company thus lost two of the five years allowed for the construction

Figure 3: Typical Arch, Monocacy Aqueduct, Chesapeake and Ohio Canal National Historical Park, near Dickerson, Maryland. L. Ewald, Jr., delineator, not dated (ca. 1936). Historic American Buildings Survey, Prints and Photographs Division, Library of Congress, HABS MD-19, Sheet 2 of 2.

It was Wright's intention to have the Resident Engineers work under the direction of Division Engineers, two in number. The First Division covered all work between Georgetown to Harper's Ferry, above Monocacy Aqueduct. The five Resident Engineers reported to the First Division Director. On August 25, 1828, this position was offered to John Martineau.[22] But Martineau also became sick and "retired" before the Monocacy Aqueduct began construction.[23] He was not replaced. Wright would exercise this supervision.

To cover the Fifth Residency with the illness of Bőye, the Resident Engineer of the Third Residency, Alfred Cruger, was given additional responsibilities for the Fourth and Fifth Residencies. Cruger was a young engineer whom Wright had brought with him from the Erie Canal. In supervising three engineering residencies, beginning in January 1830, Cruger was stretched thin.[24]

Besides his engineers, greatly reduced in effectiveness by disease, Wright also established the position of Supervisory of Masonry. The Resident and Assistant Resident Engineers were not necessarily experienced in masonry work and therefore Wright established this position. Robert Leckie was selected for this position. Leckie, born in Scotland, was an experienced mason who had worked under Benjamin Latrobe at the Marble Quarries several miles below the Monocacy Aqueduct and on the Lansford Canal in South Carolina. At the beginning of the construction of the canal, Leckie had three sons die of illness, which may have been a factor in his performance on the canal. He too, was also sick for a prolonged period during the beginning of 1830.[25] Although considered competent, Wright had some misgivings on how Leckie dealt with the men working on the canal.[26] In the design of the Monocacy Aqueduct, Wright consulted with Leckie on its plan and cost estimate.[27]

The last, but most important actor in the construction of the Monocacy Aqueduct was the contractor. On October 31, 1828, the Chesapeake and Ohio Canal Company accepted a proposal from the firm of Hovey & Hitchcock for the construction of Aqueduct No. 2 across the Monocacy river.[28] This would be the first of three firms to work on the construction of the Monocacy Aqueduct.

So at the beginning of the construction of the Monocacy Aqueduct Benjamin Wright had an extensive system for superintending the work. But because of illness of the various engineers and officers of the company, this system would not work well in supervising the work done at the Monocacy Aqueduct. The breakdown in this system of supervision would be the primary reason for the later failures at the Monocacy Aqueduct construction site.

IV. Assembling the Building Materials (1829)

The construction of the Monocacy Aqueduct required two basic commodities: stone and hydraulic cement.

The closest existing freestone quarries were at Peter's quarries at Seneca, some twenty miles below the Monocacy Aqueduct. But as Robert Leckie, Superintendent of Masonry, wrote to the Board of Directors:

> ...it would be ruinous to the company to boat the stone 20 miles against the current of the river...[29]

A search was made for a source of freestone, building stone that could be readily cut, in the vicinity of the construction site. Rock outcroppings were found on the farm of Mrs. Eliza Nelson, on the slope of Sugarloaf Mountain, approximately five miles from the site of the construction of the Monocacy Aqueduct.[30] This stone seemed adequate for building the aqueduct and by April 6, 1829, the Chesapeake and Ohio Canal Company had entered into an agreement with Mrs. Nelson to use her land for a quarry for a period of one year. In return, Mrs. Nelson received $450 from the company.[31]

The company was moving fast in these early days and time was not allowed to test the adequacy of the stone found on Mrs. Nelson's farm. Company President Charles Mercer later explained what happened:

> ...in looking along the line, for materials of stone, for the masonry of our culverts & aqueducts, we had been told that a free stone quarry might possibly be opened on Mrs. Nelson's lands, at a short distance from the canal line which would be found to supply stone of a quality suited to our masonry.
> This stone shewed itself on the bow of several hills, under a heavy mass of encumbent earth, evidently requiring much labor and expence of the Contractor who might attempt to get at it, and of its quality we had no certain evidence. Under these circumstances we did not like to have it condemned by the jury, as we might be required to pay for what the contractor, who, by our contracts, supplies his own stone, would not use or what from its imperfect quality, we might be compelled to restrain him from using.
> [The lease] gave us the use of the land and quarry for one year with some time, a few months, after its expiration for the removal of such stone as the contractor might have quarried and been unable from low-water in the river, or other cause, to remove in the year. We contracted with Mr. Hovey for the aqueduct across the Monocacy ... Mr. Hovey expended large sums of money removing the incumbent earth and opening the quarry ...[32]

Quarried stone had to be moved from Mrs. Nelson's farm to the construction site, some five miles away. Five days after they entered into an agreement with Mrs. Nelson, the company authorized the construction of a railroad, or tramway, over which the blocks of stone could be hauled to the building site.[33] There is no evidence that a railroad was constructed at this time. Several years later, a different contractor would build a railroad from a different quarry, on Joseph A. Johnson's farm, to the construction site.

Figure 4. Downstream face of the Monocacy Aqueduct. Initially built of stone from Mrs. Nelson's quarry near Sugarloaf Quarry, the aqueduct was rebuilt using the harder white stone quarried from Joseph Johnson's farm.

Figure 4: Downstream elevation, Monocacy Aqueduct, Chesapeake and Ohio Canal National Historical Park, Near Dickerson, Maryland. Jack E. Boucher, photographer, 1959. Historic American Buildings Survey, Prints and Photographs Division, Library of Congress, Prints and Photographs Division, HABS, MD,11-___,38.

At the time they entered into an agreement to quarry the stone on Mrs. Nelson's farm, they were aware of the harder white stone that existed on the farm of Mr. Joseph A. Johnson. Company President Charles Mercer wrote the resident engineer, Herman Bőye that he preferred the harder stone of Mr. Johnson's farm to the softer stone of Mrs. Nelson's:

> With respect to the white and harder stone on Joseph Johnsons [indistinct – farm?] I am strongly inclined to prefer it to the softer stone from Mrs. Nelson's quarry. Should it be used I [indistinct] that it may be done with uniformity that is after carrying up the abutment and piers of precisely equal heights from the water with the darker colored stone from Mrs. Nelsons quarries ...[34]

Although Mercer may have preferred the harder stone existing on Johnson's farm, the decision had been made that the aqueduct was to be constructed largely from the stone that came from Mrs. Nelson's farm, although not entirely. Some stone was to come from Johnson's farm. In December 1829, Chief Engineer Benjamin Wright issued orders to Assistant Resident Engineer Charles Ellet on which stone to use in which portions of the aqueduct:

> You will bear in mind that the Piers and Abutments are to be of Red or Grey Stone up to the Skewbacks, the Skewbacks are to be of White Stone.
> The Pilasters are to be of the Red Stone.
> The Ring Stone and the Arch Sheeting to be of the White Stone, or the Sheeting may be in part of Red Stone if the Contractor shall prefer.
> The Water Table to be of White Stone.
> The Parapet Walls to be of Red or White Stone, but not mixed.
> The Coping to be of White Stone ...[35]

The gray stone was the stone quarried from Mrs. Nelson's farm. The white stone was taken from Joseph Johnson's farm as was, it is believed, the red stone. What Wright is saying in this order is that the white stone, the hardest of the three stones, should be used for the structural portions of the arch – the skew backs, the ring stones and so forth. This is where the high compressive strength of the white stone would be most needed. The softer stones, the gray and red stones, were to be used in the less structural components of the aqueduct, the spandrel walls, the vertical sheeting, the pilasters, parapets and coping. From these instructions, it is clear that it was Wright's intention to use the gray stone from Mrs. Nelson's quarry for the piers and abutments.

The 1829 construction season began and Hovey used both boats on the Monocacy River and overland transportation to get the stone from Mrs. Nelson's farm to the construction site. The first cofferdam was built in the usually shallow Monocacy River in June 1829[36] and the business of pier construction begun. The stone was rough hewn at the quarry and finished at the construction site. Each stone was placed twice in place, first without the mortar to test its fit and second with the mortar. Robert Leckie provides us with a description of the process of pier construction:

C & O CANAL CO. SKETCH
CONSTRUCTION RAILWAY

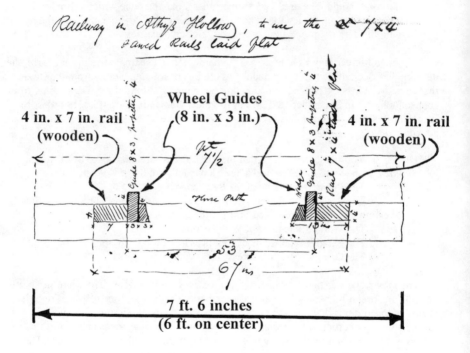

4 in. x 7 in. rail (wooden)

Wheel Guides (8 in. x 3 in.)

4 in. x 7 in. rail (wooden)

7 ft. 6 inches
(6 ft. on center)

Figure 5. Detail of Construction Railway. This drawing was developed for the Paw Paw Tunnel (Athey's Hollow), Chesapeake and Ohio Canal, but was probably very similar to that constructed at the Monocacy Aqueduct.

Figure 5. Modified and annotated drawing from National Archives, Record Group 79, Entry 214, Records of Ellwood Morris, "Railway at Athys Hollow, to use 7 x 4 Sawed Rails Laid Flat," June 21, 1838.

...You will perceive the importance of having the Piers laid in the best possible manner; the face stone must be laid solid down on their beds, with a hoisting machine, the block of stone must first be tried down dry without any mortar then hoisted and the bed put on and the stone let gently down on it, until the mortar come out all round, and the stone ly (i.e. lay) as solid as when it was in the quarry.[37] --

Besides stone, the contractor, Hovey, needed hydraulic cement for the mortar. Hydraulic cement was used in the construction of aqueducts because, unlike lime mortar, it is naturally water insoluble. In planning to build the Chesapeake and Ohio Canal, the company authorized Chief Engineer Benjamin Wright to contract with John Cocke Jr. to search the limestone region above Leesburg, Virginia, to, "...ascertain the existence, situation, quantities and probable cost of hydraulic lime."[38] The Board of Engineers study of 1826 had predicted that no limestone suitable for hydraulic cement would be found.[39] But Cocke found an excellent source of hydraulic cement, south of Shepherdstown Virginia (now West Virginia).[40] The company entered into a contract with Botelor & Reynolds to build lime kilns to burn the limestone and to grind the burnt limestone in their existing flour mill. By November 1829 Botelor and Reynolds had three lime kilns in operation burning 700 to 800 bushels per day and were grinding the burnt limestone nine hours per day producing about 500 bushels of hydraulic cement per day.[41]

The Shepherdstown cement was both inexpensive and good. The Shepherdstown operation produced hydraulic cement that was one-third to one half the cost of New York cement and one fifth of the cost of imported English cement.[42] Further, the Shepherdstown cement was superior both to New York and English hydraulic cement.[43]

But there wasn't enough Shepherdstown cement to supply all of the contractors. The Monocacy Aqueduct alone required 40,000 bushels of hydraulic cement.[44] A single masonry lock required another 3,000 bushels of hydraulic cement.[45] Contractors adopted the practice of waylaying boats bound for another contractor, so as to obtain their needed cement.[46] The company authorized additional kilns constructed at Shepherdstown at a cost of $2250.[47] This increased production but transportation problems persisted. At times the water level of the Potomac River was too low for the boat men to approach the mill at Shepherdstown to load their cement cargo.[48] Cold weather adversely affected production as so did the general scarcity of labor. At one time, the boat men refused to take their boats through the old sluice constructed by the Potomac Canal Company and forced the mill owners to deepen the tail race so as to allow them to approach the mill for their load.[49] Cement was lost when boats carrying that cement sank.[50]

In order to increase the supply of hydraulic cement, the company investigated the possibility of developing a second mill for cement production in the vicinity of the Monocacy Aqueduct. In 1829 Robert Leckie contracted with James Olcott to find limestone suitable for manufacturing hydraulic cement near the Monocacy Aqueduct.[51] Olcott found suitable limestone near Tuscaraora Creek, just north of the mouth of the Monocacy River. Leckie tested eighty batches of this material and found them all good.[52]

Small kilns to test the limestone were built on Joseph A. Johnson's land, about three quarters of a mile above the mouth of the Monocacy River.[53] Another trial kiln was built at the mouth of the Monocacy River.[54] These tests proved successful and in April 1830 the company contracted with Brackett & Guy to manufacture 40,000 bushels of hydraulic cement on Tuscarora Creek at 20 cents per bushel.[55] The Board also agreed to pay the company $50 to build a road from the new Tuscarora cement kiln to the Potomac River.[56] The hydraulic cement produced at Tuscarora Creek, however, didn't prove to be adequate, probably because the Brackett & Guy burned the limestone with wood rather than the hotter burning coal. The Board of Directors ordered that none of the Tuscarora hydraulic cement be used in the Monocacy Aqueduct.[57] Shortages of hydraulic cement continued to hinder construction of the Monocacy Aqueduct.

As 1829, the first full year of construction, was coming to an end, the Monocacy Aqueduct's contractor, Alfred Hovey, abandoned his contract.[58] All work came to an immediate halt. Before default, Hovey had 129 men on the project and 6 boys, 36 horses, 14 oxen and 9 four-wheeled carriages. After default, these numbers dropped to zero.[59]

The reasons why Hovey abandoned his contract for building the Monocacy Aqueduct are not clear, but apparently they relate to an attack made on his work by Robert Leckie and representations made about his work to the Board of Directors by Charles Ellet.[60] Just before Hovey walked off the job, Superintendent of Masonry Robert Leckie had written to Benjamin Wright about the work on the Monocacy Aqueduct, "The Masonry is so bad that it is not worth having…"[61]

V. Building and Demolishing the Piers (1830)

Losing a contractor on the most important structure on the line was a setback to the Chesapeake and Ohio Canal Company. They quickly replaced Hovey. On December 9, 1830, A.P. Osborne proposed to complete the work on the Monocacy Aqueduct at the same terms and prices which the Board had given Hovey. Osborne proposed to include Hovey's former partner, John Legg, in the work. The proposal was immediately accepted.[62]

On the same day, Chief Engineer Benjamin Wright sent a letter to Assistant Resident Engineer with instructions for how to handle the construction of the Monocacy Aqueduct under Osborne.

> The Board have this day agreed with Mr. Osbourn the bearer hereof to go on slowly with the Monocacy Aqueduct and he is to proceed as follows.
> 1st To secure all the present work so that it can pass over the Spring flood safely
> 2nd He is to commence at Mrs. Nelson's Quarry and get all the necessary stone for the remaining piers and the two abutments
> 3d He is to get all the stone for the Spandral Walls at Mrs. Nelson's quarry
> 4th He may take his choice to get the Parapet Walls between the Water Table of the Coping at Mrs. Nelson's or he may get them of the White Stone…[63]

1830 SURVEYOR'S SKETCH
MONOCACY AQUEDUCT SITE

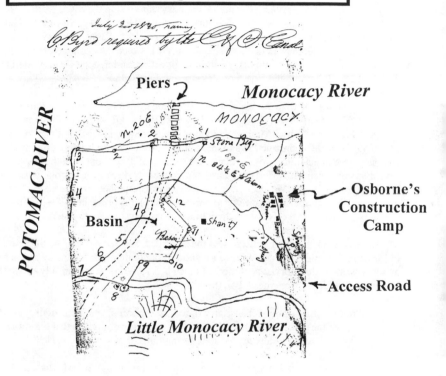

Figure 6. Surveyor's Sketch of the construction site of the Monocacy Aqueduct, 1830. This is the only known drawing of the construction site of the Monocacy Aqueduct. The second contractor's construction camp, Osborne, is shown immediately to the east of the construction site. The access road is very close to the alignment of the existing Mouth of the Monocacy Road.

Figure 6. Modified and annotated drawing from National Archives, Record Group 79, Entry 225, July 3, 1830.

By the beginning of the 1830 construction season, the company's plans had not changed – the piers were to be constructed of the gray stone from Mrs. Nelson's quarry.

It was difficult for the Chief Engineer, Benjamin Wright, to visit the construction site of the Monocacy Aqueduct – it was some forty miles distant from his office in Georgetown in an area that lacked adequate roads and he was busy supervising the entire works. He did visit the site in July 1830 and discovered that the stone, the gray stone from Mrs. Nelson's quarry, was too soft and that much of it had already cracked or split or otherwise deteriorated. The three piers that had already been constructed by contractors Hovey and Osborne would have to be demolished and rebuilt with a harder stone. He reported his findings to the President and Board of Directors on July 7, 1830.[64]

At about the same time Wright visited the Monocacy Aqueduct construction site, so did Inspector of Masonry Robert Leckie. On July 9, 1830, Leckie wrote his findings to company President Charles Mercer. He found that there was much to criticize at the Monocacy Aqueduct construction site including a scarcity of hands being employed by the contractor, the unhealthiness of the place, the number of grog shops at both the Monocacy Aqueduct construction site and the quarry and that the workers spent a great portion of their time in these shops. Leckie also criticized the lack of equipment on the construction site (there was, for example, only one wheel barrow on site), the presence of fever and augue at the aqueduct, and that the Tuscarora hydraulic cement was of doubtful quality.[65]

But what Leckie did not criticize was the stone that was being used. As Superintendent of Masonry it would have been Leckie's responsibility to detect bad stone in such a major masonry work as the Monocacy Aqueduct much before Wright's visit in July of 1830. As Company President Mercer described the situation:

> After much labor upon this quarry (i..e. Mrs. Nelson's) and constructing a part of three of the pieces of the aqueduct with it … we were apprised that the stone was imperfect, and, had already exhibited symptoms of decay.[66]

On August 14, 1830, Robert Leckie resigned from his post as Superintendent of Masonry for the Monocacy Aqueduct:

> For reasons stated in a former letter addressed to you, I must decline any further superintendence of the Monocacy Aqueduct; as under the existing circumstances, I cannot be of any efficient service to the company, and make no charge for the time I was there.[67]

Leckie recommended that A.B. McFarland take his place, an action the Board took on the same day.[68] Later in that year, Leckie would write to John P. Ingle what had happened with the gray stone from Mrs. Nelson's quarry, that when it was observed in the quarry it appeared sound but when it was quarried and put in place, exposed to the sun and frost, it would crack and split.[69]

The three piers built of bad stone were ordered by Wright to be torn down. Demolition of the three built piers on the Monocacy Aqueduct represented an enormous setback for the Chesapeake and Ohio Company. Their legislative charter called for the company to complete 100 miles of canal within the first five years. This would not be possible without the completion of the Monocacy Aqueduct. The decision to demolish the three built piers meant that the company had lost two years of construction time on what was their largest single structure to be built.

Several months previous, the company had entered into a contract with Joseph Johnson to quarry the white stone that was found on his farm. In February of 1830, Joseph Johnson had offered the quarry to the company for $500 per year and this was the price agreed upon.[70] The company thought it was a good deal:

> ... Mr. Joseph Johnson has actually supplied us with the use of the best quarry I ever saw, without limitation as to time or quantity, for all the works of the canal for $500. This sum pays him for penetrating and throwing open his fine estate, and passing and repassing with numerous carts etc. etc. by his very door to the annoyance of his amiable family and himself ... [71]

The white, hard stone from Johnson's farm would become the major source of building stone for the Monocacy Aqueduct.

The piers needed to be rebuilt and as soon as possible but Asher P. Osborne, the aqueduct's contractor, was financially strapped. On July 24, 1830, he applied to the company for an advance on the contract for the aqueduct.[72] On July 31, 1830, Wright recommended to the Board of Directors that Osborne not be given this advance.[73] Osborne then asked the Board for their permission to reassign his contract to another contractor, LeBaron, Burns & Co. of Pennsylvania. On August 21, 1830 the Board agreed.[74]

After two years of construction, the company was now on their third contractor for the Monocacy Aqueduct. The work that had been put in place by Hovey and Osborne was now being demolished. These would not be rebuilt until the following year, 1831. Adequate hydraulic cement for this construction was lacking, particularly after the failure of the Tuscarora cement. Other bad news followed. The new contractor informed the company that much of the stone that Osborne had cut was not fit for use.[75] And, to cap the plethora of bad news of the summer of 1830, Chief Engineer Benjamin Wright submitted his letter of resignation on July 31, 1830, effective October 1, 1830.[76]

Things didn't get better for the company for the rest of 1830. The new contractor, LeBaron, Burns & Co., didn't begin masonry construction operations that autumn.[77] What little work that was being done was interrupted by the onset of an early and harsh winter.[78] The company looked forward to a better year in 1831.

VI.   Completing the Aqueduct (1831-1833)

LeBaron, Burns & Co. were experienced contractors who had worked on the Pennsylvania canal. But their progress is 1831 was slow. This was a great concern for Chesapeake and Ohio Canal Company President Charles Mercer. In March 11, 1831, he wrote to the Resident Engineer, Alfred Cruger:

> I seize this occasion to urge thro you the contractors for the Aqueduct the diligent prosecution of their work. From the number of hands reported to be engaged on this work great apprehensions of its being much delayed.[79]

And again on September 10, 1831, Mercer writes Cruger:

> ... especially that you order the immediate renewal of the construction of the Aqueduct across the Monocacy ... [80]

And again on November 25, 1831:

> I was the other day at the mouth of the Monocacy, and saw, with regret, the very feeble force employed on the Aqueduct; which for want of additional centers, in the first instance, and an adequate force, to use them, to advantage, has been, by the Contractor wantonly exposed to the hazards of destruction, so far, as it is done, by the winter of ices and freshets.[81]

Despite these expressed concerns by Mercer, the contractor, LeBaron, Burns & Co., consistently had approximately 150 men working on the aqueduct and the white stone quarry, or about 50% to 100% more than the previous two contractors.[82] This didn't satisfy the company. In January 1831, for example, LeBaron, Burns & Co. had 148 men working on the aqueduct and its associated quarry. But the company estimated that 233 were needed, as broken down as follows:

> 60 men quarrying at the White Quarry
> 100 men cutting this White Stone
> 13 Four horse teams transporting the White Stone
> 33 mason, including tenders with a complement of teams drivers etc, in addition
> 2 Four horse teams hauling cement
> 1 boat and 5 men transporting sand
> 10 men procuring backing
> 10 Carpenters
> 233 Total number of men[83]

At this time the contractor was undertaking important work that didn't immediately show up as construction progress on the aqueduct. One of these projects was the construction of a railroad or tramway for hauling cut stone from the white quarry behind Joseph Johnson's house to the Monocacy Aqueduct construction site. Initially the landowners opposed constructing a road from the quarry to the construction site.[84] It was only in March 1831, when Resident Engineer Alfred Cruger entered into an agreement with landowners J.J. Harding and Mrs., J.F. Byrd, accompanied with a payment of $100, could the contractor begin constructing the wooden railway that would expedite the delivery of stone.[85] Another project was the construction of a wooden service bridge which would allow materials to be distributed efficiency to all parts of the Monocacy Aqueduct.[86]

Still work lagged. Superintendent of Masonry A.B. McFarland complained to President Mercer in April 1831:

> The Monocacy Aqueduct of all the contracts which I have hitherto superintended, is the slowest in its progress – neither can this cause be attributed to any want of skill and energy on the part of the contractors, they have a respectable gang of workmen, amounting in number to about 130, and their temporary Railroad is completed which is nearly 3 miles in length, extending from the quarries to the site of the Aqueduct – but notwithstanding that everything else is in readiness, the water continues so high that we have not been able to commence any of the foundations although several efforts to that effect, have already been made ...[87]

By April 1831 the new contractor had not yet begun work on the piers to replace the three demolished the following year. More bad news came to the company. In June 1831, the Resident Engineer, Alfred Cruger, informed the Board of Directors that the contractor was threatening to abandon the contract unless the company came up with an additional allowance for working the hard stone of the White Quarry. The company was in no position but to yield to the contractor's demands. On June 17, 1831, they ordered that additional allowances be paid the contractor for cutting the hard white stone of the White Quarry.[88] In so doing, they required the contractor to increase his work force.

It was this additional money that did the trick. Construction progress very much picked up. By August of 1831 McFarland was reporting to the Board of Directors that, "...operations of quarrying and Cutting at the Monocacy quarries – are progressing with considerable energy – about 70 workmen are still engaged..."[89]

By April 1832 the work was moving fast and McFarland was now informing the Board that, "...The Monocacy Aqueduct ...has advanced to the last arch – the center on which are about to be raised – The forces of workman are at this time quite ample ..."[90] By May of 1832, McFarland was informing the Board, "...The last arch of the Monocacy Aqueduct, it is expected will be closed in the course of the present week..."[91] The end of construction was in sight.

Construction on the Monocacy Aqueduct was finally completed in April 1833. Water was admitted to the aqueduct in October 1833.[92] With the watering of the Monocacy Aqueduct, the Chesapeake and Ohio Canal Company had met their legislative mandate to water 100 miles within five years of commencement.

VII. Summary and Conclusions

The Monocacy Aqueduct served as the largest of the eleven aqueducts for the Chesapeake and Ohio Canal for the next ninety years, until 1924 when the canal was abandoned. It still serves as a bridge over the Monocacy River for hikers, bikers, joggers, horse back riders and other visitors to the Chesapeake and Ohio Canal National Historic Park. As one of the principal features of this park the Monocacy Aqueduct is today greatly respected as one of the great icons of the American canal era. A recent public-private effort has led to a $7 million restoration project which will insure that this aqueduct will survive for at least another one hundred and seventy years.

The Monocacy Aqueduct was constructed when the American civil engineering profession was still in its infancy. Benjamin Wright, the Father of American Civil Engineering, was the undisputed designer and builder of the Monocacy Aqueduct.[93] Other early American civil engineers who were associated with the Monocacy Aqueduct include Nathan Roberts and James Geddes (who surveyed the aqueduct's initial location -- Nathan Roberts also having worked on the design of the centering), Charles Ellet, Jr., the Assistant Resident Engineer, as well as Alfred Cruger and Charles Fisk (who would later become the Chief Engineer of the Chesapeake and Ohio Canal Company). When it was completed, it was one of the largest structures of its type in the United States.

The Monocacy Aqueduct can be visited by driving to Dickerson, Maryland, and turning on to Mouth of Monocacy Road. At the end of this road is the Monocacy Aqueduct which, in the words of one of the workmen who built it, "…in point of beauty and perfection of workmanship, (it) will … compare with any work of the kind either in this Country or in Europe…"[94]

NA = National Archives

RG = Record Group

[1] See the companion paper, Robert J. Kapsch, Ph.D., George Washington, the Potomac Canal and the Beginning of American Civil Engineering: Engineering Problems and Solutions, also published in these Proceedings. These two papers explore the beginnings of American civil engineering. When George Washington began the Potomac Canal in 1785 there were no American civil engineers to be hired. Washington relied on English engineers serving as consultants. One of the reasons for the failure of the Potomac Company was because of the lack of good engineering. When the Chesapeake and Ohio Canal began construction in 1828, Benjamin Wright broke a large number of American civil engineers with him from the Erie Canal to help design and build the Chesapeake and Ohio Canal.

[2] Library of Congress, Manuscripts Division, Isaac Briggs papers. Letter, Isaac Briggs to the Board of Public Works of the Commonwealth of Virginia, "Report to the Board of Public Works … [on the route of the James and Kanawha canal …]," nd [1821?], unpaginated.

[3] U.S. House of Representatives, "Message from the President of the United States transmitting A Report From the Secretary of War With That of the Board of Engineers for Internal Improvement, concerning the proposed Chesapeake and Ohio Canal," 19th Cong. 2d sess., Document No. 10, December 7, 1826.

[4] Ibid. 41.

[5] George Armroyd, A Connected View of the Whole Internal Navigation of the United States, (Philadelphia, 1830), 219.

[6] U.S. House of Representatives, "Letter from The Secretary of War Transmitting Estimates of the Cost of Making a Canal from Cumberland to Georgetown," 20th Cong. 1st sess. Document No. 192, March 10, 1828.

[7] George Armroyd, A Connected View …, 219.

[8] U.S. House of Representatives, "Letter from The Secretary of War Transmitting Estimates of the Cost ... Document No. 192, March 10, 1828, 74. There is no explanation for the 133 foot aqueduct length estimated by Roberts and Geddes except to observe that their study consistently underestimated costs.

[9] NA-RG 79, Entry 190, Letters Received, Mercer to Wright, September 4, 1828

[10] U.S. House of Representatives, "Memorial of the Chesapeake and Ohio Canal Company," 20th Cong. 2d Sess., House Document No. 12, December 5, 1828, Appendix A, "President's Report to the Stockholders," 156.

[11] Ibid., 156.

[12] The invitation for the ground breaking ceremony read:
> The Committee of Arrangement in behalf of the Cities of the District, request the favor of you to accompany them, to be present on the morning of the Fourth of July, at the ceremony of breaking ground for the Chesapeake and Ohio Canal. Of the hour, and place of embarkation, to proceed by water to the spot, due notice will be given.

Duke University Library, Robert Leckie Papers, June 30, 1828.

[13] U.S. House of Representatives, "Memorial of the Chesapeake and Ohio Canal Company," ..., House Document No. 12, December 5, 1828, Appendix A, "President's Report to the Stockholders," 156.

[14] Ibid., 156.

[15] Chesapeake and Ohio Canal Company, *First Annual Report of The President and Directors of The Chesapeake and Ohio Canal Company Together with the Proceedings of the Stockholders at Their First Annual Meeting,* (Washington: Gales & Seaton, 1829), 22.

[16] NA-RG 79, Entry 190, Letters Received, Benjamin Wright to the President and Directors, Chesapeake and Ohio Canal Company, October, 1828. These drawings have not been located and probably have not survived.

[17] Ibid.

[18] Ibid.

[19] NA-RG 79, Entry 182, Proceedings of the President and Directors, Chesapeake and Ohio Canal Company, December 3, 1828, 122.
> Resolved, that an estimate be made by the acting member or members of the Board of Engineers, and reported to this Board of the cost which would attend the enlargement of the breadth of the Monocacy Aqueduct, so as to admit the free passage of boats moving upon it, at the same time in opposite directions.

[20] NA-RG 79, Entry 190, Letters Received, Benjamin Wright to Gen. Mercer, January 22, 1829.
> Mr. Roberts and myself have been laying down on paper our own views of the proposed Basin in Rock Creek while Gen Smith took to the Board yesterday and I presume explained to you the whole plan – We have also been engaged in forming a Plan of Centres for the Monocacy Aqueduct and giving Bills of Lumber.

[21] NA-RG 79, Entry 212, Letter Book of the Resident Engineer of the 5th Residency of the 1st Division, 1828-1831, C & O Canal Co., Board of Directors to Herman Böye, December 1, 1828.
> By direction of the President of the Ches. & Ohio Canal Co., I am requested to change your station as pointed out in the book of regulations and place you as Resident Engineer over the 5th Residency which extends from Section 64 to 84 inclusive ... and to you is assigned Mr. Charles Ellet ... and William Wallace as Rodman ..."

[22] NA-RG 79 Entry 195, "Appointment of Members of Board of Engineers," August 25, 1828, 27.

[23] NA-RG 79 Entry 194, Letters Sent, President C. F. Mercer to George Been, Chairman, November 18, 1828, Book A, 45.

[24] NA-RG 79, Entry 194, Letters Sent, J. Ingle to Alfred Cruger Esqr., January 30, 1830.
> The President & Directors ... have directed, that in addition to the duties now performed by you on the 3d and late 4th Residencies, you also act as Resident Engineer of the 5th Residency – until Mr. Böye shall so recover his health as too permit his return to the Residency ...

[25] NA – RG 79, Entry 190, Letters Received, Robert Leckie to President and Board of Directors, April 21, 1830.

[26] NA-RG 79, Entry 190, Letters Received, Benjamin Wright to President Mercer, February 24, 1830.
> Mr. Leckie is a very useful man in his way – but he corrects errors with a Mall rather than a [indistinct]. Gentle means are better than harsh in managing men in this free country, and if Mr. L would correct himself in this respect and be perfectly impartial, he would do us much good --

[27] Duke University Library, Robert Leckie Papers, Benjamin Wright to Robert Leckie, October 27, 1828.

[28] NA-RG 79, Entry 182, Proceedings of the President and Directors, 1828-1829, October 31, 1828, Vol. 1, 100. Luke Hitchcock would shortly drop out of the partnership and replaced by John Legg. NA-RG 79, Entry 182, Proceedings of the President and Directors, Chesapeake and Ohio Canal Company, January 7, 1829, 140. Alfred A. Hovey later became the sole contractor.

[29] NA-RG 79, Entry 190, Letters Received, Robert Leckie to John P. Ingle, October 19, 1830.

[30] Duke University Library, Letter, Wright to Robert Leckie, September 22, 1829. Wright wrote:
   I am not certain of the distance Mr. Hovey has to transport the above quantity of Stone from Mrs. Nelsons quarry but think it cannot be less than five miles ...

[31] NA-RG 79, Entry 194, Letters Sent, Book A, 65, John P. Ingle to Mrs. Eliza Nelson, April 6, 1829:
   I enclose you a requisition on the Treasurer of the Company for 450 $ being the amount agreed upon for the rent of your farm and the use of quarries etc. ...

[32] NA-RG 79, Entry 194, Letters Sent, Book A, 444-449, C.F. Mercer to Benjamin Price, Esq., May 29, 1832.

[33] NA-RG 79, Entry 212, Letter Book of the Resident Engineer of the 5th Residency of the 1st Division, 1828-1831, C & O Canal Co., Pr. Mercer to Mr. Bôye, April 11, 1829:
   A railroad may be made from Hovey's quarry. If made it is by agreement with Mrs. Nelson to be left for her benefit when the lease expires ...

[34] NA-RG 79, Entry 212, Letter Book of the Resident Engineer of the 5th Residency of the 1st Division, 1828-1831, C & O Canal Co., Pr. Mercer to Mr. Bôye, April 11, 1829.

[35] NA-RG 79, Entry 212, Letter Book of the Resident Engineer of the 5th Residency of the 1st Division, 1828-1831, C & O Canal Co., December 9, 1829.

[36] Harlan Unrau, "Historic Structures Report:The Monocacy Aqueduct, Historical Data, Chesapeake and Ohio Canal National Historical Park, MD.-D.C.-W.VA," (Denver, CO: Historic Preservation Team, Denver Service Center, National Park Service, January 1976), 10.

[37] NA-RG 79, Entry 212. Letter Book of the Resident Engineer of the 5th Residency of the 1st Division, 1828-1831, C. & O. Canal Co., Letter, Robert Leckie, Inspector Masonry 1st Division Ches. & Ohio Canal, to Mr. Phillips, July 28, 1829.

[38] NA-RG 79, Entry 194, Letters Sent, C. F. Mercer, Prest. Of the Ch. & O CC to John Cocke, Jr., July 28, 1828.

[39] U.S. House of Representatives, "Message from the President of the United States transmitting A Report From the Secretary of War ..., Document No. 10, December 7, 1826, 28.

[40] NA-RG 79, Entry 182, Proceedings of the President and Directors, Chesapeake and Ohio Canal Company, 1828-1829, October 11, 1828, 88.

[41] NA-RG 79, Entry 190, Letters Received, A. B. McFarland to J. P. Ingle, November 6, 1829.

[42] NA-RG 79, Entry 190, Letters Received, N. S. Roberts to Gen. Smith, October 13, 1829.

[43] NA-RG 79, Entry 190, Letters Received, Benjamin Wright to President & Directors, November 12, 1830.

[44] NA-RG 79, Entry 190, Letters Received, Robert Leckie to Charles Mercer, February 9, 1830.

[45] NA-RG 79, Entry 190, Letters Received, N. S. Roberts to Gen. Smith, October 13, 1829.

[46] NA-RG 79, Entry 194, Letters Sent, John P. Ingle to A. B. McFarland, January 8, 1830, Book A, 155. Ingle wrote of this practice:
   The practice of Contractors going on the river to stop boats with cement has been found to produce much satisfaction to Contractors and injury to the interests of the Company. ...

[47] NA-RG 79, Entry 190, Letters Received, Botelor & Reynolds to John P. Ingle, April 5, 1830.

[48] NA-RG 79, Entry 190, Letters Received, A. B. McFarland to John P. Ingle, September 27, 1829.

[49] NA-RG 79, Entry 190, Letters Received, Botelor & Reynolds to John P. Ingle, April 5, 1830. Botelor wrote of this:
   We are aware that complaints will be made for the want of Cement at our establishment, and in Justice to ourselves we think it advisable to state to you the causes which have laid the foundation for this disappointment. The Extreme cold weather in February dispersed many of our hands, and suspending the burning of lime during that Month, and when the weather moderated, and the Navigation opened, the Boatsmen refused to take their Boats aroun, and through the Shute, and take in their laden under the Mill as heretofore, and to Silence their complaints and facilitate their loading, we found it Necessary to stop our Mill for two weeks, and deepen our tail race 15 Inches,

to enable them to float out with a full load – This cold, wet, and most uncomfortable undertaking Knocked up Eleven of our hands, some of which are as yet unable to resume their labours and in addition to those interruptions we have had to contend with a flood of water, which has stopped our Mill for the last week.

[50] NA-RG 79, Entry 190, Letters Received, A. B. McFarland to John P. Ingle, November 28, 1829. McFarland wrote:

From the best information that I can collect, respecting the loss of the Boat Alabama and her cargo – It appears to have been occasioned by the mismanagement of the Boatmen – 165 of our best bags is lost with her ...

[51] NA-RG 79, Entry 182, Proceedings of the President and Directors, Chesapeake and Ohio Canal Company, 1828-1829, November 21, 1829, 403.

[52] NA-RG 79, Entry 190, Letters Received, Benjamin Wright to President and Directors, September 28, 1829.

[53] NA-RG 79, Entry 190, Letters Received. A Bill of Burning at Jct. Johnson, November 2, 1829.

[54] NA-RG 79, Entry 190, Letters Received, Purcell to President and Directors, Chesapeake and Ohio Canal Company, September 1, 1829.

[55] NA-RG 79, Entry 182, Proceedings of the President and Directors, Chesapeake and Ohio Canal Company, 1830-1831, February 3, 1830, 20-21.

[56] NA-RG 79, Entry 182, Proceedings of the President and Directors, Chesapeake and Ohio Canal Company, 1830-1831, 65.

[57] NA-RG 79, Entry 182, Proceedings of the President and Directors, Chesapeake and Ohio Canal Company, 1830-1831, September 13, 1830, 178.

[58] NA-RG 79, Entry 182, Proceedings of the President and Directors, Chesapeake and Ohio Canal Company, 1828-1829, December 7, 1829, 414.

[59] Appendix, Robert J. Kapsch, "Benjamin Wright and the Monocacy Aqueduct," *Canal History and Technology Proceedings*, Volume XIX (March 18, 2000), (Easton, PA: Canal History and Technology Press, 2000), 203.

[60] NA-RG 79, Entry 190, Letters Received, Alfred Hovey to Charles Mercer, February 11, 1831. Hovey states:

I have long delayed saying anything on the subject of the Monocacy Aqueduct, under a firm belief that Time reflection and More Experience in Operations would tend to do away the Excitement and feeling that existed at the time I left there – When Mr. Lackey [i.e. Leckie] first made the Attack, I supposed it personal with Him – The next day when I learnt that he has his Orders from the Board of Directors To declare the Job Abandoned and the great part of the men I had Employd. Heard him do so, And being informd. That the Board was act(ing) on representations made by a set of disaffected Foreigners, In connection with those of Young Elliott [i.e. Ellet]. Under such circumstances I was sure [there was] No other way for Me, but the course I took – which I should not have done had it not been for the Absolute Necissity of Attending to Business at home ... I am now sensible it would have been better for me to remain on the spot and fought it out ...

[61] Duke University Library, Robert Leckie Papers, Letter, Robert Leckie to Benjamin Wright, November 18, 1829.

[62] NA-RG 79, Entry 182, Proceedings of the President and Directors, Chesapeake and Ohio Canal Company, 1828-1829, December 9, 1829, Vol. 1, 419.

[63] NA-RG 79, Entry 212, Letter Book of the Resident Engineer of the 5th Residency, B. Wright, Engr Chief of this Canal, to Charles Ellet, Jr., December 9, 1829.

[64] NA-RG 79, Entry 182, Proceedings of the President and Directors, Chesapeake and Ohio Canal Company, 1830-1831, Vol. 2, 135.

[65] Duke University Library, Robert Leckie Papers, Letter, Robert Leckie to Company President Charles Mercer, July 9, 1830.

[66] NA-RG 79, Entry 194, Letters Sent, C. F. Mercer to Benjamin Price, Esq., May 29, 1832, Book A, 444-449.

[67] NA-RG 79, Entry 190, Letters Received, Robert Leckie to Charles Mercer, August 14, 1830.

[68] NA-RG 79, Entry 182, Proceedings of the President and Directors, Chesapeake and Ohio Canal Company, 1830-1831, August 14, 1830, Vol. 2, 162.

[69] NA-RG 79, Entry 190, Letters Received, Robert Leckie, Late Inspector of Masonry to John P. Ingle, October 19, 1830.

[70] NA-RG 79, Entry 190, Letters Received, Joseph A. Johnson to C. F. Mercer, February 17, 1830. On February 20, 1830, Mercer counter offered $400 per year. By March 3, 1830, Mercer announced to the Board that the company had agreed to provide Johnson $5-00 per year for his quarry. NA-RG 79, Entry 182, Proceedings of the President and Directors, Chesapeake and Ohio Canal Company, 1830-1831, March 3, 1830, Vol. 2, 37.

[71] NA-RG 79, Entry 194, Letters Sent, C. F. Mercer to Benjamin Price, Esq., May 29, 1832, Book A, 444-449. Johnson's house still stands, off Middleton Belt Road, and is now known as "Rock Hall."

[72] NA-RG 79, Entry 182, Proceedings of the President and Directors, Chesapeake and Ohio Canal Company, 1830-1831, July 24, 1830, Vol. 2, 149.

[73] Ibid. July 31, 1830, 153.

[74] NA-RG 79, Entry 182, Proceedings of the President and Directors, Chesapeake and Ohio Canal Company, 1830-1831, August 21, 1830, Vol. 2, 168.

[75] NA-RG 79, Entry 182, Proceedings of the President and Directors, Chesapeake and Ohio Canal Company, 19830-1831, August 21, 1830, Vol. 2, 168.

[76] NA-RG 79, Entry 182, Proceedings of the President and Directors, Chesapeake and Ohio Canal Company, 1830-1831, August 30, 1830, Vol. 2, 172. Wright stated his reasons for resignation:

> From the information I possess, of difficulties still subsisting between the Ches & Ohio Canal and the Balt & Ohio rail road [i.e. the legal suit concerning the passages at the Potomac River across the Catoctin Mountains], and the little probability of those difficulties being speedily removed; the consequence is, there is very little field for active operations for Engineers left.
>
> Under these impressions I have thought it proper and right to tender my resignation, to take effect after the first of October, or as soon thereafter as the works in Georgetown will permit my leaving.

[77] NA-RG 79, Entry 194, Letters Sent, C. F. Mercer to Byrne & Lebaron, October 28, 1830, Book A, 268. Mercer wrote:

> It is with extreme regret that I learn that no attempt has yet been made to recommence the masonry on the Monocacy Aqueduct – Without this link in the chain of the canal above Seneca, no hope can be entertained of bringing the aqueduct into use on the canal above or below it: since apart from the Potomac there is no feeder that could be resorted to but the Tuscarora which is above Monocacy.
>
> It is neither your interest nor that of the Stockhólders of the Chesapeake and Ohio Canal Company to permit an illusory hope to prevail that this work can be done sooner than you mean to complete it. I therefore now write to know of you the cause of your past unexpected delay and to obtain your candid opinion of the time at which this work will be so far completed, as to pass the canal boats through it.

[78] NA-RG 79, Entry 190, Letters Received, A. B. McFarland to Mercer, December 7, 1830.

[79] NA-RG 79, Entry 194, Letters Sent, C. F. Mercer to Alfred Cruger, Resident Engineer, March 11, 1831, Book A, 304.

[80] NA-RG 79, Entry 194, Letters Sent, C. F. Mercer to Alfred Cruger, Resident Engineer, September 10, 1831, Book A, 363-369.

[81] NA-RG 79, Entry 194, Letters Sent, C. F. Mercer to Alfred Cruger, Resident Engineer, November 25, 1831, Book A, 386.

[82] Appendix, Robert J. Kapsch, "Benjamin Wright and the Monocacy Aqueduct," *Canal History and Technology Proceedings*, … , 202-210.

[83] NA-RG 79, Entry 190, Letters Received, Cruger to Mercer, January 18, 1831.

[84] NA-RG 79, Entry 182, Proceedings of the President and Directors, Chesapeake and Ohio Canal Company, 1830-1831, January 4, 1831, Vol. 2, 251.

[85] NA-RG 79, Entry 182, Proceedings of the President and Directors, Chesapeake and Ohio Canal Company, 1830-1831, March 18, 1831, Vol. 2, 283. The railway was approximately two and a half miles long and ran east of Rock Hall, the home of Joseph Johnson. The stone carts were hauled by oxen, to the top of the hill, and then by horses and mules. See William Jarboe Grove, *History of Carrolton Manor, Frederick County, Maryland*, (Frederick, MD, Markmen & Bielfeld, 1928), 408.

[86] NA-RG 79, Entry 190, Letters Received, A. B. McFarland to President and Directors, Chesapeake and Ohio Canal Company, September 13, 1830.

[87] NA-RG 79, Entry 190, Letters Received, A. B. McFarland to President and Directors, Records of the Chesapeake and Ohio Canal Company, April 1, 1831.

[88] NA-RG 79, Entry 182, Proceedings of the President and Directors, Chesapeake and Ohio Canal Company, 1830-1831, June 17, 1831, Vol. 2, 388-389. The additional allowances included:

For quarrying ashler;	15 cents per foot
For cutting the same;	5 cents per foot
For quarrying sheeting;	23 cents per foot
For cutting the same;	10 cents per foot.
For quarrying coping;	11 cents per foot.

[89] NA-RG 79, Entry 190, Letters Received, A. B. McFarland to John P. Ingle, August 28, 1831.

[90] NA-RG 79, Entry 190, Letters Received, A. B. McFarland to Charles F. Mercer, April 28, 1832.

[91] NA-RG 79, Entry 190, Letters Received, A. B. McFarland to John P. Ingle, May 23, 1832.

[92] Harlan Unrau, *Historic Structures Report, ...* , 45-46.

[93] For unknown reasons, Wright has never been given adequate credit as the designer of the Monocacy Aqueduct. There is, however, no doubt that he designed it and oversaw much of its construction. Take, for example, the report of Army engineers John J. Abert and James Kearney in their 1831 visit to the Monocacy Aqueduct:

> Mr. Cruger, the resident engineer, showed to us the specifications of this work [i.e. the Monocacy Aqueduct], (which form a part of the contract with the builder), describing the manner in which it was to be executed, and the dimensions of the various parts. We observed, at the foot of these specifications, the name of the celebrated civil engineer, Judge Wright, who was formerly in the employ of the company.

Lt. Col. John J. Abert and Lt. Col. James Kearney, (Report of Field Survey, dated June 13, 1831 – of their 1831 inspection of the canal construction), *Chesapeake and Ohio Canal (To accompany bill H.R. No. 94)*, U.S. House of Representative Report No. 414, 23rd Cong. 1st sess., April 17, 1834, 102.

[94] NA-RG 79, Entry 190, A. B. McFarland to President and Directors, November 17, 1831.

U.S. Capitol: Recent Renovations

Newell Anderson[1]

The Capitol (Figure 1) has undergone numerous alterations, expansions, additions and renovations since the laying of the cornerstone by George Washington in 1793. Congress decided that the Capitol should be a working building rather than a museum or a monument fixed in time. As such, the Capitol has been almost continually changed and modified over time to meet the changing needs of the Congress and the People. Advances in technology have also caused and allowed modifications to the building. (Figure 2)

A Brief Chronology of the Capitol

1793 George Washington lays the cornerstone in a Masonic ceremony.

1800 Congress moves from Philadelphia. Only the north wing of the Capitol is complete.

1801 Supreme Court first meets in Capitol.

1803 President Jefferson appoints Benjamin Henry Latrobe to continue work on the building.

1807 South wing occupied by House.

1808 Latrobe begins rebuilding north wing.

1810 Senate occupies chamber in north wing; room below constructed for Supreme Court.

1814 Capitol burned by British troops.

1815-17 Latrobe rehired to restore the Capitol. He resigns over disputes about authority.

1818-29 The center building is completed. Library of Congress occupies the west central area.

1851-55 Library of Congress is damaged by fire and is rebuilt.

1851-65 Thomas U. Walter appointed "Architect of the Capitol Extension."

1853-59 Captain Montgomery C. Meigs placed in charge of construction for the Corps of Engineers.

1855 Congress adds a new cast iron dome to the Extension project.

1863 Statue of Freedom raised into place atop Dome.

1867 Present House and Senate wings completed.

1882-92 Frederick Law Olmsted Terraces are constructed.

1897 Library of Congress moves out of the Capitol west central area into the present Thomas Jefferson Building.

1900 Space vacated by the Library of Congress is in-filled with new floors.

1935 Supreme Court moves out of the Capitol into its present building.

1958-62 East Front extension.

1983-87 Restoration of West Front.

---

[1] The Structural Engineer for the Architect of the Capitol (retired), 1979 to 1997. Rensselaer Polytechnic Institute, BCE, 1958.

In its present form, the Capitol covers a ground area of approximately 174,000 square feet and has a floor area of about 719,000 square feet on five levels. The length from north to south is 751' and its greatest width is 350', including approaches. The height from the east front plaza to the top of the Statue of Freedom is 288'. (Figure 3)

The main structural system except for the roofs, Dome and recent renovations, is wall bearing, vaulted masonry. Recent renovations and roofs consist of steel, wrought iron and concrete. The Dome consists of wrought and cast iron. Some of the more recent modifications and additions are briefly described below.

Starting in 1958 the _East_ _Front_ of the Capitol was extended 32'-6". The original east front walls were built of dressed Aquia Creek sandstone, fieldstone, brick and rubble. The wall supports the Dome and varies in thickness from 3' to over 6'. (Figure 4)

The sandstone was deteriorating due to its porosity and direct exposure to the weather. The problem was solved by building an addition, and restoring the wall in selected areas. Thus, the porous sandstone exterior wall became an interior wall buttressed by the reinforced concrete addition. In addition, Congress captured additional space for office and support functions.

Starting in 1983, the _West_ _Front_ of the Capitol was restored and reinforced. The original west front wall was built of the same sandstone and materials as the east front wall; however, the wall does not support the Dome. (Figure 4)

Another difference in the _West_ _Front_ is that the three-story space behind the west central portion was occupied by the Library of Congress until 1897 and was not in-filled with floors and cross walls until 1900.

More than one thousand stainless steel tie rods of various lengths were set into the west wall, along with extensive grouting, in order to tie the west wall and the intersecting walls together as a complete system. The old fireplace flue system complicated the drilling and grouting because the flues were voids of unknown size and location. X-rays were used to locate the flues and to map the intersecting walls.

Completed in 1866, the Dome is made entirely of cast iron with wrought iron elements. Its height is approximately 199' from bearing to the top of the Statue of Freedom. The diameter of the base is approximately 100'. (Figure 5)

The 1855 legislation authorizing the design and construction of a new dome resulted from several factors. One factor was the 1851 Library of Congress fire that almost spread to the original dome structure, which was built of masonry and wood and not fireproof. Other factors included the fact that the original dome leaked, and the realization that the original dome was too small and out of scale with the rest of the building when one includes the Senate and House extensions then under construction.

Cast iron was selected as the main material because it is strong as well as fireproof. In addition, highly ornamental designs could be produced easily and cheaply, especially when compared with carved stone. Indeed, many people assume that the Dome is made of carved stone because of the ornamentation and color.

All exposed dome surfaces, inside and outside, are cast iron. Nearly 9 million pounds of ironwork shipped by train from foundries in Virginia, Maryland, New Jersey and New York would be used during construction.

The design of the Dome was a cooperative effort. Thomas U. Walter (Figure 6) was the architect. He had done extensive research into the domes of Europe. An engineer from Germany, Ottmar Sonnemann, devised the Domes' structural skeleton. Captain Montgomery C. Meigs (Figure 7) made several suggestions to simplify the design and to facilitate the construction. (Figure 8)

The basic structure consists of 36 open-web arched half-ribs evenly spaced around the circumference. The ribs are tied together with compression and tension rings and bracing. The ribs support an inner and outer dome with a system of struts and hangers. Neither Dome contributes to the structural performance of the ribs. (Figures 9 & 10)

The outer, exterior dome is detailed to accommodate anticipated thermal movements, i.e., sliding lap joints and oversized bolt holes. Such detailing is not totally consistent with good waterproofing practice but was the best compromise between two conflicting objectives, water tightness and allowance for thermal movement. As a result, water penetration has been a problem.

Starting in 1865 with the Smithsonian Institution's study of thermal performance, the performance of the Dome has been reviewed and analyzed periodically. The most recent structural analysis was accomplished by LZA Technology in 1998. LZA developed a three-dimensional, geometric mathematical model of the Dome's structural elements. A finite element analysis was done for various loading combinations. This study analysis indicated a minimum safety factor of 4 for the Dome structure.

The LZA analysis was part of an ongoing program for the renovation of the dome initiated in 1990 with the retention of Hoffmann Architects by the Architect of the Capitol. This program included lead paint abatement, cast iron repairs, bird proofing, water proofing, gutter guards, anchors for safety lines, additional catwalks, water supply and other steps to facilitate maintenance. This program is scheduled for completion in 2003.

The Statue of Freedom atop the Dome stands 19'-6" high and weighs about 15,000 pounds. The statue is hollow, was cast in bronze and is about 3/8" thick. It is supported on a cast-iron ball with the inscription *E Pluribus Unum*. (Figure 11)

During a routine painting of the Dome in 1988, a piece of bronze was found to be loose. Pinholes of daylight were observed from inside the statue. In addition, cracks in the cast-iron pedestal were observed.

The Architect of the Capitol established a screening committee to select a conservator, Washington University Technology Associates and a structural engineer, Cagley and Associates. This team was to determine and evaluate existing conditions, recommend conservation systems and methods and the means and methods to accomplish the conservation of the statue and pedestal. A report was published in October 1991, recommending removal of the statue to a ground platform for conservation. The pedestal could be repaired and conserved in place. The structural design and documents were prepared by Cagley and Associates and the statue was removed by Erickson Sky Crane on May 9, 1993, conserved and replaced to her restored pedestal on October 23, 1993. (Figures 12, 13)

The Office of the Architect of the Capitol typically has hundreds of projects at various stages of development, from conception to completion of construction. The Office also provides support services for the Legislative Branch similar to those provided by the General Services Administration for the Executive Branch. The jurisdiction includes the House and Senate office buildings, the Library of Congress, the U.S. Botanic Garden, the U.S. Supreme Court, several parking structures and a steam and chilled water plant with associated miles of tunnels and distribution system. (Figure 14)

One project recently completed is the renovation, restoration and expansion of the U.S. Botanic Garden. The original building was completed in 1933 and was the first large building in the United States to use aluminum as a structural system.

The need for the renovation became apparent in the 1970s when gardeners discovered bolt and rivet heads in the planting beds. Subsequently, it was discovered that the alloy used in the trusses exfoliates in the presence of heat and moisture. Moisture became trapped in the truss member connections, the aluminum exfoliated and expanded and thus popped the fastener heads. All of the aluminum was replaced. All of the engineering systems were upgraded to state-of-the-art levels as part of the renovation in addition to enlarging the facility and making it more visitor-friendly.

Another project just starting construction is the $368 million Capitol Visitor Center (CVC) under the east plaza. A CVC was first proposed in the 1970s. The east plaza was an ugly, black, asphalt parking lot not conforming to any plan. Tour bus loading and unloading was helter-skelter and was a source of irritation for the Capitol Hill residents. Egress to the Capitol was available at many locations and was not tightly controlled. The only direct access to the Capitol for moving materials in and out was two sidewalk lifts adjacent to the east plaza. This meant that delivery and trash/garbage trucks were present at the main tourist entrance to the Capitol. There was limited food service in the Capitol and staff had priority during the lunch hour. Rest room facilities were inadequate. There was no orientation and educational experience for the visitor. In short, visiting the Capitol could be a frustrating experience.

The CVC was designed by RTKL Associates for the Architect of the Capitol to solve all of these problems: security, circulation (people and busses), orientation and education, materials handling, food service, landscaping master plan and other support services for the visitor. Offices and support services for Congress were also included. This design resulted in a 580,000 square feet addition consisting of three levels underground and a landscaped east plaza aligned along the original design developed by Frederick Law Olmsted in the 1870s.

The construction plan presented another series of problems, i.e., circulation in the Capitol Hill area, noise control, maintenance of services, egress and fire protection/access, protection of existing Capitol foundations, time and money.

Top-down construction, perimeter slurry walls and phasing were methods selected to solve the construction problems. Construction is scheduled for completion in the fall of 2005.

These are a few of the diverse projects that have been developed through the Office of the Architect of the Capitol. These projects range from the design and construction of dog kennels for the U. S. Capitol Police, the conversion of a funeral home into a Page Residence, to the design of the Presidential Inaugural stands. This flow of projects will continue as the needs of the Congress and the People continue.

Figure 1. Earliest known photograph of the Capitol, depicting the East Front, by John Plumbe, Jr. 1846. (Architect of the Capitol)

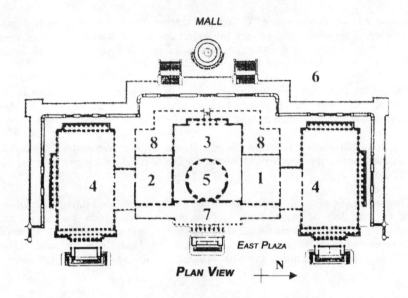

Figure 2. The numbers indicate the order in which each section of the Capitol was built. The dates below are for each section's first construction period and do not include rebuilding or repair.

1. Original north (Senate) wing, 1793-1800
2. Original south (House) wing, 1793-1807
3. Center section and Rotunda, 1818-1824
4. Present House & Senate wings and connecting corridors, 1851-1867
5. Cast-iron dome, 1855-1866
6. Terraces, 1884-1892
7. East front extension, 1958-1962
8. Courtyard infill rooms, 1991-1993

Figure 3. The height from the east front plaza to the top of the Statue of Freedom is 288'. Photo taken after the Statue of Freedom was removed for renovation in 1993. (S. Dennis, used with permission).

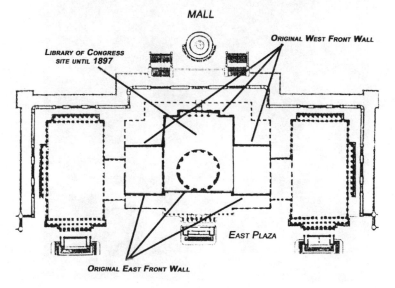

Figure 4. The original west front wall does not support the Capitol Dome. The east front wall supports the Dome.

Figure 5. Elevation of dome of U. S. Capitol. (Architect of the Capitol)

Figure 6. Thomas U. Walter, Architect of the Capitol Extension and Dome. (Architect of the Capitol)

Figure 7. As a young army captain, Montgomery C. Meigs was in charge of construction for the Corp of Engineers. During his career, Meigs had oversight for significant building projects in Washington, including the Smithsonian Institution Arts & Industries Building, the Washington Aqueduct, and Pension Building. (National Archives)

Figure 8. Dome construction during the Civil War. (Architect of the Capitol)

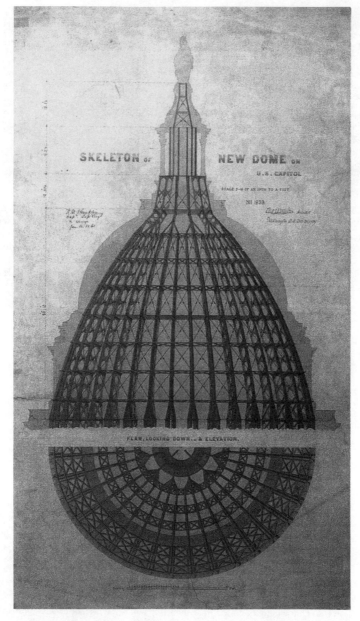

Figure 9. Skeleton of New Dome. (Architect of the Capitol)

Figure 10. Cross-section view of cast-iron dome, 1859 architectural ink and watercolor drawing by Thomas U. Walter. (Architect of the Capitol)

Figure 11. Statue of Freedom. (Architect of the Capitol)

Figure 12. Sky crane lifting Statue of Freedom during renovation. (Cagley and Associates, used with permission)

Figure 13. Aerial view of the 285 acre Capitol Complex. Buildings under the purview of the Architect of the Capitol consist of: the Capitol and grounds (including Cannon, Longworth, Rayburn, Ford House Office Buildings on right and the Russell, Dirksen, and Hart Senate Office Buildings on left), Library of Congress (Jefferson, Madison, Adams Buildings), the U.S. Supreme Court, the Capitol Police Headquarters and the U.S. Botanic Garden Conservatory. (Architect of the Capitol)

Figure 14. U.S. Botanic Garden Conservatory during reconstruction. (Architect of the Capitol)

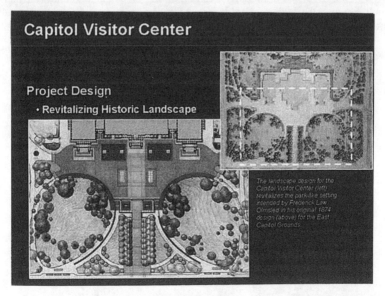

Figure 15. Proposed plans for the Capitol Visitors Center (Architect of the Capitol)

References

1. Allen, W. C. (1992). *The Dome of the United States Capitol: An Architectural History.* Washington, D.C.: U.S. Government Printing Office.

2. Becker, J. M, Haley, M. .X. (1990). Up/Down Construction – Decision Making and Performance. In: Proceedings of Design and Performance of Earth Retaining Structures; New York, Geotechnical Engineering Division, Cornell University (June).

3. Cagley and Associates (1993). "Report Summarizing the Structural Engineers' Observations During the Removal and Replacement of the Statue of Freedom Atop the United States Capitol Dome." Report to the Architect of the Capitol (November).

4. Davis, F. P. (1960). "Early Metal Space Frame Investigated." Progressive Architecture (December): pp. 164-171.

5. Hsu, S. S. (2002). Article about U.S. Capitol Visitor Center. *Washington Post* (May 28): B1, B2.

6. LZA Technology. (1998). "Structural Analysis of the Dome of the United States Capitol". Report to Alan M. Hantman, Architect of the Capitol. Washington D.C. (October).

7. Office of the Curator-Architect of the Capitol. (1992). "The United States Capitol: An Overview of the Building and its Function." (January).

8. Office of the Curator, Architect of the Capitol. (1992). "Restoration of the Statue of Freedom." Washington D. C. (February).

9. Orsak, D. (1999). "Restoration of the Statue of Freedom". *Washington Building Congress Bulletin.* Washington D. C. (December): pp. 6, 8.

10. RTKL Associates. (1991). "United States Capitol Visitor Center, Conceptual Study Submission," Report to the Architect of the Capitol. Washington D. C. (June).

11. Stover, Eric C., et. al. (2001). "Structural Analysis and Rehabilitation of the U.S. Capitol Dome." *Proceedings of the 2001 Structures Congress and Exposition.* Chang, Peter C. (Ed.) Reston, Virginia: American Society of Civil Engineers.

12. United States Botanic Garden Fact Sheet. "Conservatory Renovation Project." Architect of the Capitol. Washington D. C.

THE GEOLOGY, HISTORY, AND FOUNDATIONS
OF THE MONUMENTAL CORE
by Douglas W. Christie[1]

ABSTRACT

Many factors have influenced development of the Monumental Core in Washington, DC. Soil and groundwater conditions have influenced the foundations, and in one case the location, of the monuments and memorials. Foundation construction techniques have progressed with advances in the state of the art. The history and development of the Mall have also played an important role.

INTRODUCTION

The Monumental Core as defined for the purposes of this paper comprises the National Mall in Washington, DC, which extends from the United States Capitol on the east to the Lincoln Memorial on the west, encompassing the Washington Monument, the Reflecting Pool and the National World War II Memorial now under construction. Also included are the Korean War Veterans, the Vietnam Veterans, the Franklin Delano Roosevelt and the Thomas Jefferson Memorials.

GEOLOGY

The Washington, DC area is located within the Coastal Plain and Piedmont Physiographic Provinces. The boundary between these two provinces, known as the Fall Line, runs southwesterly from the District of Columbia-Montgomery County boundary near Silver Spring across the Potomac River north of Roosevelt Island. The Monumental Core is located within the Coastal Plain. The Coastal Plain typically contains Pleistocene terrace deposits and recent river alluvium at the lower levels, rising into exposed Cretaceous sediments on higher ground. The Piedmont Province extends from the Hudson River near Nyack, NY to a point just north of Montgomery, Alabama. It is predominantly a rolling upland developed on intensely folded and faulted metamorphic and igneous rocks. Local relief is on the order of 50 ft (15 m), with occasional greater relief near deeply cut stream valleys. Dissection is often greatest near the Fall Line. The metamorphic rocks in the Washington area include the Wissahickon Formation, the Sykesville Formation, and the Laurel Formation. The igneous rocks are more recent intrusions into the older metamorphic rocks. A thin mantle of soil covers much of the Piedmont.

[1] Associate, Mueser Rutledge Consulting Engineers, 225 West 34th Street, New York, NY 10122, phone 917-339-9300, e-mail dchristie@mrce.com.

481

Rock

Top of rock along the Mall ranges from about Elev. -220 ft (-67 m) at the Capitol to about Elev. -20 to -40 ft (-6 to -12 m) west of the Lincoln Memorial, with a localized high point of approximately Elev. -20 ft (-6 m) at the east end of the Reflecting Pool. (Darton 1951) Elevations are referenced to Mean Sea Level.   A generalized geologic section is presented on Figure 1. A portion of Darton's map is presented on Figure 2.

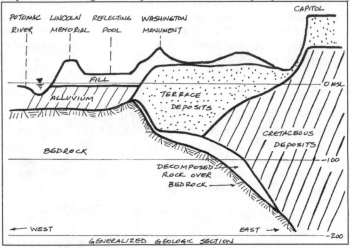

Figure 1 – Geologic Section Along Mall Looking North

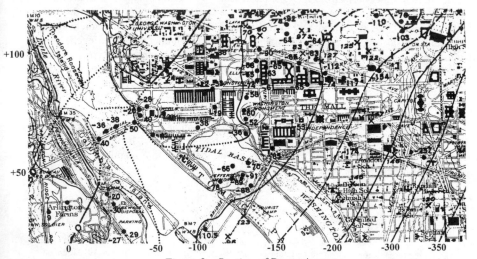

Figure 2 – Portion of Darton's map.

Cretaceous Deposits

The Cretaceous coastal plain sediments consist of a succession of wedge-shaped layers which were deposited in relatively shallow seas on the sloping bedrock surface by streams flowing eastward out of the continental interior. The interfaces between successive Cretaceous formations dip towards the southeast and the wedges thicken in the same direction. The Cretaceous sediments are lenticular on a large scale as a result of changing conditions of deposition but are much more regular in stratification than the younger overlying soils. The lowermost Cretaceous strata are grouped in the Potomac formation and consist primarily of the Patuxent arkosic sands and Patapsco clays. Erosion has removed a great thickness of the Potomac formation in downtown Washington. The Potomac formation is not present at the western portion of the Mall, but appears between 12th and 7th Streets and is approximately 250 ft (75 m) thick at the Capitol. (Mueser, Rutledge, Wentworth & Johnson, 1970)

Pleistocene Terrace Deposits

The uppermost natural sediments in the downtown Washington area comprise a succession of river terrace deposits of Pleistocene times which overlie the Cretaceous formation. A time gap of many million years is represented at the discontinuity between the two major groups of materials. These Pleistocene terraces consist of a mixture of silty and sandy clays with sands, interlayered and lensed in a complex pattern. While continental ice did not reach south to the Washington area, Pleistocene terraces were formed by debris carried in streams charged by glacial meltwater flowing from the north and northwest. The complicated alteration of soils in the terrace is a result of successive changes of sea level and rate of flow of runoff during periods of glaciation and interglacial stages. At the time of ice advance the level of the sea fell with respect to the land, stream gradients increased and sediment load decreased, resulting in a period of erosion or downcutting. During recession of the glacier inflow increased, sea level rose and comparatively coarse-grained materials were deposited. As the warming trend continued, the area was inundated and the finest grained sediments were laid down. A series of these flattop terraces at several characteristic elevations have been identified in the Washington area. These include the 25-foot terrace, the 50-foot terrace, with surface elevations between 40 and 60 ft (12 and 18 m) above sea level, and the 90-foot terrace with surface elevations between 70 and 100 ft (21 and 30 m) above sea level. Each terrace exhibits a characteristic change in gradation in a vertical profile from coarse-grained and gravelly soils at its base to sands, silts, and clays at shallower depths, corresponding to the change from low sea level at the start of ice retreat to high sea level at the warmest time of the interglacial period. Capitol Hill is at a promontory of the 50-foot terrace. The remainder of the Mall lies within the 25-foot terrace and recent deposits as illustrated in Figure 3. (Mueser, Rutledge Wentworth & Johnston, 1970)

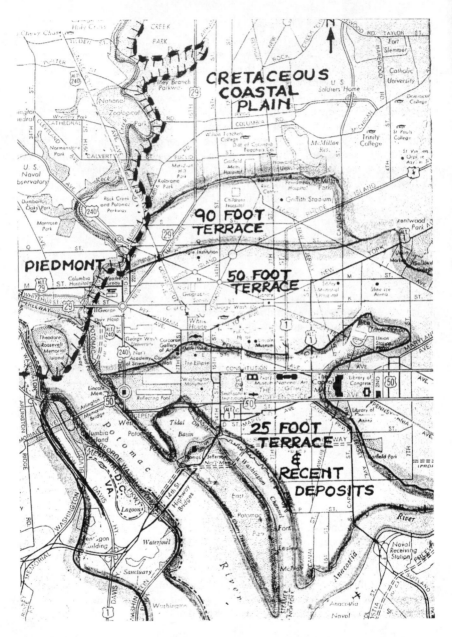

Figure 3 – Limits of Pleistocene terraces. (Courtesy of James P. Gould.)

Fill

A thin layer of urban fill is present throughout the Mall. Deeper fill is present at the west end of the Mall, west of the Washington Monument and includes dredge spoils. Fill is 20 to 25 ft (6 to 8 m) thick at the west side of the Monument, and increases to 30 to 40 ft (9 to 12 m) at the Lincoln Memorial.

A PARTIAL HISTORY OF THE NATIONAL MALL

L'Enfant's Plan

In 1791 Pierre L'Enfant proposed a plan in which he envisioned the Mall as the central axis of the Monumental Core. The Mall was to be the foremost avenue of the city, the so-called "Grand Avenue." It was to run west from the Capitol to a point directly south of the President's House where its terminus would be crowned by an equestrian statue of George Washington. According to L'Enfant's plan, the Mall was to be "four hundred feet in breadth, and about a mile in length, bordered by gardens, ending in a slope from the houses on each side." In 1792, L'Enfant was dismissed for insubordination.

During the course of the 19th century, L'Enfant's formal design for the Mall was largely forgotten. During the Civil War, the Mall grounds were used for military purposes, such as bivouacking and parading troops, slaughtering cattle and producing arms. In 1872, a 14-acre tract at 6th and B Streets was given to the Baltimore and Potomac Railroad for the construction of a depot. The railroad was also granted permission to lay tracks north to south across part of the Mall. (National Park Service, 2001s)

Downing's Plan

In 1851 Andrew Jackson Downing, a New York architect, was hired by President Fillmore to prepare a plan for development of the Mall. Downing's plan was adopted by the federal government in connection with the construction of the Smithsonian Institution and the grounds allocated to it. This plan called for English naturalistic landscaping and winding roads on the Mall, but was never fully executed. The first memorial constructed and dedicated on the Mall honored Andrew Jackson Downing. This memorial was dedicated in 1856, and is located in front of the Smithsonian Arts and Industries Building, just east of the Castle.

Canals

The canals of Washington were intended to be a major transportation artery in the scheme to build the federal city into an important port and trade center. An important part of George Washington's and Pierre L'Enfant's thinking was a canal through the middle of Washington, to link the Potomac and Anacostia Rivers. One of the main purposes of this canal was to provide a means of transporting goods to the center of the city, eliminating the need to haul them from the river. L'Enfant also expected the canal to be a beautiful waterway which would contribute to the aesthetic value of the city. L'Enfant's canal would run east from the Potomac at the mouth of the Tiber Creek nearly to the Capitol, then proceed southeast, splitting into two branches south of the Capitol. One branch would empty into the Anacostia just west of the Navy Yard, and the other branch would trend to the southwest, incorporating James Creek, which entered the Anacostia in a small bay just east of the arsenal (Greenleaf's Point). The canal routes are shown on McClelland's map presented in part as Figure 4.

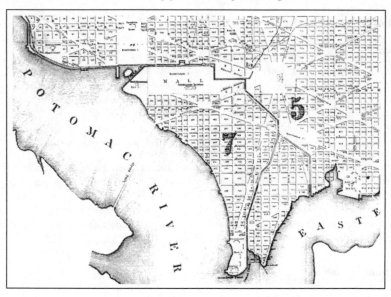

Figure 4 – Portion of McClelland's Map (1850)

The Washington City Canal was built essentially as planned except that the lower portion of the James Creek branch was not included. Construction began in the mid-1790s, but it was almost entirely constructed between 1802 and 1815, with much of the work done between 1810 and 1815. The Tiber Creek was converted into a portion of the canal from its outlet at 17th and Constitution eastward to the Capitol. Prior to its channelization, the Tiber Creek occupied a broad basin along what is now Constitution Avenue, but extending nearly to D Street NW between 13th and 14th Streets NW and a similar distance south of Constitution Avenue in the same area. It was also called the Goose Creek in this area. When it was converted to the canal, its channel was filled in where it diverged from the canal's alignment, forcing all of its flow into the canal. A portion of the canal to the west of the Capitol is shown in Figure 5.

Figure 5 – Washington City Canal and U. S. Capitol about 1858.
(Photograph from Library of Congress Collection.)

The total length of the Washington City Canal was 15,330 feet, and the original depth was about 4 feet. The Canal was used intermittently from 1816 to about 1850. It was subject to filling by sand and silt from both ends, from the Potomac and Anacostia Rivers. After about 1850 it was virtually useless, victim to sedimentation, inadequate maintenance, and the railroad. It was reputed for collecting sewage and presenting objectionable odors at low tide, as well as posing a potential health hazard. In 1871 the Board of Public Works began filling in the canal from the Rock Creek Basin to the Anacostia River. This was essentially completed by 1881 except for the last two blocks between L and N Streets South. The Constitution Avenue section was partly made into a covered sewer by building a new wall parallel to one of the canal walls, arching over the intervening space to form a conduit, and filling in the unneeded channel outside the sewer. Some of this old sewer was used in the 1970s to carry storm water and air-conditioning cooling water from Federal buildings to the Tidal Basin. The lower part of the James Creek was made into a canal in 1876, and was filled in again gradually between 1916 and 1931. Canal Street SW charts the course of the canal south of Maryland Avenue. A portion of Canal Street has since been renamed Washington Avenue. (Williams, 1972)

Former Lakes and Fill Placement at the Mall

In 1882 the US Congress appropriated funds to commence a major "Land Reclamation Project." Dredging of the Potomac River as well as the mouth of the Tiber Creek led to the creation of East and West Potomac Parks. The reclamation work, conducted by the US Army Corps of Engineers, nearly doubled the length of the Mall and created more than 700 acres (283 hectares) of new land within the city's "Federal Central Enclave." (Land Reclamation Project, 2000)

This had the effect of lengthening the Mall to approximately 2 miles (3.2 km) and extending it westward to its present boundaries by about 1900. According to a map prepared by the District of Columbia Office of the Engineer Commissioner in 1896 shown on Figure 7, the Potomac River's boundaries had been established at their present locations, the Tidal Basin adjacent to the Jefferson Memorial had been created, and the Mall had generally taken on its present shape. The piers in the present day Washington Channel along Maine Avenue reflect the original eastern shoreline of the Potomac River from the 1850s.

Two lakes were created in the fill adjacent to the Washington Monument, one to the northwest of the Monument in an area bounded by Constitution Avenue on the north and 17th Street on the west and another due west of the Monument and bounded by 17th Street on the west. They are visible in Figure 6, which also shows the partially

Figure 6 – Looking West along the Mall; Agriculture Department at left, with partially completed Washington Monument and Potomac River in background, 1866. (Photograph from Library of Congress Collection.)

completed Washington Monument. They are remnants of the old canal basin. The lakes also appear in Figure 7, a city map dated 1896. Due to concerns about the proximity of the pond to the foundations of the Washington Monument, Babcock Lake on the north side was filled in with 83,000 cu yd (63,000 m³) of earth in 1887. At the same time, 275,000 cu yd (210,000 m³) of fill was placed around the base of the Monument to create a more natural looking landscape. (Torres, 1984)

According to the 1896 map (Figure 7), the Mall was divided into seven sections as follows: The Capitol Grounds from 1st Street East to 1st Street West, the Botanical Garden from 1st to 3rd Streets West, the Public Gardens from 3rd to 6th Streets West, Armory Square from 6th to 7th Streets West, the Smithsonian Grounds from 7th to 12th

Streets West, the Agricultural Grounds from 12th to 14th Streets West, and the Monument Grounds west of 14th Street West. This layout reflects some of Downing's Plan.

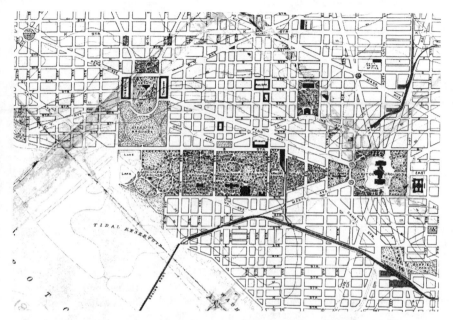

Figure 7 – Portion of DC Engineer's Map of 1896.

The McMillan Plan

In 1900 The Senate Committee of the District of Columbia was directed to prepare a comprehensive plan for the development of the entire park system in the District of Columbia. The resulting plan, known as The McMillan plan, was published in 1901. It relied heavily on L'Enfant's plan of 1792 for the Mall. One major goal was to beautify the Mall including the newly reclaimed Potomac Flats. The Committee was tasked with eliminating such undesirable elements as the railroad station, railroad tracks at grade, grazing animals, sheds, and other occupants and restoring the uninterrupted green space envisioned by L'Enfant.

The areas west of 17th Street and east of 14th Street were developed in general accordance with the McMillan Plan. At the base of the Washington Monument the McMillan plan called for a cut of up to 27 ft (8 m) to create a formal sunken garden on the west and a fill of up to 25 ft (8 m) to create a terrace on the east. Congress authorized $5 million for this work in 1928. However, studies performed under the

Independent Offices Act of 1931 suggested that this earthwork could endanger the stability of the Monument. Based on those studies, a decision was made to consider other plans, both formal and informal, for the Monument Grounds. (Improvement of the Washington Monument Grounds, 1931)

STRUCTURES ON THE MALL

United States Capitol

The United States Capitol was originally called The Federal House, and was constructed a mile southeast of the President's House on a grassy knoll known as Jenkins' Hill, the crest of a 90-foot terrace which afforded a commanding view of the area down to the rivers. The cornerstone for the Capitol Building was laid on September 18, 1793, and by November 17, 1800, the original Senate wing was completed for the first meeting of the Congress in the city of Washington. The construction of the original House wing extended over the period from 1800 to 1811. In 1814, during their occupation of Washington, the British set fire to a portion of the building that was complete, and destroyed much of the interior and the roof. The rebuilding of the original structure was undertaken and in December 1819 Congress met again at the Capitol. The central unit was begun in 1818 and the Rotunda was completed and covered by a wooden canopy by 1827.

By 1850 it became apparent that the original building was insufficient for the use of Congress and the construction of the present Senate and House wings and their connections with the original building commenced in 1851. The new house wing was completed in 1857, and the Senate wing in 1859. In 1855 the original canopy was removed to make way for the construction of the present dome which was completed in 1863. The terrace structure on the Capitol's west, north, and south sides was constructed between 1874 and 1892. An extension of the east face of the Capitol was completed in 1959. The face stone for the original building consisted of a gray sandstone from Aquia Creek, Virginia. The main foundation walls were built from a hard diorite, granite, and granite gneiss. According to tradition, rock for the foundations of the Capitol and the White House was obtained from the granite gneiss outcrop located immediately north of the position of the Lincoln Memorial on which Braddock is supposed to have landed on the way to Fort Duquesne. The Capitol building is founded on the sand and gravel of the Terrace deposits. The deeper of the main foundation walls below Elev. 70 ft (21 m) bear on sand and gravel known locally as Stratum T3.[2] The shallower of the main foundations are supported on the sand and silt of Stratum T2. The foundation walls and piers of the Capitol's terrace structure on the west side generally bear in the silt of Stratum T1. A section through the site is presented on Figure 8. Strata references on Figure 8 predate the WMATA work and do not follow that system. (Moran, Proctor, Mueser & Rutledge, 1957)

---

[2] Soil strata references in this paper follow the system established for the Washington Metropolitan Area Transit Authority soils reports and now used widely throughout the Washington area.

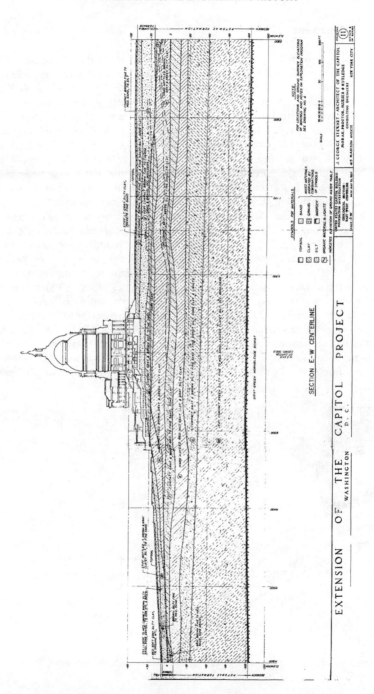

Figure 8 – East-West Section through the Capitol (Moran Proctor Mueser & Rutledge, 1957).

Washington Monument

The site originally chosen for the Washington Monument was at the intersection of the east-west axis of the Mall with a north-south line centered on the White House. Due to the poor soil conditions at the intersection, the location was shifted approximately 370 ft (113 m) east and 123 ft (38 m) south to a point where better ground conditions were present. Construction began in 1848 with private funds raised by subscription. The Monument foundations were established at approximately Elev. 15 ft (5 m) atop sand and clay soils of Pleistocene Terrace deposits of Strata T1(A) and T2. When funds ran out in 1856 work was stopped, with the Monument at a height of 152 ft (48 m). Congress was repeatedly approached for funding, but the Civil War intervened. Ultimately funding was forthcoming from Congress in the late 1870s to resume construction. At that time, concerns were raised about the ability of the spread foundations bearing on these soils to carry the load of a masonry structure in excess of 500 ft (152 m) in height. Thomas Lincoln Casey, an experienced lieutenant colonel from the US Army Corps of Engineers, designed a scheme to underpin the foundations, increasing the contact area of the foundation from 6400 sq ft (600 sq m) to 16,000 sq ft (1500 sq m). Buttresses were added to provide load transfer to the underpinning. Excavations for the underpinning extended to approximately Elev. +3 ft (1 m), the groundwater level at that time. The new foundations bear on Pleistocene Terrace Stratum T3, a very compact silty sand, with gravel, and have performed satisfactorily. Total settlements since completion of underpinning are about 7 in (180 mm). More than 60 percent of this settlement occurred as the Monument was built from 152 ft (48 m) to 555 ft (169 m) in height. A perspective through the foundations and soils is shown on Figure 9. Construction of the buttresses as part of the underpinning is shown on Figures 10 and 11.

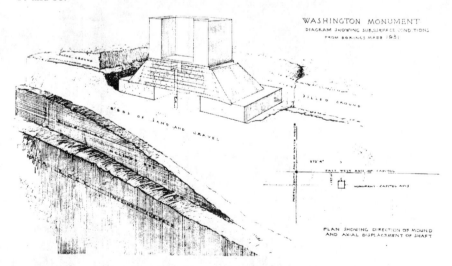

Figure 9 – Perspective view at the base of the Washington Monument (1931)

Figure 10 – Constructing buttresses at
Washington Monument (1879)

Figure 11 – Completed buttresses at
Washington Monument (1879)

World War II Memorial

Soils at the site of the World War II Memorial consist of recent fill, dredged fill, tidal marsh deposits, Cretaceous deposits, and saprolite. (US Army Corps of Engineers, 1997) The Cretaceous deposits and saprolite are relatively thin. The Memorial is designed to bear on rock at a depth of about 30 ft (9 m) below grade. The Memorial is surrounded by a structural slurry wall keyed into rock, which provides a groundwater cutoff and support for the perimeter. The interior of the Memorial will be supported on H piles driven to rock.

These methods of support were chosen due to the poor soils at the site and the need to minimize impact of construction on the memorial elm trees surrounding the site, both in terms of limiting the zone of disturbance and minimizing drawdown of groundwater and potentially drying out the tree roots. The slurry wall also addressed concerns about the effects of drawdown on the Reflecting Pool with its timber piles and slab on grade. A photograph of slurry wall construction is presented on Figure 12.

*Figure 12 – Reinforcing cage being placed for WWII
Memorial Slurry Wall (2002)*

Reflecting Pool

The Reflecting Pool was constructed in 1920 and 1921. Borings made in 1920 indicated top of rock ranging from Elev. –17 to –40 ft (-5 to -12 m). Overlying soils are probably similar to those at the World War II Memorial, except that Cretaceous deposits are presumed absent. According to drawings from 1921, piles driven to support the coping of the Pool extended to rock at approximately Elev.-30 to -40 ft
(-9 to -12 m), with a limited number of piles encountering rock as high as Elev. -25 ft (-8 m) on the northeast corner. Piles are composite, consisting of a concrete upper section placed within a steel shell and a timber lower portion. The bottom of the Pool is a slab on grade.

Rainbow Pool

The Rainbow Pool was constructed in 1920 and 1921. It was supported on composite piles similar to the Reflecting Pool, with tips on rock between approximately Elev.-17 and -28 ft (-5 to -9 m). It was removed to permit construction of the World War II Memorial, and will be reconstructed within the Memorial.

Lincoln Memorial

The Lincoln Memorial was constructed between 1914 and 1922. Test borings were made in 1913 to determine top of rock and to determine that rock was at least 2 ft (0.6 m) thick. Subsurface conditions were "sand over mud (clay) and gravel over coarse sand over mud (clay) over rock." Rock was determined to consist of blue gneiss reported in the original boring logs as having "a very hard makeup consisting of considerable quartz." Penetrating the rock was achieved by alternately driving the drill with a sledge hammer and lifting and dropping the drill. In this way two feet (0.6 m) of rock could be drilled in two to eight hours. (Boring logs for the Lincoln Memorial, 1913)

Rock is at approximately Elev. -30 to -45 ft (-9 to -14 m), and site grade is approximately Elev. 29 ft (-9 m). Ground surface at the time of construction was approximately Elev. 16 ft (5 m). The height of the Memorial above grade is approximately 100 ft (31 m).

Foundations for the main building consist of two portions. The portion below the original ground level of the park, known as the subfoundation, comprises 122 concrete piers formed within steel cylinders driven to rock. The cylinders vary in length from 49 to 65 ft (15 to 20 m) and in diameter from 3'-6" to 4'-2" (1070 to 1270 mm). They were sunk by being heavily weighted and water-jetted to a depth of absolute resistance. They were excavated by hand, the excavation extended 2 ft (600 mm) into bedrock and the cylinders filled with concrete, with a steel reinforcing cage set in each cylinder. Above the subfoundation is the upper foundation, consisting of concrete columns approximately 45 ft (14 m) high constructed atop these piers, with the column tops joined by arches cast integrally. The terrace wall and approaches were to be founded similarly in the original design, but changes were made as a result of the test borings and the wall was constructed on a shallow foundation. Due to continuing settlement of these structures, a decision was made in 1920 to underpin them to rock. The freestanding columns were shored with timbers and underpinning pits sunk to depths of 12 to 14 ft (3.7 to 4.2 m). Caissons were started by drifting from the pits to the locations of the footings for the columns and terrace wall. Caissons were excavated with picks and shovels down into rock, and dewatered by air pumps. Concrete was placed in the caissons up to within 6 in (150 mm) of the bottom of the footings. The 6-in (150 mm) spaces were subsequently dry packed. The access pits were backfilled and reinforced concrete struts constructed from the foundation wall of the main building to the freestanding columns to the terrace wall. The work was performed on every second set of columns and footings, and the intervening sets similarly underpinned when the first sets were completed. The approaches were underpinned using needle beams and girders supported on concrete piers to rock. All of the steelwork was subsequently encased in concrete. (Conklin, 1927)

In addition to the visible portions of the Memorial, a series of below-grade chambers are present extending east under the circle to the west end of the Reflecting Pool. The National Park Service formerly conducted tours of these chambers but discontinued them because the floors were slippery due to persistent dampness in the passageways. A photograph showing the Memorial under construction is shown in Figure 13. A transverse section is shown on Figure 14, and shows the portion below grade extending toward the Reflecting Pool.

Figure 13 – Lincoln Memorial under construction (1916)

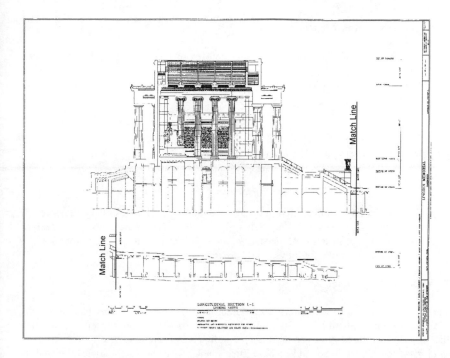

Figure 14 – Historic American Buildings Survey drawing of Lincoln Memorial (1993)

Thomas Jefferson Memorial

Ground was broken for the Thomas Jefferson Memorial in 1938 and it was completed in 1939. It is supported on piers to bedrock which extend a maximum depth of approximately 140 ft (42 m). The top of the dome is approximately 130 ft (40 m) above the roadway. The structure contains limestone from Indiana and marble from Vermont, Tennessee, Georgia, and Missouri.

Vietnam Veterans Memorial

The Vietnam Veterans Memorial is supported on concrete piles driven approximately 35 feet to rock. Ground was broken in March 1982 and the memorial was dedicated in November 1982. The black granite panels are from India.

Korean War Veterans Memorial

The Korean War Veterans Memorial is supported on H-piles driven 40 to 50 feet to rock. Ground was broken in November 1993 and the Memorial was dedicated in July 1995.

Franklin Delano Roosevelt Memorial

Ground was broken in September 1991 and the Memorial was dedicated in May 1997. The Memorial is supported on 900 steel piles driven approximately 80 ft (24 m) to rock. The Memorial consists of granite from South Dakota and Minnesota. (National Park Service, 2001b)

STRUCTURES BELOW THE MALL

Metro Station - Smithsonian

Top of rail at the Smithsonian Station on the Blue and Orange Lines along 12th Street is at approximately Elev. -8 ft (-2.4 m). This places the station in the Pleistocene Terrace deposits. Prior to construction of the station, groundwater was at approximately Elev. -5 ft (1.5 m). The station was built by cut and cover methods.

7th Street Metro Tunnel

The 7th Street Metro tunnel carrying the Green and Yellow Lines crosses beneath the Mall with invert at about Elev. -20 to -30 ft (-6 to -9 m). The deepest portion is beneath the centerline of the Mall. Decomposed rock in this area is approximately Elev. -90 ft (-27 m) or deeper, with sound schistose-gneiss bedrock at the south side of the Mall at Elev. -120 ft (-37 m). The rock is overlain by Cretaceous deposits to about Elev. -30 ft (-9 m), overlain in turn by Pleistocene Terrace deposits. Typical urban fills were encountered above the Terrace deposits, with thickness of about 8 to 10 ft (2.4 to 3 m), but as deep as 15 ft (4.5 m) near Constitution Avenue NW, presumably on the line of the Tiber Creek. Groundwater at the time of the Metro borings in 1970 was approximately Elev. -5 to -10 ft (-1.5 to -3 m). The subway consists of twin tunnels approximately 21 ft (6.4 m) in diameter each, mined with a digger shield. Primary lining was ribs and lagging, and the secondary lining was 15 in (380 mm) of reinforced concrete.

12th Street Metro Tunnel

The 12th Street Metro tunnel carrying the Blue and Orange Lines crosses the Mall with invert at approximately Elev. -10 ft (-3 m) on the south and about -30 on the north. Conditions consist of decomposed rock at about Elev. -60 to -80 ft (-18 to 24 m), overlain by Cretaceous deposits to Elev. -50 to -60 ft (-15 to -18 m). Cretaceous deposits were overlain by Pleistocene Terrace deposits overlain in turn by urban fill. The fill was typically about 10 ft (3 m) thick, but as deep as 20 ft (6 m) on the line of Constitution Avenue NW along the Tiber Creek alignment. Groundwater at the time of the Metro borings in 1968 was about Elev. -20 ft (6 m) on the north and Elev. -5 ft (-1.5 m) on the south. The subway was constructed by cut and cover methods.

Highway Tunnels

Ninth and Twelfth Street Tunnels

The Ninth and Twelfth Street Tunnels under the Mall were constructed in the late 1960s as part of the Southeast Southwest Freeway Project. They carry their respective streets beneath the Mall at relatively shallow depths.

Third Street Tunnel

The Third Street Tunnel carries Interstate Highway 395 beneath the Mall just east of Third Street West from D Street SW to D Street NW. The tunnel was built by cut-and-cover methods. It is approximately 3200 ft (1000 m) long and extends 19 ft (6 m) below local groundwater level. The tunnel opened in 1973.

REFERENCES CITED

Conklin, Edward F., (1927) *The Lincoln Memorial, Washington,* United States Government Printing Office, Washington.

Darton, N.H., (1951) *Map Showing Configuration of Bedrock Surface in the District of Columbia Region.*

Hoover, Herbert, (1933), Improvement of the Washington Monument Grounds, Communication from the President of the United States.

Moran, Proctor, Mueser & Rutledge, (1957), *Report on the Foundation Investigation of the Extension of the Capitol Project.*

Mueser, Rutledge, Wentworth & Johnston, (1970) *Final Report – Subsurface Investigation, Branch Route, Stations 3+00 (F001) to 95+70 (F002), Washington Metropolitan Area Transit Authority,* (WMATA Report No. 41).

The National Archives, (1913), Boring logs for the Lincoln Memorial.

National Park Service, (2000), *Land Reclamation Project, Creation of the National Mall,* http://www.nps.gov/nama/feature/timeline1.htm.

National Park Service, (2001a), *National Mall,* http://www.cr.nps.gov/nr/travel/wash/dc70.htm

National Park Service, (2001b), *Stones and Mortar,* http://www.nps.gov/nama/mortar/mortar.htm

Torres, Louis, (1984) *"To the immortal name and memory of George Washington,"* The *United States Army Corps of Engineers and the Construction of the Washington Monument,* Washington, D.C.: Historical Division, Office of Administrative Services, Office of the Chief of Engineers.

US Army Corps of Engineers, Baltimore District, (1997) *World War II Memorial, Washington, D.C., Final Geotechnical Report.*

Williams, Garnett P., (1972) *Washington, D.C.'s Vanishing Springs and Waterways,* US Geological Survey Circular 752.

ACKNOWLEDGEMENTS

The author wishes to acknowledge assistance from the following persons, without whom this paper would not have been possible: Hugh S. Lacy, Mueser Rutledge Consulting Engineers, for support and technical guidance; Mary Kay Lanzillotta, Hartman Cox Architects; William C. Hobson, Mueser Rutledge Consulting Engineers; James Darmody, Washington Metropolitan Area Transit Authority; Jay Padgett, GeoServices Corporation; Robert Kapsch and John Burns, National Park Service, all for technical information, and Martha Huguet and Darlene Ahl, Mueser Rutledge Consulting Engineers, for assistance in preparing the manuscript and exhibits.

Perspectives on American Civil Engineering: A View from Afar

A British perspective on American civil engineering achievement before 1840

Roland Paxton*

Abstract

Consideration of selected remarks of leading British engineers D. Stevenson and T. Telford on this theme has enabled the essential contribution of timber construction to America's early public works infrastructure to be better understood. As an example, L. Wernwag's 1810 Neshaminy River timber and iron truss bridge, of which I.K. Brunel possessed a lithograph, is discussed as an innovative achievement in bridge development.

Introduction

From a search for a perspective from contemporary British engineers, including B.H. Latrobe, W.Weston, Sir M.I. Brunel, and C.B.Vignoles, all of whom worked in the USA, and J. Rennie, T. Telford and D. Stevenson who did not, although they advised on some North American projects, only contributions by Stevenson and, to a lesser extent, Telford came to light. Both are considered authoritative and reliable commentators.

Thomas Telford (1757-1834) was at the head of the British civil engineering profession with an international reputation for innovation and excellence in road, bridge and canal engineering. His contribution is taken from the influential *Edinburgh Encyclopaedia(1)* to which he was a major contributor and a leading proprietor, and a parliamentary report*(2)*. David Stevenson (1815-1886), uncle of the writer Robert Louis Stevenson, was an up-and-coming young engineer who had had an exemplary training as a civil engineer under his father Robert Stevenson and national railway contractor William Mackenzie in the Midlands and at Edge Hill tunnel on the Liverpool and Manchester railway. He was destined within a decade to become head of the Stevenson engineering consultancy and one of Britain's leading river navigation, harbour and lighthouse engineers. From 1853-85, he was Engineer to the Northern Lighthouse Board and an international authority on river and lighthouse engineering.

Stevenson was a skilful writer producing more than 60 authoritative publications*(3)*. Those relating to this subject are, *Sketch of the Civil Engineering of North America(4)*, and papers to the Royal Scottish Society of Arts (of which he became twice President) on

----

*MBE, PhD, FICE, FRSE. Chairman ICE – PHEW. Hon. Prof. Heriot-Watt Univ., Edinburgh, EH14 4AS, Scotland; tel. 01144-131-449-5111; e-mail rolandp@civ.hw.ac.uk

"Long's frame bridge"(1839), and "Building materials of the United States of America" (1841), which were also published in the *Edinburgh New Philosophical Journal* (1841). Reference has also been made to Stevenson's pocket book, diary and autobiography.

American support for the selection of Stevenson as an appropriate source for this paper comes from Brooke Hindle who calls Stevenson's *Sketch* the best of the early European works on American engineering, adding that the earliest American surveys of native civil engineering feats were not as elaborate or well executed as British works on American technology(5). Also, from Kip Finch's scholarly critique on the *Sketch*. He wrote "one cannot fail to be impressed with the outspoken yet, fair, impartial and keen observation of this young engineer of 23 years"(6).

Stevenson's 3-month North American tour in 1837 (5 April- 8 July)

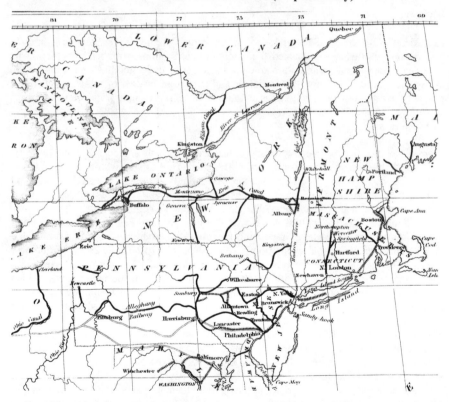

Figure 1. Map covering Stevenson's tour, 1837(3). [basically, New York – Philadelphia - Baltimore – Washington – Philadelphia – Harrisburgh - Alleghany Railway – Pittsburg – Erie – Buffalo – Niagara – Toronto – Montreal – Quebec – Montreal – Whitehall -Albany – NewYork - Boston – Lowell – Boston – Newhaven - New York, with local excursions. He traveled, 2000 miles by steamer, 450 miles by road and 590 miles by railway(7).]

Before Stevenson visited America reports were prevalent in Britain "of great American feats in Engineering and Mechanics which no one was disposed to put much faith in, the Americans having got the unenviable credit of being better at drawing 'long bows' than doing real work"*(8)*. His *Sketch* under the following heads helped to dispel this myth.

Harbors. Stevenson noted that in many ports the accommodation for large ships had been obtained at a much smaller expense than European docks and harbors of similar capacity. Economy in design and construction had been achieved by means of a quay arrangement of primitive and temporary structures, using wood from land clearance, which were fit for purpose because of their sheltered natural locations near deep water. Harbors lacked quay cranes. Ship repair facilities were primitive. The only two dry docks, the finest specimens of masonry Stevenson saw in America, were at Boston and Norfolk for naval use.

Stevenson instanced New York harbor with its tidal range of about 5 ft. as a typical example of a good American harbor. Protected from the Atlantic waves and without the aid of docks or dredges, vessels of the largest class lay afloat during low water of Spring tides, moored to the quays, and by the erection of wooden jetties the port was enlarged at low cost. For ship repair, he noted the "screw" dry dock arrangement for suspending vessels from a submersible wooden platform and thought the hydraulic version "a beautiful application of the principle of Bramah's press"*(9)*. As regards shipping, the American packet ships trading between New York and Europe were "generally allowed to be the finest class of merchant vessels at present navigating the ocean"*(10)*. Their voyages, by sail, averaged 22 days from New York to Liverpool and 32 days in return*(7)*.

Stevenson considered the harbors along the 4,000 miles of coast from the Gulf of St. Lawrence to the Mississippi, in conjunction with river, lake and canal navigation, to have had a "mighty" effect in advancing the prosperity of America*(10)*.

Lake and river navigation. In contrast to the East Coast seaports, most of the lake harbors were formed in exposed locations and were of a more permanent construction. The 1,452 ft. long Buffalo pier was formed of pitched rubble masonry without mortar. Presque-Isle had breakwaters of 3,000 ft. and 4,000 ft. in length. Dunkirk harbor breakwater had been formed during winter by erecting strong wooden cribs filling them with stones on the ice in the correct position and then breaking the ice to sink them into place. Most lake harbors had encountered wave damage and required annual repair.

Stevenson comments that steam navigation was first fully and successfully introduced into real use in America on the River Hudson in 1807 and that it was still capable of further improvement, particularly in the taking of measures to reduce the incidence of boiler explosions. He found the Western Water (Ohio–Mississippi) steamboats with their small high-pressure engines built for economy and finery of their cabins often decidedly unsafe. In contrast, the Great Lakes steamers were strongly built and similar to sea-going boats.

The Eastern Water steamboats were characterized by their small draught, light and slender construction, great speed, and use of large condensing engines. From timings made on a trip from Albany to New York in the "Rochester", Stevenson calculated its average speed at 14.97 mph with a maximum of about 16.5 mph*(11)*. Its "combination of fine lines and great engine power" produced nearly double the speed of the fastest British steamer. "The secret was that the bow of the Yankee boat was like a knife and the speed of her piston 500 ft. per minute as against the English tub's 210 ft.". Stevenson had 8 ft. long models made of the "Rochester" and the sea-going "Naragansett" and published drawings of their lines and "very soon thereafter vessels with fine lines and long stroke pillar engines were plying on the Clyde at higher speeds than in America!"*(7)*.

Canals. Stevenson considered the North American canals "stupendous" in that they enabled vessels suited to inland navigation to pass from the Gulf of St. Lawrence to the Gulf of Mexico, and from New York to Quebec or New Orleans, the latter involving a journey of 2,702 miles*(12)*. He emphasized that the most remarkable feature of these canals was their length, in which they far surpassed those of Europe, rather than their cross-sectional area which was not as large as that of many European canals, and the zeal and rapidity applied in their provision. The chief objective of their designers was to achieve works quickly with economy, safety and stability, but "although on an extensive scale...not in the same spacious style as that of older and more opulent countries"*(13)*.

Stevenson commented, "At the first view, one is struck with the temporary and apparently unfinished state of many of the American works, and is very apt before inquiring into the subject, to impute to want of ability what turns out, on investigation, to be a judicious and ingenious arrangement to suit the circumstances of a new country, of which the climate is severe...where stone is scarce and wood is plentiful, and where manual labour is very expensive"*(13)*. Although "wanting in finish and even in solidity", such works served their purpose efficiently for many years. The undressed slopes of cuttings and embankments, roughly built rubble arches, stone parapet walls coped with timber and timber canal locks did not arise from any lack of engineering skill. The use of wood in canal locks enabled their quick completion at low initial cost and provided for possible improvement later by enabling the easy, cheap and speedy transport of more durable and expensive materials and, where this occurred, the necessity of destroying substantial and costly masonry was obviated.

Stevenson also noted the important contribution of slackwater navigation to inland transport, that is, damming rivers by mounds or sluices, bypassed by locks, to increase water depth along a reach. The 363-mile Erie Canal (543 miles with branches) connecting New York via the Hudson to the Great Lakes made from 1817 to 1825 was "perhaps the most important" American public work, being the first in America to convey passengers and the longest in the world for which accurate information was available*(14)*. By 1837 its success was such that it was being widened to 70 ft. and deepened to 7 ft., by bank raising, largely by "an iron scoop drawn by two horses, which acted like a plough"*(7)*. The River Mohawk was crossed by aqueducts of 748 ft. and 1,188 ft. in length (with timber troughs on masonry supports). The 16 ft. deep Louisville and Portland Canal bypass of rapids on the Ohio had been excavated in rock for nearly all of its length. The Morris Canal built by J.D. Douglass (1790-1849) was remarkable for its use of 23 inclined planes, instead of locks, with an average lift of 58 ft. for 30-ton boats. In his table of details of 79 canals Stevenson noted that on the Chesapeake & Ohio Canal a 4 miles and 80 yd. long tunnel was required through the Alleghany Mountains*(15)*.

Roads. Stevenson found road making "very little cultivated" and most roads in a "neglected and wretched condition". He experienced the discomfort of "corduroy" or log roads over which the coach advanced by a series of jumps "calculated to shake the teeth out of the heads of the unfortunate passengers"(7). Labour for road-making was inadequate and expensive. The best roads were those of New England which were made of gravel, but no attention had been given to forming or draining them. More attention had been given to one or two lines of road, the most "remarkable" being the partly built "National Road" stretching 700 miles from Baltimore to Illinois. Experiments to obtain a durable city road had been made on Broadway, New York, including a pine pavement over which carriages passed easily and noiselessly, a practice later tried in Europe(16).

Bridges. Stevenson found that American bridges were generally constructed of wood and on a scale far surpassing those of Britain, instancing nine examples varying in length from 1500 ft. to over a mile. Although good building materials were generally plentiful, to have built stone bridges would in most cases have been too costly. Many bridges consisted of a wooden superstructure resting on stone piers. Generally they exhibited "good engineering" and he instanced the 5-arch timber bridge over the River Delaware at Trenton (1804-1806) with its 160-200 ft. bowstring spans of laminated planking and iron hangers built by Theodore Burr. Stevenson considered the bridge over the Susquehanna at Columbia with its 29 arches of 200 ft. span built on a similar principle in 1832-1834 in 6 ft. of water by Moore and Evans(7), "perhaps the most extensive arched bridge in the world, a magnificent work and its architectural effect is particularly striking"(17).

Stevenson was impressed by two timber bridges over the River Schuylkill in Philadelphia, Market Street (1801-1804) built by Timothy Palmer and the large span "Colossus" (1812-1813) by Lewis Wernwag, which he sketched on the spot (see Figure 2). Also, more generally, by Ithiel Town's Lattice and Stephen Long's frame timber bridges, both of which Stevenson promoted in British publications, by preparing designs for their use in Scotland at Norham and in India, and making a scale model of the latter.

Figure 2. Stevenson's sketch of the "Colossus" Bridge, Philadelphia in May 1837(18).

Railways. Stevenson commented on their rapid development, noting that between the opening of America's first railway in 1827 and 1837, more than 1,600 miles of railway had been built with a further 2,800 miles ongoing. He took a particular interest in the track and found that there was little uniformity of construction on the different lines. The stone blocks often used to support the rails were frequently damaged by the severe frosts. The use of wooden supports for rails was in most situations more economical than stone. They were also less liable to frost damage, made track repair easier and, being more elastic, were less damaging to locomotives and wagons. Most of the early rails and chairs were of British manufacture. American railroads were generally constructed at a much lower cost than their British counterparts because of their exemption from high land costs and compensation for damages, their execution in a less substantial and costly manner and because wood, the principle material used, was obtained at a very small cost.

5

American Railways

The table on the other side shews the distances the height overcome by means of 10 inclined planes 5 on each side of the mountain the lengths and gradients of which are given in the following table

No of Plane.	Length in Feet.	Gradient.	Height overcome.
Plane No. 1.	1607.74	One in 10.71	150 feet.
... 2.	1760.43	... 13.29	132.40 ...
... 3.	1480.25	... 11.34	130.50 ...
... 4.	2195.94	... 11.68	187.86 ...
... 5.	2628.60	... 13.03	201.64 ...
... 6.	2713.85	... 10.18	266.50 ...
... 7.	2655.01	... 10.19	260.50 ...
... 8.	3116.92	... 10.13	307.60 ...
... 9.	2720.80	... 14.35	189.50 ...
... 10.	2295.61	... 12.71	180.52 ...

The total rise and fall on the whole length of the line which is about 36 miles is 2571.19 feet, of this height 2007.02 are overcome by 10 inclined planes and 564.17 feet by slight inclinations on the other parts of the Railway. The planes are worked by an endless rope passing round grooved wheels at the top and bottom of the plane the Ropes are 7½ in in ~~diameter~~ circumference (now they are 8 inches) Two Stationary engines of 25 h.p. each are placed at the top of each plane. The water for supplying the Boilers is conveyed in wooden pipes in some places one mile

Figure 3. Alleghany Railway details from Stevenson's pocket notebook*(19)*.

Stevenson also took a particular interest in the stationary steam engine operated inclined planes used for transit where the summit level of a railway could not be attained by an gradient sufficiently gentle for the use of locomotive engines or where the formation of such inclinations would have been too costly. He gave as an example the Portage or Alleghany Railway over the Alleghany Mountains which had inclined planes "on a more extensive scale than in any other part of the world"(20). This link was part of a route extending 395 miles from Philadelphia to Pittsburg via the Columbia Railroad, the eastern division of the Pennsylvania Canal, the Alleghany Railroad and the western division of the Pennsylvania Canal. Stevenson travelled this line in 91 hours averaging 4.34 mph. The 36-mile Alleghany Railway had 10 inclined planes overcoming 2,007 ft. the longest of which carried 72 tons/hr.(see Figure 3).

The first steam locomotives in America were made in Britain, but by 1837 they were being made in great numbers in America the largest works being at Philadelphia and Lowell. "Those parts indispensable to the efficient action of the machine were very highly finished but the external parts were left in a much coarser state than in engines of British manufacture"(21).

Waterworks. Although commenting on numerous waterworks, Stevenson found those at Fairmount, Philadelphia (1819-22-36), using River Schuylkill water and operated by waterwheels, "remarkable for their efficiency and simplicity as well as their great extent, being the largest waterworks in North America"(22). New York's "gigantic" Croton scheme devised by J.D. Douglass, although underway was not completed until 1842. Stevenson was impressed with the 1,204 ft. long overfall dam executed on Ariel Cooley's plan using timber cribs filled with stone, sunk into place and then rubble-backed (23).

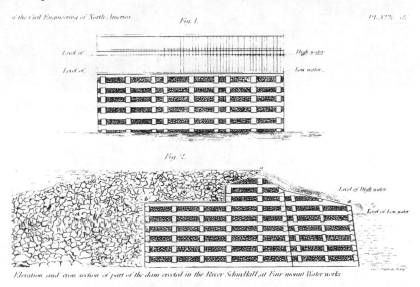

Figure 4. Fairmount Waterworks, Philadelphia – Details of dam built c.1820(4).

This construction method, very generally practised in America for creating slackwater navigations, in this instance the Schuylkill Navigation with 33 other dams and 29 locks, and for some bridge foundations in moving water, had proved successful. For the 98¾-mile city distribution system, pipes made of cast iron were used rather than wood, about one half of which were cast in America and the remainder were imported from Britain.

Figure 5. Fairmount Waterworks, Philadelphia (2002) - After tasteful conservation.

Stevenson was impressed by the extent of the water-power used on the Merrimac and other rivers. He found the mill machinery at Lowell to be excellent and referred to its beneficial effects and excellent regulation. He furnished details of 27 mills.

House Moving. Stevenson noted that the high cost of labour, a dollar a day, twice its British equivalent, encouraged Americans to adopt "many mechanical expedients which in the eyes of British engineers seem very extraordinary"*(24),* one of the most curious being the moving of entire buildings. He saw a brick house in New York measuring 50 ft. by 25 ft. moved 14½ ft. It consisted of three stories and a garret. The whole operation took about 5 weeks, but the time actually employed in moving the house was 7 hours. The sum for which Mr. Burn the house mover contracted to complete the operation was £200. In 14 years he had removed upwards of 100 houses without any accident. Stevenson also saw a wooden church with galleries and spire which was moved 1,100 ft.

Telford

Telford's authoritative treatise on world inland navigation published in the *Edinburgh Encyclopaedia(1)* includes 6 pages relating to North America giving an overview and brief descriptions of particular projects taken mainly from Gallatin's report*(25)*.

Telford considered North America "singularly favoured by nature in the immense extent and ramification of its navigable waters...forming a triangle...of which the line from New Orleans to above Pittsburg on the Ohio, 1,000 miles, may be reckoned the base, and the other sides 1,600 miles each so that the basin of navigation is nearly one half greater than that of the Volga...From the gentle inclination and the very winding course of the rivers, the branches of the Mississippi are for the most part navigable for some kind of craft almost to their heads. But by far the most important branch is the Ohio as affording a direct communication with the eastern states and, being already the channel of a great inland trade, forming a great river which flows 1,188 miles to join the Mississippi. The great extent of ship navigation in the St. Lawrence and its lakes, making altogether a distance of not less than 2,000 miles, and its lying in the direct line towards Europe, will always render this the cheapest channel of communication...By running canals over the spaces where portages and obstructions by rapids...now occur, the most distant sections of the union may be connected". He noted the American "eagerness" to develop their inland navigation and their "already numerous and extensive projects"*(1)*.

Individual elements noted by Telford included the two lower locks 18 ft. deep excavated out of solid rock on the Potomac at Great Falls (Weston 1795-1796) and an aqueduct 280 ft. long and 22 ft. above the river on the Middlesex to Boston harbour canal (1808), "the greatest work of the kind in the United States" at that time.

Telford was generally influenced by the success of James Finley's Merrimac Bridge, Newburyport, suspension bridge from the account in T. Pope's *A Treatise on Bridge Architecture* (1811), but thought that "British dexterity upon superior materials" would improve on American practice*(2)*. From 1814 Telford developed the technology to achieve the Menai suspension bridge by 1826 independently of Finley's practice*(26)*.

Wernwag's "Economy Bridges" at Neshaminy River and Frankfort Creek c.1810

The École Royale des Ponts et Chaussées' lithographed version of Wernwag's bridges (see Figure 6) possessed by I.K. Brunel*(27)*, based on an engraving published at Philadelphia in 1813*(28)*, and reissued in 1815*(29)*, attributes the "construction" of the Neshaminy River and Frankfort Creek bridges to 'Jh. Kirkbright. Surprisingly Wernwag is not acknowledged, but the Philadelphia engravings, with his signature under the design, confirm that the bridges were "built for" Josh. Kirkbright. Both bridges were on the Post Road to New York northeast of Philadelphia and included drawbridges. Their spans were from 52 ft.(draw span)-60 ft. Wernwag was confident that he could extend this principle to spans of 120 ft.(draw span)-150 ft. and that of his "Colossus" (see Figure 6) from 340 ft. 3¾ in. to 500 ft.*(30)*. This bridge also featured in the Philadelphia edition of the *Edinburgh Encyclopaedia* presumably at Telford's instigation or with his approval.

Wernwag's ingenious "Economy Bridge" elevation*(28)* (see Figure 6 – lower, the only details given), appears to show a basic understanding of the main structural forces at work, almost certainly derived from practical expertise and experience rather than theory

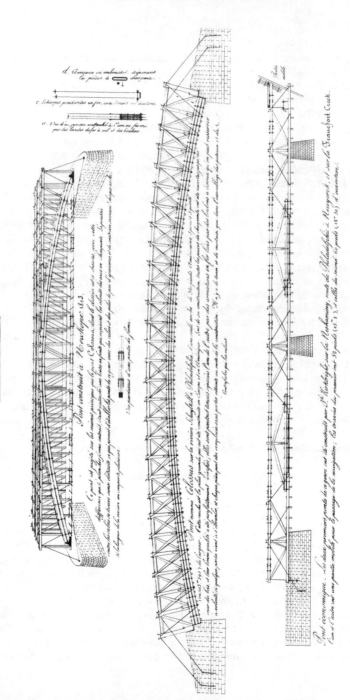

Figure 6. École Royale des Ponts et Chaussées lithograph of Wernwag bridges c.1810-1813 (1826)(27).

and, in general, represents a good and economical solution to a difficult problem(31). In erecting these bridges it seems probable that the timber and iron cantilever elements were pre-assembled and bolted into position. The portion over the abutment allowed a standard pier unit cut short to be used, providing fixity at the end. The pre-assembled suspended span portions, although somewhat clumsy in appearance, allowed the center of the span to be placed by boat or crane, followed by completion of the remainder of the top chords. In its finished state the structure formed a continuous open-webbed truss, the earliest known to the author in which iron was allowed a principal structural role. Without the interconnecting top chord its design principle is reminiscent of Scotland's Forth Bridge.

Conclusions

Chrimes rightly comments that standard histories of civil engineering make little mention of developments in the USA prior to 1850, apart from some notable bridge works, and that they are "glossing over a remarkable achievement as regards development of the transportation infrastructure"(32). As has been shown, knowledge of this achievement in advancing the prosperity of the nation had already been promoted by Telford and particularly by Stevenson who, in his historically valuable Sketch, emphasized the essential role played by low cost, high quality, timberwork. Both commended the "zeal" and "eagerness" which drove this transportation revolution. A British perspective of this achievement is, therefore, its indispensable contribution in diverse ways to the creation of the nation's vast water-borne transportation infrastructure, to the building of railways and bridges and to the provision of water for public health and powering machinery.

The means of creating this achievement differed from British practice. Although American engineers were guided by the same basic design principles as British engineers, dissimilarities in their practice arose from the different circumstances applicable, such as the nature and cost of the materials and labor employed, climate, topography and ground conditions. A need for economy, rapid execution and future design flexibility dictated timber construction, different from the situation in Britain with its developed ironworks and technologically skilled and relatively low-paid workforce. The lack of "solidity" and "finish" in American works, therefore, stemmed not from any want of knowledge or skill, but from a judicious weighing of the circumstances. In general, any work done was the minimum necessary that was fit for purpose over its design life. Passenger comfort and Aesthetics had low priority, a notable exception being the neo-classical adornments at Fairmount Waterworks. On lighthouses, overall coverage of the 5,450 miles of coast was good, but the apparatus and its regulation were below European standards. Overall, by c.1840 an indigenous, largely self-contained, American practice had developed.

Particular elements of this achievement which can be considered of international significance included, the Alleghany mountain railway despite its operational drawbacks, Finley's development of the suspension bridge exemplified in Merrimac bridge which acted as a catalyst for Telford's epoch-making improvements and the development of fast, elegant, steamboats in Britain encouraged by Stevenson. Arguably, the most outstanding, were the magnificent timber bridges of Burr, Palmer and Wernwag and, more generally, the development by c.1810 of the medium span open-webbed truss, which led progressively to Town's trusses from 1820, Long's from 1830, Howe's from 1840 and later the Warren trusses familiar to modern engineers.

References and Notes

(1)   Telford, T. "Navigation, Inland", *Edinburgh Encyclopaedia.* Edinburgh (1830).
          XV. 273-278. First published c. 1821. Also published in Philadelphia (1832).
(2)   Telford, T. *Papers relating to a bridge over the Menai Strait,* London (1819). P.P.,
          H. of C., 1819(60)V. 4.
(3)   Leslie, J. & Paxton, R. *Bright Lights – The Stevenson Engineers, 1752-1971.*
          Edinburgh (1999). 182-185.
(4)   Stevenson, D. *Sketch of the civil engineering of North America; comprising
          remarks on the harbours, river and lake navigation, lighthouses, steam-
          navigation, water-works, canals, roads, railways, bridges and other works.*
          London, John Weale (1838). Second edition (1859).
(5)   Hindle, B. *Technology in early America.* Univ. of North Carolina Press (1966).50.
(6)   Finch, J.K. *Engineering classics of James Kip Finch. Edited by Neal FitzSimons.*
          Kensington, Maryland. (1978). 111-116.
(7)   Stevenson, D. MS. Diary (1837). Nat. Lib. Scot. Acc. 10706, 224, 226.
(8)   Stevenson, D. MS. Autobiography 1824-1855, (1870). Private. 19-21.
(9)   Stevenson, D. *Sketch.* Op. cit. 30.
(10)   Ibid. 18-19
(11)   Ibid. 144.
(12)   Ibid. 187-188.
(13)   Ibid. 191-192.
(14)   Ibid. 57, 202
(15)   Ibid. 214.
(16)   Ibid. 215-222.
(17)   Ibid. 224-227.
(18)   In author's possession. Engraved in Stevenson's *Sketch.* Op. cit. 1859. 140.
(19)   Stevenson, D. MS. Engineering pocket notebook. In author's possession.
(20)   Stevenson, D. *Sketch.* Op. cit. 262.
(21)   Ibid. 258-259.
(22)   Ibid. 278.
(23)   Gibson, J M. "The Fairmount Waterworks". *Philadelphia Museum of Arts* (1988).
(24)   Stevenson, D. *Sketch.* Op. cit. 309.
(25)   United States. Treasury Department. *Report of the Secretary of the Treasury on
          the subject of public roads and canals* (1808).
(26)   Paxton R.A., "Menai Bridge (1818-1826) and its influence on suspension
          bridge development". *Trans. Newcomen Soc.* (1977-1978). **49**, 87-88.
(27)   Recueil de . . dessins ou feuilles de textes relatifs à l'art de l'ingénieur, extraits de
          la première collection…et lithographiés à
          (Paris, 1826). See Elton Engineering Books. *Cat. Number 14*, London. (1999). 24.
(28)   Copy seen at the American Philosophical Society, Philadelphia.
(29)   Nelson, L.H, *The Colossus of 1812.* ASCE (1990). 21.
(30)   *Niles' Register.* (1813), **III**, 322-323 & (1830), 25 September. 74.
(31)   The author acknowledges comments from Prof. A. Bolton and Mr. A.C. Wallace.
(32)   Chrimes. M. "American civil engineering literature in British libraries prior to
          1850". Paper presented to the Construction History Society, London (Mar. 1993).

FORTS TO FREEWAYS, PORTAGES TO PROJECT MANAGEMENT:
Canadian Civil Engineering and International Technology Transfer

Alistair MacKenzie[1]

Abstract:

Civil Engineering is an International Profession. It is a collaborative profession; we seldom keep secrets from each other and we learn from the experiences of our colleagues around the world. In the 150 years since the establishment of the American Society of Civil Engineers, the profession has benefited from the international interchange of technology and with the common border between the United States and Canada and the generally similar development needs, the interchange of technological information and of engineers between the two countries has been particularly significant. The result is that the history of Civil Engineering in both countries has many commonalities in the development of technology and in the sharing of experience and knowledge between many of the best-known Civil Engineers in the history of both countries. A study of the careers of several of these engineers provides a picture of how information and technology is transferred between countries. Sometimes, however, it has been the competing national interests of each country that has spurred the initiation of a particular project and forced a learning experience that has been of benefit in developing technological expertise.

----
[1]P.Eng. FCSCE, C. Eng. MICE, M.ASCE, Chair, National History Committee of the Canadian Society for Civil Engineering. Associate Professor, Ryerson University, 350 Victoria Street, Toronto, Ontario, Canada. Tel. 416-979-5360; amackenz@ryerson.ca

Introduction:

With the long common border between the United States and Canada, it is hardly surprising to find many examples of interchange of information and technology in Civil Engineering. This is particularly evident in Building Construction, Infrastructure Works, notably Canals, Railways and Bridges, and in Public Health Projects. International cooperation has always been a feature of relations between the two countries and from a Civil Engineering perspective this is brought into focus by considering the many mutual efforts in solving engineering problems across the border. The many bridges across the Niagara River are possibly the most visible example of this cooperation, but of as much significance may be the more recent co-operative projects between our two countries occasioned by World War 2 and later, perhaps the most outstanding of all of our cooperative ventures, the construction of the St. Lawrence Seaway.

Not to be forgotten, however, is the fact that Canada and the United States have not always been as friendly as they are today and several of Canada's historic engineering projects were built to counter either an economic or a military threat from the United States.

Pre-history:

Although not as prolific in "pre-historic" remains as many other parts of the world, North America has a significant number of earthworks carried out in the remote past, like the constructions at Cahokia, Illinois and the Great Serpent Mound in Hopewell, Ohio. The earliest engineering works in Canada carried out by the Native North Americans were the construction of dwellings, sometimes, like the Iroquoian Longhouse, substantial and ingenious. They also developed a transportation infrastructure for hunting and trading using "portages" to link the abundant lakes and rivers. The lines of communication, which these indigenous inhabitants originally developed, are followed in many instances by our present day waterway, railway and freeway systems. The first "railway" in Western Canada, a CSCE National Historic Site, does just this. In 1877, Walter Moberly, of CPR fame, built the Grand Rapids Tramway, a 3.5-mile "mule powered" railway, for the Hudson's Bay Company. The tramway followed a traditional portage route bypassing rapids on the Saskatchewan River at Grand Rapids, Manitoba.

The Arrival of the Europeans:

Although the first Europeans to arrive in North America are now believed to be Lief Ericsson's Norsemen who established a settlement at L'Ainse aux Meadows in Newfoundland around 1000AD, the first substantial works of engineering in Canada were carried out some 500 years later by French and British explorers and settlers. These works were defensive structures to protect pioneering groups and their early settlements from attack by the indigenous population or by their competing colonial aspirants. These defensive works were designed and constructed by Military

Engineers, initially French and later British as the Colonial aspirations and imperatives of these two nations ebbed and flowed.

The technologies employed were transferred from Europe and reflected the different engineering philosophies of these two countries: the theoretical and analytical approach of the French, the practical and pragmatic approach of the British. Since those times, transfer of technology has continued and in the 150 years since the founding of the American Society of Civil Engineers, both the United States and Canada have benefited from this pooling of knowledge.

The French Era:

As the French Military produced many engineering works during Canada's "French era" perhaps the best way to illustrate the contribution of these engineers to the development of Civil Engineering in North America is to consider the career of the man, perhaps the most prolific of all the early engineers, and the least known, Gaspard-Joseph Chaussegros de Léry. It was he who was to provide the longest and most effective contribution of all of the French Military engineers.

Quebec City, the oldest City in the New World, was heavily fortified in the mid 18th. century. The historic Fortifications of Quebec, the oldest National Historic Civil Engineering Site commemorated by the Canadian Society for Civil Engineering, were commenced in 1745 under the direction of de Léry, who had been appointed Chief Engineer of the King's Works in New France in 1716, a post held until his death in 1756.

He was born in Toulon, into a family experienced in engineering and architecture. His father, Gaspard, was an engineer who had been involved in several civil and military projects around Toulon. With his stepbrother Louis-Anne, Gaspard-Joseph learned his profession as an apprentice to his father. At that time this type of training was all that was available, as the French had not yet developed the excellent theoretical engineering training establishments like the École nationale des ponts et chaussées, for which they became famous.

As an army officer, de Léry took part in several campaigns in Europe including the Siege of Turin in 1706. As a result of these experiences he developed an interest in the design of fortifications and in 1714, used the knowledge gained to write a treatise on their construction. The original manuscript of this work is now in the Public Archives of Canada.

De Léry arrived in Canada in 1716 on a "temporary" mission. In 1718, this "temporary" mission became permanent on his appointment as Chief Engineer of New France. One of his first duties was to construct defensive measures for the City of Québec. The fortifications which he constructed were later to be expanded and improved upon by both French and English Military engineers. This type of defensive structure, otherwise best known through the works of Vauban, gives an insight into the Civil Engineering knowledge and expertise of the French military, and provides evidence of important advances which they had developed in the understanding of soil mechanics. These works are the first evidence of the transfer of Civil Engineering technology from the Old World to the New. De Léry continued his work on fortifications in both of the future countries of Canada and the United States, designing and constructing the forts at Chambly, Niagara and Saint Frédéric. He was also responsible for Canada's first venture into heavy industry, Les Forges du Saint-Maurice. In addition to his engineering skills, he displayed outstanding aptitude in Architecture, evidenced by his designs for Québec's Cathedral, the Château Saint-Louis, the Intendant's Palace and Montréal's Parish Church.

A talented and farsighted engineer, De Léry also studied the possibility of constructing a canal between Montréal and Lachine, to bypass the rapids, thus anticipating the eventual use of the St. Lawrence as a major transportation route.

The British Military Era:

It was the necessity for Canals that brought the first influx of British Military Engineers to Canada. The St. Lawrence and Ottawa Rivers and the Great Lakes had provided the "highway" along which the Fur Trade had developed. Indeed, one of the earliest canals anywhere in North America was a short stretch built in 1797 or 1798 on the Canadian Side of the St. Mary's River at Sault Ste. Marie by the Northwest Company, the "rival" fur traders to the better known Hudson's Bay Company. This short canal was only around 1,000 yards long, with a single lock 38 ft. by about 9 ft. with a rise of 9 ft, about half of the total rise between the lakes. The lock was destroyed during the War of 1812 and was succeeded only in 1855 by the first American Lock. The British Military were also involved in considerable fortification works, the Halifax Citadel being an example of their work. It was, however, the canal works that were to facilitate the transition from Military to Civil Engineering in Canada.

The Transition to Civil Engineering:

Canals: The War of 1812 had demonstrated the vulnerability of the waterway between the Port of Montreal and Upper Canada by the St. Lawrence River. An alternative route away from the international border was urgently required. The route up the Ottawa River, through the Rideau River and down the Cataraqui River to Kingston was adopted. To make this route navigable, three small canals were required to bypass the rapids at Long Sault on the Ottawa River, and the Rideau and Cataraqui rivers had to be canalized to form the Rideau Canal. In 1819, two companies of the British Army's Royal Staff Corps, under Lt. Col Henry du Vernet, who were working on upgrades to the Soulanges Locks near Montreal (the first locks to be constructed in North America), were ordered to start immediately on the first of the Ottawa River Canals, at Grenville. Work on the other two canals, Chute à Blondeau and Carillon were commenced in 1826 and all were in operation by 1834. The remaining part of the project, the construction of the Rideau Canal, also under the direction of British Military Engineers led by Lt. Col. John By, was commenced in 1827 and completed in 1831. Built for military purposes, the subsequent service of this network was almost entirely civilian.

It is on the Rideau Canal project that we begin to discern a change in the way in which works of engineering construction were now beginning to be carried out. Non-military engineers filled several of the most important engineering positions. Two of these were particularly experienced engineers, John McTaggart who had worked with John Rennie, and Nicol Hugh Baird who had been associated with Thomas Telford through his father, Hugh Baird. Both McTaggart and Baird brought with them considerable knowledge of construction methods and techniques. Baird had worked with his father, Hugh Baird, on the Edinburgh and Glasgow Union Canal, built between 1818 and 1821, and had further international experience in rehabilitation work in St. Petersburg, Russia. One of the more interesting aspects of the construction of this canal was the necessary adaptation these engineers learned to make to the British engineering methodologies with which they were familiar, in order to deal with the very different conditions they encountered in Canada. Baird was later to work on the Trent-Severn Waterway in Ontario and on this project was to demonstrate excellent Project Management skills to compliment his engineering abilities.

The proliferation of Canadian Canals along the St. Lawrence River was a result of the need to establish a transportation corridor to compete with the Erie Canal, which threatened to draw commerce away from Canada by providing an alternative route to the interior of the Continent. The Erie Canal, started in 1817 and completed in 1825, is frequently cited as the first training "facility" for American Civil Engineers and it provides an excellent early example of transfer of technology. British engineers were contributors to its success and Canadian engineers as well as American engineers benefited greatly from the development of professional and technical knowledge that was derived from this great work. Thomas Coltrin Keefer, first President of the CSCE and later a President of ASCE, gained his first engineering experience on the Erie Canal.

The most significant waterway project to affect both countries is without question the construction of the St. Lawrence Seaway, opened in June 1959. This project is an outstanding example of international cooperation and the sharing of technological and managerial expertise.

Railways: Canadians have a special interest in Railroad history, as Canada is the only country in the world that owes its very existence to the successful completion of that major Civil Engineering Railway Project, the construction of the trans-continental CPR. The interchange of people and ideas in Civil Engineering is vividly illustrated in North American Railroad History. One man above all is pre-eminent as the great railway "Baron" of North America, James Jerome Hill. Born in Rockwood, Ontario, Hill decided at age 18 to explore the Orient, but got no further than St. Paul Minnesota. His first job was as a steamboat clerk, but by dint of hard work and self-study he rapidly became one of St. Paul's most successful businessmen. His career took a big step forward when together with two Canadians, Donald A. Smith and George Stephen he acquired the St. Paul and Pacific Railroad. Turning this inefficient railroad into a profitable venture was to project these three people to the forefront when the Canadian Government looked for entrepreneurs to construct the Canadian Pacific Railway. As the CPR got underway and crisis followed crisis, Hill became increasingly more interested in pushing a railroad of his own through U.S. territory to the Pacific. He was, however, instrumental in acquiring for the CPR the services of an American railroad man, William Cornelius Van Horne, who as CPR General Manager pushed this huge project through to success. From the time of Van Horne's appointment, Hill's involvement and influence receded and he left the company in 1883, by which time Hill and Van Horne had essentially become rivals. As Pierre Berton writes: "Hill, the Canadian turned American, and Van Horne, the American turned Canadian, would both push their railroads through to the Pacific". Further American involvement in the construction of the CPR was considerable. In the West, the principal contractor was an American, Walter Onderdonk, on the prairies another American company Langdon and Shepard were the prime contractors, and the very difficult section north of Lake Superior was in the hands of yet another American Company, the New Jersey based North American Railway Contracting Company.

The successful learning of railroading techniques and the transfer of technology in the railroad world are well illustrated through the experiences of John F. Stevens. Stevens, a future ASCE President, was another of J.J. Hill's protégés who in 1882 began work with Langdon and Shepard on the prairie section of the CPR. He was later employed directly by the CPR Company as an Assistant Engineer in the mountains of Western Canada. During this time he had faced almost every challenge possible for a railroad engineer and was able to apply this knowledge effectively when many years later he was to face one of the biggest challenges ever to face any engineer. When the CPR was completed he initially returned to work for J.J. Hill on the Great Northern Railroad, becoming Chief Engineer in 1895. However, his place in engineering history was assured when in 1905, Theodore Roosevelt appointed him Chief Engineer of the Panama Canal. In the two years that he worked on the canal, he reorganized the venture to such a degree that his methodology was followed until the canal was complete. In "The Path Between the Seas", David McCulloch says of Stevens and his methods: "John Stevens ushered in what was to be known on the Isthmus as the railroad era. And it is one of the ironies of the story that the unseen guiding spirit as the canal got underway was James J. Hill. Stevens was not only Hill's man, but he would run the work the way Hill ran the Great Northern. Indeed, the building of the Panama Canal was among other things one of the greatest of all triumphs in American railroad engineering." Perhaps Canadian experience contributed to this success?

Before leaving the Canadian Pacific Railway, it may be noted that one of the most significant innovations which increasingly impacts our lives as travel times dwindle across the globe was the introduction of Universal Time and the establishments of time "zones". The man responsible for initiating this methodology was the original Chief Engineer of CPR, Sir Sandford Fleming.

Another engineer who learned in, and contributed to, both countries and who combined Canal and Railroad expertise was Walter Shanly. Shanly was one of the first notable Canadian engineers to receive his engineering training in Canada. Born in Ireland, he was 19 years old when the family immigrated to Canada. As Shanly is an excellent example of an engineer who made significant contributions to Civil Engineering in both the United States and Canada, it is worth looking at his career in some depth. His history is of interest not only because of its international context, but also because it extends our knowledge of how engineers of that era learned their engineering skills "on the job".

His father had come to Canada with the intention of becoming a Farmer and had acquired property near London, Ontario. Their neighbours were a Civil Engineer, Hamilton Killaly and his family. The two families socialized and became very friendly. Walter showed a great interest in Killaly's earlier Civil Engineering career and when after two years of farming, Killaly decided to resume his career, Walter decided that he, too, would like to pursue a career in Civil Engineering. Killaly promised to help him as soon as he had secured his own position and when in 1840 Killaly was appointed chairman of the Board of Works for the United Province of Canada, he took Walter with him to Montreal. At that time, the only method of becoming a civil engineer was to be "sponsored" by an established engineer, as there were no formal training schemes.

Walter's first appointment was to the Chambly Canal, one of the elements of an "international" waterway connecting the St. Lawrence to Lake Champlain. A short spell as an Inspector of Roads was followed by an appointment to the Beauharnois Canal, at Valleyfield on the St. Lawrence. On this Project Walter Shanly began to acquire both technical and managerial expertise. Such was his application that by 1845 he was left in sole charge of the Project. From there, he went to work on a section of the Welland Canal where he introduced his younger brother Frank to civil engineering.

When the Welland Canal work was finished in 1848, both brothers were out of work and Frank went to the United States. He succeeded in getting employment on the Northern Railroad, being built from Ogdensberg to Lake Champlain, and shortly thereafter asked Walter to join him. With his previous experience, Walter was given a senior appointment and once again had to learn "on the job". This time he had to adapt the engineering knowledge gained from his canal experiences and attempt to apply it to railroads. His management knowledge proved to be particularly useful and he applied strict professional methodology in dealing with contractors, in contrast to many of his contemporaries who were, in his opinion, much too close to contractors to enable them to carry out professional duties in an ethical and impartial manner.

Returning to Canada in 1851, he was appointed Chief Engineer on the Bytown and Prescott Railway and the following year, Chief Engineer of the Toronto and Guelph Railway

Shanly's strict professional approach to the ethics of Civil Engineering construction contracting was to undergo a severe test shortly after this. In 1853 the Toronto and Guelph Railway became the Western Division of the Grand Trunk Railway and Walter was appointed Chief Engineer of that Division. The Contractor on this section was C. S. Gzowski and Company and their successful bid had been based partly on working for bonds in the railway. Shortly thereafter the Grand Trunk tried and failed to come to an agreement with the Great Western Railway to take over the already built Niagara Falls to Windsor line as their western extension. Following this failure, they decided to extend the Toronto - Guelph line to Sarnia, paralleling the Great Western route. A deal was made with Gzowski and Company to carry out this work and Walter Shanly found that he was no longer an "independent" engineer but was to work for the contractor. Unhappy with this arrangement, Walter thereafter became more involved in management than in engineering. On completion of the Grand Trunk Railway he was appointed General Manager and Chief Engineer, but was shortly thereafter downgraded to General Traffic Manager, and resigned in 1862. He successfully ran for office in the Provincial election of 1863 and after Confederation, continued his political career, being elected to the Federal Parliament in 1867.

His great engineering achievement in Canada, the Western Division of the Grand Trunk Railway had made him one of the best-known engineers in Canada. He was to gain further fame internationally by carrying out a bigger and much more difficult project in the United States, the completion of the Hoosac Tunnel in North Western Massachusetts. Walter teamed up with brother Frank to act as contractors for this project. The five-mile tunnel through the Hoosac Mountain had been under construction since 1852 with limited success. Innovations, which the Shanlys brought to the project, included the first use of nitroglycerine in tunneling and the first use of compressed air drills in a North American Tunnel (these having been previously used successfully in the Mont Cenis Tunnel in the Alps). Walter Shanly also applied with great success the "Project Management" techniques of Time and Cost Control and Human Resource Management that he had learned on his earlier projects in both Canada and the United States. With outstanding effort, dedication and superb organizational skills the project was completed in 1874.

Combining his duties as an engineer and a Member of Parliament, Shanly introduced the incorporation bill of the Canadian Society of Civil Engineers in the House of Commons in April of 1887. He was nominated as President of the Society following Thomas Keefer's term but declined.

An International railway project that has been jointly recognized by ASCE and CSCE is the White Pass and Yukon Railway. This narrow gauge (3 feet) railway, which runs from Skagway, Alaska to Whitehorse, Yukon, was again a collaborative effort between Engineers and Contractors from America, Canada and Britain. Involved were Sir Thomas Tancred from Britain, Erastus Corning Hawkins from the U.S., Engineer John Hislop and Contractor Michael Heaney from Canada. During the Second World War, the line was taken over by the U.S. Army to serve the construction of the Alaska Highway and the "staging" airfields.

Ship Railways: One of the most interesting of Canada's National Historic Civil Engineering Sites commemorates an uncompleted Project, yet one that had considerable international implications. This was to be a Ship Railway across the Isthmus of Chignecto, the 17-mile wide strip of land separating the Northumberland Strait and the Bay of Fundy. Started in 1887, it was about 90% complete when it was abandoned due to financial and political problems. The project was the brainchild of Engineer Henry G.C. Ketchum, a Canadian born engineer with international Civil Engineering experience acquired in both Brazil and Britain.

As some method of crossing the Isthmus would cut the travel distance between the St. Lawrence and the Maritime Ports by some 500 miles, proposals for a Canal across there had been made since the early colonial times. The earliest recorded suggestion was made in 1686, during the French era. A later proposal came from Colonel Robert Morris of the Royal Engineers in 1783, but technical difficulties, related to the extreme tidal differences at both ends, coupled with potentially very high costs had always prevented any action.

It could be argued that the Native Americans originally developed the concept of Ship Railways on this continent. Evidence has been found on some "portage" routes that they laid log roads on which to drag particularly heavy canoe loads, rather than use the more traditional method of unloading the canoes and carrying both canoe and load. However, much later, in the mechanical age of the mid to late 1800's, Ship Railways were being seen as viable alternatives to Canals. The British Engineer, Sir James William Brunlees (President of the Institution of Civil Engineers, 1882/3), had in 1859, proposed a ship railway as an alternative to the Suez Canal and in 1873, Sir John Fowler was asked to look at a similar project to transport steamers past the First Cataract of the Nile. In North America, James Buchanan Eads had, in March of 1880, presented to the U.S. Congress his proposal to build a ship railway across the Tehuantepec Isthmus of Mexico as a cheaper and more efficient alternative to the Panama Canal. Both Ketchum and Eads acknowledged their debt to Brunlees earlier work and the technology was seen to be sound although there were no major examples in existence. It is, perhaps ironic that it was to be the same man, Ferdinand de Lessops, who, in effect, killed both Brunlees' and Ead's projects by his work on the Suez and Panama Canals.

In his earlier time in Britain, Ketchum had worked with Sir Benjamin Baker, and had persuaded him to become a partner in the Chignecto project. Everything seemed to be in place to ensure a satisfactory result, and indeed, there appeared to be a common perception at the time throughout the engineering community world wide that had the Chignecto Marine Railway been satisfactorily completed, it would have led to the construction of other major schemes, notably Ead's Tehuantepec project. The eventual failure of the project through lack of political will may well have put an end to what might have been an exciting future for this type of transportation project. Around this time, as a "feeder" to the Tehuantepec scheme, a further Ship Railway of around 100 miles was being proposed across Florida from Jacksonville to the Gulf of Mexico. Other proposals included a 170 mile Ship Railway between Toledo on Lake Erie and Michigan City on Lake Michigan, and further one, 75 miles long, between Toronto on Lake Ontario and Georgian Bay on Lake Huron. International transfer of technology was very evident on these projects, although in the end, the concept proved to be another excellent idea bypassed by other technological developments.

In addition to his work on the CMR, Ketchum is a significant historic figure in Canadian Civil Engineering history as, in 1862, he was the first graduate of Canada's first Civil Engineering program at King's College, now the University of New Brunswick.

Bridges: Perhaps no area of Canada is more influenced by the proximity of the United States than Southern Ontario. This region, the most heavily populated in Canada, has strong economic ties to adjacent New York State and enjoys similar social and cultural interests. All of these relationships are facilitated by the ease of cross border transportation and communication between the two regions, due to the many bridges across the Niagara River. The present day bridges, and their predecessors, provide a long and interesting history of cooperation and technology transfer. The first bridge projects on the Niagara River were "joint ventures" between the United States and Canada. William Hamilton Merritt, who is best known for his efforts in first promoting and later constructing the Welland Canal, was the first to initiate a survey to determine a possible site for a bridge. Merritt was born in Bedford, New Jersey, but moved with his parents to St. Catherines, Ontario when he was only 3, (therefore Canadians have claimed him ever since). He had partnered with Charles Beebe Stuart, an engineer who had, amongst many other achievements, been Municipal Engineer of Rochester, and knew the area and its political climate very well. The partners set up parallel bridge companies in the United States and Canada and invited proposals for a bridge. The two principal contenders were Charles Ellet and John A. Roebling. As is well known, it was Charles Ellet who was successful and commenced work by organizing his famous "kite" competition to initiate the spanning of the river with the first ropes and then cables. From the time of the completion of Ellet's suspension bridge in 1848, to the present day, the list of engineers who designed and built the many bridges that succeeded Ellet's, reads like a role of honour of famous American and Canadian Engineers. After Ellet came Edward Serrill's bridge in 1851 at Queenston, John A. Roebling's Road and Rail Suspension Bridge in 1855, and Samuel Keefer's Clifton Suspension Bridge in 1869. Further structures were Leffert Lefferts Buck's Whirlpool Rapids Bridge in 1897 and his Falls View Bridge in 1898 on the site of Keefer's earlier structure; these were followed by Richard Buck's (no relation apparently) Lewiston and Queenston Suspension Bridge in 1899. All of these bridges were in the general vicinity of the Niagara Falls, but an other notable bridge was the International Railway bridge between Fort Erie, Ontario and Buffalo, NY, designed and constructed by Casimir Gzwoski and the Phoenix Bridge Company, again a US/Canadian joint effort.

Bridges with a joint US and Canadian content have not been without their share of problems. The Cornwall Bridge, spanning the St. Lawrence between Cornwall Island, Ontario and Massena NY collapsed on September 6, 1898 as the result of a problem foundation. The history of the problems that beset one of the other notable US/Canadian joint ventures, the Quebec Bridge, is too well known to merit further elaboration.

Whilst discussing bridges, it may be appropriate to include an outstanding example of technology transfer between Britain and Canada in the design and construction of the Victoria Bridge in Montreal in 1853. The engineer responsible was the Chief Engineer of the Grand Trunk Railway, Alexander M. Ross and the bridge used Robert Stephenson's "tubular" technology, in its superstructure and T.C. Keefer's "ice breaker" pier design. The "tubular" design was first applied on Stephenson's historic Britannia Bridge but as Dr. Petroski has pointed out, this design was an example of a technological innovation that was an "economic and environmental failure". The "tubular" structure on the Victoria Bridge was removed when the bridge was widened in 1897 and replaced by more conventional trusses.

Ports and Harbours: Competition between the respective interests of the United States and Canada led to the development of a deep-water port and a major graving dock on Canada's west coast. The Port of Victoria, BC, owes its existence to Canadian concern that the opening of the Panama Canal would significantly change the pattern of trade in the Pacific and have an adverse effect on Canada's economy. The resulting Breakwater and Docks constructed at Ogden Point in 1917 was an interesting project that combined the efforts of Canadian engineers, British Contractors and Britain's Royal Navy.

On the other side of Victoria harbour is a project initiated as a result of divergent national priorities and the British Navy's interests. During the late 1800's, the British Navy urgently needed dry dock facilities on the West Coast. The only dry dock to which they had access at that time was the floating dock at Mare Island, San Francisco, opened in 1854. A second and larger dock was opened in 1868, but at that time, following the Civil War, relations between the United States and Britain (and therefore Canada) were somewhat tense and the British Navy could therefore not rely on being able to access a dock on what was potentially enemy territory. The only other dock of acceptable size in the Pacific was in Melbourne, Australia although smaller docks existed in Valpariso, Hong Kong and Singapore. To solve the problem, the Naden Dry Dock was constructed in 1879 in Esquimalt, BC, by a combination of Canadian and British Engineers and Contractors. The Mare Island graving dock was filled in during the 1950's, and thus the Naden Dock is the oldest surviving dry dock on the West coast of the Americas and is now a Canadian National Historic Civil Engineering Site.

Public Health: The first site commemorated by the Canadian Society for Civil Engineering as a National Historic Civil Engineering Site was the Pumphouse of the Hamilton, Ontario water supply project designed and constructed in 1859 by Thomas Coltrin Keefer, first president of CSCE and later to be President of ASCE. Keefer had a long and distinguished career as an engineer, an author and an educator. His work on this particular Water Supply project, illustrates the way in which technological information was shared across the border. The Board of Water Commissioners for the City of Hamilton was very conscious of the need for a supply of pure water for the City as a result of the Cholera epidemics that had regularly broken out in Hamilton as in many other North American Cities. Although the scheme proposed by Keefer appeared to be quite satisfactory, the Board wanted confirmation that the project was to use the very best practices in design and construction. In the United States, the work done on the New York City water supply had made John B. Jervis and Alfred W. Craven, renowned international figures. These two eminent engineers were therefore invited to review Keefer's plans. They fully approved the Keefer scheme, noting only a minor recommendation regarding the "staging" of the works. Such international cooperation illustrates the very best practices in sharing Civil Engineering knowledge and expertise.

World War 2:

One of the most intense project efforts of World War 2 was carried out far from the live theatres of war. The Alaska Highway is commemorated as an International Historic Civil Engineering Landmark by both ASCE and CSCE. The construction of the 1520 miles (2450 km) of this highway, over the very short period of 8 months and 12 days in 1942, was an outstanding feat of engineering and construction expertise. Once again, international cooperation was a key to success. Originally named the Canadian – Alaskan Military Highway, and nicknamed the "Alcan" Highway, it was officially named the Alaska Highway in 1943. A more purely Canadian wartime venture was the design and construction of 88 airfields and their infrastructure in 1941 and 1942, to "house" the British Commonwealth Air Training Plan. An unexpected bonus of this project was that it provided the basis of Canada's post war air transportation infrastructure.

Conclusion:

This brief and selective look at the interrelationships in Civil Engineering between the United States and Canada goes some way to illustrating the interdependence of the resources and expertise of many different countries in producing the world in which we live. The development of Civil Engineering in Canada owes much to French, British and American Engineers. Canada too has contributed significantly to the development of American Civil Engineering. Transfer of technological information between our two countries and the rest of the world continues to the present day, and for North Americans, the creation of NAFTA and the NAACE promises even closer ties and greater cooperation in Civil Engineering matters between member states in the future.

References:

Andrews, Mark (1998) *"For King and Country: Lieutenant Colonel John By, R.E. Indefatigable Civil-Military Engineer"*, Heritage Merrickville Foundation, Merrickville, Ontario.

Angus, James T. (1988) *"A Respectable Ditch: A History of the Trent – Severn Waterway 1833-1920"*, McGill-Queen's University Press, Montreal.

Ball, Norman Ed. (1988) *"Building Canada: A History of Public Works"*, University of Toronto Press, Toronto.

Berton, Pierre (1970) *"The National Dream: The Great Railway 1871-1881"*, McClelland and Stewart, Toronto.

Berton, Pierre (1971) *"The Last Spike: The Great Railway 1881-1885"*, McClelland and Stewart, Toronto.

Charbonneau, André (1987) *"Gaspard-Joseph de Léry (1682-1756): The First Canadian Engineer"*, Proceedings CSCE Conference, Montréal, Québec.

Disher, J.W. and Smith E.A.W. (2002) *"By Design: The Role of the Engineer in the History of the Hamilton Burlington Area"*, H.E.I., Inc. Hamilton, Ontario.

Hart, Peter (1997) *"A Civil Society: A Brief Personal History of the Canadian Society for Civil Engineering"*, CSCE. Montreal.

Ircha, M.C. (1989) *"The Chignecto Ship Railway: An Innovative Approach"*, Proceedings CSCE Conference, St. John's, Newfoundland.

Legget, Robert F. *"Ottawa Waterway: Gateway to a Continent"*, (Toronto 1975)

McCullough, David (1997) *"The Path between the Seas: The Creation of the Panama Canal 1870-1914"*, Simon & Schuster, New York.

Petroski, Henry (1994) *"Design Paradigms: Case Histories of Error and Judgement in Engineering"*, Cambridge University Press, New York.

Stamp, Robert M. (1992) *"Bridging the Border: The Structures of Canadian-American Relations"*, Dundurn Press, Toronto

Underwood, Jay (1995) *"Ketchum's Folly"*, Lancelot Press, Hantsport, N.S.

White, Richard (1999) *"Gentlemen Engineers: The Working Lives of Frank and Walter Shanly"*, University of Toronto Press, Toronto.

ALASKA: BUILDING "THE GREAT LAND"

By Howard P. Thomas, P.E., Fellow, ASCE, and Kelly S. Merrill, P.E., Member, ASCE

INTRODUCTION

Prior to statehood in 1959, Alaska was a territory. Comprising a 656,000-square-mile area characterized by vast and remote wilderness with major climatic and terrain challenges, civil engineers and civil engineering have played a major role in Alaska's development. To commemorate the Alaska Section's 50th anniversary in 2001 and the Society's 150th anniversary in 2002, the Alaska Section of ASCE decided, as a volunteer effort, to assemble a PowerPoint® presentation portraying highlights of the state's civil engineering history over the past 100 years or more.

A major purpose of the project was to inform and encourage young people contemplating an engineering career and, to date, the presentation has been well received in a number of schools in the state. For convenience, the presentation is subdivided into Mining, Rail, Military, Highways, Dams, Oil & Gas, Ports & Harbors, Aviation, and Public Health and Sanitation. It also includes segments on the 1964 Earthquake, Alaska's two Outstanding Civil Engineering Achievement (OCEA) award-winning projects, and several proposed projects that were either abandoned or have not yet been constructed. Assembled from a variety of sources including members' project files, many of the photos have never before been published.

The following presents highlights of the presentation including several of Alaska's "engineering heroes" down through the years. A bibliography at the end provides additional recommended reading.

MINING

Prominent Alaska mines are depicted in Figure 1.

The development of Alaska started with the "Gold Rush" in the late 19th century. The "Gold Rush" actually started in the Yukon Territory of Canada and progressed westward to Nome and Fairbanks. However, because of their relative accessibility, mines in the Juneau area were among the first to open in the late 1800s.

The miners' interest wasn't limited to gold. Mining Engineer Stephen Birch (1872-1940) (Figure 2) was one of the first to discover a world-class copper deposit at Kennecott. Graduated from Columbia University School of Mines with a degree in Civil Engineering, Birch first came to Alaska in 1898. A man of both vision and practicality, Birch was a risk-taker who later formed the Kennecott Copper Company which flourished under his leadership. He eventually became a millionaire. According to AIME, "Birch had the ability, unusual in a mining

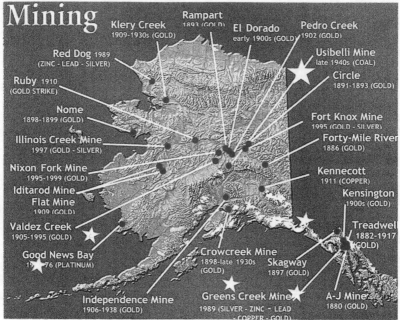

Figure 1. Mining History of Alaska

engineer, to sell himself the Presidency of Kennecott Mines Co and the financial acumen to acquire control of the great copper mines in the US and Chile."

Opened in 1911, the Kennecott project required constructing a 195-mile railroad to tidewater at Cordova, a phenomenal project in itself. The six-story-high concentrator mill building (Figure 3) was constructed using 200,000 board feet of lumber hauled in from the Puget Sound area. The mine was closed in 1938 after producing $200 to $300 million of copper. Designated a National Historic

Figure 2.
Stephen Birch, 1872-1940

Figure 3. Kennecott Copper Mine - 1911

Landmark in 1978, the National Park Service acquired many of the significant buildings and lands of the historic town of Kennecott in 1998.

RAILROADS

Alaska's few railroads are shown on Figure 4.

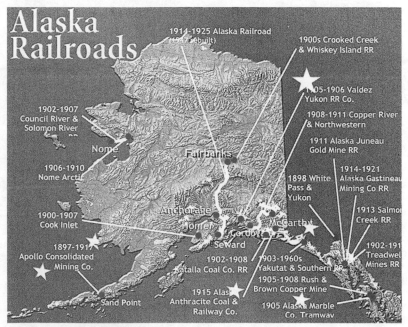

Figure 4. Alaska's Railroads

Construction of railroads accompanied many of the early mining projects. Most of these were short, narrow-gauge railroads not designed for passenger service. The first railroad built on Alaska soil was the White Pass and Yukon Railroad, a passenger railroad, constructed in only two years starting in 1898. Extending from Skagway toward the gold fields of the Yukon, the line was quite literally carved into the rocky mountainsides outside of Skagway. The tools of the day were manpower, horses, black powder, makeshift wooden ladders, wheelbarrows and sheer willpower.

Figure 5 depicts the famous White Pass & Yukon railroad bridge over Dead Horse Gulch. At the time it was built, it was the highest railroad bridge in the world. The gulch was named after the many horses that perished in this area during the construction of the bridge.

Master builder of the railroad was Mike Heney and the Chief Engineer was Erastus C. Hawkins. Heney and Hawkins were a remarkable team. Heney grew up

Figure 5. White Pass & Yukon Railroad 1898-1900

in Ontario, Canada, with little formal education, ran away from home at 14 to work on the Canadian Pacific Railroad. He was outgoing and likeable with an endless fund of stories. Brave, resourceful and determined, he was full of energy and was referred to as "Big Mike" and "The Irish Prince." Never married, the challenge of Alaska was the love of Heney's life.

Erastus Hawkins (Figure 6) provided the "brains" of the team and Heney the "'brawn." Having started with a New York engineering firm at the age of 19, Hawkins more than made up for lack of a formal degree with diligent self-study and learning on the job. Fair, honest and detail-oriented, Hawkins was described as "a bundle of energy and stamina who could outwalk, outride, and outlast most of his contemporaries." Ten-thousand-foot-high Mt. Hawkins in the Chugach Range and a large island in Prince William Sound near Cordova are named for him.

**Figure 6.
Erastus Corning Hawkins
1860-1912**

The White Pass and Yukon Railroad still operates today, offering passengers a very scenic trip from Skagway to Lake Bennett (originally, the railroad extended all the way to Whitehorse).

As mentioned above, the Kennecott project required construction of a 195-mile-long railroad along what had long been considered an almost impassable route following the Copper River (Figure 7). Once again, Heney was the Contractor and Hawkins was the Chief Engineer (later General Manager). An especially challenging feature of the route of the Copper River and Northwestern Railroad was the necessary crossing of the river between the Miles and Childs Glaciers. Termed the "Million Dollar Bridge" (Figure 8), no hydrologic data were available to the designers. In addition, the Miles Glacier discharges a ceaseless barrage of icebergs from early spring to late Autumn.

**Figure 7. CR&NW
Locomotive Derailment**

**Figure 8.
Million Dollar Bridge -
Copper River 1910**

Constructed under incredible hardship in 1908-1910, timber pneumatic caissons similar to those used in the Eads Bridge in St. Louis and the Brooklyn Bridge in New York were used to excavate 50 ft below the river level to allow concrete pier construction. During the winter of 1909-1910, the crews worked around the clock to beat spring breakup and the last steel bolt had just been placed when the ice moved and the falsework collapsed. Called "one of the great engineering feats of all time," the cost of the 1500-ft-long, 4-span bridge actually turned out to be $1.4 million ($50 to $100 million in 2002 dollars). The bridge was decked and converted to a road bridge 20 years after the railroad was shut down. However, the 1964 earthquake caused one of the piers to shift and one end of the No.4 span to drop. A "temporary" ramp, still in place today, was constructed to allow limited continued traffic over the bridge.

In 1915, President Woodrow Wilson established the Alaska Engineering Commission to head up the design and construction of the Alaska Railroad to Fairbanks. By that time, a private railway had already been constructed from Seward to Portage. Now Alaska's largest city, Anchorage was established as a construction camp (tent city) near the mouth of Ship Creek. The 466-mile route of the railroad traverses discontinuous permafrost and areas of slope instability which continue to pose maintenance challenges to this day. Construction peaked in 1917 with 4,500 workers. The railroad's completion in 1923 was commemorated by President Warren G. Harding driving a golden spike at Nenana near the railroad bridge which, at the time, with its 700-ft span, was the second-longest, single-span railroad bridge in the US.

MILITARY

Alaska's military bases are shown in Figure 9.

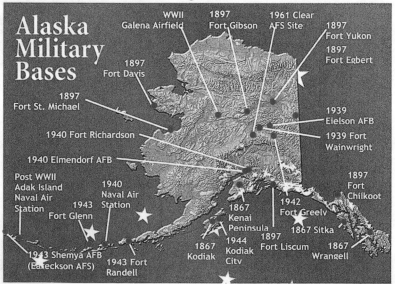

Figure 9. Alaska's Military Bases

The U.S. military established remote outposts in the southern portion of Alaska as early as 1867 when the territory was purchased from Russia. Thirty years later, during the gold rush, several forts were established in the Interior on the Yukon River and on Norton Sound. But it wasn't until World War II that the U.S. military established a major presence in Alaska. Between 1940 and 42, major garrisons and airfields were constructed with an impetus from the lend-lease program under which aircraft were provided to Russia. These included the present Ft. Wainwright and Eielson AFB near Fairbanks.

Referred to as the "Father of Military Construction in Alaska," General Benjamin Talley (1903-1998) (Figure 10) was a career military engineer who became an Alaska legend. His career spanned nearly half a century. Described as a capable, energetic and enthusiastic leader, Talley supervised construction of Elmendorf Air Force Base (known as Ft. Richardson at the time) near Anchorage in 1940, readied Alaska for defense from Japanese invasion in 1940, oversaw all military construction in Alaska during WW II, and oversaw reconstruction of Anchorage and southcentral Alaska after the 1964 earthquake as a civilian engineering manager. In each decade from the 1930s through the '80s, Talley accomplished more than most military officers would dream about in a lifetime. Awarded the Distinguished Service Cross for his accomplishments, he and his second wife Virginia retired to Anchor Point, Alaska.

Between 1953 and 58, Distant Early Warning (DEWline) radar detection sites were constructed at very remote sites along the Beaufort Sea Coast and Aleutian Chain. The objective of these was early warning of surprise attack by Soviet bombers. A joint American-Canadian defense project, over 50 sites were established from the western Aleutians to Baffin Island, Canada. Twenty two of these were constructed in Alaska. From 1955 to 1961, the so-called "White Alice" communication network with 31 stations was constructed. However, by the early 1960s, satellite

Figure 10. General Benjamin B. Talley, 1903-1998

technology had eclipsed the usefulness of this network. Today (in 2002), a $7-billion National Missile Defense System with facilities on Shemya Island and near Delta Junction is being deployed in Alaska.

HIGHWAYS

Comprising only 2300 miles of paved roads, Alaska's road system is a limited one, neither interconnected nor covering all parts of the state (Figure 11). Vast portions of the state are not served by roads. "Bush" Alaska is characterized by remote, scattered villages, frequently off the road system with access limited to air and sometimes water (summers only). Not only is the cost of constructing modern roads in Alaska high because of the remoteness, the terrain, and the severe climate, but so is the cost of maintaining them.

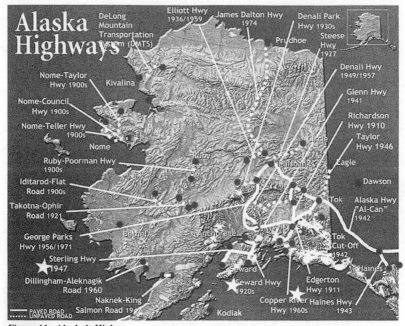

Figure 11. Alaska's Highways

The Alaska Road Commission was established by President Teddy Roosevelt in 1905. The ARC designed and maintained a network of trails for mushers, trappers and miners in the early years. It was only in later years they built secondary dirt roads suitable for wagons. West Point graduate Wilds Preston Richardson (1861-1929) had impressed Roosevelt with his Alaska military duty during the gold rush years and was appointed the first President of the ARC. Richardson emphasized priority to upgrading the Valdez-Fairbanks trail which later came to be named for him, as was Ft. Richardson near Anchorage.

Figure 12. Alaska Highway Construction - Tokhim River July 1942

With the impetus from WW II, the
1,400-mile Alaska Highway extending
from Dawson Creek, British Columbia,
to Delta Junction, Alaska, was pushed
through as a joint US-Canadian project
in 1942 (Figure 12). The original road
was built by the Corps of Engineers as a
pioneer trail without much regard to
grades. The Bureau of Public Roads took
over the following year and upgraded it
to secondary road standards. The
Commanding Officer of the remarkable
project was Brigadier General "Patty"
O'Connor (Figure 13), born in Bay City,
Michigan, in 1885. After attending the
University of Michigan and Notre
Dame, O'Connor graduated from West
Point in 1907. The Alaska Highway
represented a momentous development
in the history of the Far North.
Constructed by the Corps of Engineers
in just eight months, it secured Alaska's
connection to the "Lower 48" during
World War II. Afterward, it made
Alaska more accessible to civilians.
Today, popular with tourists, the Alaska
Highway is paved throughout its length.

Figure 13.
James A. "Patty" O'Connor, Born 1885

Figure 14. Yukon River Bridge - 1970s

Bridge construction has also been a
challenge in Alaska. Until the oil
pipeline came, there was no bridge
across the mighty Yukon River which
divides the state north and south.
Completed in 1972, the "Ed Patton"
Bridge (named after the pipeline
consortium's President) is a half-mile-
long, steel box-girder bridge which
carries both the "Dalton Highway" and
the pipeline itself (Figure 14).
Incorporating innovative approaches to
use of low temperature steel and seismic

Figure 15. Sitka Harbor Bridge - 1973

analysis, the bridge's piers are also designed for large ice forces with anchors
grouted into the bedrock to resist overturning. The following year, a 1250-ft
cable-stay bridge was constructed in Sitka; this was the first vehicular cable-stay
bridge built in the U.S (Figure 15).

Ever since the Copper River and Northwestern Railroad head shut down in 1938, there had been an interest in converting the old railbed to a highway. Fifteen million dollars was spent to construct the Copper River Highway from 1953 to 1972 and, at that point, only about 34 miles of original railbed remained to be converted to road. Environmentalists then succeeded in stalling the project and it remains uncompleted to this day (2002).

Figure 16. Prudhoe Bay West Dock Causeway Bridge

In the 1990s, oil field traffic involving movement of large prefabricated building modules and drill rigs required bridges with unprecedented live-load capacities. The Prudhoe Bay West Dock Causeway Bridge on the North Slope was designed for a 2,300-ton live load (Figure 16). Part of the state's paved road system, the modern Canyon Creek Bridge on the Kenai Peninsula received a state OCEA award in 1999 (Figure 17).

Figure 17. Canyon Creek Bridge - Seward Hwy - 1990s

DAMS

In August 1967, the Fairbanks area received 6 inches of rain in 6 days, resulting in the worst flood in Fairbanks history (Figure 18). The water on downtown

Figure 18. Chena River Flood - Fairbanks 1967

streets was 6 ft deep, seven people drowned, and 15,000 were left homeless. This flood led to many state and national improvements, including helping to convince Congress to establish a national flood insurance program. It also resulted in the

Chena Lakes Flood Control Project, which included the 7-mile-long Moose Creek Dam and its associated floodway. The Salmon Creek Dam (Figure 19), built in Juneau in 1913 to provide hydropower and water to meet industrial and community needs, was the highest concrete arch dam in the world at the time it was constructed.

Formation of the Alaska Power Authority in 1976, now called the "Alaska Energy Authority," led to construction of a number of hydroelectric dams (also power interties) in the state during the 1980s. These included the Bradley Lake Dam near Homer completed in 1991 (Figure 20). With an installed capacity of 126 megawatts and built at a cost of $300 million, including two parallel 20-mile transmission lines, the Bradley Lake site had been selected by the Corps of Engineers as the most cost-effective of 19 potential southcentral Alaskan sites. The project included a 13-ft-diameter power tunnel bored from the lake to a powerhouse on Kachemak Bay below.

OIL AND GAS

When people think of Alaska oil and gas, most think of the colossal Prudhoe Bay field, and few realize that oil exploration and development in Alaska actually dates back to the beginning of the 20th century. Oil was first discovered at Katalla near Cordova in 1896 when a gold prospector named Tom White accidentally fell into a pool covered with oil seepage. In 1902, the Alaska Development Company brought in Alaska's first oil gusher at Katalla (Figure 21). But the wells ended up producing less than 20 barrels a day and all fuel produced from the petroleum was used locally. The dream ended in 1933 when a Christmas Eve fire destroyed the Katalla oil refinery.

Figure 19. Salmon Creek Dam - Juneau 1913

Because of the very high, up-front costs and almost-overwhelming logistics, about all of the oil exploration and development in Alaska has been by the multi-national oil companies. The first major field was discovered at Swanson River on the Kenai Peninsula south of Anchorage in 1957. After numerous "dry holes," William Bishop, a Richfield

Figure 20. Bradley Lake Dam - Kachemak

Figure 21. Katalla Oil Fields - 1902

geologist, picked the discovery well site based on limited seismic data. Alaska's statehood in 1959 was attributed in part to development of a viable oil industry based on the Swanson River oil fields.

But it was the Prudhoe Bay strike in February 1968, which captured the interest of the nation. With an estimated 22 billion barrels of oil and 30 trillion cubic feet of natural gas, Prudhoe Bay is by far the largest oil field ever found in North America. The field began production when the Trans-Alaska Pipeline System (TAPS) was completed in 1977. It has dominated the state's economy and has helped to provide an improved standard of living for a generation of Alaskans since then. Although it has been overshadowed by the Prudhoe Bay field, the nearby Kuparuk field which started producing in 1981 is a very large oil field in its own right.

PORTS AND HARBORS

Alaska has more than 6,000 miles of coastline, more than the rest of the U.S. combined. Ports and harbors provide essential access points for transportation routes and extraction of mineral, timber and commercial fishing resources. Unalaska on the Aleutian Chain is the

Figure 22. Sheetpile Dock Under Construction - North Slope

gateway to the Bering Sea region and is one of the most productive seafood processing ports in the nation. Conventional wharf structures have seen much use but, since 1980, the open-cell sheetpile concept, conceived by an Alaska firm, has been widely used in Alaska and elsewhere for docks, bridge abutments, weirs and erosion-protection facilities (Figure 22).

AVIATION

Aviation has played an important part in the development of Alaska. Alaska airfield development started in the 1920s. Initially, these were essentially bush operations using grass landing strips. Most of the early, rural airfield runways were constructed for small, light aircraft. With World War II and the lend-lease program in the early 1940s, construction of airfields significantly expanded. Frozen ground thaw degradation at Ft. Wainwright (then Ladd Field) and Eielson AFB were reported by Karl Terzaghi in 1952. In 1948, Congress authorized funds to build two "intercontinental" airports at Anchorage and Fairbanks. Construction began in 1949 and both airports opened for business in 1951.

One of the designers of the Anchorage Airport was Irene Ryan (1909-1997) (Figure 23). A charter member of the Alaska Section of ASCE, Irene grew up in the "oil patches" of Oklahoma and Texas. The first woman graduate of the New Mexico School of Mines, Irene was also the first woman to solo a plane in Alaska (in 1932). Very versatile, she practiced as a consulting civil, mining, geological and petroleum engineer. Later elected to the State Senate, Irene also served as Alaska's first female cabinet member. In a word, she wasn't intimidated by working in a man's world. Today, the "Ted Stevens Anchorage International Airport" (as it is now called) (Figure 24) has become the fifth largest cargo airport in the world and is in the midst of a multi-million-dollar terminal redevelopment program. And, right next door is Lake Hood, the world's largest and busiest seaplane base, accommodating more than 800 takeoffs and landings on a peak summer day!

Figure 23.
Irene Ryan, 1909-1997

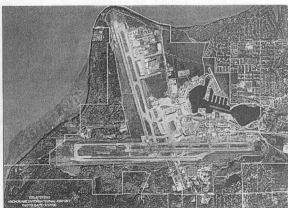

Figure 24. Anchorage International Airport - 2000

PUBLIC HEALTH AND SANITATION

Delivery of good public health and sanitation have been major challenges in Alaska, especially in the bush villages, due to remoteness, climate and permafrost. Seldom were the villages laid out in a compact, orderly pattern as would be amenable to a local sewer/water system. "Honey bucket" systems, in which each household collects and delivers its human wastes to a sewage lagoon, continue in many villages. Costs of installing and maintaining a modern flush toilet system in a small village have often proven to be prohibitive. However, upgrades are underway in many of the villages today, thanks especially to Village Safe Water (a department of ADEC), the Alaska Native Tribal Health Consortium and the Denali Commission, a state-federal partnership established by Alaska's senior Senator Ted Stevens in 1998 as a mechanism for funding worthy infrastructure projects in the bush communities.

In the mid-80s, at Senator Stevens' instigation, the Defense Environmental Restoration Account/Program was established to remedy the "sins of the past" and clean up old military sites throughout the state, including both active installations and FUDS (Formerly-Used Defense Sites). "Barrel roundups" were common. The barrels were mostly empty, some had rusted through and released their contents, and others still contained residual fuels and hazardous wastes. Some sites have required characterization and remediation of soil and groundwater contamination. Others, because of their remoteness, required little or no active remediation. Especially because of logistics, cleanup costs have tended to be high.

Amos "Joe" Alter (1916-2000) was a public health and cold regions engineering pioneer (Figure 25). Appointed "Territorial Sanitary Engineer" by Gov. Ernest Gruening, Joe served in this position for 21 years. A charter member of the Alaska Section of ASCE, Joe was state section President in 1956 and was the Alaska Section's first and only District Director, so far. Also a founder of ASCE's Technical Council on Cold Regions Engineering, Joe was an ASCE Honorary Member.

Figure 25.
Amos "Joe" Alter, 1916-2000

Figure 26. Eklutna Water Treatment Plant

Figure 27. Asplund Water Treatment Plant - Anchorage

Modern award-winning facilities constructed in the Anchorage area include the Eklutna Water Project (Figure 26) and the Asplund Wastewater Treatment Plant (Figure 27).

1964 EARTHQUAKE

The largest earthquake ever recorded in North America, the Great Alaska Earthquake struck with a Moment Magnitude of 9.2 on Good Friday, March 27, 1964, at 5:36 pm local time. With its source deep beneath the earth's surface in the Prince William Sound area, duration of shaking was reported by eyewitnesses to be 4 to 7 minutes. As a result of the long-period cyclic motions, major bluff failures occurred in the downtown and west Anchorage areas (Figure 28). The earthquake and tsunami claimed 125 lives and caused $311 million in property loss (in 1964 dollars). Most of the dollar damages were in the Anchorage area but most of the life loss was caused by the tsunami.

Figure 28. Good Friday Earthquake - 4th Avenue 1964

TRANS-ALASKA OIL PIPELINE

In addition to several finalists, Alaska has had two OCEA award-winning projects. These were the Trans-Alaska Oil Pipeline System (TAPS) in 1978 and the Whittier Access Project in 2001. Mammoth in its scope and undertaking, the 48-inch-diameter, 800-mile-long pipeline was built to unprecedented environmental standards. It was completed in 38 months by 28,000 workers in

June 1977. Having
pumped 11 billion
barrels of oil so far,
enough to fill 14,000
tankers, the pipeline
has provided nearly
25 percent of the
nation's
domestically-
produced crude for
the past 25 years.
This in turn has
provided revenues to
fuel state services
and has provided

Figure 29. Trans-Alaska Pipeline

jobs for a generation of Alaskans. Civil
engineering innovations on the project included
the elevated pipeline construction mode over
thaw-unstable permafrost and the 62,000 Vertical
Support Members (VSMs) equipped with
thermosyphons that extract heat from the ground
during the cold winter months and thereby
maintain a frozen condition of the ground around
each support (Figure 29).

Engineering Manager on the design and
construction of the project was Harold R. Peyton
(Figure 30). A pioneer in the cold regions
engineering field, Dr. Peyton was one of the
originators of the first academic course in Arctic

Figure 30.
Dr. Harold Ray Peyton

Engineering, now a requirement for professional
registration in the state. He also designed and
developed the first university arctic cold laboratory
and served as principal consultant on ice forces for
the first permanent oil platform in Cook Inlet. ASCE
has named a national award for cold regions
engineering after Dr. Peyton, who was also a State
Section Charter member.

A contemporary of Peyton was engineering educator
Elbert F. ("Eb") Rice (Figure 31). Known as "Mr.
Arctic Engineering," Dr. Rice taught civil
engineering at University of Alaska Fairbanks from
1952 to 1982. Known for his unique wit, humor and
concern for individuals, Eb bridged the gap between
theory and practical, down-to-earth construction in
the Northland. He also took delight in assuming any

Figure 31.
Dr. Elbert F. "Eb" Rice,
1923-1982

side of a discussion to promote thought. In addition to being an ASCE State Section Charter Member, Rice was State Section President in 1960.

WHITTIER ACCESS PROJECT

A gateway to the Prince William Sound area, until last year, the town of Whittier, Alaska, had been walled in by mountains and icefields, its only land access via a railroad line constructed through two rock tunnels in 1942 as a wartime project. The Whittier Access project (Figure 32), constructed in 1999-2000, included converting the longer of the two tunnels (2.5 miles long) into a dual-use single-lane highway/rail facility. In addition to the conversion, the 5.2-mile project includes construction of two bridges and a new 500-ft-long two-lane tunnel for the highway. The dual-use tunnel is the longest highway tunnel in North America and the longest combined rail-highway use tunnel in the world. In addition, Whittier is the first US tunnel with a ventilation system combining jet and portal fans, the first tunnel with a unique computerized traffic control system, and the first designed to operate in -40° F and winds up to 150 mph.

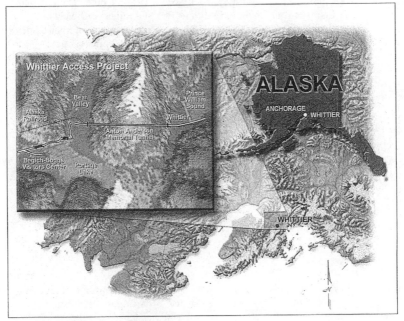

Figure 32. Whittier Access Project

PROPOSED PROJECTS

Alaska has always evoked creative ideas promoted by risk-takers and, over the years, numerous projects, some of them very imaginative, have been proposed. Some of these never saw the light of day but others are still under consideration.

Starting with TAPS, projects subject to requirements of the 1972 National Environmental Policy Act (NEPA) have received intense environmental scrutiny. As a result, public policy and permitting issues often prevail over engineering issues.

Bering Strait Tunnel

The idea of connecting North America with Russia (Eastern Siberia) via the Alaska Bering Straits was first suggested many years ago. A 42-mile tunnel profile complete with airshafts at the Diomede Islands was presented in a 1906 international proposal. The eastern portal would be in the vicinity of the village of Wales on the Seward Peninsula. Many have suggested that it should be a rail tunnel, but there are no railroads (or highways) within hundreds of miles of either terminus. Including the approaches, this would be an extremely expensive project.

Rampart Dam

The Rampart Dam was an ambitious hydropower scheme proposed in 1959 to provide power equivalent to twice the capacity of the Grand Coulee Dam. A dam constructed at Rampart Canyon on the Yukon River 100 miles northwest of Fairbanks would have flooded the Yukon Flats, creating a reservoir larger than Lake Erie. John F. Kennedy promoted the project in his 1960 presidential campaign and Irene Ryan, mentioned previously, served on the Rampart Economic Advisory Board created by the Corps of Engineers. The project was abandoned by 1978 as the need for power and environmental impacts were questioned.

Susitna Hydroelectric Project

Another large hydroelectric project was the Susitna Hydroelectric Project first proposed in 1960. It would have been a two-dam system, including an 885-foot earth-filled structure at Watana, and a 645-foot concrete arch dam at Devil Canyon. The Corps of Engineers wrote an Environmental Impact Statement but many Alaskans had reservations about the project for environmental and cost reasons. Like with the Rampart Project, opponents also voiced concern that the project would generate far more power than Alaska needed. As project cost estimates reached $5.3 billion, the state decided to drop the project by 1985.

Project Chariot

Following World War II, peaceful uses of atomic energy were promoted under a national program called "Swords into Plowshares." Starting in 1958, Edward Teller of the Atomic Energy Commission personally espoused a project to excavate a harbor in northwest Alaska using nuclear explosives. The site selected was Cape Thompson on the Chukchi Sea, 30 miles southeast of Point Hope. The original plan was to excavate a full-scale harbor using two 200-kiloton nuclear explosions, detonated 150 ft below ground. Ostensibly to open up the remote area for mineral extraction, some questioned the need for a port on the Chukchi Sea. A revised plan was prepared for a one-tenth-scale experimental harbor. The project was abandoned in 1963 when the Nuclear Test Ban Treaty dampened efforts on the project and political support had evaporated as concerns over potential health effects grew.

Other Projects

The state capitol of Juneau is not on the state's road system and many Alaskans feel isolated from their legislature and state government. In 1976, the citizens voted to move the state capitol from Juneau to a site at least 30 miles from Anchorage or Fairbanks. A site at Willow, north of Anchorage, was selected. A master plan and concept designs were developed and a price tag of $2.8 billion was estimated to complete the move. A vote needed to fund the move was defeated in 1982.

Projects currently under consideration include bridging the 2-to-4-mile-wide Knik Arm (with its 38.9-ft tidal range, the second largest in North America) near Anchorage to reduce travel time to Port MacKenzie, constructing a bridge over the Tongass Narrows in Ketchikan (where one currently has to take a ferry boat to get from the town to the airport), and constructing a gas pipeline from Prudhoe Bay (there is presently no way to get natural gas to market). Gas pipeline proposals being considered include two routes to the Lower-48 via Canada (one following the Alaska Highway from Delta Junction and the other following the Beaufort Sea coastline to the MacKenzie Delta) and an export project including construction of an LNG plant at Valdez and special tankers for transport to Asia.

ACKNOWLEDGEMENTS

The authors would like to express their appreciation to Anne Brooks, Tom Wolf and Jim Rooney, the other members of our ASCE Alaska Section committee, Chandra Johnson, who prepared all of the graphics including the several pop-up maps for the PowerPoint® presentation, the many Alaska Section members who donated photos, Frank Norris of the National Park Service for a historical review, Claude Vining of the Alaska District Corps of Engineers, and CH2M HILL, for their much-appreciated administrative support.

BIBLIOGRAPHY

Alaska's Builders–50 Years of Construction in the 49th State, Associated General Contractors of Alaska, 1998.

Alter, Amos J. *Alaska's Engineering Heritage "Bridge from Past to Future"* Alaska Section, American Society of Civil Engineers, 1976. 140 pp.

Beach, Rex. *The Iron Trail.* Published by Rex Beach, 1912.

Coates, Peter A. *The Trans-Alaska Pipeline Controversy.* Lehigh University Press, 1991.

Cohen, Stan. *Rails Across the Tundra, A Historical Album of the Alaska Railroad.* Pictorial Histories Publishing Company, Missoula Montana, March 1980. 104 pp.

Cohen, Stan, *The White Pass and Yukon Route, A Pictorial History.* Pictorial Histories Publishing Company, Missoula, Montana, March 1980. 104 pp.

Crittenden, Katharine Carson, *Get Mears! Frederick Mears: Builder of the Alaska Railroad.* Binford & Mort Publishing, 2002.

Herron, Edward A. *Alaska's Railroad Builder–Mike Heney,* Julian Messner, Inc., 1960.

Janson, Lone E. *The Copper Spike.* Alaska Northwest Publishing, Anchorage, 1975. 175 pp.

Mighetto, Lisa and Homstad, Carla. *Engineering in the Far North–A History of the US Army Engineer District,* Historical Research Associates, Inc., 1997.

Minter, Roy. *The White Pass, Gateway to the Klondike,* with foreword by Pierre Berton. University of Alaska Press, 1987. 366 pp.

Naske, Claus M. *Paving Alaska's Trails-The Work of the Alaska Road Commission.* University Press of America, 1986.

Quinn, Alfred O. *Iron Rails to Alaskan Copper–The Epic Triumph of Erastus Corning Hawkins,* Maverick Publications, Inc., Bend, Oregon, 1995.

Soberg, Ralph. *Bridging Alaska–from the Big Delta to the Kenai.* Hardscratch Press, 1991.

Swalling, A. I. *Oh, to be Twenty Again–and Twins!* A&M Publishing, 1999.

Tower, Elizabeth A. *Big Mike Heney, Irish Prince of the Iron Trails, Builder of the White Pass and Yukon and Copper River Northwestern Railways.* 60 pp.

Wilson, William H. *Railroad in the Clouds,* 1977.

Subject Index

Page number refers to the first page of paper